Chemical Vapour Deposition

Growth processes on an atomic level

Online at: https://doi.org/10.1088/978-0-7503-3107-4

Chemical Vapour Deposition

Growth processes on an atomic level

Karin Larsson
Department of Chemistry-Angstrom Laboratory, Uppsala University, Lagerhyddsvagen, Uppsala, Sweden

IOP Publishing, Bristol, UK

ISBN 978-0-7503-3107-4 (ebook)
ISBN 978-0-7503-3105-0 (print)
ISBN 978-0-7503-3108-1 (myPrint)
ISBN 978-0-7503-3106-7 (mobi)

DOI 10.1088/978-0-7503-3107-4

Version: 20221201

IOP ebooks

British Library Cataloguing-in-Publication Data: A catalogue record for this book is available from the British Library.

Published by IOP Publishing, wholly owned by The Institute of Physics, London

IOP Publishing, No.2 The Distillery, Glassfields, Avon Street, Bristol, BS2 0GR, UK

US Office: IOP Publishing, Inc., 190 North Independence Mall West, Suite 601, Philadelphia, PA 19106, USA

Multimedia content is available for this book from http://doi.org/10.1088/978-0-7503-3107-4.

Contents

Preface

The earliest uses of chemical vapour deposition (CVD) for the deposition of metals can be dated back to the 1900th century. The CVD growth of the tungsten metal was reported as early as 1855. It is one of the earliest uses of CVD for the growth of metals. However, the CVD process came to develop quite slowly in the next-coming years. It was not until the end of World War II that the CVD process began to be used on a large scale. The reason was that the researchers had started to realize the potential of deposition of thin film coatings. The emphasize in the 1960s was on the CVD of refractory metals, particularly tungsten. This was followed by a gradual shift to other refractory metals and composites. The development of CVD for semiconductor/microelectronics applications became the dominant interest in the 1970s (with a primary focus on silicon). The development of machine-tool coatings was also important in this decade. The interest broadened to cover a multitude of materials in the 1980s.

The importance of CVD for many industries, as well as for the academy, has been growing ever since. There has been an increasing need for improvement of existing, as well as development of new, CVD technologies. This need has been most urgent when other thin film deposition techniques have shown to be problematic, or even inadequate. Many CVD techniques have, thus, been developed in the search for the most suitable deposition method for a specific application.

The main advantage with the CVD methods of today is their simplicity and cost-effective growth of very thin layers with a high material quality. However, it is not only the range of CVD methods that is expanding, so is the number of applications. Examples of applications are electrical, opto-electrical, optical, mechanical, and chemical, which can be found in the electronic, optical, tooling, and chemical industry. In addition to the use of CVD for a specific application, CVD is today used to produce thin film coatings that can be used for several different applications. For instance, it has become possible to produce coatings that consists of several different material layers (i.e., multilayered coatings), and the resulting coating will thereby become a multifunctional material (with one functionality for each layer).

Different analysis techniques have been developed by which it is possible to characterize the geometrical structure and chemical composition in the surface region of a CVD grown thin film. A small nucleus of techniques, generally viewed as the work horse of the field, include (i) x-ray photoelectron spectroscopy (XPS), (ii) Auger electron spectroscopy (AES), (iii) low-energy electron diffraction (LEED), (iv) scanning probe microscopy, (v) ultraviolet photoemission, (vi) vibrational spectroscopy (reflection–absorption infrared and high-resolution electron energy loss), (vii) thermal desorption spectroscopy (TDS), and (viii) molecular beam spectroscopy (MBS). There is another microscopy technique in which a beam of electrons is transmitted through a material to form an image; transmission electron microscopy (TEM).

Even though these characterization methods are of an overall high quality, there are still some problems connected with the analysis of CVD growth processes. For instance, an atomic-level resolution is very difficult (or even not possible) to achieve for a specific surface geometrical structure. This is especially the case with the in-depth resolution, and for both the geometrical structure and chemical composition. In addition, there are also some major problems when trying to analyse the CVD growth process over time, or to analyse the growing film in the CVD reactor. For the latter type of analysis, in-situ characterization is required that is both very difficult and expensive to arrange. Even though these types of difficulties will by time become diminished by further development in the field, some major problems will not. And that has to do with the principles of the CVD process itself. For instance, by changing one parameter (like the substrate temperature), another (or several others) will be changed at the same time (like type and concentration of active growth species right above the growing thin film surface). Hence, if it is of an interest to study the effect of just one parameter on the CVD growth of a certain material, then one must find another way than to use the more traditional characterization methods. Luckily, this possibility is nowadays achievable by the introduction of fast computers and high-level theoretical programs. In fact, an increasing reliance has been placed upon theoretical modelling of the complex CVD processes in the growth of various types of materials.

Since its beginning in the 1970s, simulation of CVD processes and equipment modelling have come a long way. The simulation of CVD processes was, initially, categorized in three different types: detailed CVD chemistry modelling, CVD reactor modelling and CVD microscopic modelling. The detailed chemistry modelling aims at identifying homogeneous and heterogeneous reaction mechanisms, in combination with simple one- dimensional reactor models. On the other hand, hydrodynamic models are used to study the flow and transport phenomena in the CVD reactor modelling. These calculations are based on relatively simple models of the CVD chemistry. Finally, the microscopic modelling describes the interaction between the molecular transport and deposition within sub-micron trenches and contact holes on wafers for semiconductor manufacturing.

Since the late 1990s, there has been an integration of reactor scale models with theoretical multiscale models. This integration has been enabled by the very fast development of computer performances. The CVD reaction processes occur at multiple time and length scales. Thus, there is the need to link the macroscale to the micro- or atomic scale. Multiscale modelling is a promising route towards the prediction of microstructure and properties of materials that have been grown by using CVD methods. There are two types of multiscale modelling. One is referred to as hierarchical, or sequential, modelling, in which more accurate calculations are first taking place for a smaller scale model. The results from these calculations will then be used as input in the next level (and less accurate) method for calculations on a larger scale model. The other type is referred to as concurrent multiscale calculations, which considers multiple scales simultaneously. For both types of multiscale modelling, different theoretical methods are (with their appropriate time and length scale resolutions) used to describe specific regions of the material.

Ab initio methods, such as density functional theory (DFT), are the computational chemistry techniques that can provide the necessary input data to the upper scale methods. The DFT methods can, in fact, provide the key insights about the basic mechanisms of the atomic assembly. As such, the question of bridging the electronic models to the atomistic models has during the last decades received an increased attention. For instance, the many activation barriers that are present during the CVD surface growth processes can, by using DFT calculations, be efficiently used as input data in kinetic Monte Carlo modelling of the CVD film growth. Important advances in the predictive capabilities of multiscale modelling would most probably benefit from a wider use of first-principle calculations. Further integration of quantum calculations into multiscale modelling and simulations of CVD growth of thin film is perhaps the most important, but also the most difficult, challenge for the next years to come.

The purpose with the present book is not to be a detailed textbook of quantum mechanical methods (like DFT). It is not either a textbook presenting the most common chemical vapour deposition methods. Also, the purpose is not to be a review over recent modelling efforts in the field of CVD growth of various types of functional materials. On the contrary, the purpose is to show the methodology behind the modelling and simulation of CVD growth processes. More precisely, the book will give advice about what to avoid, and how to proceed, for successful theoretical results. Examples area given with some specific material types, but the methodology is generally applicable for any solid-state material. In this context, the book gives instructions for the practical handling in the course of the calculations.

This type of training in theoretical calculations and simulations are extremely useful (i) for the design of new experimental CVD setups, (ii) for the development of new materials with specific properties, and (iii) for interpretation of experimental results. It is also highly valuable when experimental methods cannot give the information needed.

The book is written for a wide range of readers, ranging from undergraduate and graduate students that are learning about the CVD technique, to people with a larger experience from the usage of CVD.

Author biography

Karin Larsson

Karin Larsson is a Professor Emerita in Inorganic Chemistry, Department of Chemistry-Angstrom Laboratory, Uppsala University, Sweden. She has experience from material characterization, CVD synthesis, and theoretical simulations.

Karin Larsson received her PhD in Chemistry in 1988. She studied the dynamics of especially water molecules in solid hydrates by using solid state NMR spectroscopy. This was the first time Karin Larsson came across quantum mechanics since the theory behind NMR has been developed from this type of theory. She thereafter continued within the synthesis world by growing diamond thin films using hot-filament and laser-assisted chemical vapour deposition (CVD) techniques. She was then not only performing the synthesis experiments, but also planned and constructed the CVD machines. Karin Larsson became very interested in the atomic-level details in the CVD growth processes, and, therefore, switched research area for the second time in her career. She now became active in the field of first principle calculations, using density functional theory (DFT) calculations quantum-mechanical based molecular dynamic (MD) simulations. She became then especially interested in the interpretation, understanding and prediction/modification of the following processes; (i) CVD growth processes, (ii) interfacial processes for renewable energy applications, and (iii) interfacial processes for e.g. bone regeneration (incl. biofunctionalization of surfaces). The quantum mechanics skills obtained during her PhD period, and the CVD growth experience received during her PostDoc period, was then of great use. These are research activities that she has been very active in for more than 30 years by now.

Karin Larsson has been very active by publishing in highly high-renown international journals (such as J. Phys. Chem., Phys. Rev. B, Thin Solid Films, Phys. Rev. Lett., etc.), and in the supervision of a large number of Master Students, PhD students, and PostDoc Fellows. She has also coordinated large International Projects as well as EU-financed International ones. Her main scientific philosophy was then to combine synthesis, characterization, and theoretical tools in the research projects.

Karin Larsson has also during all years been very active with teaching, both for Undergraduate and Graduate Students. She has also developed most of these courses herself. The courses include: (i) Instrumental Methods Within Materials Chemistry, (ii) Inorganic Surface Chemistry and Corrosion, (iii) Surface Chemistry; Surface Science and Technology, (iv) Materials Design, (v) Theoretical Methods for Surface and Materials Studies, (vi) Theoretical Materials Design, and (vii) Chemical Bonding: From MO to Bond Theory.

Karin Larsson did also become an Elected member of IVA (Royal Swedish Academy of Engineering Sciences), division V (Mining and Minerals), in 2009.

Chapter 1

Introduction

The surface of a material experiences various types of interactions with its environment (e.g. wear, corrosion, radiation, electrical or magnetic fields, etc). It must, therefore, have the properties needed to withstand these negative interactions. Examples of desirable surface properties are reflectivity, semi-conductivity, high thermal conductivity, and erosion resistance. Despite all these necessary precautions, it might also be necessary to improve the material surface performances for specific applications. Industrial products are, therefore, often coated with various types of thin films. This is, especially, the situation for modern electronic products, where the thin film quality is of largest importance for the overall electronic performances. For example, the performances of integrated circuits in computers have been shown to be completely dependent on the quality of the coating films. Another example is the solar cells, where the quality of the coatings has been shown to determine the efficiency of the energy conversion efficiencies. The deposition of a thin film coating on another material produces a composite material and, as such, allows it to have surface properties that can be very different from those of the corresponding bulk materials [1].

1.1 Chemical vapour phase deposition

1.1.1 Thinfilm coating technologies

Industrial products are commonly coated with various types of thin films. This is especially the situation for modern electronic products, where the coatings are of an outmost importance for the overall electronic performances. As an example, the quality of such thin films has been shown to determine the performance of integrated circuits in computers. Solar cells are another example, where the quality of the coating films determines the efficiency of the energy conversion efficiencies. Table 1.1 summarizes the surface properties that can be obtained, or modified, by the use of coatings.

Many different types of thin film technologies have been developed, which all can be divided into three major technologies (see table 1.2). The first one considers the

doi:10.1088/978-0-7503-3107-4ch1

Table 1.1. Materials properties affected by the coating, with their plausible application areas.

Application Areas	Properties
Electrical	Crystal lattice control Dielectric constant Magnetism Resistivity Superconductivity
Optical	Emissivity Photoconductivity Reflectivity Refraction Selective absorption
Mechanical	Adhesion Ductility Friction Hardness Strength Toughness Wear
Chemical	Catalysis Corrosion Diffusion Electrochemical Oxidation
Porosity	Pore size Pore volume Surface area

pasting of already prepared thin films on solid substrates. The pasting of gold leaf on solids is, historically, a well-known technique. The second technology is the formation of a thin film on a solid material by using chemical reactions between the solid and active gas phase species. A layer of a different material (compared to the solid) can then be formed. The thermal oxidation of crystalline silicon wafers, to form a silicon dioxide (SiO_2) layer at the surface, is a good example of this technology. Finally, the third type of technology is the formation of thin films by depositing gaseous species onto the substrate. These types of epitaxy, ion plating, and ion implantation are all examples of PVD methods in which there is a net transfer of growth precursors from a solid reservoir to the growing film. This film is formed as the product of chemical reactions that take place in the gas phase (involving one or more gaseous species). The chemical vapour deposition (CVD) method competes directly with the PVD method, but it is also used in conjunction

Table 1.2. Thin film technology.

Thin film technology	Subgroups	Subsubgroups	Examples of applications
Pasting of films (prepared elsewhere) on substrates			Pasting of gold leaf Pasting of thin Si (e.g. for making SOI)
Conversion of surface materials			Thermal SiO_2 on Si Surface modification
Deposition of thin films on substrates	Deposition by reactions completed only on substrates	Wet process	Painting Plating Printing Ink-jet
		Dry process	Thermal CVD Vacuum evaporation
	Thin film deposition by reactions using active species generated elsewhere and transported to substrates	Plasma process	PECVD Sputtering Ion plating Reactive plasma deposition (RPD)
		Hot filament process	HF-CVD Initiated (i)CVD
		Photoassisted decomposition	Photo-CVD

and many of the newer processes are, actually, hybrids of the two technologies (such as plasma-enhanced CVD or activated sputtering).

The CVD method is atomistic to its nature. This means that the species that reaches the film surface are atoms or molecules, or a combination of these two types of growth species. The deposition will thereafter take place by chemical reactions onto the film surface. The CVD method can, in turn, be divided into three groups. Group one uses a plasma for the activation of the source gas in the CVD chamber. A common name of this type of method is plasma-enhanced CVD (PECVD). The molecules in the source gases are in this method decomposed by the collisions with energetic electrons that are generated in plasma. The second group of CVD methods uses catalytic cracking reactions to generate activated species from the gas source molecules. A heated metal wire is often used for this purpose and the method is often called hot filament assisted CVD (HFCVD). However, the thermal-induced activation can also take place by heating the whole CVD chamber from outside (i.e. by using an external oven that covers the main part of the reactor). This type of CVD-process is therefore called a thermal-assisted CVD (TACVD). The third CVD group uses the energy of radiation (by using a laser) for the activations of the source gas molecules and is, therefore, generally called laser assisted CVD (LACVD). The costs of the CVD techniques are generally quite high. However, this is mostly acceptable since the use of these techniques are most often the only way to assemble the structures necessary to achieve the needed performance.

1.1.2 Common chemical vapour deposition techniques

CVD is a highly versatile process where activated gaseous reactants undergo chemical surface reactions in the deposition of a thin solid material onto another solid material (i.e. the substrate). (It can thus be distinguished from physical vapour deposition (PVD) processes where no surface chemical reactions take place.) The source gas is introduced into the CVD reaction chamber, and therein activated to form very reactive growth species. Figure 1.1 demonstrates various plausible chemical reactions occurring in the CVD reaction chamber—both in the gas phase and on the substrate (or growing film). If the reaction occurs in the gas phase (i.e. homogeneous reactions), it will lead to the formation of fine particles that finally attaches to the growing film. This is most often not the desired product. On the other hand, the chemical reactions on a heated substrate (i.e. heterogeneous reactions) produce a dense film, which is most often the desired outcome of the CVD process. A combination of homogeneous and heterogeneous reactions tends to produce a porous film [2].

As presented above, the CVD reactions are initiated by activating the sources gas in the CVD chamber. The different activation methods use heat, plasma, or light, and their respective CVD process is denoted as HFCVD (or TACVD), PECVD, and LACVD. Other forms of notations can also be found in the literature. The incoming source gas will be thermally cracked onto the surface of a heated metal filament coil when using HFCVD. This process is, therefore, sometimes referred to as catalytically assisted CVD (CACVD). Several fragments of the source gas will thereby be formed,

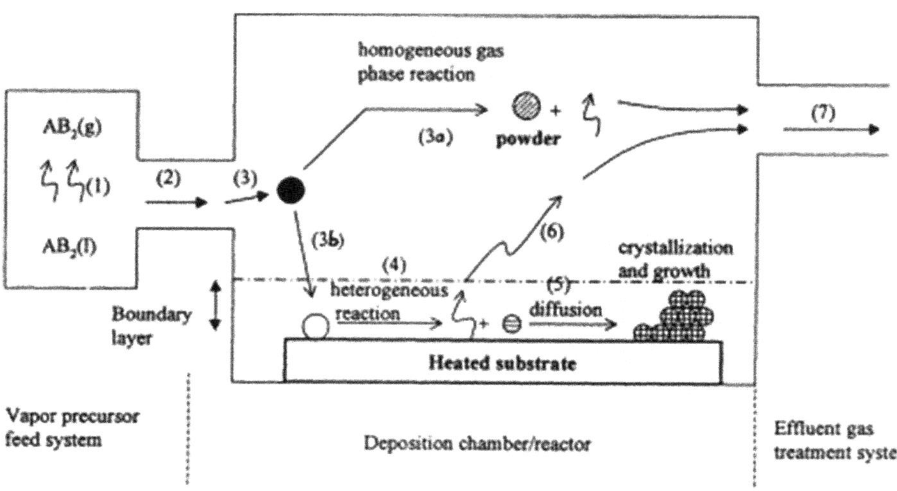

Figure 1.1. A schematic illustration of the key CVD steps in the deposition process. [2], reproduced with permission © Elsevier.

which are all in their ground state. (A filament coil made of Ta is generally used in the deposition of diamond.) It is, in addition, most often needed to separately heat the substrate, but to a lower temperature. Contrary to the situation with focused heating in the HFCVD process, the complete CVD chamber will be heated to a uniform temperature in the TACVD process. The drawback with this method is that not only will the source gas and substrate/film be heated, but also the chamber walls. There is, hence, an apparent risk that deposition also will take place on these walls. Unwanted materials, such as flakes from the chamber walls, might then get loose and attach to the growing film coating. PECVD is a technique in which electrical energy, rather than thermal energy, is used to initiate homogeneous reactions in the gas phase. Chemically active ions and radicals are thereby formed in the generated plasma, which can participate in heterogeneous reactions on the surface of the growing film (or substrate). In the LACVD method, the source gas will be directly activated by a focused laser beam. The generation of active fragments can, thereby, take by two different means. The source molecules will either be cracked by the temperature of the laser beam or undergo excitations that are induced by the laser frequency. Thus, the use of a laser, in the activation of the source gas, makes it possible to tailor-make the formation of specific fragments in the activation process.

Various groups of CVD techniques have, hence, been developed that use different types of gas phase activation methods. However, these groups can be further divided into subgroups depending on specific types of experimental parameters, such as source gas types (also called precursors), pressure, schemes of gas introduction into the chamber, etc. For example, metalorganic CVD (MOCVD) uses a metalorganic precursor. In the broader definition, MOCVD involves precursors containing metal–oxygen bonds (e.g. metal–alkoxides), metal–nitrogen bonds (e.g. metal alkylamiddes), and even metal hydrides (e.g. trimethylamine alane). Furthermore, metal–organic vapour phase epitaxy (MOVPE), also called organometallic vapour phase epitaxy (OMVPE), is an MOCVD process that produces heteroepitaxial single crystalline films on single crystalline substrates from metal–organic precursors.

The deposition can also be conducted in an atmospheric (i.e. 'normal') pressure, or low pressure, with the following notations: normal pressure CVD and low-pressure CVD, respectively. It is also common to deposit at ultrahigh vacuum (UHV). There are different UHV CVD processes, such as molecular beam epitaxy (MBE) and chemical beam epitaxy (CBE). This latter method uses volatile metal–organic precursors together with gaseous co-precursors. The closely related technique of metal–organic MBE (MOMBE) uses volatile metal–organic precursors and gaseous co-precursors that are derived from the solid element. The chemical reactions in CBE and MOMBE occur only on the substrate, with no significant gas-phase reactions. Hence, no gas-phase reaction will be involved in the deposition of heteroepitaxial thin films. There is also a type of CVD process where the gaseous reactants are introduced separately into the reactor—atomic layer deposition (ALD). The reactants are then sequentially introduced into the reaction chamber, with an inert purging gas in between. As an alternative, the reactant pulses can be separated by evacuation. The ALD technique exists with both thermal and plasma-assisted gas phase activation.

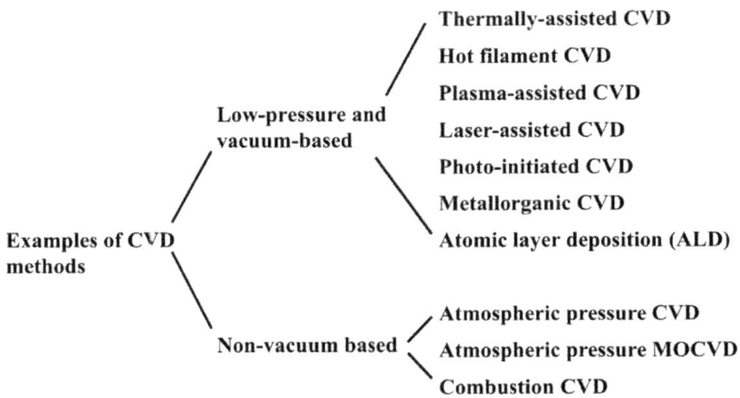

Figure 1.2. CVD and examples of its variants.

The above presented CVD methods do not vary only from the process parameters, but also from scalability. The scalability is a factor that largely affects the properties of thecoating thin film. It might seem very difficult to choose amongst all these CVD technologies. There are, though, some guidelines. That is, the final selection of a specific CVD technology must depend on the requirement of the resulting CVD product. There are three different types of criteria that must be met. One is the capability to deposit the required type of coating and thickness. The second criterium is the capability to coat the component uniformly with respect to both size and shape. The third criteria is the need for a cost-effective deposition (in terms of costs for the substrate, coating material, coating technique, equipment down time, etc).

An overview of the vacuum-based and non-vacuum-based CVD methods is given in figure 1.2, and a more detailed overview is presented in chapter 2. Furthermore, a thorough review of the different CVD technologies is given in [2].

1.1.3 History of CVD

The earliest uses of CVD for the deposition of metals can be dated back to the 1900th century. In fact, the CVD growth of the tungsten metal was reported as early as 1855. It is one of the earliest uses of CVD for the growth of metals. A gas mixture containing WCl_6, in an excess of H_2 was used for the deposition. As reviewed in the pioneering work by Powell *et al* [3], the first practical use of CVD was developed in the 1880s. Carbon and metal deposition where then used to strengthen the filaments in incandescent lamps. However, the CVD process came to develop quite slowly over the next 50 years. The first reports on the CVD growth of silicon, by using hydrogen reduction of $SiCl_4$, can be dated to 1909 [4] and 1927 [5]. These activities, in addition to other CVD growth activities in the field of Si-based photocells [6] and rectifiers [7], became the ground for the wide-spread use of thin silicon films in the electronics industry.

It was not until the end of World War II that the CVD process began to be used on a large scale. The reason was that the researchers had started to realize the potential of deposition of thin film coatings. The emphasize in the 1960s was on the CVD of refractory metals, particularly tungsten. This was followed by a gradual shift to other refractory metals and composites. The development of CVD for semiconductor/microelectronics applications became the dominant interest in the 1970s (with a primary focus on silicon). The development of machine-tool coatings was also important in this decade. The interest broadened to cover a multitude of materials in the 1980s. Furthermore, the modification of silica gel, with boron oxide, was made by using hot-filament CVD in the early 1990s [8].

The importance of CVD for many industries, as well as for the academy, has been growing ever since. There has been an increasing need for improvement of existing, as well as development of new, CVD technologies. This need has been most urgent when other thin film deposition techniques have shown to be problematic, or even inadequate. Many CVD techniques have, thus, been developed in the search for the most suitable deposition method for a specific application. The CVD technique has, therefore, since the start, undergone major developments and improvements.

Some important dates in the development of CVD techniques are listed below:

- 1960: introduction of the terms 'CVD' and 'PVD'
- 1960: introduction of CVD in the semiconductor industry
- 1960: the concept of CVD TiC coating of cemented carbide tools is being introduced
- 1968: start of industrial use of CVD coated cemented carbides
- 1968: first development of metallo-organic CVD for the deposition of gallium arsenide
- 1974: introduction of plasma-enhanced CVD
- 1976: first applications of low-pressure CVD in electronics
- 1979: introduction of the hot filament chemical vapor deposition (HFCVD) technique for the synthesis of amorphous silicon films from silane gas at low substrate temperatures and high deposition rate
- 1983: introduction of electron cyclotron resonance for plasma CVD
- 1993: ALD examples appear in the scientific literature

For a more detailed review of the CVD development, see [9, 10].

1.1.4 Use of CVD today

CVD is a constantly expanding and improving technology, which has been proven by the recent developments in CVD methods like metallo-organic, plasma and laser CVD. The main advantage with the CVD methods used today is their simplicity and cost-effective growth of very thin layers of materials with a high material quality. However, it is not only the range of CVD methods that is expanding; so is the number of applications. Examples of applications are electrical, opto-electrical, optical, mechanical and chemical, which can be found in industries, such as the electronic industry, optical industry, tool industry, and chemical industry. In addition to the use of CVD for a specific application, CVD is today used to produce thin film coatings that can be used for several different applications (from microelectronics to gas turbine engines) [11]. For instance, it has become possible to produce coatings that consists of several different material layers (i.e. multilayered coatings), and the resulting coating will thereby become a multifunctional material (one functionality for each layer).

CVD methods are also, in addition to the growth of two-dimensional coatings, used to grow nanoscale particles [12]. It is also possible to tailor the structure of thin films and nanoparticles on an atomic level. The creation of new materials with exceptional functionality such as magnetoresistance [13], tunable optical emission/absorption [14], high efficiency photovoltaic conversion [15], and ultralow thermal conductivity [16] will thereby be enabled. These developments have led to revolutionary devices such as giant magnetoresistance (GMR) sensors and magnetic memories [17], quantum dot lasers and detectors [18], and thermal/chemical protection systems for components operating in hostile environments (e.g. aircraft engine combustion chambers) [19]. The new CVD growth possibilities have been shown to be very beneficial for modern semiconductor devices, and then also for the microelectronic industry [11]. For example, it is now possible to shrink the microelectronic circuitry into the nanoscale regime [20]. It is also possible to create thin films that (i) can withstand high temperatures, (ii) show superconductivity and non-linear optical behaviour, and (iii) show field dependent permittivity/permeability. Furthermore, it is possible to synthesize materials with high thermoelectric figures of merits [21]. Also, the CVD growth of electron spin engineered materials, including quantum dot arrays with magnetic atoms, is very promising for the opening of a route to quantum communications and/or computing [22].

In addition to the possibility to deposit a multifunctional and/or a nanostructured material, it has now become possible to evenly coat a large surface area. This possibility has opened up the possibility to use CVD in many application areas. Semiconducting processes, membrane functionalization, microelectrochemical membranes (MEMS), the design of nanostructured electrodes for photovoltaics and energy storage, the development of novel catalysis, and development of sensors are some of these areas of applications [23–28]. Other interesting applications include thermal, chemical, and mechanical protective coatings [29]. The possibility to perform CVD growth at atmospheric pressure makes it possible to work with in-line glass manufacturing processes. The production of 'self-cleaning' glass is an example of large-scale deposition, where a thin film of transparent TiO_2 is being deposited on glass. The dirt on this glass will thereafter break down in the sunlight by the usage of photocatalysis [30, 31].

The microelectronic industry has a large need for very thin high purity films with precise control of uniformity, composition, and doping. CVD is today a widely used micro-fabrication process in growth of materials in various forms (such as monocrystalline, polycrystalline, and amorphous coatings, as well as powders and fibers). It is also used for heteroepitaxial growth on substrates. Moreover, the shrinking dimensions of the electronic devices set high demands on the materials to be used. With CVD it is possible to produce almost any metal and non-metal element (such as carbon (in various forms), silicon, germanium, and tungsten), as well as compounds (such as carbides, nitrides, oxides, oxynitrides, carbon fluorocarbons, and intermetallics). Table 1.3 presents general classes of CVD-grown film (and specific examples of materials), together with their applications and typical precursors used.

Table 1.3. Selection of films grown by CVD, their applications and typical precursors used.

General class of material	Specific material	Applications	Typical CVD precursors
Dielectric oxides	TiO_2, ZrO_2, HfO_2, $HfSi_xO_y$	High-κ gate dielectric layers in CMOS technology	Metal-β-diketonates[1] + O_2/O_3
	Ln_2O_3, $LnSi_xO_y$, $LnAlO_3$	Dielectric capacitor layers in DRAMs	Metal alkoxides[2] + O_2/O_3
	SiO_2, Ta_2O_5, Nb_2O_5	Dielectric capacitor layers in DRAMs	Metal alkoxides[2] + O_2/O_3
	SiO_2, Ta_2O_5, Nb_2O_5	Optical coatings	Metal alkylamides[3] + O_2/O_3
			Organometallics[4] + O_2/O_3
Ferroelectric oxides	$SrTiO_3$, $(Ba,Sr)TiO_3$, $Pb(Zr,Ti)O_3$, $SrBi_2(Ta_xNb_{1-x})_2O_9$, $Bi_4Ti_3O_{12}$, $Pb(Sc,Ta)O_3$, $Pb(Mg,Nb)O_3$	DRAMs, NVFERAM, computer memories, infrared detectors, microelectron-mechanical devices, transducers, ceramic capacitors	Metal-β-diketonates[5] + O_2 Metal alkoxides[6] + O_2
Ferrites	$(Ni,Zn)Fe_2O_4$, $(Mn,Zn)Fe_2O_4$,	Recording media, high frequency read heads	Metal-β-diketonates[7] + O_2
Super-conducting oxides	$YBa_2Cu_2O_{7-x}$, Bi–Sr–Ca–Cu–O	Josephson junctions, bolometers, squids	Metal-β-diketonates[8] + O_2
Conducting oxides	$(La,Sr)CoO_3$, $(La,Mn)O_3$, RuO_2, $SrRuO_3$	Ferroelectric capacitor electrodes	Metal-β-diketonates[9] + O_2
Low-emissivity and conducting oxides	F-doped SnO_2 Sn-doped In_2O_2	Architecual coatings on flat glass (solar control, anti-reflective), display coatings on glass	$[SnCl_4]/[NH_4F]/O_2[InCl_3]/$ $[SnCl_4]/O_2$, $[In(thd)_3]/[Bu_2Sn$ (acetate)$]/O_2$, [In $(acac)_3]/$ $[Sn(acac)_2]/O_2$
Electrochromic and photochromic oxides	WO_3 MoO_3	Architecual coatings on flat glass (displays)	$[WCl_6]/O_2$, $[W(CO)_6]/O_2$ $[Mo(CO)_6]/O_2$
Thermo-chromic oxides	VO_2	Architectural coatings on flat glass	$[V(Cl)_4]/O_2$, $[V(OCl)_3]/O_2$
Self-cleaning coatings	TiO_2	Architectural coatings on flat glass	$[Ti(Cl)_4]/[Ti(OPr^i)_4]/O_2$

(*Continued*)

Table 1.3. (*Continued*)

General class of material	Specific material	Applications	Typical CVD precursors
Garnets	$Y_3Fe_5O_{12}$	Microwave elements, magneto-optic recording	Metal-β-diketonates + oxygen
Elemental semiconductors	Si	Microelectronic devices	$[SiH_4]$, $[Si_2H_6]$, $[SiH_xCl_{4-x}]$ (x=0–3)
	Doped Si	Semiconductor devices	$[SiH_4]$/$[PH_3]$/$[AsH_3]$/$[B_2H_6]$
	Ge		$[GeH_4]$
	Diamond and diamond-like carbon		$[C_nH_{2n+2}]$
III/V Compound	GaAs	Solar cells, LEDs	Group III trialkyls[10] + Group V
Semi-conductors	GaAs/AlGaAs	Heterostructure lasers, solar cells, HEMTS, HBTs, FETs	trihydrides[11] or organometallics[12]
	GaP	Red LEDs, photocathodes	
III/V Compound	GaAsP, InGaP	Red LEDs	Group III trialkyls[10] + Group V
Semi-conductors	AlGaInP	Yellow/green LEDs	trihydrides[11] or organometallics[12]
	GaN	Blue LEDs	(cont.)
(cont.)	InGaN	Green LEDs	
	InGaP/AlGaInP	Red laser pointers	
	InP	Gunn diodes, radar devices	
	InP/InGaAs	Detectors in optical fibre technology	
	InGaAsP	Emitters in optical fibre technology	
	GaSb/AlGaSb	Thermal imaging devices, environmental sensors	
	ZnS	Blue phosphors, TFELs	
	ZnSe, ZnSSe,	Blue/green LEDs and lasers	
	ZnMgSSe	Solar cells	
	ZnCdS, CdS/CdTe	Infrared detector, thermal imaging systems	
	CdTe, CdHgTe		

II/VI Compound Semi-conductors		Semiconductor devices	Group II dialkyls[13] + Group VI dihydrides[14] or Group VI dialkyls[15]
Metals	Al	Interconnects in microelectronic devices, metallized polymers, gas diffusion barriers, optical coatings, and adhesion aids	$[Bu^i_3Al]$, $[AlH_3NR_3]$
Metals (cont.)	W	Metallization in transistors, interconnects in ICs, wear and erosion protection	$[WF_6]$, $[WCl_6]/H_2$; $[WCO_6]$; W-alkyls
	Cu	Interconnects in ICs	[Cu(II)(β-diketonate)]/H_2; [(hfac)Cu(I) L] (L= $[PMe_3]$, [1,5-COD], alkyne, VTMS)/H_2
	Au, Ag	Metallization in ULSI	$[Me_2Au$(β-diketonate)]; [Ag(β-diketonate)]
	Pt, Pd, Ni	Metal contacts in microelectronic devices, catalysts, protective and decorative coatings	$[Pt(acac)_2]$, $[PtMe_3]$; (1,5-COD) $[PtMe_2]$; $[Pd(allyl)_2]$; [CpPd(allyl)]; $[Ni(CO)_4]$; $[NiCp_2]$; $[Ni(MeCp)_2]$
	Ti	Adhesion layers, metal foils, corrosion resistant coatings	$[TiI_4]$, $[TiBr_4]/H_2$
	Cr	Corrosion protection	$[CrF_2]$, $[CrCl_2]$, $[CrCO_6]$, $[CrPh_2]/H_2$
	Mo	IC contact and gate metallization, wear resistant coatings, infrared reflector, laser mirror coating	$[MoF_5]$, $[MoCl_5]$, $[Mo(CO)_6]$
	Ru	IC contact and gate metallization, diffusion barriers	$[Ru_3CO_{12}]$, $[RuCp_2]$, $[Ru(acac)_2]/H_2$
	Ta	Capacitors, resistors, corrosion resistant coatings	$[TaCl_5]$, $[TaF_5]$, $[Ta(CO)_5]/H_2$
Metal nitrides	AlN	Surface acoustic wave devices, packaging materials in microelectronic devices	$[AlBr_3]/[NH_3]$, $[AlMe_3]/[NH_3]$; $[Al(NMe_2)_3]/[NH_3]$; $[AlMe_3(NH_3)]$ adduct
	Si_3N_4	Chemical passivation, encapsulation of silicon bipolar and MOS devices	$[SiH_4]/[NH_3]$; $[Si_2Cl_6]/[NH_3]$; $[Si(NMe_2)]/[N_{4-n}H_n]$
	TiN	Wear-resistant and friction reducing coatings, transparent optical films, energy efficient windows, decorative coatings, low resistant contacts, diffusion barriers, and gate eletrodes in microelectronic devices	$[TiCl_4]/[NH_3]$, $[TiI_4]/[NH_3]$; $[Ti(NR_2)_4]/[NH_3]$

(*Continued*)

Table 1.3. (*Continued*)

General class of material	Specific material	Applications	Typical CVD precursors
	ZrN, HfN	Hard coatings for machine tools, diffusion barriers and gate electrodes in microelectronic devices	$[M(NR_2)_4]/[NH_3]$; $[M(NEt_2)_4]/[Me_2NNH_2]$; $[M(NR_2)_4/N_2]$ (plasma-assisted)
		Metallization in ULSI	
		Metal contacts in microelectronic devices, catalysts, protective and decorative coatings	
		Adhesion layers, metal foils, corrosion resistant coatings	
		Corrosion protection	
		IC contact and gate metallization, wear resistant coatings, infrared reflector, laser mirror coating	
Metal nitrides (cont.)	TaN, NbN	Diffusion barriers for Cu in ICs; gate electrodes	$[TaCl_5]/[NH_3]$ $[TaBr_5]/[NH_3]$; $[Ta(NMe_2)_5]/[NH_3]$, $[Ta(NEt_2)_5]/[NH_3]$; $[Ta(NEt_2)_3(Nbu^t)]/[NH_3]$; $[Ta(NEt_2)_3(NCMe_2Et)]/[NH_3]$ $[NbCl_5]/[NH_3]$; $[NbCl_5]/[Me_2NNH_2]$; $[Nb(NR_2)_5]/[NH_3]$, $[Nb(NR_2)_4]/[NH_3]$
	WN	Diffusion barriers, gate electrodes in microelectronic devices	$[WCl_6]/[NH_3]$, $[WF_6]/[NH_3]$; $[W(CO)_6]/[NH_3]$; $[W(NBu^t)_2(NR_2)_2]/[NH_3]$; $[(RN)WCl_4(NCR)]/[NH_3]$; $[W_2(NMe_2)_6]/[NH_3]$
	MoN		$[MoF_6]/[NH_3]$, $[Mo(CO)_6]/[NH_3]$; $[Mo(NMe_2)_4]/[NH_3]$; $[Mo(NBu^t)_2(NMe_2)_2]/[NH_3]$
Metal carbides	TiC	Hard coating for cutting, milling, forming and stamping tools	$[TiCl_4]/[CH_4]$; $[Cp_2TiCl_2]/[H_2]$
	ZrC	Coating for nuclear fuel paricles	$[ZrCl_4]$, $[ZrBr_4]/[CH_4]$ or $[C_3H_{11}]$

	HfC	Oxidation-resistant coating for composites, coating for superalloys	$[HfCl_4]/[CH_4]$
		Metallization in ULSI	
		Metal contacts in microelectronic devices, catalysts, protective and decorative coatings	
Metal carbides (cont.)	Cr_7C_3, Cr_3C_2	Corrosion- and wear-resistant coatings	$[CrCl_3]/n$-butane; $[Cr(C_6H_5\text{-}i\text{-}Pr)_2]$; $[CrBu^t_2]$
	WC, W_2C, W_3C	Tool coatings, catalysts	$[WF_6]$/cyclopropane; $[WCl_6]/[CH_4]$
	V-carbide	Corrosion- and wear-resistant coatings	$[VCl_4]/[CH_4]$; $[VCp_2]$
	Ta-carbide, Nb-carbide	Protective coatings	$[TaCl_5]/[MeCl]/[H_2]$; $[NbCl_5]/[CCl_4]/[H_2]$
	SiC	Blue LEDs, heat sinks, protective coatings	$[MeSiCl_3]/[H_2]$
	GeC	Optical coatings	$[C_nH_{2n+2}]/[GeH_4]$

[1] e.g. $[M(\beta\text{-dik})_4]$ (M = Zr, Hf), $Ln(OR)_3$, $Ln(\beta\text{-dik})_3$ (Ln = lanthanide) + O_2/O_3

[2] e.g. $[M(OR)_4]$ (M = Si, Ti, Zr, Hf), $[M(OR)_5]$ (M = Ta, Nb) + O_2/O_3

[3] $[M(NR_2)_4]$ (M = Ti, Zr, Hf) + O_2/O_3

[4] e.g. $[Cp_2MMe_2]$, $[(RCp)_2MM_2]$ (M = Zr, Hf) + O_2/O_3; $[LaCp_3]$, $[Ln(RCp)_3]$ + O_2/O_3

[5] $[Ba(thd)_2]$, $[Sr(thd)_2]$, $[Ti(OPr^i)(thd)_2]$, $[Bi(thd)_3]$; $[Pb(thd)_2]$, $[Sc(thd)_3]$, $[Mg(thd)_2]$ +O_2

[6] $[Bi(OR)_3]$, $[Ta(OR)_5]$, $[Nb(OR)_5]$ + O_2

[7] $[Ni(thd)_2]$, $[Zn(thd)_2]$, $[Mn(thd)_2]$, $[Fe(thd)_3]$ + O_2

[8] $[Y(thd)_3]$, $[Ba(thd)]$, $[Sr(thd)_2]$, $[Cu(thd)_2]$, $[Ca(thd)_2]$ + O_2

[9] $[Ru_3(CO)_{12}]$, $[Ru(\beta\text{-dik})_3]$, $[Ru(RCp)_2]$ (R = H, Et)

[10] $[R_3M]$ (M = Ga, In, Al; R = Me, Et)

[11] $[AsH_3]$, $[PH_3]$, $[NH_3]$

[12] $[Bu^tAsH_2]$, $[Bu^tPH_2]$, $[R_3Sb]$

[13] $[R_2M]$ (M = Zn, Cd; R = Me, Et)

[14] $[H_2Se]$, $[H_2S]$

[15] $[Et_2Se]$, $[Et_2S]$, $[Et_2Te]$, $[Pr_2Te]$

1.1.5 Pros and cons with CVD technologies

CVD has several important advantages that make it the preferred process in many cases. These can be summarized as follows [2]:

- CVD methods can be used for the growth of a wide range of materials, including nanoparticles, 1D nanomaterials, 2D materials (such as films) and 3D structures.
- Uniform films can be produced with good reproducibility and adhesion strength (to the substrate), and with a reasonably high deposition rate.
- CVD (especially TACVD) can be used to coat complex-shaped components and deposit films with good conformal coverage.
- The CVD method has the ability to control the crystal structure and orientation, as well as the surface morphology, by controlling the CVD process parameters. The deposition rate can be adjusted when using CVD. From low-deposition rates, for the growth of thin films, to the deposition of thick protective coatings (in some cases centimeters thick).
- The processing costs are reasonable for the conventional CVD technique.
- There is the flexibility to use a wide range of chemical precursors such as halides, hydrides, and organometallics, which enables the deposition of a large spectrum of materials.
- The CVD methods use relatively low deposition temperatures, which, for example, enable the deposition of refractory materials at a temperature much lower than their melting temperature.
- The CVD equipment is relatively simple. It does not require ultra-high vacuum, and can be adapted to many process variations.
- ALD holds a special position among the CVD techniques. This is because it offers the possibility to grow thin films in a controlled, self-limiting manner. Three-dimensional configurations can, thereby, usually be coated with relative ease.

However, the CVD methods also have several disadvantages:

- A major disadvantage is that it generally works at temperatures above 600 °C where the thermal stability of the substrate may limit is applicability. However, the development of special CVD technologies (such as plasma CVD and metallo-organic CVD) partially offsets this problem.
- The CVD method requires the use of chemical precursors with high vapour pressures, which are often hazardous and sometimes even toxic. The by-products of these precursors are also toxic and must be neutralized, which may be an expensive process. However, these drawbacks have been minimized by using, e.g. electrostatic spray-assisted vapour deposition (ESAVD) (which uses more environmentally friendly precursors).
- It is difficult to deposit multicomponent materials with well-controlled stoichiometry when using a mixture of different types of precursors. This is the case when the different precursors have different vaporisation rates.

However, this limitation does not exist when using a single-source chemical precursor.

- The use of more sophisticated reactor systems, such as low pressure or UHV CVD, PECVD, and LACVD, tends to increase the costs of the deposited material. However, there are also other types of more simple CVD methods, such as flame-assisted CVD (FACVD), that can be used where the production cost is an issue.

1.2 Overview of thin film characterization techniques

Different analysis techniques have been developed through which it is possible to characterize the geometrical structure and chemical composition in the surface region of a CVD grown thin film. There is, though, a small nucleus of techniques that is generally viewed as the work horse of the field. These ones include (i) x-ray photoelectron spectroscopy (XPS), (ii) Auger electron spectroscopy (AES), (iii) low-energy electron diffraction (LEED), (iv) scanning probe microscopy, (v) ultraviolet photoemission, (vi) vibrational spectroscopy (reflection-absorption infrared and high-resolution electron energy loss), (vii) thermal desorption spectroscopy (TDS), and (viii) molecular beam spectroscopy (MBS) [32]. The first two techniques ((i) and (ii)) are used to gain information about the surface chemical composition, and the third and fourth ((iii) and (iv)) are used for the determination of surface structures. Moreover, information about chemical bond formation and surface reactivity can be achieved by methods (v) and (vi), whilst information about adsorbate bond strengths and surface reaction rates can be achieved by (vii) and (viii), respectively. The majority of these techniques have proven to be very useful in the study of thin film growth. There is another microscopy technique in which a beam of electrons is transmitted through a material to form an image; transmission electron microsopy (TEM). This material is most often an ultrathin sheet less than 100 nm thick (TEM is, therefore, a very suitable analysis method for CVD thin films). Moreover, TEM can capture very fine details in the material—even as small as a single column of atoms. It is, therefore, a major analytical method in both physical, chemical and biological sciences, and finds applications in various areas of materials sciences (such as nanotechnology and semiconductor research).

Even though these characterization methods are of an overall high quality, there are still some problems connected with the analysis of CVD growth processes. For instance, an atomic-level resolution is very difficult (or even not possible) to achieve for a specific surface geometrical structure. This is especially the case with the in-depth resolution. This is the situation for both the geometrical structure and chemical composition. Just to mention one example, it is not possible to experimentally identify detailed variations in chemical composition over a specific surface area. In addition, there are also some major problems when trying to analyse the CVD growth process over time, or to analyse the growing film in the CVD reactor. For the latter type of analysis, *in situ* characterization is required that is both very difficult and expensive to arrange. Even though these types of difficulties will over time become diminished by further development in the field, some major problems will not. This has to do with the principles of the CVD process itself (for details, see chapter 3). For instance, by changing one parameter (like the substrate temperature), another (or several others) will be changed at the same time (like type and concentration of active growth species right above the growing thin film surface). Hence, if it is of interest to study the effect of just one parameter on the CVD growth of a certain material, then one must find another way than to use the more traditional characterization methods. Luckily, this possibility is nowadays achievable by the introduction of fast computers and high-level theoretical programs. In fact, an increasing reliance has been placed upon theoretical modelling of the complex CVD processes in the growth of various types of materials.

1.3 Theoretical modelling and simulations

1.3.1 The beginning of CVD modelling

Since the end of the 1970s, there has been an increasing need for improvement of existing, as well as development of new, CVD processes and technologies. Trial-and-error approaches were for this purpose used, which turned out to be very inefficient and costly. A more systematic approach was later recognized, which led to the development and use of predictive modelling tools (e.g. CVD reactor modelling and modelling of the chemistry processes in a CVD reactor (based on elementary reactions)). Eversteyn *et al*, as early as 1970, predicted film uniformity by using a simplified analytic description of transport phenomena in a CVD reactor [33]. However, more realistic descriptions of hydrodynamics, transport phenomena and chemistry were presented in the late 1970s [34–36]. In fact, this period shows the beginning of the computational era for both CVD processes and equipment modelling.

1.3.2 Development of CVD modelling techniques

Since its beginning in the 1970s, simulation of CVD processes and equipment modelling has come a long way. The simulation of CVD processes was, initially, categorized in three different types: detailed CVD chemistry modelling, CVD reactor modelling and CVD microscopic modelling. Detailed chemistry modelling aims to identify homogeneous and heterogeneous reaction mechanisms, in combination with simple one-dimensional reactor models [37–57]. On the other hand, hydrodynamic models are used to study the flow and transport phenomena in the CVD reactor modelling. These calculations are based on relatively simple models of the CVD chemistry. The majority of the CVD reactor modelling studies address classical reactor configurations [37, 58–87]. Finally, the microscopic modelling describes the interaction between the molecular transport and deposition within sub-micron trenches and contact holes on wafers for semiconductor manufacturing [88–92].

Since the late 1990s, there has been an integration of reactor scale models with theoretical multiscale models [91–97]. This integration has been enabled by the very fast development of computer performances. It has also become possible to combine detailed descriptions of multi-dimensional transport models with detailed chemistry

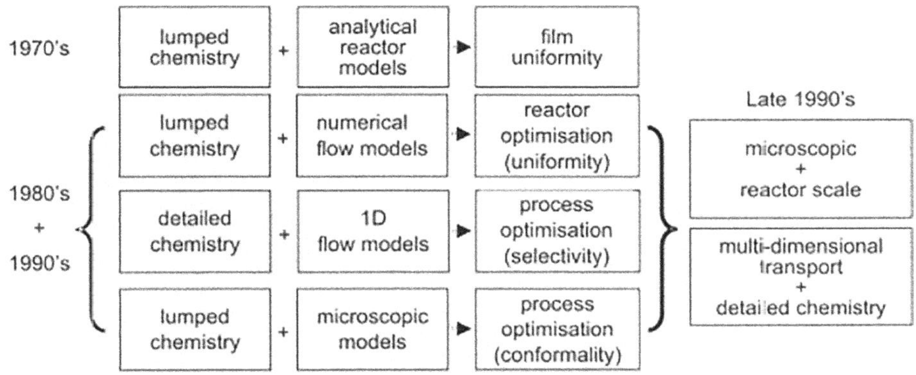

Figure 1.3. History of CVD model development and its use in CVD processing. [109], reproduced with permission © Elsevier.

models [46, 59, 98–101]. A multiscale modelling method for simulating and analysing the growth of CVD thin films has, thereby, emerged. Computational chemistry techniques have to an increasing extent been used in the development of this CVD growth model. These theoretical developments are schematically illustrated in figure 1.3. More detailed presentations of the development of CVD simulation models (in the past decades) can be found in [102–104].

1.3.3 CVD modelling and simulations of today

The CVD reaction processes are very complex, involving processes that are strongly coupled with each other. These processes are occurring at multiple time and length scales. Fast chemical reactions, such as surface migration (by a single jumping of an adsorbed species between neighbouring surface sites), have characteristic duration times of 10^{-12}–10^{-15} s and involve displacements of approximately 10^{-10} m. In addition, most CVD experiments involve gas phase reactions that produce numerous chemical species, and there are strong and complex interactions between the growing thin film surface (or substrate) and this reactive gas mixture. This complexity puts large demands on the CVD modelling and simulation techniques. For instance, there is the need to link the macroscale to the micro- or atomic scale, which is a very difficult task. Many numerical models have been developed to simulate the CVD processes, but most of them have only focused on a particular scale [105–107]. Theoretical methods like molecular dynamic simulations [108] and Monte Carlo simulations [110] were used for this purpose. A theoretical method that accurately describes all chemical CVD processes is expected to implement equations governing electronic interactions and molecular motion over various time and length scales.

The complexity level of the present CVD process modelling, and simulations, is impressive, especially when compared to the methods used in the early days of CVD modelling. This is mainly due to the large progress in the development of advanced methods and numerical techniques, in addition to the enormous increase in computer capacities. Considerable efforts are, however, still being made to develop

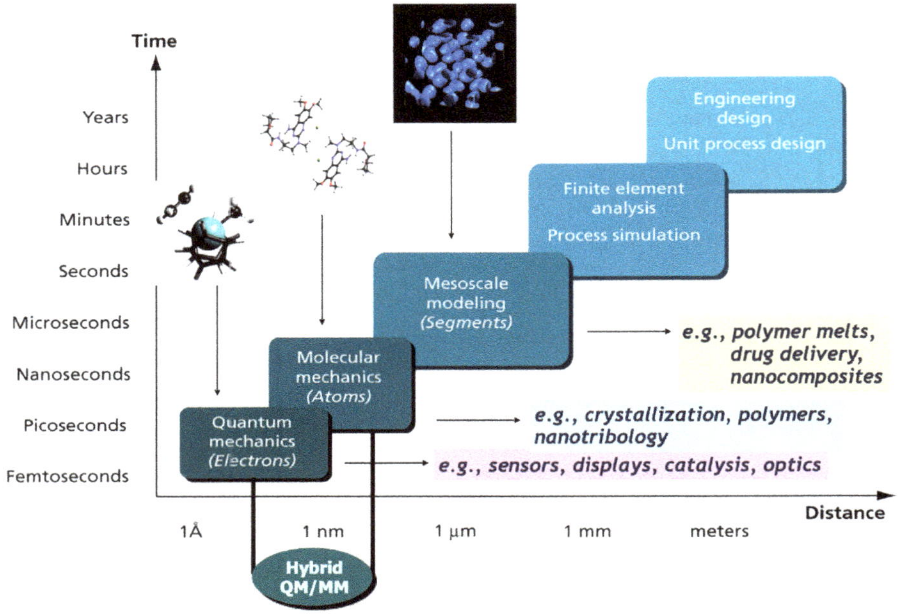

Figure 1.4. Characteristic time and length scales for various groups of theoretical methods and material types (BIOVIA, Inc.).

new algorithms through which it will be possible to combine models [111–113] and to increase the speed of calculations [114].

Multiscale modelling is a promising route towards the prediction of microstructure and properties of materials that have been grown by using CVD methods [115]. This is, however, a tremendous scientific and economic challenge. There are two types of multiscale modelling. One is referred to as hierarchical, or sequential, modelling, in which more accurate calculations are first taking place for a smaller scale model. The results from these calculations will then be used as input in the next level (and less accurate) method for calculations on a larger scale model. The other type is referred to as concurrent multiscale calculations, which considers multiple scales simultaneously. For both of these two types of multiscale modelling, different theoretical methods are (with their appropriate time and length scale resolutions) used to describe specific regions of the material [116]. Figure 1.4 give examples of theoretical methods for specific time and length scales, together with materials phenomena that can be studied. Examples of modelling and simulations of CVD growth mechanisms include silicon [115, 116], CND [117, 118], graphene [118], and ceramic materials (SiC, TiC, etc) [118].

1.3.4 Impact of CVD modeling and simulations

Computational modelling has had a large impact on the field of CVD. Not only in the way that the CVD growth processes have been developed, but also in how the equipment has been developed. However, the knowledge about the reaction mechanisms and kinetics is still far from sufficient for many important CVD processes. To experimentally try to gain more knowledge about these processes is both time-consuming and expensive. Thus, there are no chemistry models available for many newly introduced CVD processes. Moreover, whereas the current multi-scale theoretical methods largely succeed in relating the macroscopic process conditions to macroscopic properties of the deposited films, little progress has been in making a connection to film properties at the atomic scale (and their influence on the functional properties of the film). Further development of computational chemistry techniques, especially for surface reactions, is therefore very important. It will, if successful, make it possible to combine the detailed description of transport phenomena with the reaction chemistry, thereby forming a single computational model.

Ab initio methods, such as density functional theory (DFT), are the computational chemistry techniques that can provide the necessary input data to the upper scale methods. The DFT methods can, in fact, provide the key insights about the basic mechanisms of the atomic assembly. As such, the question of bridging the electronic models to the atomistic models has, during the last few decades, received increasing attention [119–123]. For example, DFT methods have been used to calculate energy barriers and rate constants, which, further on, have been used in upper scale models, such as atomistic molecular dynamic models and Monte Carlo models [119].

New forms of interatomic potentials, being based on the embedded atom method, Tersoff and bond order potentials, can be combined with molecular dynamics to investigate many aspects of the CVD processes. Moreover, the many activation barriers that are present during the CVD surface growth processes can, by using DFT calculations, be efficiently used as input data in kinetic Monte Carlo modelling of the CVD film growth. Examples of earlier work in this area include modelling of silicon nitride CVD [124], and selective metal–organic CVD of AlGaAs films on patterned substrates [95]. This illustrates the interest of *ab initio* calculations for the study of gas-phase and surface processes. Important advances in the predictive capabilities of multiscale modelling would most probably benefit from a wider use of first-principle calculations. Further integration of quantum calculations into multi-scale modelling and simulations of CVD growth of thin film is perhaps the most important, but also the most difficult, challenge for the years to come.

The purpose of the present book is not to be a detailed textbook of quantum mechanical methods methods (like DFT). It is also not a textbook presenting the most common chemical vapour deposition methods. Furthermore, the purpose is not to be a review over the recent modelling efforts in the field of CVD growth of various types of functional materials. On the contrary, the purpose is to show the methodology behind the modelling and simulation of CVD growth processes. More precisely, the book will give advice about what to avoid, and how to proceed, for a successful theoretical result. Example areas are given with some specific material types, but the methodology is generally applicable for any solid-state material. In this context, the book can be used as a very useful book of instructions for the practical handling of theoretical methods in the course of the calculations.

References

[1] Picreaux S and Pope. L 1986 Taylored surface modifications by ion implantation and laser treatment *Science* **226** 615–22

[2] Choy K 2003 Chemical vapour deposition of coatings *Prog. Mater Sci.* **48** 55–170

[3] Powell C F, Oxley J H and Blocher J M 1966 *Vapor Deposition* (New York: Wiley)

[4] Pring J N and Fielding W 1909 CLXXI—the preparation at high temperatures of some refractory metals from their chlorides *J. Chem. Soc.* **95** 1497–506

[5] Hoelbling R 1927 Über die herstellung und einige eigenschaften von reinem metallischen silicium *Z. Angew. Chem.* **40** 655–9

[6] Teal G K, Fischer J R and Treptov A W 1946 A new bridge photo-cell employing a photo conductive effect in silicon. Some properties of high purity silicon *J. Appl. Phys.* **17** 1879

[7] Storks K H and Teal G K 1948 Electrical translating materials and method of making them *US Patent* **2** 441–603

[8] Sato S, Sakurai H, Urabe K and Isumi Y 1985 Vapor-phase Beckmann rearrangement of cyclohexanone oxime over silica-boria catalyst prepared by chemical vapor deposition method *Chem. Lett.* **14** 277–8

[9] Jones A C and Hitchman M L 2009 *Chemical Vapour Deposition* (Cambridge: The Royal Society of Chemistry)

[10] Choy K L 2019 *Chemical Vapour Deposition (CVD)* (London: CRC Press)

[11] Bunshah R 1994 *Handbook of Deposition Technologies of Films and Coatings: Science Technology and Applications* (Blue Ridge, NJ: Noyes Publications)

[12] Shull R D and Sanchez J M 1993 *Nanophases and Nanocrystalline Structures* (Warrendale, OH: TMS)

[13] Daughton J and Granley J 1999 Giant magnetoresistance devices move *Ind. Phys.* **5** 22–5

[14] Painter O, Lee R K, Schere A, Yariv A, O'Brien J D, Dapkus P D and Kim I 1999 Two-dimensional photonic band-gap defect mode laser *Science* **284** 1819–21

[15] Shah A, Torres P, Tscharner R, Wyrsch N and Keppner H 1999 Photovoltaic technology: the case for thin-film solar cells *Science* **285** 692–8

[16] Fairbanks J W and Hecht R J 1987 The durability and performance of coatings in gas turbine and diesel engines *Mater. Sci. Eng.* **88** 321–30

[17] Parker M A, Coffey K R, Howard J K, Tsang C H, Fontana R E and Hylton T L 1996 Overview of progress in giant magnetoresistive sensors based on NiFe/Ag multilayers *IEEE Trans. Magn.* **32** 142–8

[18] Pan D and Towe E 1998 Conduction intersubband (In, Ga) As/GaAs quantum dot infrared photodetectors *Electron. Lett.* **34** 1883–4

[19] Hillery R V 1996 *Coatings for High-Temperature Structural Materials: Trends and Opportunities* (Washington, DC: National Academy Press)

[20] SIA 1997 *The National Technology Roadmap for Semiconductor* (Washington, DC: Semiconductor Industry Association)

[21] DiSalvo F J 1999 Thermoelectric cooling and power generation *Science* **285** 703–6

[22] Awschalom D D and Kikkawa J M 1999 Electron spin and optical coherence in semiconductors *Phys. Today* **6** 33–8

[23] Niinisto L, Paivasaari J, Niinisto J, Putkonen M and Nieminen M 2004 Advanced electronic and optoelectronic materials by atomic layer deposition: an overview with special emphasis on recent progress in processing of high-*k* dielectrics and other oxide materials *Phys. Status Solidi* A **201** 1443–52

[24] Knes M, Niesch K and Niinisto L 2007 Synthesis and surface engineering of complex nanostructures by atomic layer deposition *Adv. Mater.* **19** 3425–38
[25] Zaera F 2013 Nanostructured materials for applications in heterogeneous catalysis *Chem. Soc. Rev.* **42** 2746–62
[26] Stein A, Wilson B E and Rudisill S G 2013 Design and functionality of colloidal-crystal templated materials—chemical applications of inverse opals *Chem. Soc. Rev.* **42** 2763–803
[27] Marichy C, Bechelany M and Pinna N 2012 Atomic layer deposition of nanostructure materials for energy and environmental applications *Adv. Mater.* **24** 1017–32
[28] Donders M E, Oudenhoven J F M, Baggetto L, Knoops H C M, Van der Sanden M C M, Kessels W M M and Notten. P H L 2010 All-solid-state batteries: a challenging route towards 3D integration *Atomic Layer Deposition Applications 6, ECS Trans.* **33** 213–22
[29] Wachtman J B and Haber R A 1992 *Ceramic Films and Coatings* (Blue Ridge, NJ: Noyes Publications)
[30] Mills A, Hill D, Crow M and Hodgen S 2005 Thick titania films for semiconductor photocatalysis *J. Appl. Electrochem.* **35** 641–53
[31] Mills A, Hill D, Bhopal S and O'Neill S A 2003 Thick titanium dioxide films for semiconductor photocatalysis *J. Photochem. Photobiol. A: Chem.* **13** 56–60
[32] Leng Y 2013 *Materials Characterization: Introduction to Microscopic and Spectroscopic Methods* (New York: Wiley)
[33] Eversteyn F C, Severin P J W, van den Brekel C H J and Peek H L 1970 A stagnant layer model for the epitaxial growth of silicon from silane in a horizontal reactor *J. Electrochem. Soc.* **117** 925–31
[34] Wahl G 1977 Hydrodynamic description of CVD processes *Thin Solid Films* **40** 13–26
[35] Jensen K F and Graves D B 1983 Modeling and analysis of low pressure CVD reactors *J. Electrochem. Soc.* **130** 1950–7
[36] Coltrin M E, Kee R J and Miller J A 1984 A mathematical model of the coupled fluid mechanics and chemical kinetics in a chemical vapor deposition reactor *J. Electrochem. Soc.* **131** 425–33
[37] Coltrin M E, Kee R J and Evans G H 1989 A mathematical model of the fluid mechanics and gas-phase chemistry in a rotating disk chemical vapor deposition reactor *J. Electrochem. Soc.* **136** 819–28
[38] Allendorf M D and Kee R J 1991 A model of silicon carbide chemical vapor deposition *J. Electrochem. Soc.* **138** 841–52
[39] Okkerse M, de Croon M H J M, Kleijn C R, Marin G B and van den Akker. H E A 1998 A surface and a gas-phase mechanism for the description of growth on the diamond (100) surface in an oxy-acetylene torch reactor *J. Appl. Phys.* **84** 6387–98
[40] Arora R and Pollard R 1991 A mathematical model for chemical vapor deposition processes influenced by surface reaction kinetics: application to low-pressure deposition of tungsten *J. Electrochem. Soc.* **138** 1523–37
[41] Lengyel I and Jensen K F 2000 A chemical mechanism for *in situ* boron doping during silicon chemical vapor deposition *Thin Solid Films* **365** 231–51
[42] Coltrin M E, Ho P, Moffat H K and Buss R J 2000 Chemical kinetics in chemical vapor deposition: growth of silicon dioxide from tetraethoxysilane (TEOS) *Thin Solid Films* **365** 251–63
[43] Veneroni A and Masi M 2006 Gas-phase and surface kinetics of epitaxial silicon carbide growth involving chlorine-containing species *Chem. Vapor Deposition* **12** 562–8

[44] Liu B, Hicks R F and Zinck J J 1992 Chemistry of photo-assisted organometallic vapor-phase epitaxy of cadnium telluride *J. Crystal Growth* **123** 500–18
[45] Tirtowidjojo M and Pollard R 1988 Elementary processes and rate-limiting factors in MOVPE of GaAs *J. Crystal Growth* **77** 108–14
[46] Mountziaris T J, Kalyanasundaram S and Ingle. N K 1993 A reaction-transport model of GaAs growth by metalorganic chemical vapor deposition using trimethyl-gallium and tertiary-butyl-arsine *J. Crystal Growth* **131** 283–99
[47] Kim E J and Gill W N 1994 Modeling of CVD of silicon dioxide using TEOS and Ozone in a single-wafer reactor *J. Electrochem. Soc.* **141** 3462–72
[48] Southwell R P and Seebauer E G 1996 A predictive kinetic model for the chemical vapor deposition of TiSi2 *J. Electrochem. Soc.* **143** 1726–36
[49] Nitodas S F and Sotirchos S V 2002 Development and validation of a mathematical model for the chemical vapor deposition of alumina from mixtures of aluminum trichloride, carbon dioxide, and hydrogen *J. Electrochem. Soc.* **149** C120–9
[50] Wang Y-F and Pollard R 1994 A method for predicting the adsorption energetics of diatomic molecules on metal surfaces *Surf. Sci.* **302** 223–34
[51] Birakayala N and Evans E A 2002 A reduced reaction model for carbon CVD/CVI processes *Carbon* **40** 675–83
[52] Meeks E, Kee R J, Dandy D S and Coltrin M E 1993 Computational simulation of diamond chemical vapor deposition in premixed $C_2H_2/O_2/H_2$ and CH_4O_2-strained flames *Comb. Flame* **92** 144–60
[53] Simka H, Hierlemann M, Utz M and Jensen K F 1996 Computational chemistry predictions of kinetics and major reaction pathways for Germane gas-phase reactions *J. Electrochem. Soc.* **143** 2646–53
[54] Sato K, Haruta H and Kumashiro. Y 1997 *Ab initio* molecular-orbital study on the surface reactions of methane and silane plasma chemical vapor deposition *Phys. Rev.* B **55** 15467
[55] Frenklach M 1998 Simulation of surface reactions *Pure Appl. Chem* **70** 477–84
[56] Metzger R, Werner C and Spitzer. A 2000 Adsorption of precursor molecules in the CVD process of (Ba;Sr) titanate simulated by *ab-initio* calculations *Thin Solid Films* **365** 242–50
[57] Maruno S, Furukawa T, Nakahata T and Abe. Y 2001 A chemical mechanism for determining the influence of boron on silicon epitaxial growth *Jpn. J. Appl. Phys* **40** 6202–7
[58] van de Ven J, Rutten G J M, Raaymakers M J and Giling. L J 1986 Gas phase depletion and flow dynamics in horizontal MOCVD reactors *J. Crystal Growth* **76** 352–71
[59] Kleijn. C R 2000 Computational modeling of transport phenomena and detailed chemistry in chemical vapor deposition—a benchmark solution *Thin Solid Films* **365** 294–306
[60] Kleijn C R, Hoogendoorn C J, Hasper A, Holleman J and Middelhoek. J 1991 Transport phenomena in tungsten LPCVD in a single-wafer reactor *J. Electrochem. Soc.* **138** 509–16
[61] Moffat H K and Jensen K F 1988 Three-dimensional flow effects in silicon CVD in horizontal reactors *J. Electrochem. Soc.* **135** 459–70
[62] Visser E P, Kleijn C R, Govers C A M, Hoogendoorn C J and Giling. L J 1989 Return flows in horizontal MOCVD reactors studied with the use of TiO_2 particle injection and numerical calculations *J. Crystal Growth* **94** 929–46
[63] Kleijn C R and Hoogendoorn. C J 1991 A study of 2-and 3-D transport phenomena in horizontal chemical vapor deposition reactors *Chem. Eng. Sci.* **46** 321–34
[64] Kommu S, Wilson G M and Khomami. B 2000 A theoretical/experimental study of silicon epitaxy in horizontal single-wafer chemical vapor deposition reactors *J. Electrochem. Soc.* **147** 1538–50

[65] Meziere J, Ucar M, Blanquet E, Pons M, Ferret P and Di Cioccio L 2004 Modeling and simulation of SiC CVD in the horizontal hot-wall reactor concept *J. Crystal Growth* **267** 436–51

[66] Nishizawa S and Pons. M 2006 Growth and doping modeling of SiC-CVD in a horizontal hot-wall reactor *Chem. Vapor Deposition* **12** 516–22

[67] Cheng T S and Hsiao. M C 2006 Computation of three-dimensional flow and thermal fields in a model horizontal chemical vapor deposition reactor *J. Crystal Growth* **293** 475–84

[68] Pollard R and Newman J 1980 Silicon deposition on a rotating disk *J. Electrochem. Soc.* **127** 744–52

[69] Fotiadis D I, Kremer A M, McKenna D R and Jensen. K F 1987 Complex flow phenomena in vertical MOCVD reactors: effects on deposition uniformity and interface abruptness *J. Crystal Growth* **85** 154–64

[70] Evans G and Greif. R 1987 A numerical model of the flow and heat transfer in a rotating disk chemical vapor seposition reactor *J. Heat Transf.* **109** 928–35

[71] Wang Y B, Chaussavoine C and Teyssandier. F 1993 2D modelling of a non-confined circular impinging jet reactor; Si chemical vapour deposition *J. Crystal Growth* **126** 373–95

[72] Hwang J Y, Park C, Huang M and Anderson. T 2005 Investigation of mass transport phenomena in an upflow cold-wall CVD reactor by gas phase Raman spectroscopy and modeling *J. Crystal Growth* **279** 521–30

[73] Mukinovic M, Brenner G, Khanderi J, Spollmann S, Fischer R A, Tafipolsky M, Cadenbach T and Schmid. R 2005 A multiscale simulation approach for the MOCVD of GaN using a single-molecule precursor in a vertical stagnation flow reactor *Chem. Vapor Deposition* **11** 306–16

[74] Oh I-H, Takoudis C G and Neudeck. G W 1991 Mathematical modeling of epitaxial silicon growth in pancake chemical vapor deposition reactors *J. Electrochem. Soc.* **138** 554–66

[75] Peskin A P and Hardin G R 1998 Gallium arsenide growth in a pancake MOCVD reactor *J. Crystal Growth* **186** 494–510

[76] Yang L, Farouk B and Mahajan R L 1992 Three-dimensional predictions of silicon deposition in a barrel type CVD reactor *J. Electrochem. Soc.* **139** 2666–73

[77] De Paola E and Duverneuil P 1998 Simulation of silicon deposition from SiHCl3 in a CVD barrel reactor at atmospheric pressure *Comput. Chem. Eng.* **22** S683–6

[78] Di Stanislao M, Valente G, Fascella S, Spampinato C, Masi M and Carra S 2002 Multi-model hierarchy approach to simulate barrel reactors for epitaxial silicon deposition *J. Phys. IV* **12** 121–8

[79] Werner C, Ilg M and Uram. K 1996 Three-dimensional equipment modeling for chemical vapor deposition *J. Vac. Sci. Technol.* A **14** 1147–51

[80] Bergunde T, Dauelsberg M, Makarov Y N, Kadinski L, Weyers M, Schmitz D, Strauch G and Jurgensen. H 1997 Heat transfer and mass transport in a multiwafer MOVPE reactor: modelling and experimental studies *J. Crystal Growth* **170** 66–71

[81] Dauelsberg M, Kadinski L, Makarov Y N, Bergunde T, Strauch G and Weyers. M 2000 Modeling and experimental verification of transport and deposition behavior during MOVPE of $Ga_{1-x}In_xP$ in the planetary reactor *J. Crystal Growth* **208** 85–92

[82] Hopfmann C, Werner C and Ulacia. J I 1991 Numerical analysis of fluid flow and nonuniformities in a polysilicon LPCVD batch reactor *Appl. Surf. Sci.* **52** 169–87

[83] Houf W G, Grcar J F and Breiland. W G 1993 A model for low pressure chemical vapor deposition in a hot-wall tubular reactor *Mat. Sci. Eng.* B **17** 163–71

[84] Schoof G J, Kleijn C R, Van den Akker H E A, Oosterlaken T G M, Terhorst H J C M and Huussen F 2002 Simulation and validation of SiO LPCVD from TEOS in a vertical 300 mm multi-wafer reactor *J. Phys. IV* **12** 51–61

[85] Kleijn. C R 1991 A mathematical model of the hydrodynamics and gas-phase reactions in silicon LPCVD in a single-wafer reactor *J. Electrochem. Soc.* **138** 2190–200

[86] Soderberg. A 1992 Investigation of buried etch stop layer in silicon made by nitrogen implantation *J. Electrochem. Soc.* **139** 561–6

[87] Habuka H, Nagoya T, Katayama M, Shimada M and Okuyama K 1995 Modeling of epitaxial silicon thin-film growth on a rotating substrate in a horizontal single-wafer reactor *J. Electrochem. Soc.* **142** 4272–8

[88] Raupp G B and Cale T S 1989 Step coverage prediction in low-pressure chemical vapor deposition *Chem. Mater.* **1** 207–14

[89] Hasper A, Kleijn C R, Holleman J, Middelhoek J and Hoogendoorn C J 1991 Modeling and optimization of the step coverage of tungsten LPCVD in trenches and contact holes *J. Electrochem. Soc.* **138** 1728–38

[90] Cale T S and Raupp G B 1990 Free molecular transport and deposition in cylindrical features. *J. Vac. Sci. Technol.* B **8** 649–55

[91] Coronell D G and Jensen K F 1994 Simulation of rarefied gas transport and profile evolution in nonplanar substrate chemical vapor deposition *J. Electrochem. Soc.* **141** 2545–51

[92] Akiyama Y and Imaishi N 1995 Applicability of one-dimensional diffusion model for step coverage analysis—comparison with a simple Monte Carlo method *Appl. Phys. Lett.* **67** 620–2

[93] Merchant T P, Gobbert M K, Cale T S and Borucki L J 2000 Multiple scale integrated modeling of deposition processes *Thin Solid Films* **365** 368–75

[94] Cale T S, Bloomfield M O, Richards D F, Jansen K E and Gobbert. M K 2002 Integrate multiscale process simulations *Comp. Mat. Sci.* **23** 3–14

[95] Cavalotti C, Nemirovskaya M and Jensen K F 2003 A multiscale study of the selective MOVPE of AlxGa1−xAs in the presence of HCl *J. Crystal Growth* **248** 411–6

[96] Jensen K F, Rodgers S T and Venkataramani R 1998 Multiscale modeling of thin film growth *Curr. Opin. Solid State Mater.* **3** 562–9

[97] Rodgers S T and Jensen K F 1998 Multiscale modeling of chemical vapor deposition *J. Appl. Phys.* **83** 524–31

[98] Mankelevich Y A, Rakhimov A T and Suetin N V 1996 Two-dimensional simulation of a hot-filament chemical vapor deposition reactor *Diamond Relat. Mater.* **5** 888–94

[99] Kuijlaars K J, Kleijn C R and van den Akker H E A 1998 Simulation of selective tungsten chemical vapour deposition *Mater. Sci. Semicond. Process.* **1** 43–54

[100] Ingle N K, Theodoropoulos C, Mountziaris T J, Wexler R M and Smith F T J 1996 Reaction kinetics and transport phenomena underlying the low-pressure metalorganic chemical vapor deposition of GaAs *J. Crystal Growth* **167** 543–56

[101] Ern A, Giovangigli V and Smooke M D 1997 Detailed modeling of three-dimensional chemical vapor deposition *J. Crystal Growth* **180** 670–9

[102] Kleijn C R 1995 Chemical vapor deposition processes ed M Meyyap-pan *Computational Modeling in Semiconductor Processing* (Boston, MA: Artech House)

[103] Kleijn C R, Kuijlaars K J, Okkerse M, Van H, Santen and van den Akker H E A 1999 Some recent developments in chemical vapor deposition process and equipment modeling *J. Phys. IV* **9** 117–32

[104] Masi M, Di Stanislao M and Veneroni A 2003 Fluid-dynamics during vapor epitaxy and modeling *Prog. Crystal Growth Charact. Mater.* **47** 239–70

[105] Jensen K F 1994 X *Handbook of Crystal Growth* vol 3A, ed D T J Hurle (Amsterdam: Elsevier)

[106] Gilmer G H 1994 X *Handbook of Crystal Growth* vol 3A, ed D T J Hurle (Amsterdam: Elsevier)

[107] IslamRaja M M, Capelli M A, McVittie J P and Saraswat K C 1991 A 3-dimensional model for low-pressure chemical-vapor-deposition step coverage in trenches and circular vias *J. Appl. Phys.* **70** 7137–40

[108] Zhou X W, Johnson R A and Wadley H N 1997 A molecular dynamics study of nickel vapor deposition: temperature, incident angle, and adatom energy effects *Acta Mater.* **45** 1513–24

[109] Kleijn C R, Dorsman R, Kuijlaars K J, Okkerse M and van Santen H 2007 Multi-scale modeling of chemical vapor deposition processes for thin film technology *J. Cryst. Growth* **303** 362–80

[110] Yang Y, Zhou X W, Johnson R A and Wadley H N 2001 Monte Carlo simulation of hyperthermal physical vapor deposition *Acta Mater.* **49** 3321–32

[111] Mae K 2002 Molecular dynamics aided kinetic Monte Carlo simulations of thin film growth of Ag on Mo(1 1 0) with structural evolution *Surf. Sci.* **482–85** 860–5

[112] Haslam A J, Moldovan D, Phillpot S R, Wolf D and Gleiter H 2002 Combined atomistic and mesoscale simulation of grain growth in nanocrystalline thin films *Comput. Mater. Sci.* **23** 15–32

[113] Voter A F 1997 Hyperdynamics: accelerated molecular dynamics of infrequent events *Phys. Rev. Lett.* **78** 3908

[114] Shu J W, Lu Q, Wong W O and Huang H 2002 Parallelization strategies for Monte Carlo simulations of thin film deposition *Comp. Phys. Comm.* **144** 34–45

[115] Rabe D 1998 *Computational Materials Science* (Weinheim: Wiley-VCH)

[116] Karakasidis T E and Charitidis C A 2007 Multiscale modeling in nanomaterials science *Mater. Sci. Eng.* C **27** 1082–9

[117] Raji K, Thomas S and Sobhan C B 2011 A chemical kinetic model for chemical vapor deposition of carbon nanotubes *Appl. Surf. Sci.* **257** 10562–70

[118] Elliott J A, Shibuta Y, Amara H, Bichara C and Neyts E C 2013 Atomistic modelling of CVD synthesis of carbon nanotubes and graphene *Nanoscale* **5** 6662–76

[119] Oleinik I I, Pettifor D G, Sutton A P and Butler J E 2000 Theoretical study of chemical reactions on CVD diamond surfaces *Diam. Relat. Mater.* **9** 241–5

[120] Rugerrone P, Kley A and Scheffler M 1997 Microscopic aspects of homoepitaxial growth *Prog. Surf. Sci.* **54** 331–40

[121] Pettifor D G II, Oleinik D, Nguyen-Mahn and Vitek V 2002 Bond-order potentials: bridging the electronic to atomistic modelling hierarchies *Comput. Mater. Sci.* **23** 33–7

[122] Wimmer E and Wolf W 2000 *Ab initio* methods as an integral part of microelectronic materials modeling: status and perspectives *Mater. Sci. Semicond. Process.* **3** 3–11

[123] Esteve A, Djafari-Rouhani M, Faurous P and Esteve D 2000 Modeling of the silicon (100) thermal oxidation: from quantum to macroscopic formulation *Mater. Sci. Semicond. Process.* **3** 47–57

[124] Bagatur'yans A A, Novoselov K P, Safonov A A, Savchenko L L, Cole J V and Korkin A A 2000 Atomistic modeling of chemical vapor deposition: silicon nitride CVD from dichlorosilane and ammonia *Mater. Sci. Semicond. Process.* **3** 23–9

IOP Publishing

Chemical Vapour Deposition
Growth processes on an atomic level
Karin Larsson

Chapter 2

Common CVD reactor setups

Chemical vapour deposition (CVD) is a highly versatile process where activated gaseous reactants undergo chemical surface reactions in the deposition of a thin solid material onto another solid material (i.e. the substrate). The source gas is introduced into the CVD reaction chamber, and therein activated to form very reactive growth species. The different activation methods use heat, plasma, or light, and their respective CVD process is denoted as HFCVD (or TACVD), PECVD, and LACVD. The incoming source gas will be thermally cracked onto the surface of a heated metal filament coil when using HFCVD. Several fragments of the source gas will thereby be formed, which are all in their ground state. It is, in addition, most often needed to separately heat the substrate, but to a lower temperature. Contrary to the situation with focused heating in the HFCVD process, the complete CVD chamber will be heated to a uniform temperature in the TACVD process. The drawback with this method is that not only will the source gas and substrate/film be heated, but also the chamber walls. There is, hence, an apparent risk that deposition also will take place on these walls. PECVD is a technique in which electrical energy, rather than thermal energy, is used to initiate homogeneous reactions in the gas phase. Chemically active ions and radicals are thereby formed in the generated plasma, which can participate in heterogeneous reactions on the surface of the growing film (or substrate). In the LACVD method, the source gas will be directly activated by a focused laser beam. The generation of active fragments can, thereby, take by two different means. The source molecules will either be cracked by the temperature of the laser beam or undergo excitations that are induced by the laser frequency. Thus, the use of a laser in the activation of the source gas makes it possible to tailor-make the formation of specific fragments in the activation process.

In addition to gas phase activation, the various CVD reactors can be separated by the use of different pressures, gas flow, etc.

doi:10.1088/978-0-7503-3107-4ch2

2.1 General

The most common CVD reactors have some common features [1–7]. Gas phase species are introduced into the reactor by using a vacuum pump at the outlet of the reactor (see figure 2.1). This pump is removing not only the gas that hasn't been used in the reactor, but also desorption products from the growing film. Thus, produced volatile by-products will be removed by the gas flow through the reaction chamber, which is then cleaned in a scrubber. There is, thus, a pre-set flow of a gas mixture above a substrate onto which the final product—the thin film material—will be deposited (see figure 2.2). This gas mixture will also be held at a pre-set total pressure

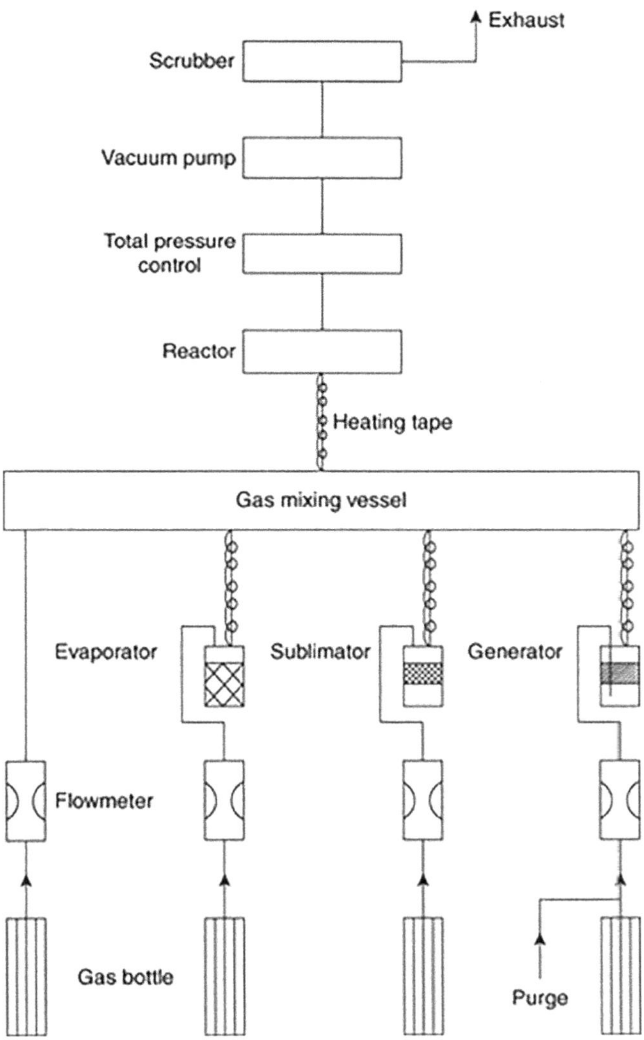

Figure 2.1. Schematic showing the main components in a CVD equipment. [1], reproduced with permission © Elsevier.

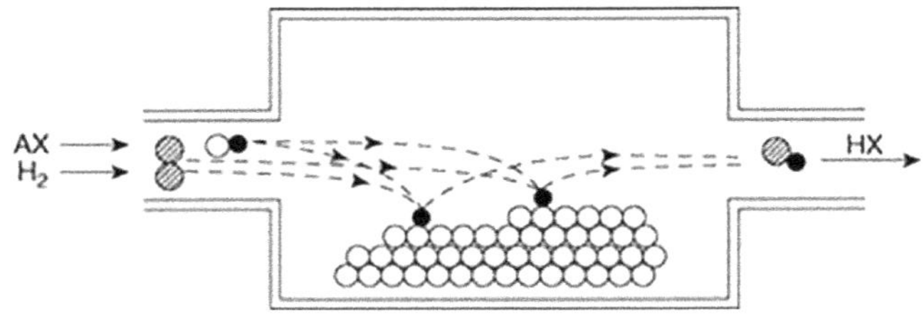

Figure 2.2. Heterogeneous surface reaction in the CVD reactor. [1], reproduced with permission © Elsevier.

in the reactor, where the relative concentrations of gas phase species determine their specific partial pressures. In addition, the gas mixture will be activated just above this film, with the purpose to create the most reactive growth species. For practical reasons, this gas mixture can contain a non-reactive carrier gas when a very high partial pressure of gaseous reactants has to be avoided. In the case of CVD growth of diamond, a hydrogen (H_2) gas has instead been used (e.g. with 98% H_2(g) and 2% CH_3(g)). Moreover, the film will be deposited on a substrate (e.g. Si) at a specific temperature, *T*, which can be applied to the substrate (and, thereby, the film) by different means. In summary, various experimental parameters must be pre-set for a CVD experiment (also called operating conditions). These ones contain (at least) the gas phase flow, total pressure, substrate temperature, etc. In addition, the general construction of a CVD reactor can be made in many different ways [1–7]. For a specific CVD construction, with a specific gas phase activation technique, the various parameters are most often interdependent, i.e. by varying one it will most probably affect one, or several, of the others. For instance, an increased total pressure will directly affect the temperature of the growing thin film material.

2.2 Classification of CVD reactors

A CVD setup can be classified according to (i) type of reactants, (ii) level of total pressure, and (iii) type of gas phase activation. For each type of class, the most common examples will be rather briefly presented here. A more detailed description of these classes is given in [1–6].

2.2.1 Reactant types

A CVD setup can be classified according to type, and state (i.e. gaseous, liquid, or solid), of the reactants. For instance, metalorganic chemical vapour deposition (MOCVD) uses metalorganic precursors, which are most often in their liquid state at room temperature (see below).

As shown in table 2.1, the reactant compounds that are used as CVD precursors can be classified as inorganic compounds and organic metallic compounds. The inorganic compounds can be further classified into hydrides and halides. A more detailed description of the various precursors is presented in table 1.3 in section 1.1.4. The hydrides and halides have high vapour pressures and are in their gas, or liquid, state at room temperature. They can, therefore, easily be supplied into the CVD reactor in their gaseous state (see section 2.2 for the liquid material precursor).

Table 2.1. Typical reactant material used in CVD.

Inorganic compound	Hydride	SiH_4, PH_3, B_2H_6, AsH_3
	Halide	SiH_2Cl_2, $SiCl_4$, $TiCl_4$
Organic metallic compound	Alkoxy compound	TEOS, TEPO, TMB, PETa
	Alkyl compound	TMP, TMAl
	Complex compound	$Sr(DPM)_2$, $Bi(DPM)_3$

Alkoxy, alkyl, and more complex compounds are mainly used as organic metallic compounds. All of them show low vapour pressure, and are in their liquid or solid state at room temperature. They must, therefore, be vapourized when used as a precursor in a CVD reactor. There is, thusly, an issue regarding how efficiently these organic metallic compounds can be vapourized and supplied to the CVD reactor.

2.2.1.1 Vapourization of liquid precursors

The reactants in a CVD reactor are in their gaseous state. However, they do not necessarily come from a gas tube. They can, instead, be transformed from their corresponding liquid (or solid) state or be the product of chemical reactions (that all takes place just at the inlets of the CVD reactor). The liquids must, however, be very volatile, and the solids must easily undergo sublimation. The liquid or solid is then dissolved in a convenient solvent. Another alternative is to use chemical reactions just prior to the gas inlet. This method is necessary when a short-lived gaseous reactant is needed for the CVD growth of a specific material. Another reason might be that the gas is extremely difficult or dangerous to handle when stored (e.g. explosive or poisonous). However, these types of reactant productions put a large demand on the construction of the CVD setup.

2.2.1.1.1 The bubbling method

The use of an initially liquid precursor is very common, and one frequently used method is the *bubbling method.* The configuration of this method is illustrated in figure 2.3. The liquid material is mixed with a carrier gas (Ar in figure 2.3) until its saturated state is reached. Furthermore, the liquid material is maintained at a constant temperature and pressure, and the flow rate of the carrier gas is controlled to send a constant amount of material to the CVD chamber. As can be seen in figure 2.3, the resulting gaseous precursor is then eventually mixed with a carrier gas and other gaseous precursors and transported to the substrate as in classical CVD. High growth rates can be reached by using this technique. However, a variation in heat of vaporization, or bubbler pressure, will cause changes in the supply amount of the material. Therefore, various types of controls, such as a mass flow meter (MFM) or a heat conduction type concentration sensor (TCD), must be installed to compensate for these possible variations in supply amounts. However, the control accuracy is limited, so the bubbling method has not recently been used that frequently. Another reason is the materials of low vapour pressure, for which a high flow rate of supply cannot be achieved. Other vaporization methods have,

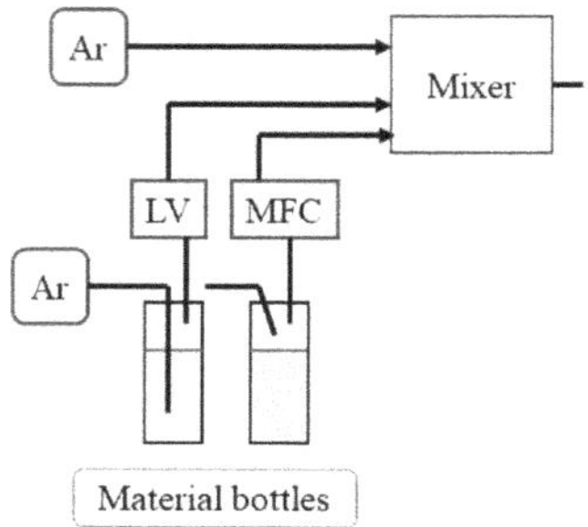

Figure 2.3. Schematic diagram of the creation of gaseous precursors by using the bubbling method. The left material bottle contains a liquid and the right a solid material. LV: laser velocimeter; MFC: mass flow control; Ar: argon.

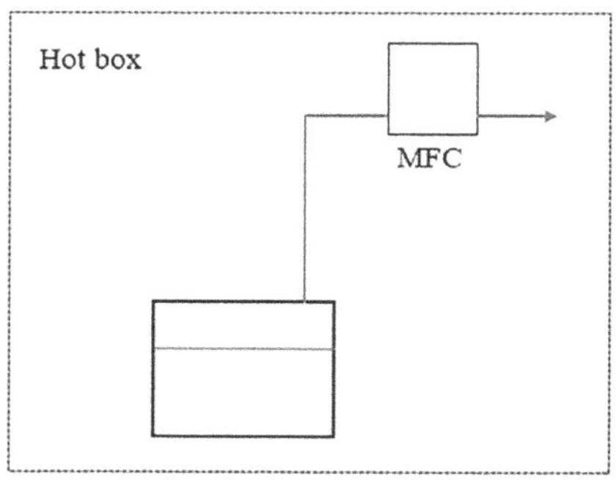

Figure 2.4. Baking vapourization method.

therefore, been used (the baking and injection method) out of which the injection method has become the mainstream method.

2.2.1.1.2 The baking method

The configuration of the baking method is illustrated in figure 2.4. The liquid precursor is heated to increase its vapour pressure above the liquid material. The flow rate of the formed vapour, into the CVD reactor, is, thereafter, controlled by using a mass flow controller (MFC). Thus, the supply of gaseous precursors into the CVD reactor is in this method directly controlled by the MFC, which is also heated to the applied temperature (see figure 2.4). Thus, the flow rate of gaseous species can be accurately controlled. Another positive aspect with this method is that the interfusion of impurities can be minimized. This specific type of vapourization method is frequently used to control the tetraetoxysilan (TEOS), an alkoxy, compound of silicon.

However, this method also has some drawbacks. Firstly, it is not suitable for materials of low vapour pressure and also not for atmospheric pressure CVD processes. Secondly, the equipment is also expensive, and, thirdly, the entire raw material supply system must be kept at a constant temperature.

2.2.1.1.3 The injection method

The injection method directly controls the flow rate of the liquid precursor and carries out the vaporization and supply into the CVD reactor of only a specific amount. This method can be further classified into the *column method* (where the liquid is vapourized in a column) and the *direct injection valve method* (where the material is vapourized directly in a control valve).

The *column method* is illustrated in figure 2.5. The carrier gas and the liquid precursor are mixed and vapourized in the heated column. The column includes metallic particles with the purpose to increase the contact area of the liquid for an efficient vapourization thereof. It is then only necessary to heat the column and the subsequent line. However, the drawback is that cavitation in the liquid can cause an unstable generation of vapour.

The *direct injection method* (also called the *liquid injection method*) vapourizes the liquid by mixing the liquid with a carrier gas while heating the flow control valve for the liquid. Since this type of vapourizer also serves as the control valve, it can be constructed in a small size and is also not expensive. The gas–liquid interface is formed at the seat of the control valve, which enables a rapid rising and falling of the gaseous precursor supply into the CVD reactor. A schematic of the *direct injection method* is illustrated in figure 2.6.

2.2.2 Total pressure

In addition, the CVD reactors can be classified according to level of total pressure:

- Atmospheric pressure CVD (APCVD)—CVD at atmospheric pressure
- Low-pressure CVD (LPCVD)—CVD at sub-atmospheric pressure (e.g. at 40 Torr).
- Ultrahigh vacuum CVD (UHVCVD)—CVD at very low pressure (typically $<10^{-6}$ Pa ($\approx 10^{-8}$ Torr).

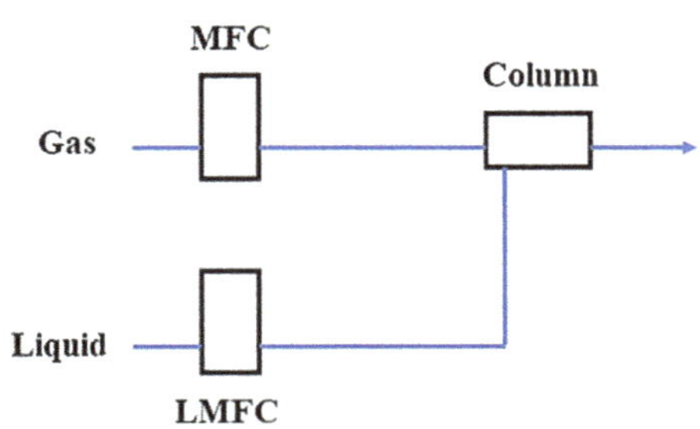

Figure 2.5. Column method.

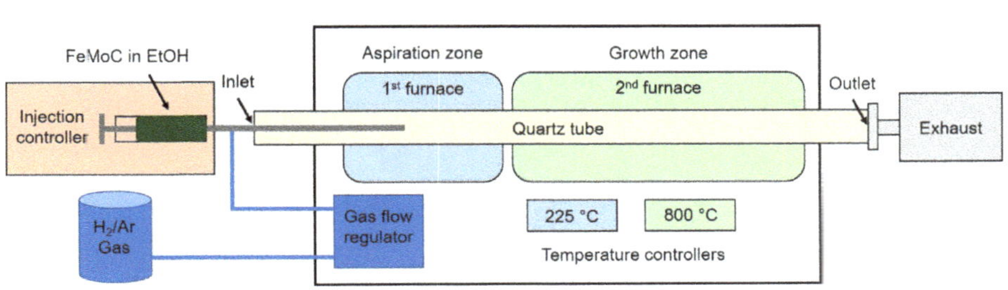

Figure 2.6. Schematic of a direct liquid injection chemical vapor deposition furnace [8].

Reduced pressures tend to reduce unwanted gas-phase reactions and, thereby, improve the thin film quality. On the other hand, a reduced pressure will prolong the CVD deposition time (i.e. causing a lower deposition rate). The majority of modern CVD reactors are either working under LPCVD or UHVCVD conditions.

2.2.3 Gas phase activation

There are various possibilities for gas phase activation, and their main types can be summarized in the following main categories;

- Thermally-assisted CVD (TACVD)
- Hot filament CVD (HFCVD)
- Plasma-assisted CVD (PACVD)
- Laser-assisted CVD (LACVD)
- Photo-initiated CVD (PICVD)
- Combustion CVD (CCVD)

Also, a combination of these gas phase activation techniques is often used (e.g. HF-assisted activation in combination with laser-assisted activation [9]). Moreover, instead of using a gas mixture in the reactor, it has been shown to be possible to sequentially introduce the different reactants into the reactor. A carefully controlled layer-by-layer growth of different elements will then be formed on the substrate (see figure 2.7). This method is, thus, called atomic layer deposition (ALD). Another name is atomic layer epitaxy (ALE) since it is most often used for heteroepitaxial growth of one material onto another. Different types of gas phase activation techniques can be used for this specific CVD technique [10]. One example of an ALE growth is the deposition of copper from copper(I) chloride and H by means of ALE [11]. (This process is described in detail in chapter 7.) This ALE process consists of four steps that are repeated until a film of desired thickness has been formed: (1) CuCl is first supplied, and the molecules are adsorbed on the substrate surface. (2) This is followed by a purging with an inert gas to remove excess precursors, leaving one monolayer of adsorbed CuCl. (3) To remove the chlorine atoms from the surface, hydrogen is added to form HCl (which is desorbed and

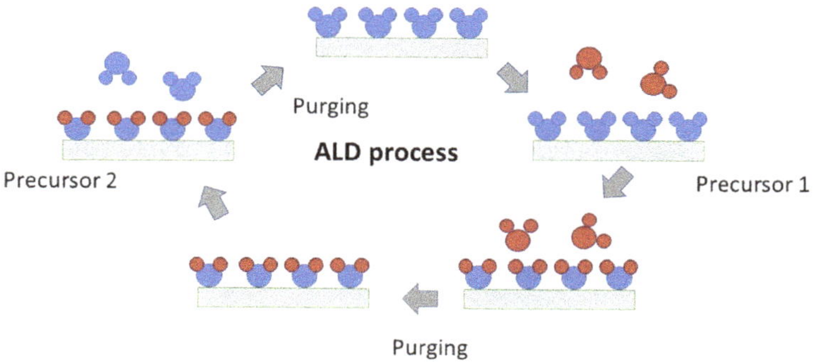

Figure 2.7. Schematic of an ALD process.

removed from the deposition chamber by pumping). (4) Finally, the reactor is purged with an inert gas (i.e. Ar) again to remove the HCl formed, as well as the excess hydrogen.

The ALD method has several features that make it advantageous over conventional CVD methods [12]:

- The growth process is self-limiting, which provides a simple thickness control, a good reproducibility, and the capability to produce sharp interfaces and superlattices.
- There is an alternate supply of precursors that favours the use of precursors that are otherwise highly reactive towards each other, thus enabling an efficient precursor utilization and a lowering of the deposition temperature.
- There is a wide processing temperature window, which makes it possible to make multilayer structures in a continuous process.

The major drawback is, however, the low growth rate. Also, it cannot be used for the deposition of an elemental material (like diamond). A horizontal, low-pressure hot-wall (i.e. thermal-assisted) ALD reactor is demonstrated in figure 2.8.

By using laser-assisted activation in the ALD process, the deposition temperature can generally be decreased and/or the deposition rate increased [9]. A specially designed combined laser- and thermal-assisted ALD reactor is presented in figure 2.9, and the corresponding pulsing scheme is demonstrated in table 2.2. It was used for the deposition of BN (see chapter 8). The horizontal reaction chamber consists of two fused silica tubes. To be more specific, the BBr_3 precursor was introduced via the inner tube (diameter 28 mm), which ended just before the deposition zone, and NH_3 entered directly into the outer reactor tube (diameter 40 mm). Moreover, N_2 (with a purity of 99.999%) was used as the carrier gas. For the photolysis of NH_3 and BBr_3, an ArF excimer laser ($\lambda = 193$ nm) was used, at a repetition rate of 20 Hz. The laser beam was focused above and parallel to the substrate surface by a cylindrical plano-convex lens (CVI) with a focal length of 500 mm. Laser-induced light emission was observed perpendicular to the laser beam by using an optical detection system consisting of a Czerny–Turner-type of grating spectrograph, a charge-coupled device (CCD) detector, and an optical multichannel analyzer.

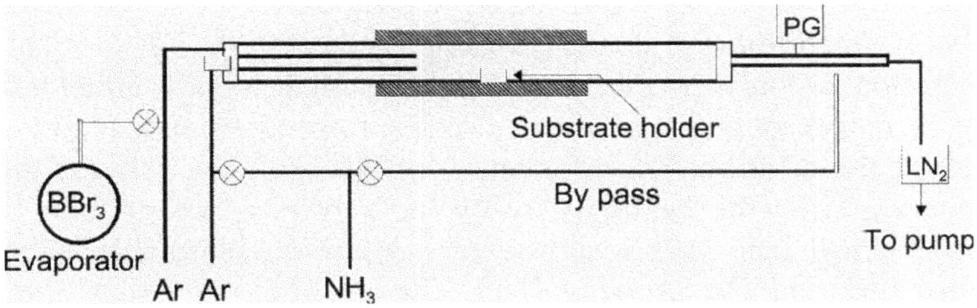

Figure 2.8. Horizontal thermal-activated ALD system. An oven is externally heating the reactor around the substrate holder. [10], reproduced with permission © Elsevier.

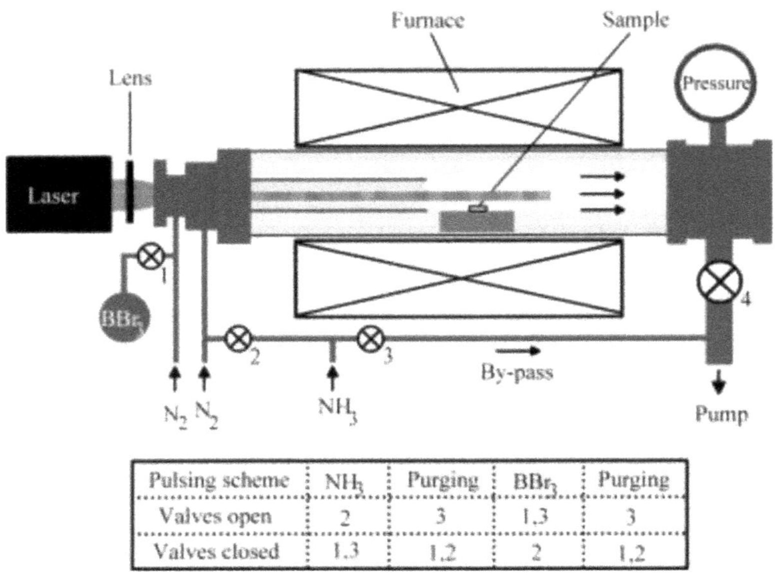

Figure 2.9. Horizontal thermally- and laser-activated ALD system. A needle valve (4) was used to keep a constant pressure in the reaction chamber holder. [9], reproduced with permission © Elsevier.

Table 2.2. Pulsing scheme for the ALD-system presented in figure 2.8.

Pulsing scheme(s)	NH_3	Purging	BBr_3	Purging
Valves open	2	3	1.3	3
Valves closed	1.3	1.2	2	1.2

2.2.3.1 Thermally-assisted CVD

The thermally-assisted CVD reactors are classified by type of substrate heating: cold-wall CVD or hot-wall CVD. The chamber is heated by an external power source in the cold-wall CVD, while the substrate is heated by radiation from the heated chamber walls (called hot-wall CVD). The temperature is then uniform in the whole chamber. However, as a drawback, deposition may also be formed on the chamber walls, in addition to unwanted gas phase reactions. On the contrary, in a cold-wall reactor it is only the substrate that is directly heated by either induction or by passing current through the substrate itself. A third alternative is to use a heater in contact with the substrate (i.e. from beneath). The chamber walls are then kept at room temperature, so the heating of only the substrate, rather than the gas or chamber walls, will help reduce unwanted gas-phase reactions that can lead to particle formation on the growing film. The possibility to vary the substrate temperature is also very important for the control of surface kinetic control. Examples of hot-wall and cold-wall CVD setups are demonstrated in figure 2.10.

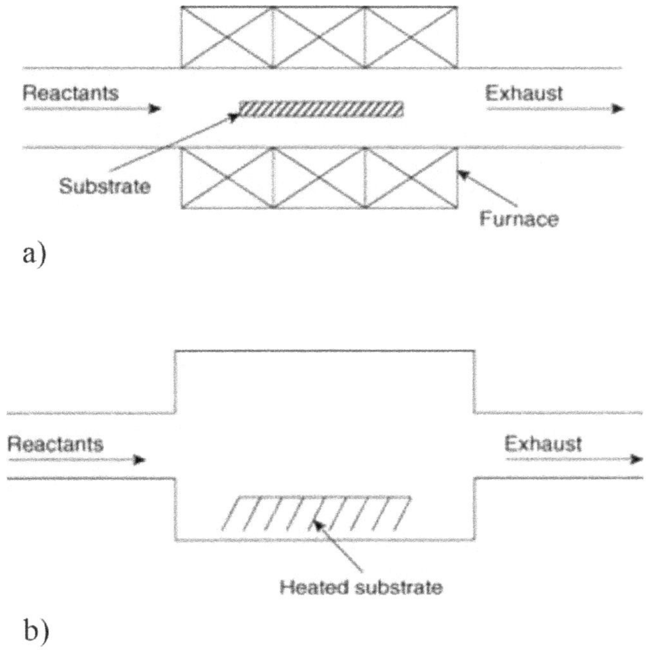

Figure 2.10. Presentations of (a) hot-wall and (b) cold-wall CVD reactors. [1], reproduced with permission © Elsevier.

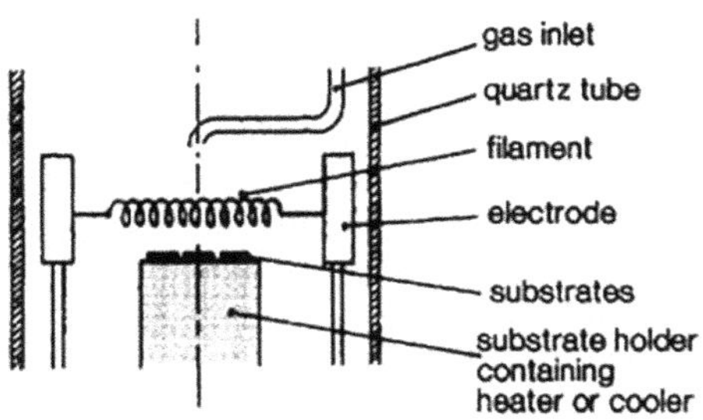

Figure 2.11. Presentation of a hot-filament cold-wall reactor. [13], reproduced with permission © Elsevier.

2.2.3.2 Hot filament-assisted CVD

Hot filament CVD (HFCVD)—also called catalytic CVD (cat-CVD)—relies on a heated coiled wire to decompose the precursor reactants present in the gas mixture and deposit a film on the substrate surface kept near the filament at lower temperatures (see figure 2.11). The filament temperature and substrate temperature are then independently controlled. This allows for a construction where the substrate is kept at a colder temperature (for improved adsorption rates) and a necessary higher temperature (for the decomposition of precursors to activated growth species) is used for the filament. The HFCVD technique is limited only by the size of the reactor and lends itself to be incorporated into continuous roll-to-roll industrial fabrication processes (see figure 2.12). For example, for the synthesis of carbon

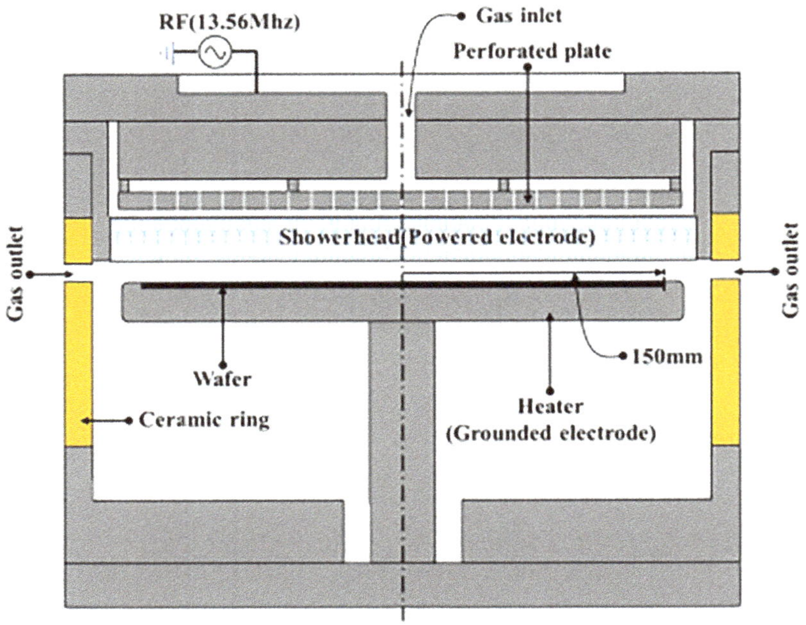

Figure 2.12. Schematic of a PECVD reactor with a showerhead structure for amorphous carbon deposition. [14], reproduced with permission © Elsevier.

materials (e.g. diamond, carbon nanotubes, graphene), a gas mixture consisting of a hydrocarbon (e.g. methane, acetylene) diluted in hydrogen is decomposed utilizing a refractory metal filament (such as tungsten, tantalum, or rhenium) that is resistively heated and maintained to 1800 °C–2300 °C (as measured with a dual wavelength pyrometer). The filament temperature plays an essential role and influences the type and quality of the materials that are grown. For the situation with carbon deposition, the higher the filament temperature, the faster the filaments become carburized. The metal carbides are brittle, resulting in relatively short filament lifetimes and unsteady growth conditions.

For the cold-wall reactor, the filament assembly is usually kept 5–10 mm above the heated substrate. The filament wire diameter, length, and geometry must be optimized according to the intended film size, as well as the filament temperature and gas pressure required by the specific CVD reaction. Moreover, the filament is held fixed by two parallel electrodes made of a refractory metal, typically molybdenum, to avoid evaporation of the electrode material onto the growing film. Coil-shaped filaments are often used since they enhance the total surface area for the heterogeneous cracking process.

However, they tend to bend downward over time during, thereby causing unsteady growth conditions. A filament arrangement consisting of multiple parallel stretched filaments is usually a better choice, since it enhances the uniformity and total area of the deposited film. However, the multiple filament arrangement requires very high-current power supplies.

The substrate temperature that is required for the synthesis of a particular material is determined by the residence time (see section 3.3.2) and mobility (see section 6.7.4) of reactant molecules and radicals that adsorb at the substrate surface. In the HFCVD reactor, the substrate temperature is intrinsically affected by the filament temperature and power, filament-to-substrate distance, and total pressure. An active substrate temperature control is, therefore, in most cases needed to be able to optimize the quality of the CVD grown films. While the filament temperature is dictated by the energy required to break the precursor gas molecules, the filament-to-substrate distance and total pressure are determined by the mean-free path of the necessary fragments that form at, or near, the filament (to avoid further reactions to unwanted products in the gas phase).

Besides determining the required filament and substrate temperatures, filament-to-substrate distance, and total pressure, the choice of substrate material and gas composition are also critical parameters to obtain the intended film type and quality. This is something that will be discussed more in detail for the CVD growth of c-BN in chapters 8–10. The right gas-phase chemical environment and substrate temperature are conditions that are necessary, but not sufficient, for the CVD growth of the intended thin film material. After the right substrate has been chosen, the final film quality and growth rate will primarily be determined by the gas phase composition.

2.2.3.3 Plasma-assisted CVD

The plasma-assisted CVD methods (also called plasma-enhanced CVD) use a plasma to enhance the chemical reactivity of the precursors. A plasma is any gas in which a significant percentage of the atoms or molecules is ionized. The fractional ionization in plasmas used for deposition varies from about 10^{-4} in typical capacitive discharges, to as high as 5%–10% in high density inductive plasmas. Moreover, processing plasmas are typically operated at pressures of a few mTorr to a few Torr, although arc discharges and inductive plasmas can be ignited at atmospheric pressure. Plasmas with low fractional ionization are of great interest for materials processing. The reason is that the electrons are so light compared to atoms and molecules that the energy exchange between the electrons and the neutral gas particle is very inefficient. Therefore, the electrons can be maintained at a very high equivalent temperatures—tens of thousands of K—while the neutral atoms remain at their ambient temperature. These energetic electrons can induce many processes that would otherwise be very improbable at low temperatures, such as dissociation of precursor molecules and the creation of large quantities of free radicals.

A second benefit of deposition within a discharge arises from the fact that electrons are more mobile than ions. Consequently, the plasma is normally more positive than any object it is in contact with, as otherwise a large flux of electrons would flow from the plasma to the object. The voltage between the plasma and, e.g. a substrate surface it is in contact with is normally dropped across a thin sheath region (see figure 2.12). Ionized atoms or molecules that diffuse to the edge of the sheath region feel an electrostatic force and are accelerated towards the neighboring substrate surface. Thus, all surfaces that are exposed to a plasma will become bombarded by energetic ions. The potential across the sheath above an electrically-isolated substrate (the floating potential) is typically of size 10–20 V. However, much higher sheath potentials can be achieved by adjusting the reactor geometry and configuration. Thus, deposited films will be exposed to energetic ion bombardment during deposition. This bombardment can lead to increases in density of the film, and help remove contaminants, improving the film's electrical and mechanical properties.

It is, thus, clear that fast and very reactive radicals, and ions, can be formed when creating a plasma of a gas mixture. Hence, many CVD setups uses a plasma to activate the gaseous phase. It, therefore, allows deposition at lower temperatures, which is often critical for the situation with temperature sensitive substrate materials. In fact, the PECVD processing temperature is as low as 200 °C to 400 °C. Thus, PECVD is commonly used for situations where the substrate, or deposited films, would otherwise suffer from degradation if subjected to the higher temperatures required by, e.g. TACVD. This is very important for the manufacture of semiconductors and the deposition of organic coatings. More generally, by varying the plasma we can add additional control to the properties of the deposited film.

Remote plasma-enhanced CVD (RPECVD) is similar to PECVD, except that the substrate is not positioned directly in the plasma discharge region. By removing the substrate from the plasma region it allows for a reduced processing temperature (down to room temperature).

In comparison with TACVD and HFCVD, and their gas phase activation by using a high temperature or a heated filament at a specific distance above the substrate, there are different plasma types, and, therefore, also different plasma-assisted CVD setups (most commonly called plasma-enhanced—PECVD). The plasma is generally created by using

- direct current (DC),
- radiofrequency (RF),
- microwave frequency (MW),
- arc, or
- pulsed DC current.

2.2.3.3.1 Direct current (DC) plasma

A simple direct-current (DC) discharge can be readily created at a few Torr between two conductive electrodes and may be suitable for the deposition of conductive materials (see figure 2.13). However, insulating films will quickly extinguish this discharge as they are deposited. This type of plasma is, thus, not suitable for the deposition of non-doped diamond films.

2.2.3.3.2 Alternating-current (AC) plasma

It is more common to excite a capacitive discharge by applying an alternating-current (AC) or radio-frequency (RF) signal between an electrode and the conductive walls of a reactor chamber, or between two cylindrical conductive electrodes facing one another. The latter configuration is known as a parallel plate reactor. Frequencies of a few tens of Hz to a few thousand Hz will produce time-varying plasmas that are repeatedly initiated and extinguished. Frequencies of tens of kilohertz to tens of megahertz result in reasonably time-independent discharges.

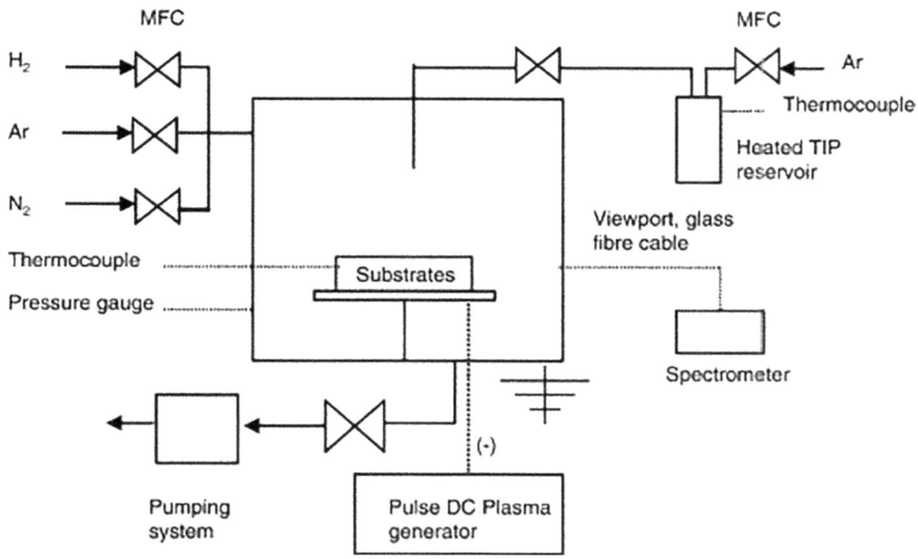

Figure 2.13. Presentation of a PECVD reactor with a DC plasma. [15], reproduced with permission © Elsevier.

Excitation frequencies in the low-frequency (LF) range, usually around 100 kHz, require several hundred volts to sustain the discharge. These large voltages lead to high energy ion bombardment of surfaces. High-frequency plasmas are often excited at the standard 13.56 MHz frequency, which is widely available for industrial use. At high frequencies, the displacement current from the sheath movement and scattering from the sheath, will assist in the ionization. Thus, a lower voltage is sufficient to achieve the higher plasma densities. Hence, one can adjust the chemistry and ion bombardment in the deposition by just changing the frequency of excitation.

Capacitive plasmas are usually very lightly ionized, which results in limited dissociation of precursors and low deposition rates. Much denser plasmas can be created by using inductive discharges, in which an inductive coil is excited with a high-frequency signal. This will induce an electric field within the discharge, and the electrons become accelerated in the plasma itself rather than just at the sheath edge.

2.2.3.3.3 Radio frequency (RF) plasma

In PECVD, the RF power is mostly used for the deposition of materials with highly insulating properties (e.g. in depositing diamond-like carbon (DLC) films). The reason is that the substrate is required to be powered with an alternating pulse. A matching box is usually used alongside the RF generator to reduce the reflected power. Furthermore, the power setting differs based on the material deposited. For the intentional control of the CVD conditions, the substrate bias voltage (being one of the important parameters for deposition) is not independent of other conditions but mostly determined by the RF power density. Since the substrate bias influences the energy of the bombarding ions onto the growing film, it is expected to be controlled independently. As can be seen in figure 2.14, the precursor gas is fed to the

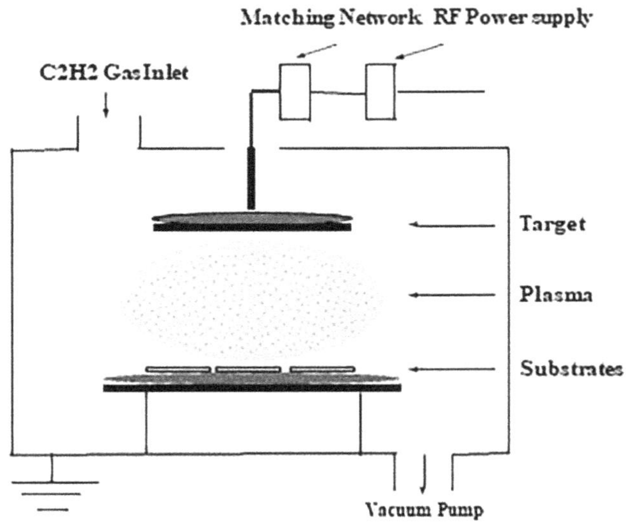

Figure 2.14. Presentation of a PECVD reactor with an RF plasma. [16], reproduced with permission © Elsevier.

RF PECVD chamber by using a showerhead arrangement to ensure uniformity of precursor concentration over the deposited film. The so-called direct exposure RF PECVD systems typically employ the showerhead as an electrode for the introduction of RF energy to create the plasma. As for the other types of PECVD systems, the precursors entering the plasma undergoes electron-molecule collisions, thereby producing high energy excited molecules and molecular fragments that can adsorb on the substrate surface and deposit the film. The substrate sits on a heated platen and the by-product gases are exhausted through ports below the wafer level.

2.2.3.3.4 Microwave frequency (MW) plasma

The production of a microwave frequency (MW) plasma in the PECVD reactor, also known as glow-discharge, is demonstrated in figure 2.15. By utilizing two electrodes in the reaction chamber, a high frequency power is applied to the gas to be ionized. When the microwave frequency is used, plasma is produced by the application of a high frequency electric field (of 2.45 GHz) at a low pressure of a carrier gas (such as oxygen, argon, or nitrogen). At such high frequencies, electrons (due to their small masses) can accelerate toward the higher energy levels. However, this is not the case for the heavier ions. Therefore, the temperature of electrons increases, but it remains low for the heavier ions. The electrons with their high energy levels are, thus, responsible for collision with the precursor gas molecules, to dissociate them and make them reactive for further plasma polymerization. This technique has been used for deposition of conductive films from a metal target.

2.2.3.3.5 Arc plasma

The arc plasmas typically operate at moderate frequencies of 1 MHz lower than the microwave plasma sources. However, arc plasmas are extremely hot and require

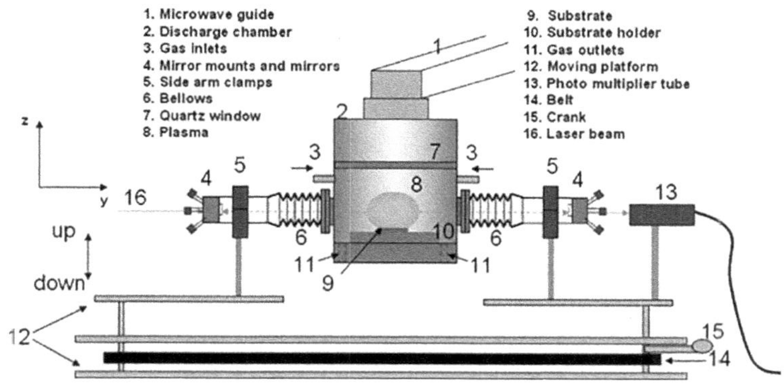

Figure 2.15. Demonstration of an MW PECVD (also called MWCVD) reactor. [17], reproduced with permission © Elsevier.

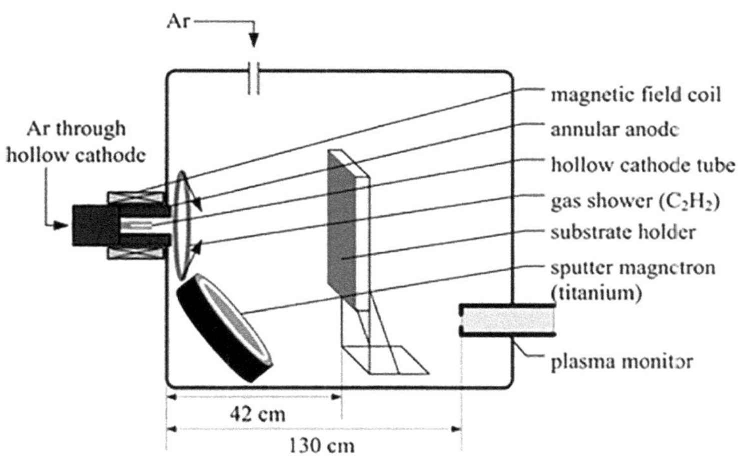

Figure 2.16. Side view on the vacuum chamber in an arc PECVD reactor. The arrow at the gas shower indicate the acetylene flow direction into the external plasma. [18], reproduced with permission © Elsevier.

higher power, so they are not usually utilized. Some efforts to avoid these limitations include a limiting discharge time (which is the time applied for the electrons to accelerate in the electric field), the use of pulsed voltages, or by using a dielectric material between the electrodes to reduce the accumulation of electrons (see figure 2.16). Arc plasma PECVD over the MW PECVD and TACVD techniques (due to the latter method's decreased production of atomic hydrogen species) [7]. However, due to the high temperature of the arc plasma deposition, a substrate cooling is necessary. But this is difficult process, so the generation of a high temperature in AP PECVD is generally regarded to be a disadvantage of this method. However, the high deposition rate and obtained film thickness are some of the advantages.

The AP PECVD technique is mainly used for the deposition of diamond where the process occurs at a higher pressure as compared with MW PECVD. Therefore, the number of collisions increases, and, consequently, the temperature is higher. Due to the high temperatures, a total dissociation of hydrogen molecules occurs and has, therefore, proven to be advantageous.

2.2.3.3.6 Pulsed DC plasma

A pulsed DC PECVD reactor is considered to be an alternative to the RF PECVD reactor. It is supposed to be a simple and cost-effective deposition system that can be used to improve the film properties (as compared to widely used RF PECVD system [20]. In the pulsed DC PECVD system (see figure 2.17), a pulsed DC power supply is a significant component for the reason that it may, by varying the pulse parameters, be possible to change (i) the gas phase chemistry, (ii) energy of ionic bombardment onto the substrate, (iii) electron density, and (iv) electron temperature to meet the requirement of the highly selective deposition process [21, 22]. However, most of the models of commercially available pulsed DC power supplies are expensive and custom-made, which has led to less flexibility in controlling the pulse conditions.

Combustion-CVD: Combustion chemical vapor deposition (CCVD), also called flame pyrolysis, is an open-atmosphere, flame-based technique for the deposition of high-quality thin films (e.g. functional layers) and nanomaterials (see figure 2.18). When using combustion-CVD, the precursors will be mixed into a combustion gas. The precursors mainly contain metalorganic compounds like, e.g. silanes or siloxanes. By applying the combustion energy on the precursors, very reactive compounds will be formed. They will be deposited on the substrate surface, forming firmly bonded silicate coatings with a thickness of 5–50 nm. The surfaces of both metals and plastic materials can be activated by a combustion treatment. Moreover,

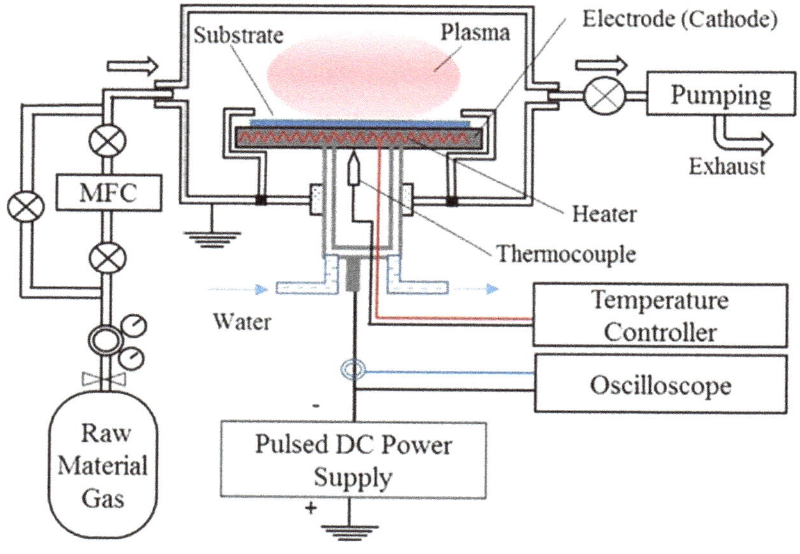

Figure 2.17. Demonstration of a pulsed DC PECVD reactor. [19], reproduced with permission © Elsevier.

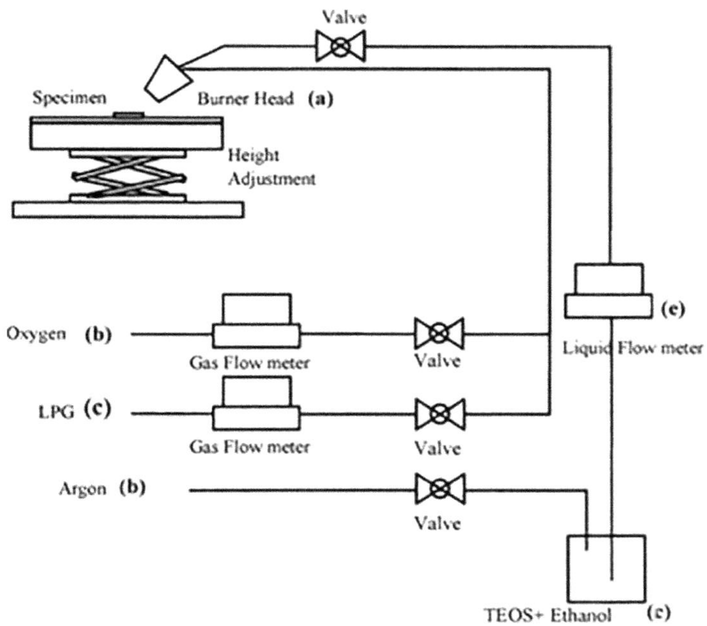

Figure 2.18. Demonstration of a combustion CVD system. [23], reproduced with permission © Elsevier.

due to the nature of the combustion, the upper molecule layers in the deposited films will become oxidized forming carbonyl, carboxyl, and hydroxyl groups. These groups will then react with succeeding coatings, with reactions based on van der Waals forces between polar atomic groups. A weak point of these reactive groups is their poor resistance to hydrolyzation. Furthermore, the increase in surface activity is reversible, and will fall back to the original situation within a short period of time (30–120 min depending on the polymer).

2.2.3.4 Photo-induced CVD

The photo-induced CVD process (PICVD) uses UV light to create activated growth species (see figure 2.19). It is similar to plasma processing, given that plasmas are strong emitters of UV radiation. PICVD can, under certain conditions, operate at or near atmospheric pressure.

Many low-temperature processing CVD techniques have previously been studied. Photo-induced CVD (PICVD) was found to be a very promising technique since the processed surface is not subjected to damaging ion bombardment that can occur in plasma-assisted CVD systems. In the earlier years, most work in this area employed high-power and costly excimer laser systems, or cheaper but much lower power incoherent lamp sources [24, 25]. However, novel excimer lamp-based systems became developed later on, which provided high powers, higher efficiencies, and extended availability of wavelengths (as compared to the more traditional lamp sources). The intense UV light is generated in these devices by using dielectric barrier discharges to form excimer molecules of rare gases and rare-gas halides. This has opened up the

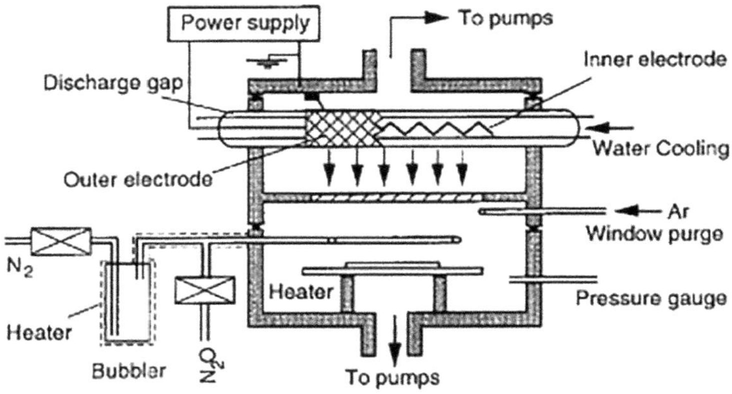

Figure 2.19. Demonstration of a photo-induced CVD (PICVD) reactor. [26], reproduced with permission © Elsevier.

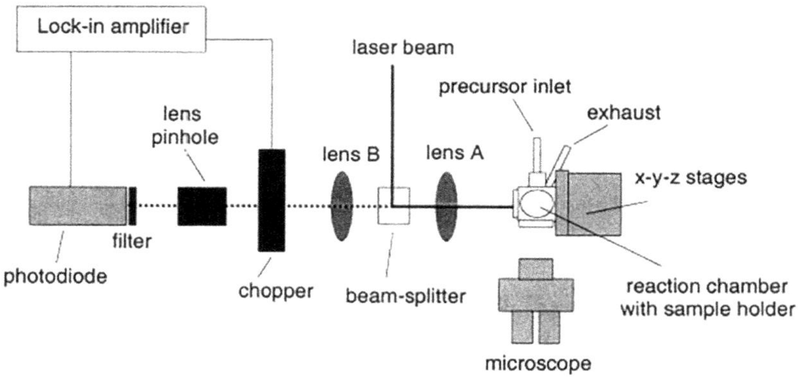

Figure 2.20. Demonstration of a laser-assisted CVD (LACVD) setup for localized heating of the substrate. [29], reproduced with permission © Elsevier.

opportunity for relatively inexpensive and large area photo-processing. Such processes have already been successfully used to deposit silicon dioxide and silicon nitride films at low temperatures by irradiating silane, nitrous oxide, and ammonia mixtures [27, 28].

2.2.3.5 Laser-assisted CVD

The laser-assisted CVD (LACVD) process uses lasers as an activation technique. It can either use the laser to heat spots or lines on a substrate (for, e.g. semiconductor applications) (see figure 2.20). However, it can also use the laser to decompose the growth precursors in the gas phase just above the growing material film [9, 29]. The laser is then aligned with the substrate surface, but at a specific distance in between (see figure 2.7). This specific use of a laser is discussed in detail in section 2.2.3.3.

For the use of a thermally activated LACVD deposition, the thermalization of the laser energy is fast compared to the surface reactions on the growing film. The laser acts, if focused, as a localized heat source and the gaseous reactants should not absorb the laser light (i.e. the chosen wavelength should not excite the gas phase). However, if the photons cause dissociation or excitation of the molecules through absorption of the laser light, the process will be photochemically activated. The laser beam can be positioned parallel or perpendicular with respect to the substrate surface. The parallel beam configuration is only used in photolytic LACVD, where the laser beam is absorbed partially by the gas phase (see figure 2.9). Photolytic activation allows for low temperature deposition of thin films and is, thus, similar to PECVD. On the other hand, the perpendicular laser configuration is used to deposit lines, rods, and more complicated three-dimensional structures of different materials (see figure 2.20). As illustrated in figure 2.21, the different growth reaction steps occur only within the laser-heated zone.

The main differences between TACVD and thermal LACVD are the higher growth rate and the smaller size of the deposition area for thermal LACVD. The higher growth rate is due to the effective, nearly three-dimensional mass transport of the gaseous species to the deposition zone. Since laser beams can be focused, small areas can be heated and a typical laser beam in the visible region can be focused down to a spot size below 1 μm in diameter. These differences can have a great impact on the growth process, which makes a direct comparison between TACVD and thermal LACVD sometimes difficult. Theoretically, any material that can be deposited as a thin film by conventional CVD can be deposited as lines, dots, or three-dimensional structures by thermal LACVD. Lines are deposited by moving the focused laser beam along the substrate surface during deposition. The shape of the substrate can be arbitrary, i.e. planar or three-dimensional. This technique is generally referred to as direct writing.

It is also possible to grow rods and other three-dimensional structures by using thermal LACVD. The laser focus should then be moved away from the substrate surface when the deposition has been initiated (see figure 2.22).

The ability to grow rods by LACVD can be used for [31]:

(1) Investigation of process chemistry and kinetics before depositing more complex structures.
(2) Deposition of new materials under extreme physical conditions (e.g. for high temperatures, high pressures, within electric or magnetic fields, etc).

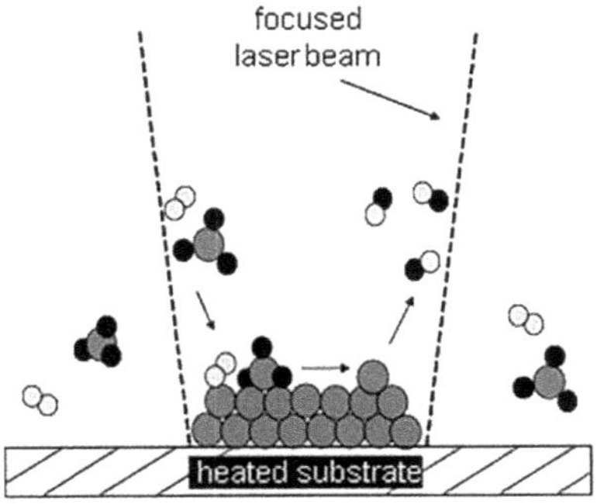

Figure 2.21. Thermal LACVD growth processes for a perpendicular laser beam configuration.

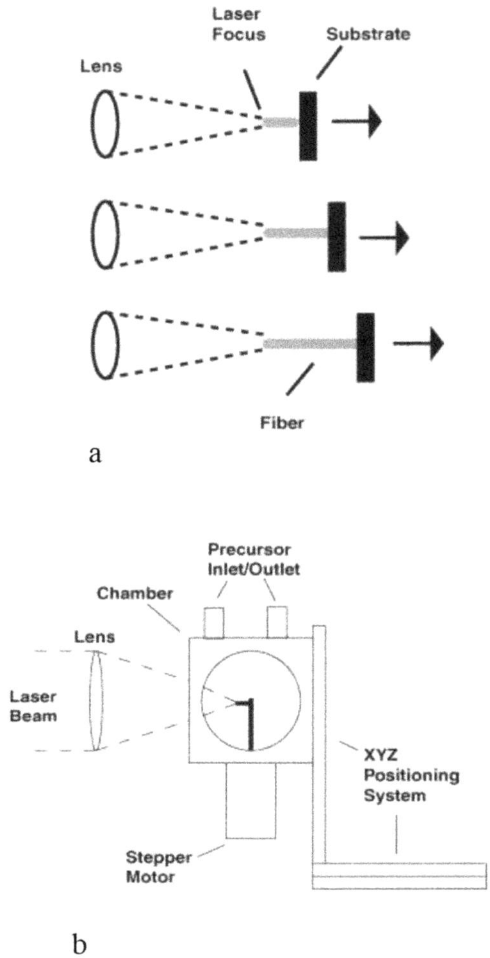

Figure 2.22. Deposition of a rod by moving the substrate away from the laser focus with the same speed as the growth rate of the material. [30], reproduced with permission © Elsevier.

(3) Fabrication of single crystals with high purity from a container-less cold atmosphere.
(4) Rapid prototyping of three-dimensional structures.

An SEM image of a single crystalline rod can be seen in figure 2.23.

To grow more complex three-dimensional structures such as springs, the laser focus must be moved away from the substrate surface in a more complicated manner (see figure 2.22(b)). In this case, the laser beam has to be focused at the end of a fiber, which is offset from an axis of rotation. Once growth begins, rotation around the axis starts, and the structure is translated slowly perpendicular to the laser beam. The translation speed determines the pitch of the spring. By using this method, different sizes and shapes of carbon springs have been grown from ethene (see figure 2.24) [30].

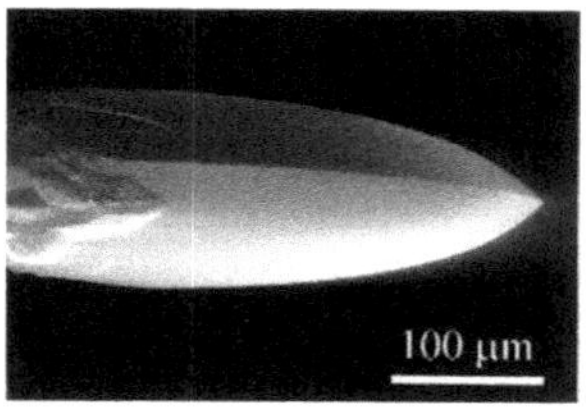

Figure 2.23. SEM image of a typical single crystalline rod. [29], reproduced with permission © Elsevier.

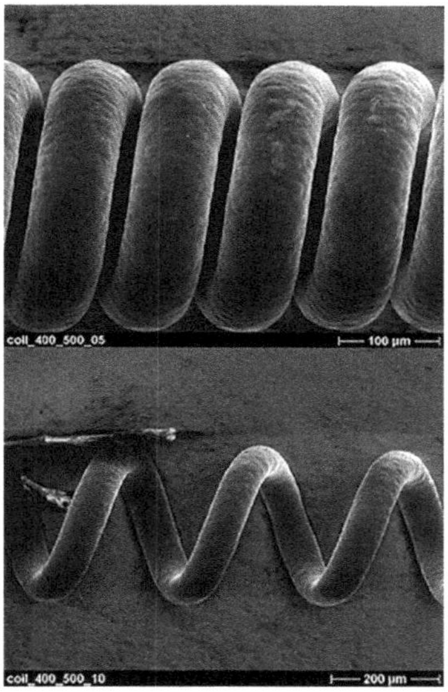

Figure 2.24. SEM photos of two springs deposited at 400 mW laser power and 500 mbar ethylene.

References

[1] Carlsson J-O and Martin P M 2010 *Handbook of Deposition Technologies for Films and Coatings: Science, Applications and Technology* 3rd edn (Norwich, NY: William Andrew) 314–361

[2] Choy K L 2019 *Chemical Vapour Deposition (CVD)* (London: CRC Press)

[3] Jones A C and Hitchman M L 2009 *Chemical Vapour Deposition* (Cambridge: The Royal Society of Chemistry)

[4] Choy K-L 2021 *Chemical Vapour Deposition (CVD): Advances, Technology and Applications* (Boca Raton, FL: CRC Press)

[5] Fischer R A 2005 *Precursor Chemistry of Advanced Materials: CVD, ALD, and Nanoparticles* 1st edn (Berlin: Springer-Verlag GmbH)

[6] Bunshah R 1994 *Handbook of Deposition Technologies of Films and Coatings: Science Technology and Applications* (Blue Ridge, NJ: Noyes Publications)

[7] Pierson H O 1999 *Handbook of Chemical Vapor Deposition (CVD)* 2nd edn (Norwich, NY: William Andrew)

[8] Esquenazi G L, Brinson B and Barron A R 2018 Catalytic growth of carbon nanotubes by direct liquid injection CVD using the nanocluster $[H_xPMo_{12}O_{40}\subset H_4Mo_{72}Fe_{30}(O_2CMe)_{15}O_{254}(H_2O)_{98-y}(EtOH)_y]$ *C-J. Carbon Res.* **4** 17
[9] Olander J, Ottosson M, Heszler P, Carlsson J-O and Larsson K 2005 Laser-assisted atomic layer deposition of boron nitride thin films *Chem. Vap. Deposition* **11** 330–7
[10] Ma°rlid B, Ottosson M, Pettersson U, Larsson K and Carlsson J-O 2002 Atomic layer deposition of BN thin films *Thin Solid Films* **402** 167–71
[11] Ma°rtensson P and Carlsson J-O 1997 Atomic layer deposition of copper on tantalum *Chem. Vap. Deposition* **3** 45–50
[12] Ehrlich D J and Tsao J Y 1989 *Laser Microfabrication; Thin Film Processes and Lithography* (Boston, MA: Academic)
[13] Haubner R and Lux B 1993 Diamond growth by hot-filament chemical vapor deposition: state of the art *Diam. Relat. Mater.* **2** 1277–94
[14] Dong G L, Seung K S, Han H S and Ko S 2019 The effect of hole density variation in the PECVD reactor showerhead on the deposition of amorphous carbon layer *Vacuum* **163** 37–44
[15] Mathur S and Kuhn P 2006 CVD of titanium oxide coatings: comparative evaluation of thermal and plasma assisted processes *Surf. Coat. Technol.* **201** 807–14
[16] Gholamali H, Shafiekhani A, Darabi E and Elahi S M Ag/Au alloy LSPR engineering by co-deposition of RF-sputtering and RF-PECVD *Appl. Organometal. Chem.* **32** 4316
[17] Ma J, Richley J C, Ashfold M N R and Mankelevich Y A 2008 Probing the plasma chemistry in a microwave reactor used for diamond chemical vapor deposition by cavity ring down spectroscopy *J. Appl. Phys.* **104** 103305
[18] Zimmermann B, Fietzke F, Klostermann H, Lehmann J, Munnik F and Möller W 2012 High-rate deposition of amorphous hydrogenated carbon films by hollow cathode arc PECVD *Surf. Coat. Technol.* **212** 67–71
[19] Mamuna A A, Furutaa H and Hattaa A 2018 Pulsed DC plasma CVD system for the deposition of DLC films *Mater. Today Commun.* **14** 40–6
[20] Scholl R A 1998 Assymetric bipolar pulsed power: a new power technology *Surf. Coat. Technol.* **98** 823–7
[21] Samukawa S and Mieno T 1996 Pulse-time modulated plasma discharge for highly selective, highly anisotropic and charge-free etching *Plasma Sources Sci. Technol* **5** 132–9
[22] Lieberman M A and Ashida S 1996 Global models of pulse-power-modulated high-density low-pressure discharges *Plasma Sources Sci. Technol.* **5** 145–60
[23] Oh S K, Ryu S, Kim Y H, Kim Y K M T and Kim D S 2009 High-temperature oxidation behavior of SiO_2 protective layer deposited on IN738LC superalloy by combustion CVD (CCVD) *Appl. Surf. Sci.* **256** 124–9
[24] Tanimoto S, Matsui M, Kamisako K, Kuroiwa K and Tarui Y 1992 Investigation on leakage current reduction of photo-CVD tantalum oxide films accomplished by active oxygen annealing *J. Electrochem. Soc.* **139** 320–30
[25] Nishimura Y, Tokunaga K and Tsuji M 1993 Deposition of tantalum oxide films by ArF excimer laser chemical vapour deposition *Thin Solid Films* **226** 144–8
[26] Zhang J-Y, Lim B and Boyd I W 1998 Thin tantalum pentoxide films deposited by photo-induced CVD *Thin Solid Films* **336** 340–3
[27] Yamagishi K and Tarui Y 1986 Photo-CVD of tantalum oxide film from pentamethoxy tantalum for VLSI dynamic memories *Jpn. J. Appl. Phys.* **25** L306–10

[28] Bergonzo P and Boyd I W 1993 Low pressure photodeposition of silicon nitride films using a xenon excimer lamp *Appl. Phys. Lett.* **63** 1757–9
[29] Bjorklund K L, Lu J, Heszler P and Boman M 2002 Kinetics, thermodynamics and microstructure of tungsten rods grown by thermal laser CVD *Thin Solid Films* **416** 41–8
[30] Williams K L, Köhler J and Boman M 2006 Fabrication and mechanical characterization of LCVD-deposited carbon micro-springs *Sensors Actuators* A **130–131** 358–64
[31] Nordine P C, de la Veaux S C and Wallenberger F T 1993 Silicon fibers produced by high-pressure LCVD *Appl. Phys.* A **57** 97–100

Chapter 3

CVD processes on an atomic level

Industrial products are commonly coated with various types of thin films. This is especially the situation for modern electronic products, where the coatings are of an outmost importance for the overall electronic performances. As an example, the quality of such thin films has been shown to determine the performance of integrated circuits in computers. Solar cells are another example, where the quality of the coating films determines the efficiency of the energy conversion efficiencies. Many different types of thin film technologies have been developed and chemical vapor deposition (CVD) is one of these. The thin film is then formed by the deposition of gaseous species onto the substrate. The CVD method is atomistic to its nature. This means that the species that reach the film surface are atoms or molecules, or a combination of these two types of growth species. The deposition will thereafter take place by chemical reactions onto the film surface. It is thereby possible to use theoretical modelling in the identification and modification of the CVD growth mechanisms.

3.1 Introduction

There are different types of so-called 'controls' to choose from when using the CVD technology; diffusion, thermodynamic, and surface kinetic control. The diffusion control is a physical control where the CVD growth rate is completely determined by the diffusion rate of the gas phase reactants through the boundary layer (see figure 3.1). On the contrary, the thermodynamic and surface kinetic controls are chemical controls that are completely governed by the total reaction energy and energy barrier, respectively, of the CVD growth reaction. The growth rate is thereby determined by experimental parameters, such as gas flow, pressure, temperature, etc.

It is generally possible to choose between the different types of controls by simply varying the temperature in the spatial region around the growing film (also called the substrate temperature, T_{sub}). As can be seen in figure 3.2, there are two distinct breakages in the slope of the growth rate as a function of T_{sub}. For very high T_{sub}

doi:10.1088/978-0-7503-3107-4ch3

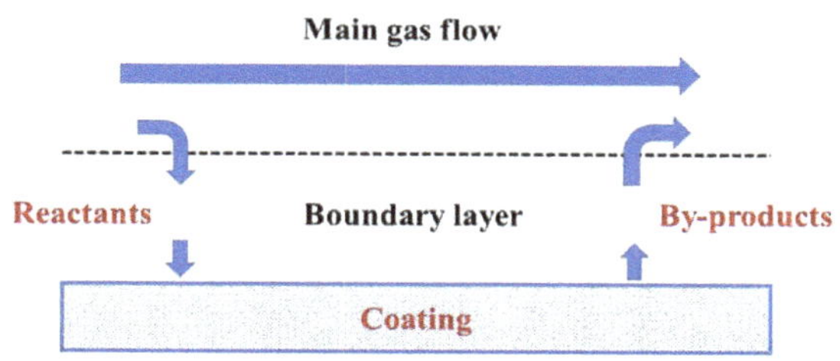

Figure 3.1. Schematic showing the deposited coating in a gaseous CVD environment. The reactant gas flow is decreasing from maximum (in the main flow) to approximately zero (at the coating-gas interface) when passing over the boundary layer.

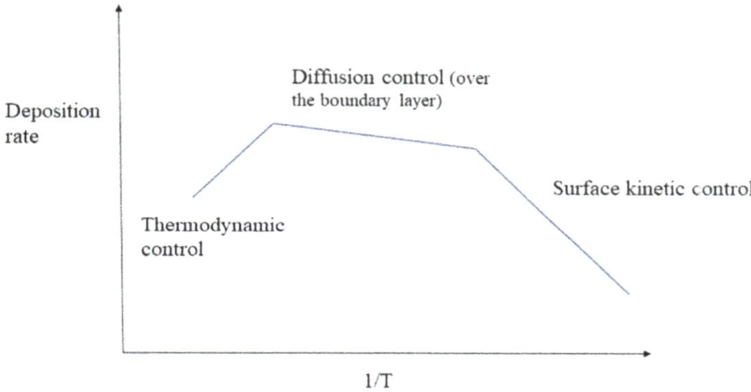

Figure 3.2. Deposition rate as a function of temperature over the three different control regions (i.e. limitation regions).

values, all energy barriers in the growth mechanism will be overcome. Also, the diffusion of reactants through the diffusion layer (i.e. boundary layer) will be satisfactorily high. Hence, the CVD growth process will be under thermodynamic control. This means that the growth process is completely governed by the total reaction energy of the chemical growth process on the growing thin film surface (or substrate). For this situation it will be possible to steer the growth by simply varying the type and partial pressures of reactants in the gas phase (in the close proximity to the growing film). This means that for an exothermic reaction process, the growth rate is only dependent on the supply of gaseous reactants into the reactor.

On the other hand, for a medium-sized T_{sub} value the diffusion of reactants through the diffusion layer will become too low, while there are still no problems with the energy barriers on the surface. Thus, there is a deficiency in gaseous reactants reaching the substrate-gaseous interface (i.e. as a result of the sluggish diffusion over the boundary layer). This boundary layer is therefore most often called the depletion layer. The reactor is now working under diffusion control (which is a very difficult control to handle). For an even lower T_{sub} value, the total energy barrier of the surface growth process will become a hindrance and the growth is now under a surface kinetic control. This is a very desirable control, since it will now be possible to steer the complete CVD growth process by controlling each

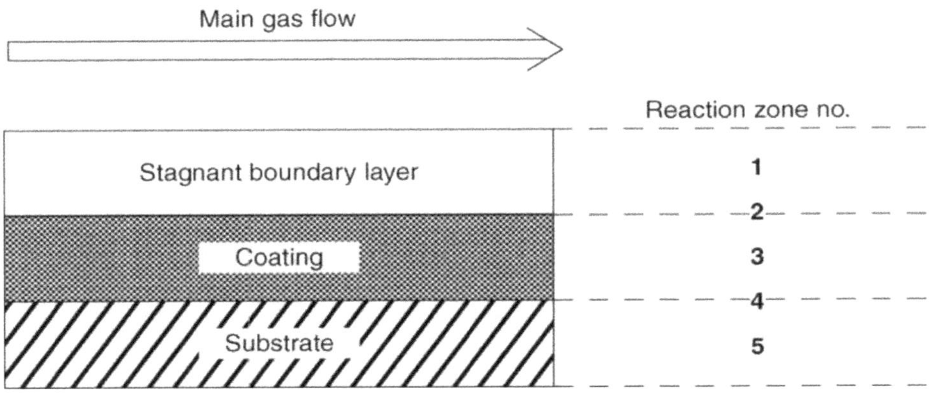

Figure 3.3. Important reaction zones in CVD. [1], reproduced with permission © Elsevier.

individual elementary step in the growth mechanism. It is, then, possible to control the complete growth mechanism, and, thereby, the final result. This type of control, with its surface chemical processes, has been discussed in section 3.1.

It is not only the chemical processes that take place on the growing thin film surface that are of importance during CVD deposition. In fact, the overall chemical processes in the proximity to the growing thin film can be divided into six different areas (see figure 3.3). The most important areas, for the chemical growth processes, are the substrate/thin film interface (4), the thin film/gas interface (2), and the gas phase just above the growing thin film (1). Other areas are the substrate material (5) and the deposited thin film material (3), where solid-state reaction can take place at increased temperatures. In addition, there may be gas phase reactions in the main gas flow. Only the substrate/thin film interface and the thin film/gas interface will be considered in the present book. The reason is that other types of theoretical methods must be used in studying the chemical reactions in the other areas.

3.2 Chemical reactions in the substrate/thin film interface

3.2.1 General

The substrate is very important for both the CVD growth of thin films, as well as for heterogeneous catalytic reactions. In fact, the substrate is most often the part of a chemical process that needs improved or new surface properties. The main difference between CVD growth and heterogeneous catalysis is that the product in the CVD growth process will remain on the surface, while the product in heterogeneous catalysis will become desorbed from the surface. (Ideally, the catalyst material should not be destroyed or used in the catalytic reaction.) These heterogeneous reaction types put different demands on the substrate.

For the situation with heterogeneous catalysis, the substrate should be (i) stable, (ii) chemically inert, and (iii) be able to uphold a large area. Example of substrate materials for catalytic reactions are Al_2O_3, SiO_2, and C [2].

There are, in many aspects, completely different demands on the substrate material for the use in CVD growth. Suitable CVD substrates can, therefore, be divided into various groups, depending on the material, with its properties, that is to be produced. The following groups of substrates can be identified:

1. One or more properties of the substrate need to be improved (i.e. by depositing a thin film to increase the corrosion resistance of the substrate). For the need to improve several properties (e.g. corrosion resistance, hardness, etc) several types of thin films are generally deposited on the substrate (i.e. in a sandwiched structure). It is for this purpose extremely important that there is a very good adhesion between the substrate and the adlayer, and between the different adlayers.
2. The role by the substrate is to form another material, which will afterwards either be removed from the surface (e.g. in the formation of 2D graphene) or stay ontop of the substrate (e.g. in the formation of a device such as a FET).

The substrate is, thus, extremely important for a CVD growth of a specific material. For example, it must control the growth of a desired material, and with a desired structure (e.g. a monocrystalline phase). In addition, the demand can also be to control the growth of a specific crystalline orientation (e.g. a (111) surface of a crystalline material). There are also many other examples, including the control of a specific doping concentration and type. Thus, the CVD substrates can be further divided into several sub-groups:

1. A common type of substrate forms an inter-layer between itself and the desired film. This takes place either with reactions between the substrate and the gaseous species, or subsequently solid phase reactions at increased temperatures. An example is the CVD growth of diamond on a Si substrate, with a SiC interlayer.
2. The substrate is less reactant, and the film can be directly formed on the substrate.
3. An epitaxial growth of a material puts specific demands on the substrate. One very important example is the lattice mismatch between the substrate and the desired CVD material. For instance, monocrystalline diamond is, for this purpose, grown on top of a natural diamond substrate with an identical crystalline orientation. For the situation where the lattice mismatch is a bit too large, it is very common to use an interlayer with an improved lattice matching to both the substrate and the desired thin film material. In fact, several interlayers may be necessary for this purpose.

To exemplify the use of a CVD thin layer to improve the properties of the substrate, let us take the deposition of diamond as an example. Diamond is the hardest known material, and as such, it has the potential to increase the hardness of another type of material (here defined as the substrate). This improvement has frequently been used for various types of cutting tools. However, the main demand, for this type of application, is that the diamond thin film is strongly attached to the substrate (to diminish the risk of peeling in a harsh environment). Since the steels are the most commonly used and cost-effective structural materials in modern industry, CVD coating diamond films on steel substrates are extremely important, combining the unique surface properties of diamond with the superior toughness and strength of the core steel substrates, and will open up many new applications in the industry. However, CVD diamond deposition on steel substrates continues to be a persistent problem. There are, unfortunately, large problems when trying to deposit diamond on steel. The reason is that diamond will be 'dissolved' by the steel. More precisely, the high carbon diffusion in iron induces a severe internal carburization, and voluminous graphite will be precipitated from the substrate. In order to hinder the catalytic graphite formation, various methods have been applied for the adherent diamond film deposition, such as pre-imposed various interlayers or multi-interlayers, special controls of the deposition process, the approaches of substrate alloying, and so on. However, it has been found that adherent diamond films can be directly deposited on Al alloying steel substrates [3].

One example of the usage of a substrate to grow a desired phase of a material is the usage of diamond monocrystalline substrates in the heteroepitaxial CVD growth of cubic BN (c-BN). For the present example, the main demand on the substrate is (i) that it has an identical crystalline structure as the deposited material, and (ii) that the lattice mismatch (between the substrate and the film) is very small. Large strains in the deposited film, and/or peelings thereof, will otherwise occur. For the present example, diamond and c-BN have very similar crystalline structures, and their lattice mismatch is as low as 0.89 %. The simulations would be much more difficult for a lattice mismatch larger than that. See section 3.2.2 for further discussions in this topic.

3.2.2 Substrate influence on the CVD growth of a specific material phase

The example in the present chapter deals with the influence of a diamond(100) substrate on the growth of c-BN(100). As presented above, this is an example of a situation with a very small lattice mismatch (of only 0.89%), and the purpose with this chapter is to demonstrate a theoretical methodology to use in studying this type of CVD growth. The following steps are then essential to start with:

1. You must prove that the lattice mismatch does not affect the construction of your supercell model!
2. You must also be clear about the purpose of your calculations! What is most important to study? What is interesting to study?

An example of step 2 is the following. Let us say that your primary goal is to construct a stable c-BN adlayer, on the diamond substrate, that consists of 10 atomic layers. This is your most important goal. Then you have to make a plan over how to build this c-BN adlayer, while keeping in mind that you shall simulate, for instance, an ALD processes. That is, this c-BN adlayer shall be grown layer-by-layer.

In addition, you may also find it interesting to investigate the following aspects of the ALD growth:

- How many B and N layers will be affected by the substrate? In other words, how thick must the c-BN film be to not experience the substrate? This is

information that is of the largest importance for both the growth processes and surface processes of c-BN (incl. surface reactivities).
- How will the surface reconstruction evolve during the ALD process? This is information that is very important for both growth and electronic properties.
- How will surface termination affect the ALD growth process and c-BN surface properties?
- Is there a combined effect of surface termination and substrate type?

There is a description below of how to prove that the actual lattice mismatch (even though very small) does not affect the construction of a periodic supercell in modelling the diamond/c-BN system.

3.2.2.1 Evaluation of a super-cell approach

The following methodology will here be used to outline the effects of a lattice mismatch on the use of a supercell model of diamond/c-BN:

1. In short, the fastest way is to build a very thick c-BN model. The thickness must, in fact, be so large that the middle of the model has bulk properties. In practice, you optimize the structure for fixed cell parameters. You, thereafter, analyze the pDOS of the middle atomic layers and compare with the pDOS for bulk c-BN. If they don't match, increase the thickness of the c-BN slab. For an identity in pDOS, you have reached a c-BN slab thickness that models everything from bulk c-BN to a c-BN surface.
2. You take the resulting model from (1) and simply replace the cell parameter with the corresponding parameters for a diamond model. The reason for this is that when you model a diamond/c-BN system you must choose the diamond cell parameters. This is because the real diamond substrate is so much thicker than the CVD/ALD grown c-BN film, which means that the c-BN film will adapt much easier to the diamond substrate than the other way around. You, thereafter, optimize the structure for fixed parameters.
3. For the two final c-BN models (i.e. the results of (1) and (2)), calculate and compare the pDOS spectra of the upper surface atomic layers (e.g. one B and one N layer). The difference in pDOS spectra gives you information about the effect by the lattice mismatch on the surface reactivity and those properties that rely on the surface electronic structure.
4. If the difference in pDOS is of no practical importance, you can rely on the supercell approach in modelling the diamond/c-BN interface. If not, you must look for other modelling methods (see section 9.1.3 for further information).

This strategy will be presented here more in detail. For a c-BN model with the c-BN's cell parameters, a supercell with the finally obtained slab thickness can be seen in figure 3.4(a), and the pDOS spectra of the two middle atomic B and N layers are shown in figure 3.5(a). In addition, a supercell model of bulk c-BN is shown in figure 3.4(b), and the corresponding DOS spectra (i.e. the pDOS spectra for all atoms in the supercell) are shown in figure 3.5(b). As can be seen in figure 3.5, there is a relative translation of the DOS spectra since the upper valence band edge is positioned at 0 eV. This is natural since we are comparing different structures. However, it is the spectra themselves that should be compared. As is obvious from figure 3.5, they are not identical but we couldn't come closer in identity by using a

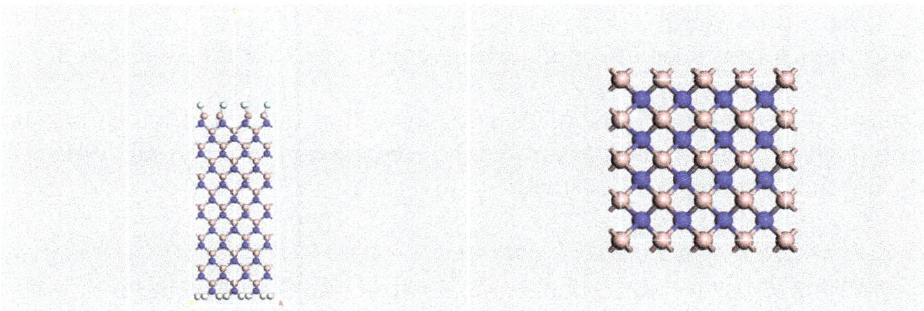

Figure 3.4. Supercell models of (a) the final slab thickness of c-BN, and (b) bulk c-BN. The lower and upper surface of the c-BN slab is terminated with H and F species, respectively. The cell parameters (i.e. x and y values for the slab, and x, y, and z for the bulk cell) for the slab model are identical to the tabulated values for crystalline c-BN.

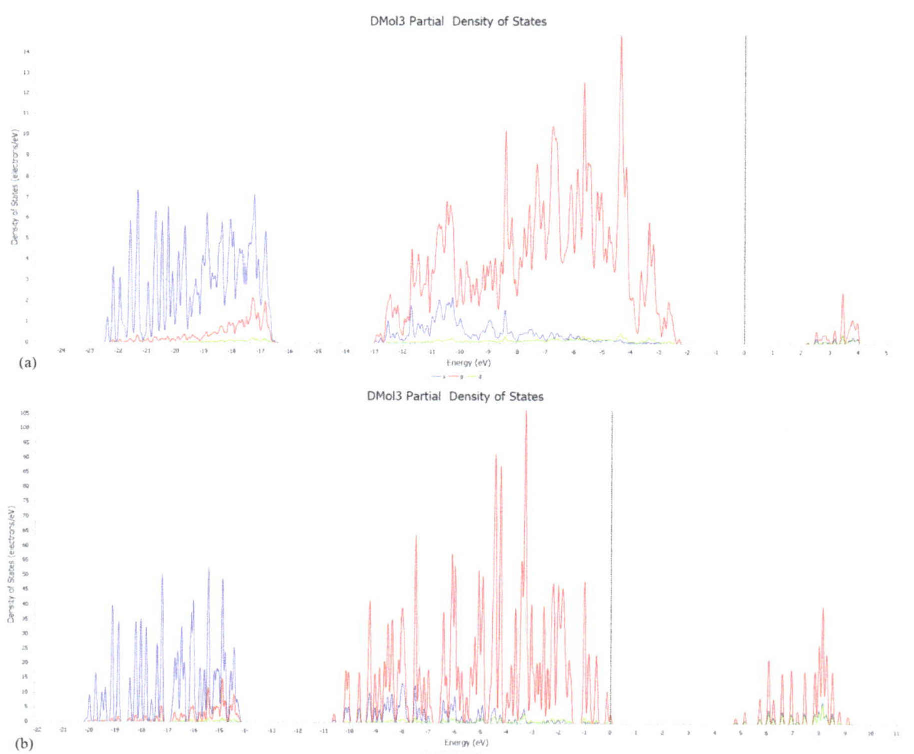

Figure 3.5. (a) pDOS and (b) DOS spectra for the final slab thickness of c-BN and bulk c-BN, respectively. The corresponding geometrical structures are shown in figure 3.4.

larger c-BN slab thickness. It must, though, be stressed that the similarity in spectra is quite satisfactory. However, the small difference in appearance is most probably due to the fact that we have used different *x*- and *y*-axis in constructing the models. Thus, we have different levels of restrictions in the models in that the two cell edges along the *x*-, *y*-, or *z*-axis are identical. A strong advice is, therefore, to continue the study by using identical *x*- and *y*-axes. Anyway, we have found the critical thickness where the inner part of the c-BN slab mimics bulk c-BN and we can move either upwards or downwards to come to a surface with its specific pDOS spectra (and, thereby, surface properties).

3.2.2.2 Critical thickness of the diamond substrate

By using a supercell approach in the construction of a diamond/c-BN model, the first step to take is to make sure that the c-BN layer is not affected by the lower surface of the diamond slab. As will be further discussed in chapter 6, a supercell is, during the quantum mechanical calculations, infinitely translated in the *x*-, *y*- and *z*-directions. This means that both the model of the diamond substrate and the c-BN adlayer are modelled as infinite slabs (in the *x*- and *y*-directions), with a vacuum layer in between. The following methodology can be used to find the most optimal diamond slab thickness.

Several supercell models are, at first, constructed by starting from a four-layer thick diamond slab and, thereafter, by a stepwise increase in thickness (in increments of one C atomic layer). Geometry optimizations should be performed for each of these thicknesses, and the resulting structures are shown in figure 3.6, and the results from pDOS analysis of the N atomic layer that is binding to the diamond substrate (see figure 3.7). A two-atomic thick BN adlayer has been used for these calculations, with N binding to the diamond surface and the B surface layer terminated by F species. The results from a pDOS calculation can be regarded as a fingerprint for the electronic and geometrical structure of the atoms that have been analyzed. As can be seen in figure 3.7, there is a peak at 0 eV (i.e. the upper valence band edge) for a four-layer thick diamond model, which cannot be seen for thicker structures. Moreover, the pDOS structure just below and above 0 eV is changing slightly for a diamond thickness up to eight atomic C layers, after which it does not change anymore. It must be stressed here that both the theoretical parameters used in the calculations and the supercell size are the same for all diamond thicknesses. Thus, the position of the upper valence band edge is not expected to vary for the different structures. There are, hence, variations in the pDOS for this type of system, which are not visible to the eye when studying the corresponding model structures (see figure 3.6). It is very clear from these pDOS results that a diamond thickness of at least eight atomic C layers is necessary to use in modelling a realistic diamond substrate, and its effect on the diamond/c-BN interface. Thus, this size must be used for all further calculations on a diamond/c-BN system, where diamond is assumed to function as a substrate.

3.2.2.3 Layer-by-layer growth of c-BN on a diamond substrate

Let us now assume that we simulate a heteroepitaxial ALD (i.e. layer-by-layer) growth of c-BN(100) onto a diamond(100) substrate. As will be discussed more in

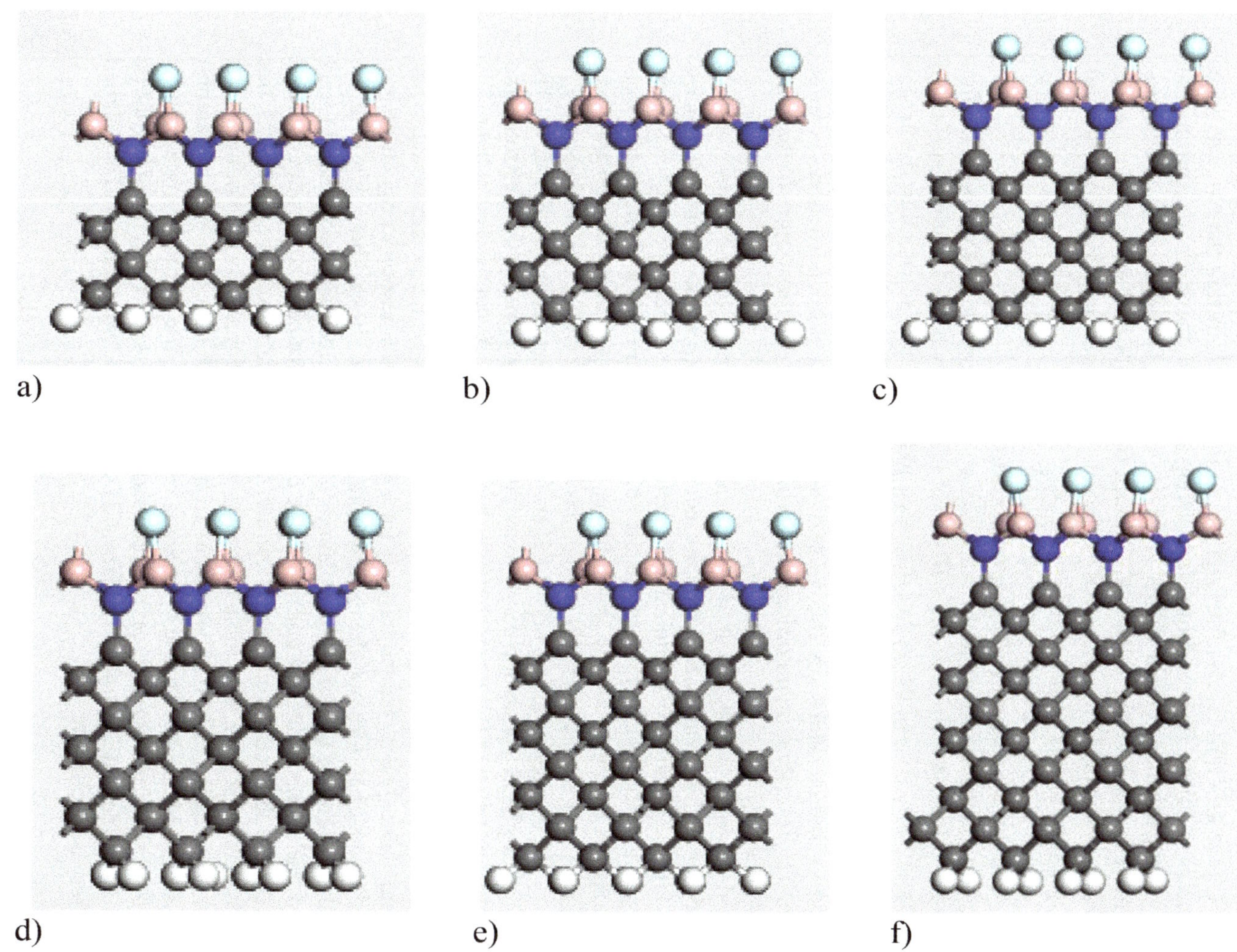

Figure 3.6. A two-layer thick BN adlayer attached to a diamond(100) slab of various thicknesses: (a) 4, (b) 5, (c) 6, (d) 7, (e) 8, and (f) 11 C atomic layers. An N layer is binding to the diamond surface, and the **B** surface is terminated with F species.

detail in chapter 8, a c-BN(100) material consists of alternative B and N layers. Let us also assume that the N- and B-covered surfaces are H- and F-terminated, respectively. Another assumption is to initiate the c-BN(100) growth by starting with an N layer (being binding to the diamond(100) surface). The reason for these choices

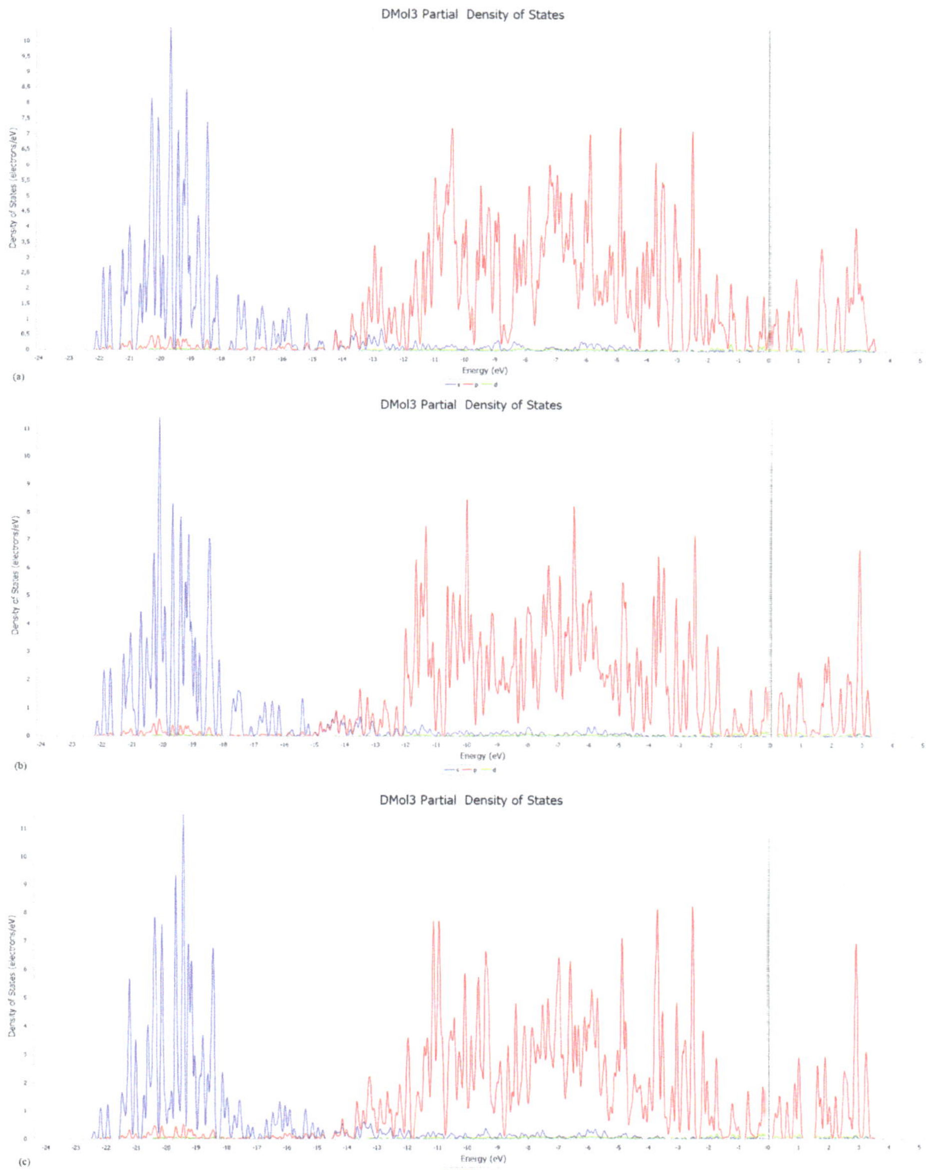

Figure 3.7. pDOS spectra of the N atomic layer in the structures presented in figure 3.6: (a) 4, (b) 5, (c) 6, (d) 7, (e) 8, and (f) 11 C atomic layers. An N layer is binding to the diamond surface, and the **B** surface is terminated with F species.

Figure 3.7. (Continued.)

will be further discussed in section 8.5. When forming the first layer of N atoms on the diamond(100) surface, one has to start with just one N atom.

The methodology now is to attach this single N atom to every possible adsorption site, geometry optimize the system, and finally calculate the total energy. As is usual for surface models, all atoms, except for the lowest positioned ones, should then be optimized. The lowest surface is also H-terminated. The reason is that we have to simulate a continuation into bulk material, and this cannot be done if the lower atoms are also allowed to relax and are radical ones (to be further discussed in section 5.8). When the most stable adsorption site has been found (i.e. with the most negative total energy) it is possible to start building the first atomic N layer. This is not problematic if there is a small enough lattice mismatch between the two materials. This is the situation for the present example with diamond and c-BN (see section 3.2.1) so the atomic N layer can be easily constructed just by starting from the preferred adsorption site. For those materials with a larger lattice mismatch, more calculations have to be performed (see above in the present chapter). The most stable positions of the N atoms in the first N layer (i.e. binding to the diamond C atoms) are demonstrated in figure 3.8. It is clear that these N atoms prefer to adsorb on top on the diamond surface C atoms, and not in the expected position for a natural continuation of the diamond structure (which is a bridge position).

The second atomic layer, which is a B layer, is thereafter constructed by continuing with a cubic BN lattice, followed by geometry optimization (see figure 3.9). This is also the situation for all next-coming layers in the present example (see figures 3.10–3.17). In addition, the pDOS of the interface will be calculated, and analyzed, for all c-BN thicknesses. Thus, both geometrical and electronic structures will be analyzed for the c-BN surface region. Moreover, the electronic structure of the c-BN surface region will be compared with the surface region of pure c-BN. The purpose of the latter type of analysis is to decide on the

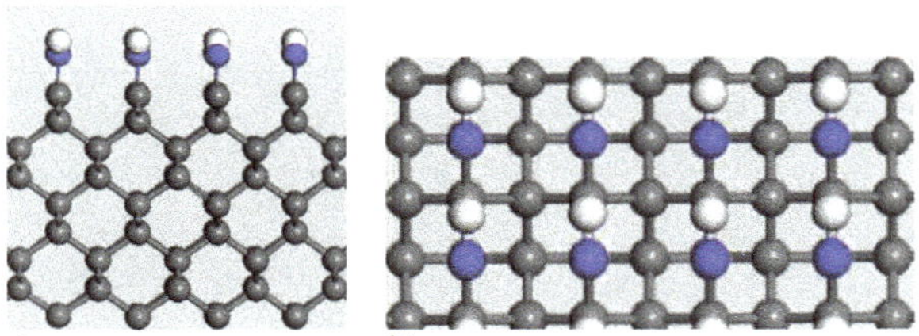

Figure 3.8. (a) Side view and (b) top view of a geometry optimized structure of a diamond(100)/NH interface.

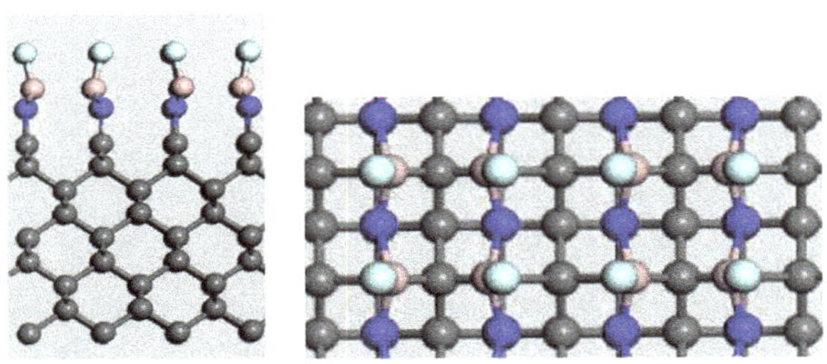

Figure 3.9. (a) Side and (b) top views of a diamond(100)/NBF interface.

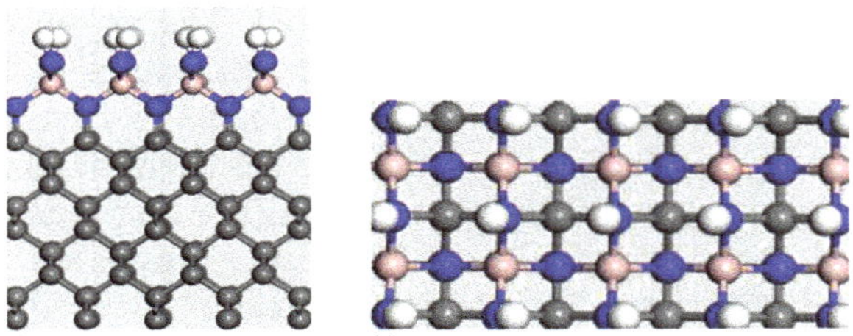

a) b)

Figure 3.10. (a) Side and (b) top views of a diamond(100)/NBNH interface.

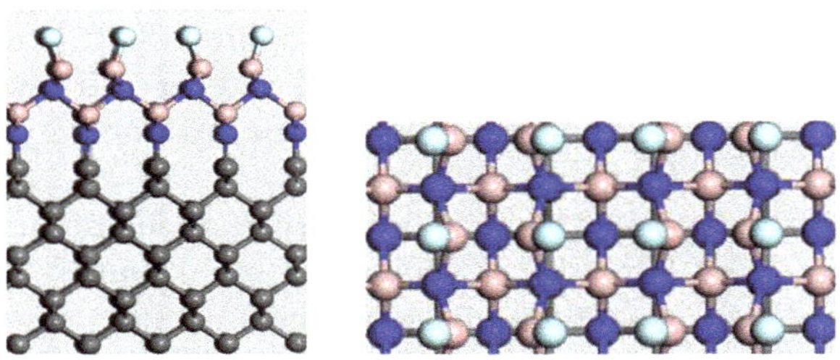

a) b)

Figure 3.11. (a) Side and (b) top views of a diamond(100)/$(NB)_2F$ interface.

a) b)

Figure 3.12. (a) Side and (b) top views of a diamond(100)/$(NB)_2NH$ interface.

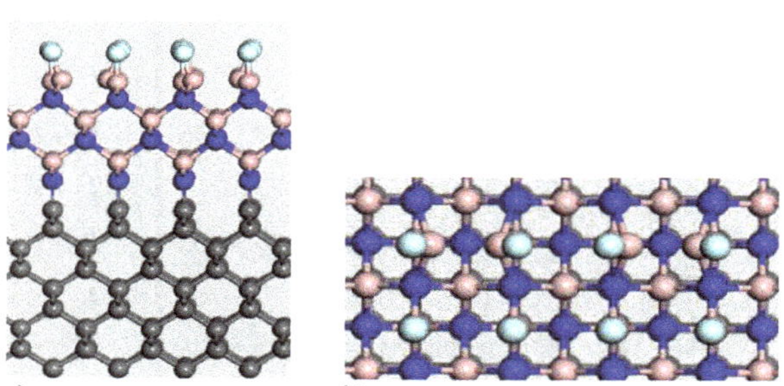

a) b)

Figure 3.13. (a) Side and (b) top views of a diamond(100)/$(NB)_3F$ interface.

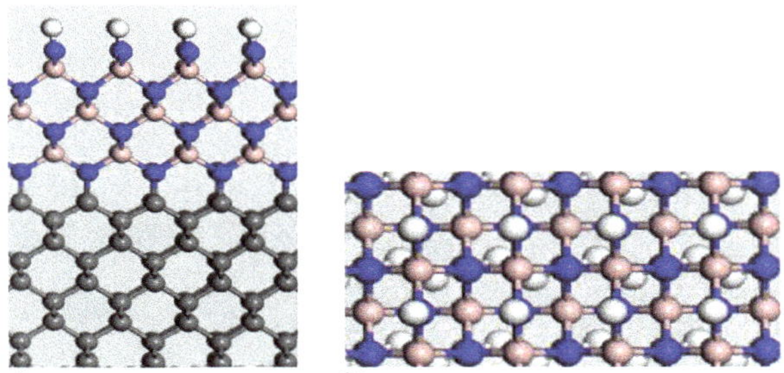

a) b)

Figure 3.14. (a) Side and (b) top views of a diamond(100)/$(NB)_3NH$ interface.

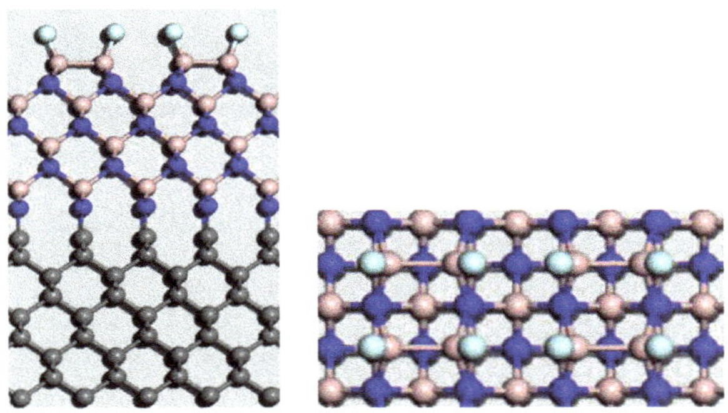

a) b)

Figure 3.15. (a) Side and (b) top views of a diamond(100)/$(NB)_4H$ interface.

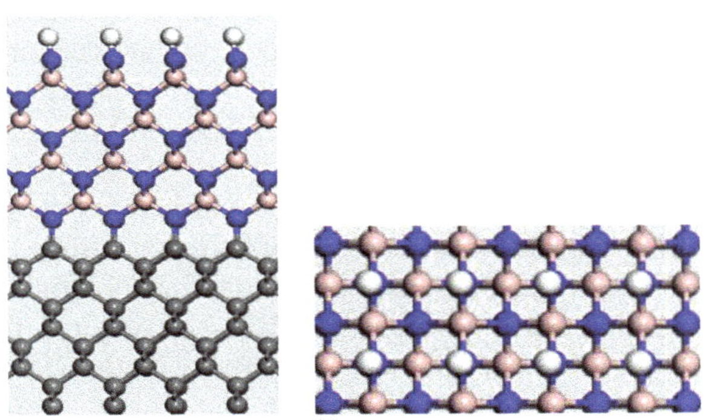

a) b)

Figure 3.16. (a) Side and (b) top views of a diamond(100)/$(NB)_4NH$ interface.

thinnest c-BN layer, above which the c-BN does not experience the diamond substrate anymore.

In modeling a (100) surface of pure c-BN, a thick enough c-BN slab has to be built. The requirement is that the middle atomic layers (in the z-direction) have an identical electronic structure as bulk c-BN. As was shown above in the present chapter, this is accomplished by using a 24 atomic layer thick c-BN(100) model (see

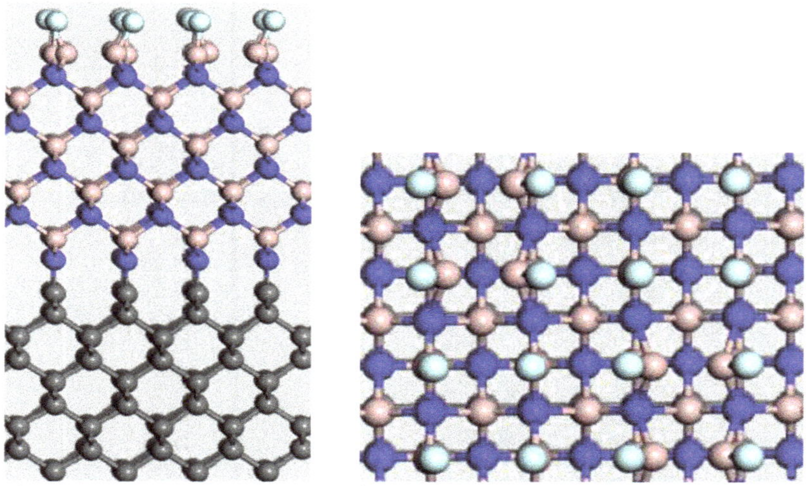

Figure 3.17. (a) Side and (b) top views of a diamond(100)/$(NB)_5H$ interface.

figure 3.4). The reason for this demand on the model is that one has to make sure that the upper c-BN surface does not experience the lower surface of the model. As for real c-BN(100) surfaces, the geometrical structure goes from a bulk-like structure to a surface geometry when starting from this middle position of the model and move upwards along the z-axis. As can be seen in figure 3.5, the pDOS spectrum of the middle atom layer in the thick c-BN slab is close to identical to the pDOS spectrum of bulk c-BN. We can, therefore, conclude that the surface of this c-BN slab correctly describes a realistic c-BN surface.

By comparing the pDOS spectrum of the F-terminated surface B atom in figure 3.4(a) with the corresponding pDOS spectra of the surfaces of the c-BN adlayers of various thickness (see figure 3.18), it will be possible to find the critical thickness of the c-BN adlayer. (The same comparison has also been done here for a very thick c-BN slab (of the same thickness) with surface N atoms (see figure 3.19).) In other words, the thickness for which the c-BN adlayer does not experience the underlying diamond substrate. It is instead identical to the (100) surface of bulk c-BN. In addition, it is not only the critical size that will be obtained by analyzing the c-BN surface pDOS spectra (for various c-BN adlayer thicknesses). One also gets information about the evolution of surface electron structure and reactivity when increasing the thickness of the c-BN slab from a single atomic layer on the diamond substrate and up to the critical c-BN thickness (i.e. in a layer-by-layer growth).

Let us start by comparing the pDOS spectra for the c-BN adlayers with N-covered surfaces (see figure 3.18). It is obvious that the pDOS spectra for the monolayer thick c-BN slab looks completely different when compared with the other slab thicknesses (compare figure 3.18(a) with figures 3.18(b)–(e)). However, the latter pDOS spectra are very similar. The unfilled peak at approximately 0.2 eV

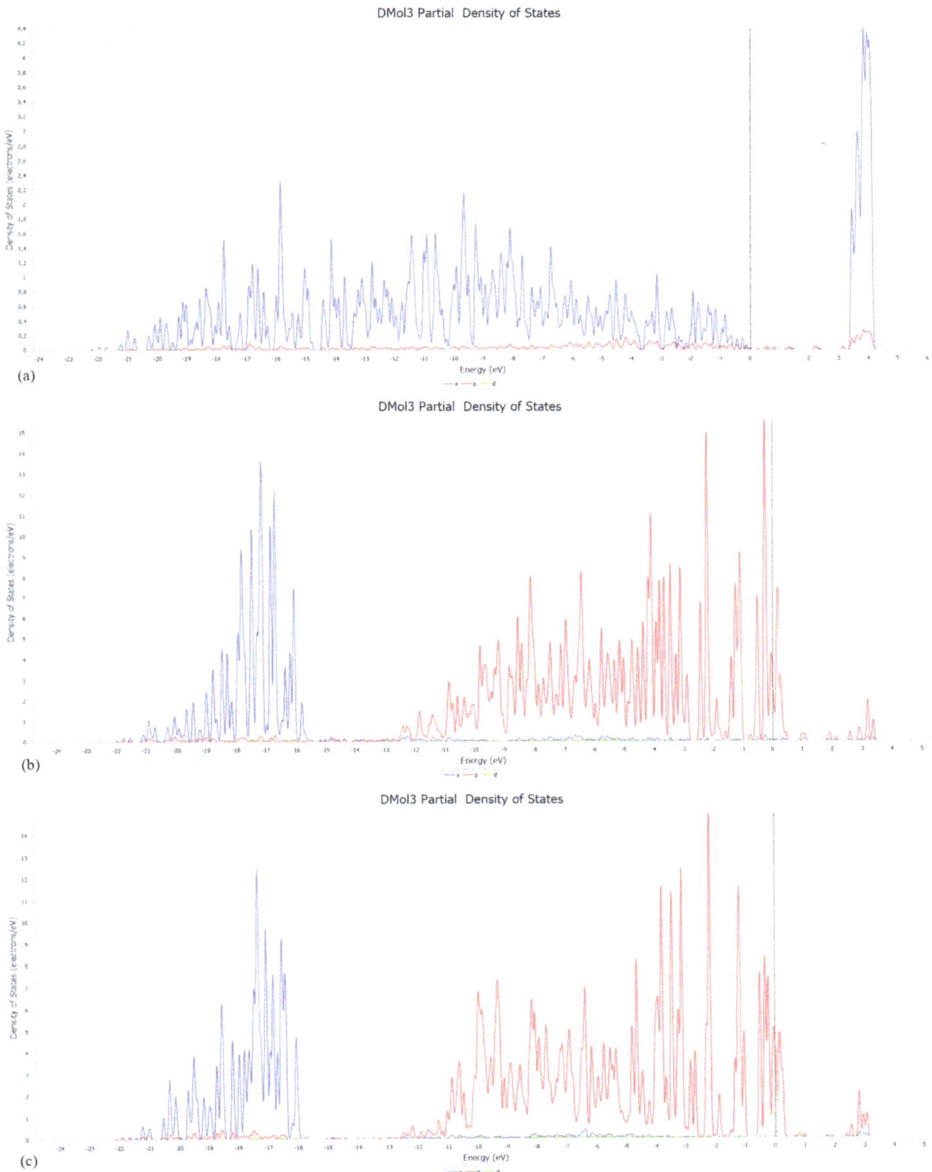

Figure 3.18. pDOS spectra of the surface N atomic layer in variously thick BN adlayers attached to a diamond (100) substrate; (a) 1, (b) 3, (c) 5, (d) 7, and (e) 9 atomic thick BN adlayer. These spectra are to be compared with the (f) spectra for the surface N atomic layer on a 24 atomic thick c-BN slab (see figure 3.4(a)). Blue: s-orbitals; red: p-orbitals; green; d-orbitals.

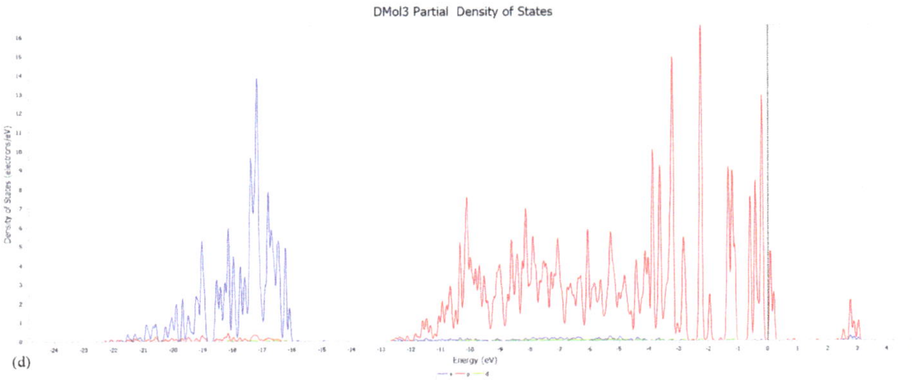

(d)

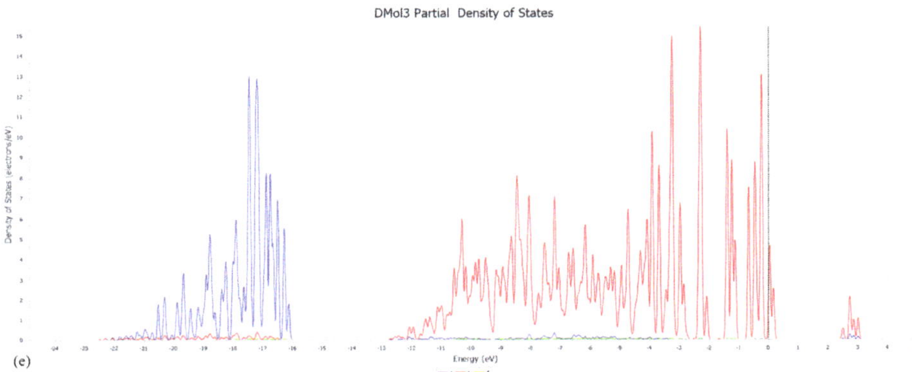

(e)

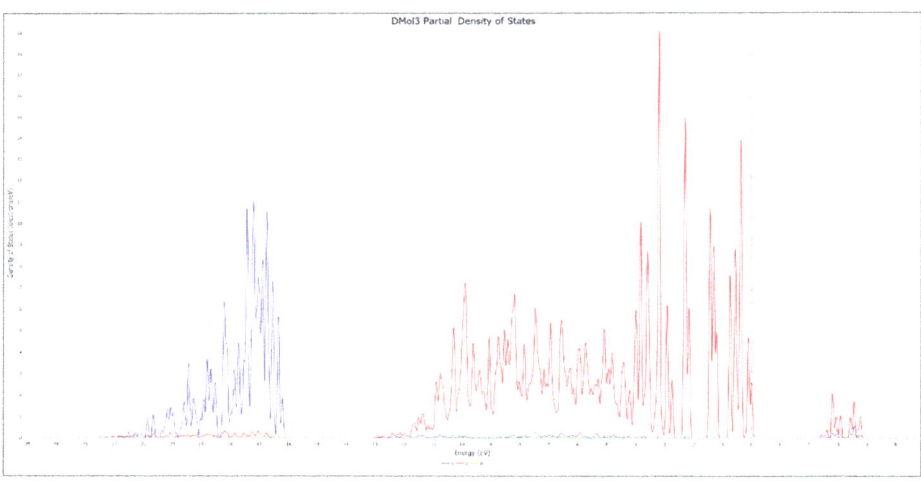

(f)

Figure 3.18. (Continued.)

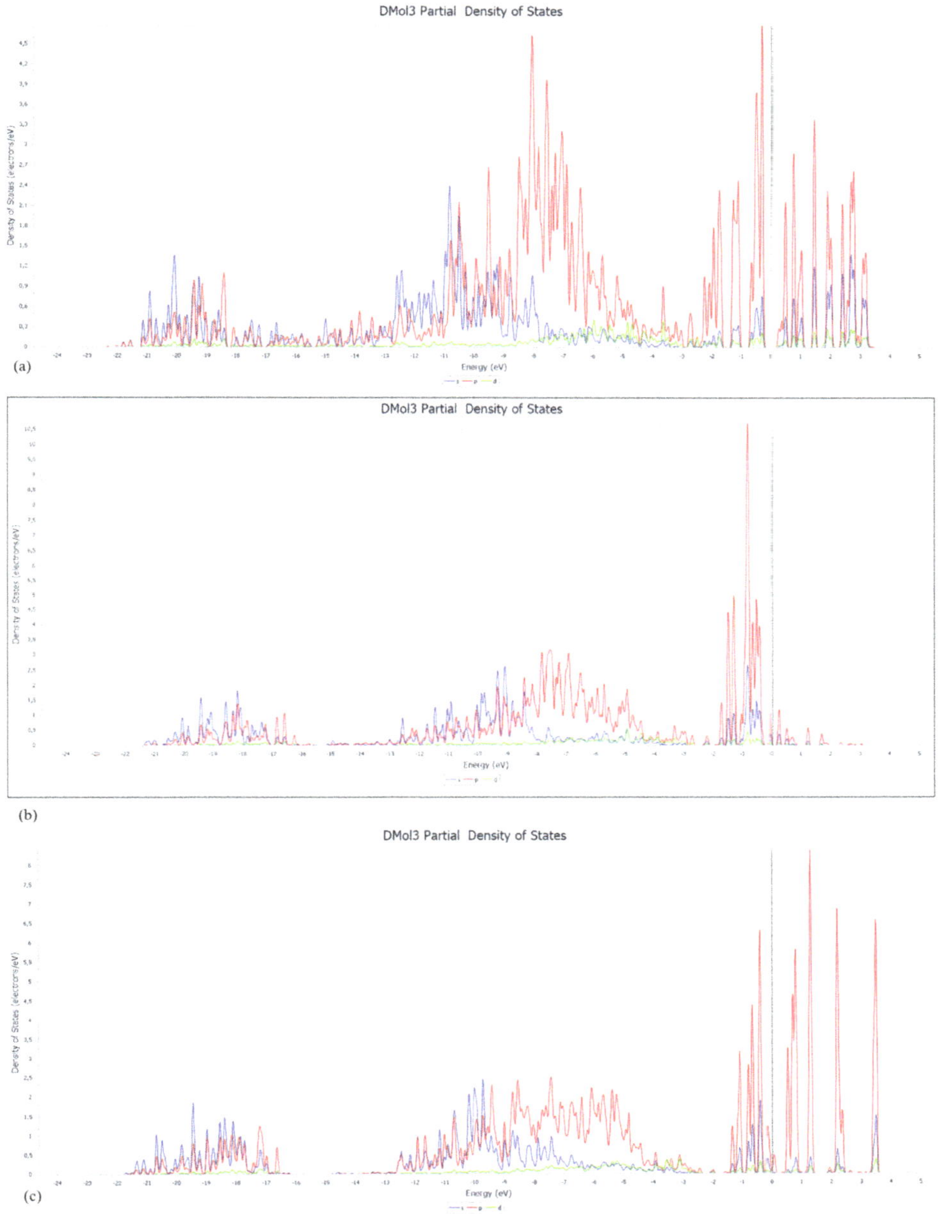

Figure 3.19. pDOS spectra of the surface B atomic layer in variously thick BN adlayers attached to a diamond (100) substrate; (a) 2, (b) 4, (c) 6, (d) 8, and (e) 10 atomic thick BN adlayer. These spectra are to be compared with the (f) spectra for the surface B atomic layer on a 24 atomic thick c-BN slab. Blue: s-orbitals; red: p-orbitals; green; d-orbitals.

Figure 3.19. (Continued.)

(figure 3.18(b)) starts splitting up in figure 3.18(c) and remains in its final state when going from figure 3.18(d) to figure 3.18(e). In addition, the smaller peaks between 0.2 and 2.5 eV are being removed. Furthermore, the pDOS spectra in figure 3.18(e) is very similar to the one for the freestanding very thick c-BN slab (see figure 3.18(f)).

(The transfer of pDOS spectra of the free-standing c-BN slab in figure 3.18(f), relative to the spectra for the diamond(100)/BN model, can be explained by the use of different models.) It is, thus, from these pDOS spectra possible to conclude that the critical thickness of an attached c-BN layer (onto a diamond(100) substrate) corresponds to an eight atomic layer thick c-BN slab. A c-BN slab with a larger thickness should, hence, have a surface reactivity comparable to a free-standing slab. Also, it would be possible to model a larger thickness of c-BN by only including the c-BN adlayer (i.e. excluding the diamond substrate), which will ultimately reduce both computer time and needed computer capacity. However, it is absolutely necessary to do the same comparison for also the B-covered c-BN slabs.

The situation is, however, completely different for the B-covered surfaces (see figure 3.19). It is obvious that the pDOS spectra for the variously thick c-BN slabs are strongly coupled to their respective surface reconstruction. The pDOS spectra for the common 2 × 1 reconstruction, to various textens, are presented in figures 3.19(a), (b), and (d). The very weak tendency for 2 × 1-reconstruction (see figure 3.9) towards the strongest one (see figure 3.15) is demonstrated by the different pDOS spectra in figures 3.19(a), (b), and (d). As was the situation for the N-covered slabs, there is a large difference between the thinnest c-BN slab (i.e. the two-atomic layer c-BN slab) and the next in size (i.e. the four-atomic thick slab). However, this latter slab shows a pDOS pattern that is not that different compared to the eight-atomic thick c-BN slab. On the contrary, the six-atomic thick c-BN slab shows a completely different pDOS pattern, which can be explained by the different surface reconstruction (see figure 3.13). The alternating non-reconstructed B–B pairs are represented here by a non-hybridized p-peak at approximately 0.1 eV (see figure 3.19(c)). The 10-atomic thick c-BN slab does also show a completely different pDOS pattern as compared to the two-, four, and eight-atomic thick ones, and resembles to some extent the pDOS pattern for the alternating 2 × 1-reconstructed six-atomic thick slab. However, the non-hybridized p-peak is absent for this very thick slab (see figure 3.19(e)) (there has not been any attempt to further analyze these differences in the present example), which can be explained by the different surface reconstruction (see figure 3.13). The alternating non-reconstructed B–B pairs are represented here by a non-hybridized p-peak at approximately 0.1 eV (see figure 3.19(c)). However, the most important result from the present example is that the pDOS spectra for the 2 × 1-reconstructed B-surface in the eight-atomic layer thick c-BN slab, is very similar to the one for the corresponding free-standing very thick c-BN slab (see figures 3.18(d) and (f)). Both of these surfaces have identical types of surface reconstruction, which makes a comparison possible. It is, thus, also possible from these pDOS spectra to conclude that the critical thickness of an attached c-BN layer (onto a diamond(100) substrate) corresponds to an eight-atomic thick c-BN slab (as was the situation for the N-covered c-BN slabs).

These calculations have shown that a thickness of eight atomic layers is a critical thickness for a heteroepitaxial c-BN slab (attached to a diamond(100) substrate), above which the c-BN material does not experience the substrate material anymore.

What is also of the largest interest to study is the evolution of c-BN surface geometry when starting from the first N atomic layer on the diamond substrate, and continuing up to the critical thickness of c-BN. As can be seen in figures 3.8–3.17, the most stable geometrical structure of the surface atoms differed considerably for the different thicknesses. As presented above, the first N atomic layer was found to bind ontop on the diamond surface C atoms (figure 3.8). For each of these c-BN thicknesses, various initial surface geometries were tested and used in the optimization calculations, and the most stable outcome from these calculations are presented in figures 3.8–3.17. It is, thus, most likely that these final geometry structures also form at CVD (or preferably ALD) conditions.

In studying the evolution of surface geometry for various BN thicknesses, let us start with surface B atoms. For a two-atomic layer thick BN adlayer, the surface B atom will strive towards a 2×1 reconstruction. The tendency for a pairwise attraction is clearly seen in figure 3.9. This tendency is even stronger for a four-layer thick BN adlayer (see figure 3.11). However, something different has occurred for a six-layer thick adlayer, were alternate pair-wise attractions can be observed in the B–B rows. The other surface B atoms are not affected by its neighbours (see figure 3.13). On the other, the surface B atoms go back to very strong pair-wise attractions (to 100%) for an eight-atomic thick BN adlayer. In fact, it is apparent that bonds have now been formed (see figure 3.15). Moreover, the surface go over to another type of surface structure for a ten-atomic thick BN adlayer, which will then remain for even larger BN adlayers. As can be seen in figure 3.17, in each B–B row, two neighbouring B–B pairs show interatomic attraction, while the following two do not. And then it continues like that.

The variations for the corresponding N surfaces are, however, not that severe. In fact, it stays in an almost perfect cubic position for all BN adlayer thicknesses studied (see figures 3.8, 3.10, 3.12, 3.14, and 3.16).

The variations in surface geometry can be explained by the specific properties to elemental N and B. The N atoms have five valence electrons, but only binds to three other atoms (2 B and one H) in these structures. Hence, it can form a paired electron, which is a quite stable situation. This is the most possible explanation to the present observation. The B atom, on the other hand, is a very peculiar atom. It has only three valence electrons, and can, therefore, only bind to three other atoms. However, if it receives an electron pair from, e.g. a binding F species, then it can bind to a fourth species. This is something we can observe to various degrees in figures 3.9, 3.11, 3.13, 3.15, and 3.17. However, further investigations are needed to clarify these observations in detail.

We have found here that both the geometrical and electronic structure will vary over a specific adlayer thickness of c-BN onto a diamond substrate. This variation in material properties is most likely important for not only the general surface reactivity and electronic properties of c-BN, but also very important for its growth mechanism. Thus, this might be the explanation why the vapour phase growth of c-BN differs when starting from the nucleation on a substrate to a continuous growth of a thick c-BN film. Furthermore, a deeper knowledge about this critical c-BN thickness, with its individual atomic layers, can be used to design c-BN thin films with desired properties. And, ultimately, it can be used to grow phase-pure c-BN.

The methodology presented here is not restricted to only c-BN growth on a diamond substrate. In fact, it can be used for any material, substrate type, and for different kinds of surface terminating species. The effect by the diamond substrate on the phase-pure growth of c-BN will be further discussed in chapter 9.

3.3 Chemical reactions in the thin film/gas interface

3.3.1 Introduction

The chemical CVD growth processes in the interphase between the growing thin film and the gaseous environment is very complex, and also generally far from equilibrium. Thus, the CVD growth mechanism is constituted by a large number of 'elementary' reactions (see figure 3.20). The most important ones are adsorption of growth precursors, chemical reactions on the surface (either with another adsorbate or with a gaseous species), desorption of rest products, surface migration to, e.g. a kink or ledge (i.e. the final incorporation into the material lattice). These examples of 'elementary' reactions are just classes of reactions that can be further divided into several elementary reactions depending on the specific CVD growth process and other experimental conditions.

1. Transport of precursors to the vicinity of the substrate surface
2. Transport of precursors to the substrate surface
3. Adsorption of precursors on the substrate surface
4. Diffusion of precursors on the substrate surface
5. Chemical surface reactions
6. Nucleation
7. Desorption of gaseous reaction products
8. Transport of the gaseous reaction products away from the reaction zone

The heterogeneous chemical processes that take place on a solid surface during CVD growth of a specific material are very similar to the processes that take place during heterogeneous catalysis. The main difference is the circumstance that a product is formed on a substrate during CVD, while the substrate is a catalytic material in the heterogeneous catalysis. For the latter situation, the substrate (or

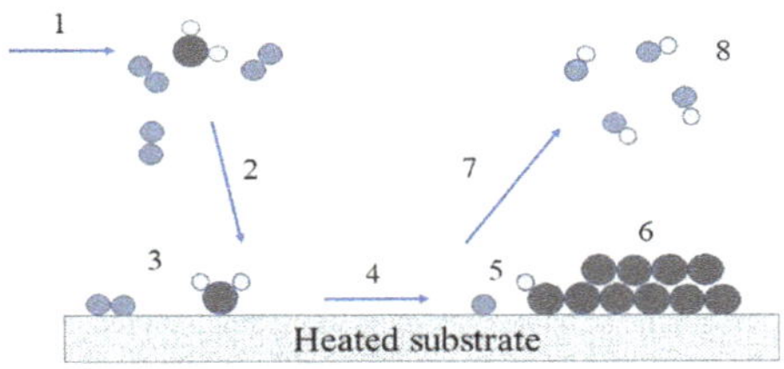

Figure 3.20. Most important classes of elementary reactions during CVD growth.

catalyst) should not either be consumed or destroyed, and any product must be desorbed from the surface. Due to the overall similarities, also heterogeneous catalysis can be modelled by using the methodologies and methods presented in the present book. However, the process that will be explicitly dealt with here is the CVD (or ALD) growth of solid thin film materials.

3.3.2 Adsorption of growth species

One important prerequisite for CVD growth to occur is that the growth precursor, or any of its fragments, finally becomes chemisorbed to both the substrate and the growing film. The concept of chemisorption means that the adsorbate binds with a covalent, or ionic, bond to the surface. The other alternative is physisorption, where the adsorbate binds to the surface with very weak van der Waals interactions. An example of chemisorption is the adsorption of H and CH_3 radicals to a radical surface site on a solid-state surface (see figures 3.21(a) and (b), respectively). Furthermore, the adsorption of a non-radical CH_4 molecule is an example of van der Waals interactions (see figure 3.22).

However, during ordinary CVD of thin film materials it is most common that several layers of gaseous species (like growth species) onto the growing film. The first layer is most commonly chemisorbed to the surface, and the following ones are physisorbed (see figure 3.23 for a general example). Let us now assume that a surface radical site has been formed on this latter type of surface. The empty site can, thereafter, be filled in two ways; (i) either by a direct chemisorption from the gas phase, or (ii) by a chemisorption of a species from any of the physisorbed layers. The species in these latter types of layers are very mobile since they only experience weak van der Waals (VdW) interactions with the surface (of the strengths $0 < \text{VdW} \lesssim 15\,\text{kJ mol}^{-1}$). The mobility of the individual species in these VdW-layers are, thus, highly mobile like in a fluid, and can therefore very easily migrate to a neighbouring radical surface site. There is therefore a competition between a direct chemisorption from the gas phase and an indirect chemisorption from the VdW-layers. As stated in section 3.1, these surface

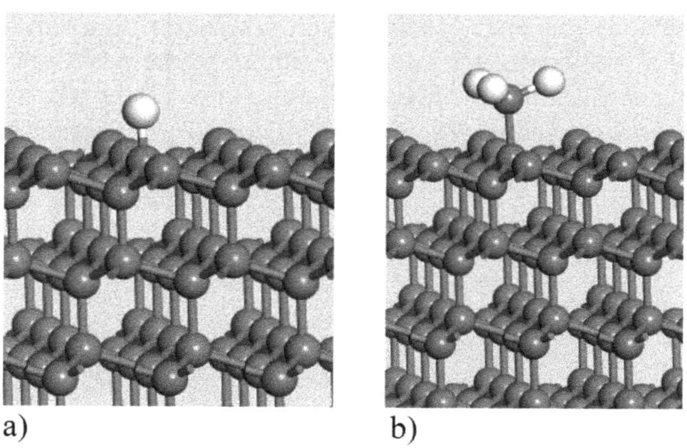

Figure 3.21. Chemisorption of radical (a) H and (b) CH_3 species to a diamond(111) surface.

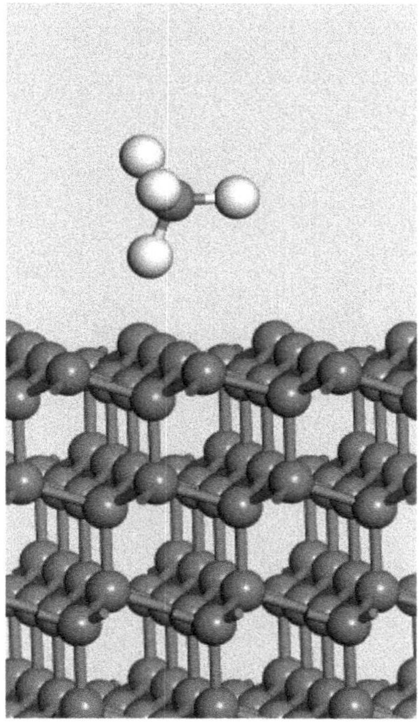

Figure 3.22. Physisorption of a non-radical CH_4 species to a diamond(111) surface.

processes are extremely important when working under the condition of surface kinetic control, which is the type of control that should preferably be used for a precise growth of a thin film material.

Let us now take one step back and study the collisions of gas phase species, i, onto the surface. Equation (3.1) gives an expression of the corresponding collision rate.

$$\nu_{i,\,\text{collision}} = \frac{p_i}{\sqrt{2\pi k T m_i}} \tag{3.1}$$

where p_i is the partial pressure of i in the gas phase, k is the Boltzmann constant, T is the temperature, and m_i is the mass of i.

Furthermore, the adsorption coefficient, s (also called the sticking coefficient), is defined as the probability for chemisorption. Thus, if it takes three collisions before the gaseous species, i, is finally chemisorbed to the surface, s_i is equal to ¾ (or 0.75). A direct chemisorption from the gas phase onto a surface radical site is thus: $0 < s < 1$. The values $s = 0$ and 1 are extremely rare. In fact, the chemisorption of the very small and very reactive H radical has an s value of approximately 0.8 for the majority of surface material types. So, it is expected that larger and/or less reactive gas phase species have smaller s values.

We know from the collision rate expression (equation (3.1)), and from the related adsorption rate expression (equation (3.2)), that a larger sticking coefficient gives a higher adsorption rate. We have,

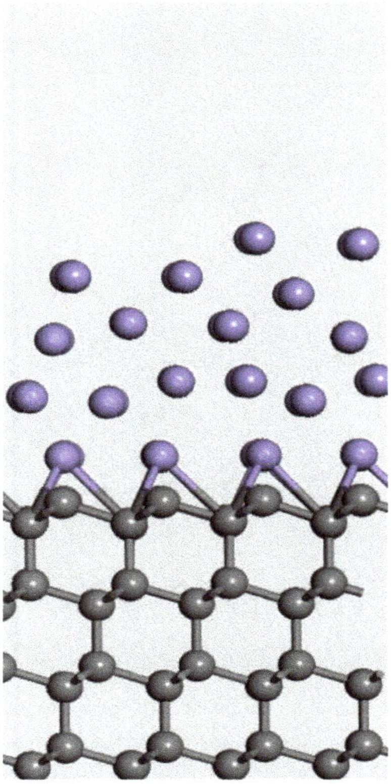

Figure 3.23. Adsorption of an arbitrary species on a diamond(111) surface. The first atomic layer is chemisorbed to the surface, and the others are physisorbed.

$$\nu_{i,\ \mathrm{adsorption}} = \frac{p_i s_i}{\sqrt{2\pi k T m_i}} \tag{3.2}$$

thus, two alternatives. Either to improve the reactivity of the surface and/or increase the reactivity of the growth species (and we should preferably try to do both). The sticking coefficient can be expressed as

$$s/s_0 = 1 - \theta \tag{3.3}$$

where θ is the surface coverage of a specific adsorbate, s_0 is the corresponding sticking coefficient at $\theta = 0$, and s has a value between 0 and 1.

Let us first have a look at the very small and reactive H radical (which is the smallest known species). Let us also assume that we shall adsorb this H species to a diamond(111) surface. Thus, mainly covalent surface-H bonds will be formed. As can be seen in figure 6.8(a), the adsorption energy as a function of surface coverage shows a linear relation. This means that not even the last H adsorbate will experience any repulsive interactions from its neighbouring adsorbates. Thus, for this ideal situation, equation (3.3) will show an ideal linear relation (see figure 3.24).

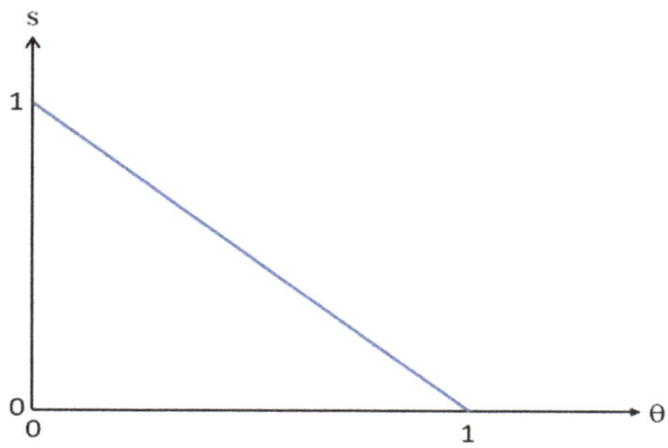

Figure 3.24. Linear relation of the sticking coefficient, s, as a function of the surface coverage, θ.

Let us instead study the adsorption of methyl species, CH_3, to this specific surface. It is a radical molecule, and as such reactive. It is also expected to bind covalently with the diamond surface. It has an umbrella-like form when approaching the surface, and finally becomes chemisorbed (see figure 3.21(b)). We can intuitively understand that it will become progressively much more difficult for this molecule to chemisorb to the diamond surface as the surface coverage increases. The equation for the sticking coefficient is no longer ideal. It will now take the following form (see figure 3.26). **s/s₀** will decrease exponentially when approaching a 100% surface coverage. In fact, it is not possible to completely terminate a diamond(111) surface with CH_3 species. The sterical repulsions between the adsorbates are too large.

It is apparent that equation (3.3) needs to be modified. This can be done by replacing s_0 with an Arrhenius expression:

$$s = s_o(1 - \theta) = s_0' e^{-\Delta E(\text{barr, ads})/kT}(1 - \theta) \tag{3.4}$$

where s_0' a pre-exponential factor, $-\Delta E$(barr, ads) is the energy barrier for the adsorption, k is the Boltzmann constant, and T is the temperature. The pre-exponential factor can take different forms, but this will not be dealt with in the present book. However, it is easily understandable that bulky adsorbates like CH_3 will have severe difficulty in reaching the surface with neighbouring adsorbates since the inter-adsorbate repulsions will be to large. This is the same as stating that the barrier energy will be too high.

The situation described in figure 3.24 is quite rare since this will be the form of s for an adsorption barrier energy that is equal to zero. The situation described in figure 3.25 is more common, but to various extents. However, the most common situation, with several layers of growth species adsorbed on the surface (see figure 3.23), will look like figure 3.26. Thus, the final s-value, s_{tot}, is the added value of a direct chemisorption directly from the gas phase, s_{direct}, and the value of an indirect chemisorption from the weakly bounded van der Waals layers, $s_{indirect}$.

As presented above, the atoms in the van der Waals layers are highly mobile. Therefore, an equilibrium situation will soon evolve where it takes on average τ seconds until the physisorbed species have become desorbed into the gas phase;

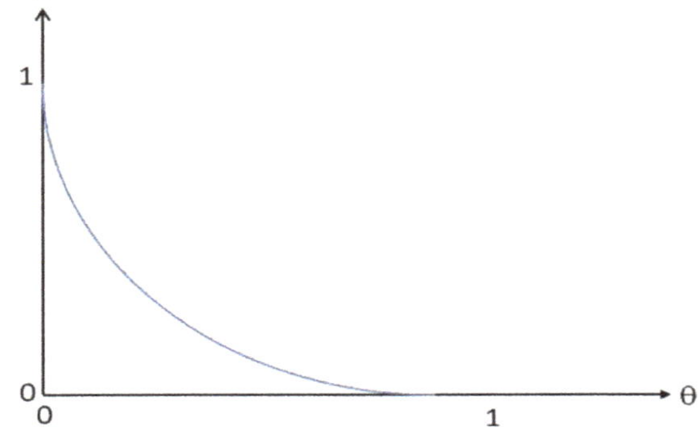

Figure 3.25. Exponentially decreasing sticking coefficient, s, as a function of the surface coverage, θ.

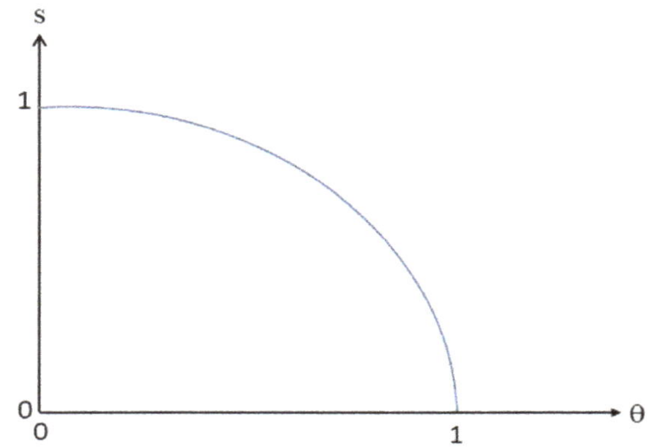

Figure 3.26. Realistic situation for the sticking coefficient, s, as a function of the surface coverage, θ.

$$i(\mathrm{VdW}) \rightleftarrows i(g) \tag{3.5}$$

where i(VdW) is the van der Waals-bounded adsorbate and i(g) is the gas phase form of the adsorbate. This specific time is called the residence time, τ, and can be expressed as follows:

$$\tau = \tau_0 \mathrm{e}^{-\Delta E(\text{fysisorption})/kT} \tag{3.6}$$

where τ_0 is the pre-exponential factor, ΔE(fysisorption) is the fysisorption energy, k is the Boltzman constant, and T is the surface temperature. As a conclusion, the surface temperature shall be kept as low as possible for a long residence time on the surface. This will, in turn, increase the total sticking coeffient, s_{tot}. This is in fact a third parameter to play with, in addition to the reactivities of the adsorbate and surface. Thus, if we want a fast chemisorption of a specific adsorbate to the surface, then we shall try to increase its reactivity (if possible), and the reactivity of the surface. At the same time, we shall lower the temperature of the surface. The effect of these three factors is clear from equation (3.6). The opposite is true if we want to avoid chemisorption, which may be the situation for careful surface characterizations (where the chemisorption of gaseous and unwanted species should be as low as possible). Choose a high surface temperature and try to make the surface as non-reactive as possible.

However, it is not only the temperature, surface reactivity and gas phase reactivity that governs the degree of chemisorbed species onto the surface (i.e. surface coverage), but also the partial pressure of the reactive gas phase species, i. Thus, in determining the amount of gas species adsorbed onto a surface, different types of adsorption isotherms have to be used. There are two main types of adsorption isotherms, out of which one is applicable at low pressure (Langmuir adsorption isotherm) and one at higher pressures (BET isotherm). The main difference between these two adsorption isotherms is that only a monolayer is assumed to be formed when using the Langmuir adsorption isotherm, but that several layers can be formed when using the BET isotherm (see figure 3.27).

Let us start by deriving an expression for the Langmuir adsorption isotherm:

$$\theta_i = \frac{K_i p_i}{1 + K_i p_i} \tag{3.7}$$

where θ_i is the surface coverage of the chemisorbed species i, K_i is the equilibrium constant for the chemisorption of i, and p_i is the partial pressure of the gas phase species i above the growth material film. For this task one has to make the following four assumptions:

1. The gas is ideal.
2. Only one monolayer of chemisorbed i species will be formed.
3. The sticking coefficient for the chemisorption of i is equal to one.
4. All adsorption sites are identical.
5. The chemisorption of i is an equilibrium process.

Since the adsorption is assumed to be in equilibrium, the adsorption process can be written as:

$$\text{surface(rad)} + i \rightleftarrows i\text{surface} - i \tag{3.8}$$

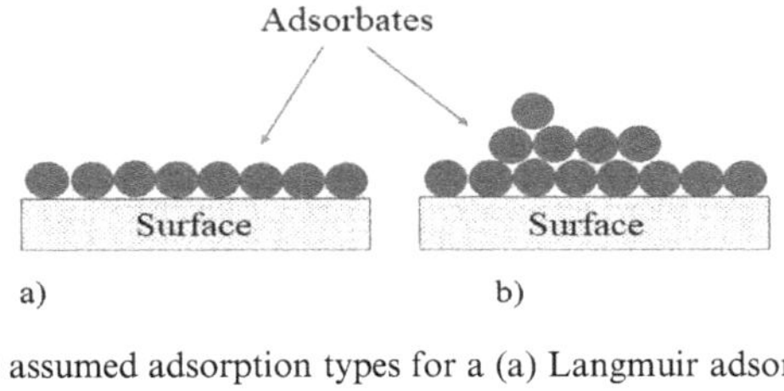

Figure 3.27. Sketch over the assumed adsorption types for a (a) Langmuir adsorption isotherm, and (b) BET isotherm.

The reaction rate to the right (v_r) is then equal to the reaction rate to the left (v_l). Since these reaction rates can be expressed as

$$v_{r,\,i} = k_{r,\,i}\theta_{\text{rad}}p_i \tag{3.9}$$

$$v_{l,\,i} = k_{l,\,i}\theta_i \tag{3.10}$$

The following is then true:

$$k_{r,\,i}\theta_{\text{rad}}p_i = k_{l,\,i}\theta_i \tag{3.11}$$

and

$$K_i = \frac{k_{r,\,i}}{k_{l,\,i}} \tag{3.12}$$

where K_i is the equilibrium constant for the adsorption of i. It is then possible to rewrite equation (3.11) as:

$$K_i = \frac{\theta_i}{\theta_{\text{rad}}p_i} \tag{3.13}$$

If we now assume that the x-species is the only type of adsorbate:

$$\theta_{\text{rad}} = 1 - \theta_i \tag{3.14}$$

$$K_i = \frac{\theta_i}{(1 - \theta_i)p_i} \tag{3.15}$$

Equation (3.15) can finally be rewritten as equation (3.7):

$$\theta_i = \frac{K_i p_i}{1 + K_i p_i}$$

For a very low partial pressure (i.e. $K_i p_i \ll 1) \Longrightarrow \theta_i = K_i p_i$. In other words, there is a linear relation between θ_i and p_i with the slope K_i. On the other hand, for a very large partial pressure (i.e. $K_i p_i \gg 1) \Longrightarrow \theta_i = 1$ (see figure 3.28). In other words, 100% of the identical adsorption sites have been terminated with i species.

Let us now have a look at the equilibrium constant, K_i. Since the adsorption process is a chemical reaction, K_i can be expressed as:

$$K_i = C\,\mathrm{e}^{-\Delta \mathrm{E(i\ adsorption)}/RT} \tag{3.16}$$

where C is a pre-exponential constant, ΔE is the adsorption energy for the terminating with i, R is the gas constant, and T is the temperature. Since the adsorption energy is negative (i.e. under the assumption of an exothermic adsorption process), an increase in T will cause a decrease in K_i and vice versa for a decrease in T. This is expected since there will be a larger degree of desorption at higher temperatures, and a larger degree of adsorption at lower temperatures. This variation in slope (i.e. K_i) with temperature is demonstrated for the Langmuir adsorption isotherm shown in figure 3.28.

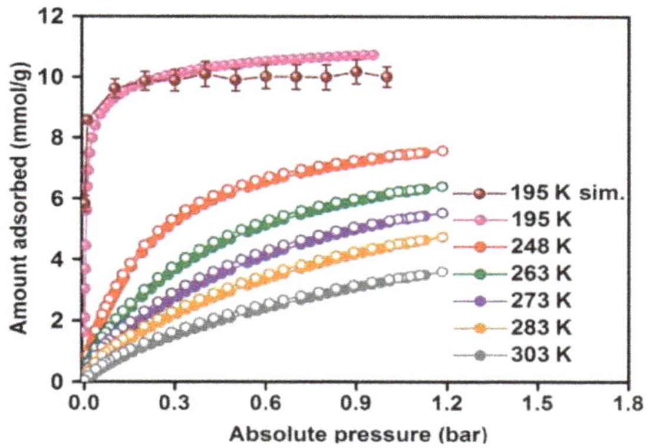

Figure 3.28. Langmuir adsorption isotherms for CO_2 adsorption at six different temperatures [4].

As can easily understood from the assumptions made in the development of the expression for the Langmuir adsorption isotherm, there are limitations of this type of isotherm:

1. The adsorbed gas has to behave ideally in the vapor phase. This condition can be fulfilled at low pressure conditions only.
2. The assumption of a monolayer formation is also only possible under a low pressure condition. Under a high pressure condition the assumption breaks down as gas molecules becomes physisorbed to the firstly adsorbed monolayer. The BET theory is more appropriate for this situation (see below).
3. Another assumption was that all the sites on the solid surface are equal and have an equal affinity towards the adsorbate. However, this is only true for a perfectly planar material surface, without any defects or other type of surface morphologies (like steps and kinks).
4. The assumption of an equilibrium in the adsorption process is most often not the real situation. However, the probability for equilibrium is larger at a lower gas pressure.

From these facts we can conclude that the Langmuir equation is valid under low pressure conditions.

A similar expression as equation (3.7) can be constructed for the BET adsorption isotherm (see equation (3.17)). Several atomic layers are then adsorbed onto the surface, with the first one chemisorbed to the surface (or at least more strongly bonded than the other layers) and the others physisorbed with very weak van der Waals interactions.

$$\theta_i = \frac{c_i p_i}{(p_0 - p_i)(p_0 + p_i(c_i - 1))} \tag{3.17}$$

where θ_i is the surface coverage of i, c is the BET-constant, p_i is the partial pressure of gas phase species i, and p_0 is the vapour pressure of the corresponding liquid phase. The following assumptions have been made in the construction of the expression:

1. All adsorption sites are identical and well-defined (one per adsorbate).
2. An adsorbate in a lower layer can act as a single adsorption site for an adsorbate in an upper layer.
3. The uppermost adsorbate layer is in equilibrium with the gas phase (i.e. identical adsorption and desorption rates).
4. Energy, of the size of adsorption energies, must be applied for desorption of adsorbates to take place.
5. The first adsorption layer binds much stronger than the others.

The derivation of equation (3.17) is not presented here, but the interested reader can find more information in [3]. This is also the situation for the various types of BET isotherms described below.

As illustrated in figure 3.29, there are different types of possible BET adsorption isotherms:

Type I: This reversible isotherm is a pseudo-Langmuir isotherm since it depicts a monolayer type of adsorption. A type I isotherm is obtained when $p_i/p_o < 1$ and $c_i > 1$ in the BET equation (equation (3.17)), which is related to the adsorption energy of the first monolayer and varies from solid to solid. Microporous solids, with pore diameters less than 2 nm, give this type of isotherm. These materials have relatively small external surfaces, and examples thereof are activated carbon, molecular sieve zeolites, COFs/MOFs, and certain porous oxides. The limiting uptake is governed by the accessible micropore volume rather than by the internal surface area.

Type II: This reversible isotherm is very different compared to the Langmuir adsorption isotherm. The flatter region in the middle represents the formation of a monolayer. A type II isotherm is obtained when $c_i > 1$ in the BET equation

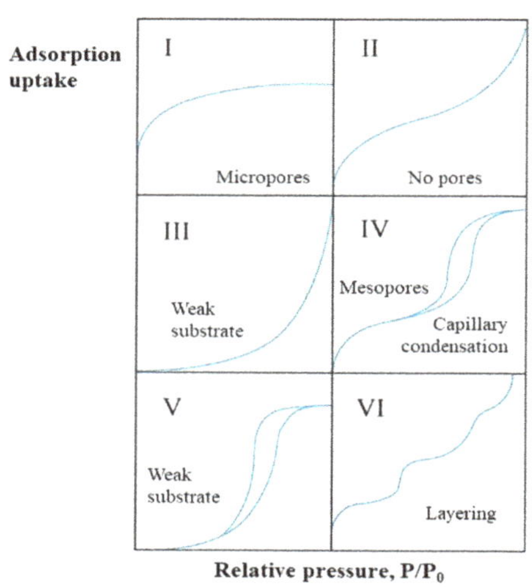

Figure 3.29. BET adsorption isotherms of various types.

(equation (3.17)). This is the most common isotherm obtained when using the BET technique for a non-porous or microporous material, representing an unrestricted monolayer-to-multilayer adsorption. The beginning of the almost linear middle section of the isotherm, is often taken to indicate the stage at which monolayer coverage is complete and multilayer adsorption about to begin.

Type III: The reversible Type III isotherm is obtained when the c <1, showing an unrestricted multilayer formation process. It is formed when the lateral interactions between the adsorbed molecules are strong in comparison to the interactions between the surface and the adsorbate.

Type IV: This isotherm occurs when capillary condensation takes place. The gas condenses in the tiny capillary pores of the solid at pressures below the saturation pressure of the gas. At the lower pressure regions, there is the formation of a monolayer followed by a formation of multilayers. Materials with pore diameters between 2 and 50 nm, gives this type of isotherm. The characteristic feature of the Type IV isotherm is its hysteresis loop, which is associated with capillary condensation taking place in the mesopores, and the limiting uptake over a range of high p_i/P_0.

Type V: Type V isotherm is very similar to the type IV isotherm. The initial part of the isotherm is attributed to a monolayer-to-multilayer type of adsorption, following the same path as the corresponding part of the Type II isotherm. The distinction between Types IV and V is analogous to that between Types II and III.

Type VI: The Type VI isotherm, in which the sharpness of the steps depends on the system and the temperature, represents stepwise multilayer adsorption onto a uniform non-porous surface. The step-height represents the monolayer capacity for each adsorbed layer and, in the simplest case, remains nearly constant for two or three adsorbed layers.

The BET adsorption isotherm is widely used in the field of materials science for the determination of surface areas of solid materials [5]. Gaseous molecules, like N_2, are generally used for this purpose. The reason is that nitrogen is available in high purity and has a strong interaction with most solid surfaces.

3.3.3 CVD growth mechanisms

There are two main types of CVD mechanisms that differ in the way the reactants react with each other. They are called Langmuir–Hinshelwood (L–H) and Eley–Rideal (E–R). In the L–H mechanism, two reactants become chemisorbed onto the surface, and thereafter react with each other (see figure 3.30). In the other mechanism, E–R, one of the reactants is first chemisorbed to the surface, where it, thereafter, reacts with the second gaseous reactant (see figure 3.31). For each of

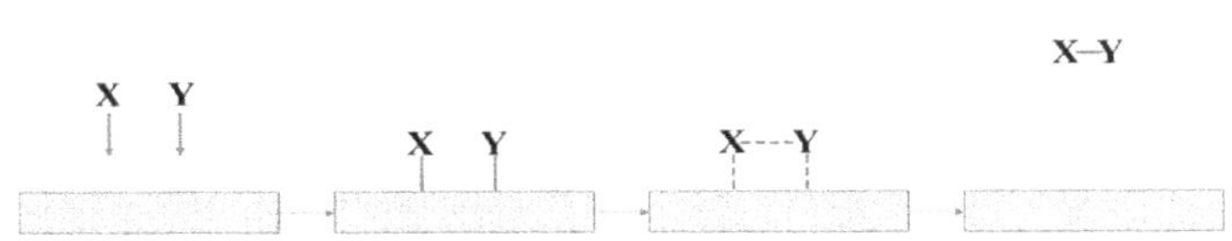

Figure 3.30. Presentation of a Langmuir–Hinschelwood mechanism.

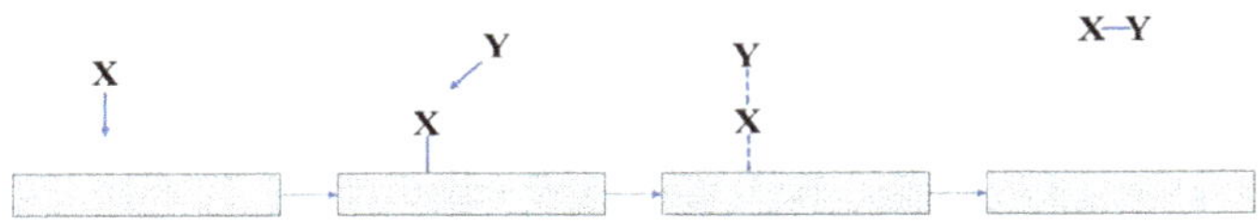

Figure 3.31. Presentation of an Eley–Rideal mechanism.

these mechanisms, there are all of the major types of heterogeneous reactions that can take place on a solid surface during CVD growth of a material; adsorption, desorption, migration, surface reaction, incorporation into the final lattice, etc. Each of these reaction types will be exemplified in detail in chapters 5–9.

It is possible to differentiate between the different CVD growth mechanisms by studying the kinetics. In addition, we get information about the overall growth rate. However, we must first define a plausible growth mechanism. Let us assume here that we want to determine the type of growth mechanism for the following total reaction mechanism:

$$2CuCl(g) + H_2(g) \rightarrow 2Cu(s) + HCl(g)$$

It is a chemical reaction so we have to, at first, find plausible elementary steps. Let us first assume that we have a reaction of type L–H (S denotes a surface site):

(1) $2CuCl(g) + 2S(rad) \rightarrow 2CuCl(ads)(separate)$
(2) $2CuCl(ads)(separate) \rightarrow 2CuCl(ads)(neighbours)$
(3) $2CuCl(ads)(neighbours) + H_2(g) \rightarrow 2CuClH(ads, TS)$
(4) $2CuClH(ads, TS) \rightarrow 2CuClH(ads)(neighbours)$
(5) $2CuClH(ads)(neighbours) \rightarrow 2Cu(s) + 2HCl(g) + 2S(rad)$

We must then make sure that the sum of these elementary steps gives the total reaction. By adding all species to the left of the expressions (1) to (5), and the ones to the right, we get

$$2CuCl(g) + H_2(g) \rightarrow 2Cu(s) + 2HCl(g)$$

We shall now continue to identify the slowest elementary step in the mechanism (i.e. the rate-determining step). This 'bottle-neck' is most often the first very slow step (in case there are several). Adsorption processes are often very fast, so step (1) can be ruled out. However, surface migration is most often very slow, so let us assume that (2) is the rate-determining step in the suggested L–H mechanism above.

Generally, the rate of an elementary step (aA+bB → C) can be expressed as

$$\nu = k\frac{[A]^a[B]^b}{[C]^c} \tag{3.18}$$

where k is the rate constant, and [A], [B], and [C] are the concentrations of A, B, and C, respectively. Hence, in our example the following will hold:

$$\nu_{tot} = \nu_2 = k_2\frac{\Theta^2_{2CuCl(ads)(neighbours)}}{\Theta^2_{2CuCl(ads)(separate)}} \tag{3.19}$$

However, we have now to continue to find a relation between ν_{tot} and the reactants. The first step in the mechanism, (1), is very fast compared to (2), and is, therefore, expected to reach an equilibrium before the rate-determining step has proceeded to any considerable extent. We can then use the equilibrium expressions for (1):

$$K_1 = \frac{\Theta_{CuCl(ads)}}{p_{CuCl(g)}\Theta_{rad}} \tag{3.20a}$$

$$K_1^2 = \frac{\Theta^2_{2CuCl(ads)(separate)}}{p^2_{CuCl(g)}\Theta^2_{rad}} \tag{3.20b}$$

where $\Theta_{CuCl(ads)(separate)}$ (being identical to $\Theta_{CuCl(ads)}$) is the surface coverage of CuCl species, and $p_{CuCl(g)}$ is the partial pressure of CuCl(g) above the surface. We can also, for simplicity, assume that the following reactions ((3)–(5)) have reached equilibrium. Thus, the following equilibrium expressions can be established for (3), (4), and (5), respectively:

$$K_3 = \frac{\Theta^2_{2CuClH(ads)(TS)}}{\Theta^2_{2CuCl(ads)(neighbours)}\, p_{H_2(g)}} \tag{3.21}$$

$$K_4 = \frac{\Theta^2_{2CuClH(ads)(neighbours)}}{\Theta^2_{2CuClH(ads)(TS)}} \tag{3.22}$$

$$K_5 = \frac{p_{HCl(g)}\Theta_{rad}}{\Theta_{CuClH(ads)}} \tag{3.23a}$$

$$K_5^2 = \frac{p^2_{HCl(g)}\Theta^2_{rad}}{\Theta^2_{2CuClH(ads)(neighbours)}} \tag{3.23b}$$

where equations (3.23a) and (3.23b) are assumed here to describe the same equilibrium. We must remember here that the activity of a solid material (i.e. concentration) is equal to 1, so the concentration expression for Cu(s) is not explicitly marked in equation (3.23).

We have now five equations with five unknown parameters, so it will be possible to find an expression of equation (3.19) as a function of the reactant concentrations. We, therefore, continue with the following mathematical exercise:

$$\text{Equation (3.20b)} \Rightarrow \Theta^2_{CuCl(ads)(separate)} = K_1^2 p^2_{CuCl(g)}\Theta^2_{rad} \tag{3.24}$$

$$\text{Equation (3.21)} \Rightarrow \Theta^2_{2CuCl(ads)(neighbours)} = \frac{\Theta^2_{2CuClH(ads)(TS)}}{K_3\, pH_2(g)} \tag{3.25}$$

$$\text{Equation (3.22)} \Rrightarrow \Theta^2_{2\text{CuClH(ads)(TS)}} = \frac{\Theta^2_{2\text{CuClH(ads)(neighbours)}}}{K_4} \quad (3.26)$$

$$\text{Equations (3.25 and 3.26)} \Rrightarrow \Theta^2_{2\text{CuCl(ads)(neighbours)}} = \frac{\Theta^2_{2\text{CuClH(ads)(neighbours)}}}{K_3K_4\ \text{pH}_2\text{(g)}} \quad (3.27)$$

$$\text{Equations (3.23b and 3.27)} \Rrightarrow \Theta^2_{2\text{CuCl(ads)(neighbours)}} = \frac{\text{p}^2_{\text{HCl(g)}}\Theta^2_{\text{rad}}}{K_3K_4\ K_5^2\text{pH}_2\text{(g)}} \quad (3.28)$$

$$\Theta_{\text{tot}} = \Theta_{2\text{CuCl(ads)(neighbours)}} + \Theta_{2\text{CuCl(ads)(separate)}} + \Theta_{2\text{CuClH(ads)(separate)}} + \Theta_{2\text{CuClH(ads)(neighbours)}} + \Theta_{\text{rad}} = 1 \quad (3.29)$$

Let us now assume that we are working under a very low pressure, which means in practice that $\Theta_{\text{rad}} \approx 1$. The final expression for the total growth rate is then:

$$\text{Equations (3.19, 3.24, 3.28 and 3.29)} \Rrightarrow \nu_{\text{tot}} = k_2 \frac{\text{p}^2_{\text{HCl(g)}}}{K_3K_4\ K_5^2K_1^2\text{p}^2_{\text{CuCl(g)}}\text{p}_{\text{H}_2\text{(g)}}} \quad (3.30)$$

As a conclusion, under the assumption of a Langmuir–Hinschelwood type of mechanism, the final expression for the overall CVD growth rate, as a function of reactant concentrations (i.e. partial pressures in the case of gas phase species), can be expressed as equation (3.30).

The next step is to assume that the CVD reaction mechanism is of type E–R, and redo the mathematical exercise performed above. The reaction mechanism is then assumed to look like this:

(1) CuCl(g) + S(rad) → CuCl(ads)
(2) $1/2H_2$(g) + S(rad) → H(ads)
(3) CuCl(ads) + H(ads) → CuCl(ads)(neighbouring H) + H(ads) (neighbouring CuCl)
(4) CuCl(ads)(neighbouring H) + H(ads)(neighbouring CuCl) → CuClH(ads) (TS)
(5) CuClH(ads)(TS) → CuClH(ads) + S(rad)
(6) CuClH(ads) → Cu(s) + HC*l*(g) + S(rad)

We must again make sure that the sum of these elementary steps gives the total reaction.

By adding all species to the left of expressions (1) to (6), and the ones to the right, we get:

$$\text{CuCl(g)} + 1/2\text{H}_2\text{(g)} \rightarrow \text{Cu(s)} + \text{HCl(g)}$$

We shall now continue by identifying the rate-determining step, which most probably is the elementary step (3) (since it involves the migration of both CuCl and

H on the surface). Thus, let us assume that (3) is the rate-determining step in the suggested E–R mechanism above. Hence, in our example the following will hold:

$$\nu_{\text{tot}} = \nu_3 = k_3 \frac{\Theta\text{CuCl(ads)(neighbouring H)}\Theta\text{H(ads)(neighbouring CuCl)}}{\Theta\text{CuCl(ads)}\Theta_{\text{H(ads)}}} \tag{3.31}$$

However, we now have to continue to find a relation between ν_{tot} and the reactants. Furthermore, the first and second steps in the mechanism, (1) and (2), are very fast compared to (3), and are, therefore, expected to reach an equilibrium before the rate-determining step has proceeded to any considerable extent. We can, thus, use the equilibrium expressions for both (1) and (2):

$$K_1 = \frac{\Theta_{\text{CuCl(ads)}}}{\text{p}_{\text{CuCl(g)}}\Theta_{\text{rad}}} \tag{3.32}$$

$$K_2 = \frac{\Theta_{\text{H(ads)}}}{\text{p}^{1/2}_{\text{H}_2\text{(g)}}\Theta_{\text{rad}}} \tag{3.33}$$

where $\Theta_{\text{CuCl(ads)}}$ and $\Theta_{\text{H(ads)}}$ are the surface coverages of CuCl and H species, respectivey. Also, $\text{p}_{\text{CuCl(g)}}$ and $\text{p}_{\text{H}_2\text{(g)}}$ are the partial pressure of CuCl(g) and H_2, respectively. We can also, for simplicity, assume that the following reactions ((4)-(6)) have reached equilibrium. Thus, the following equilibrium expressions can be established for (4), (5), and (6), respectively:

$$K_4 = \frac{\Theta_{\text{CuClH(ads)(TS)}}}{\Theta_{\text{CuCl(ads)(neighbouring H)}}\,\Theta_{\text{H(ads)(neighbouring CuCl)}}} \tag{3.34}$$

$$K_5 = \frac{\Theta_{\text{CuClH(ads)}}\,\Theta_{\text{rad}}}{\Theta_{\text{CuClH(ads)(TS)}}} \tag{3.35}$$

$$K_6 = \frac{\text{p}_{\text{HCl(g)}}\Theta_{\text{rad}}}{\Theta_{\text{CuClH(ads)}}} \tag{3.36}$$

We have now six equations with six unknown parameters, so it will be possible to find an expression of equation (3.31) as a function of the reactant concentrations. We, therefore, continue with the following mathematical exercise:

$$\text{Equation (3.32)} \Rrightarrow \Theta_{\text{CuCl(ads)}} = K_1\text{p}_{\text{CuCl(g)}}\Theta_{\text{rad}} \tag{3.37}$$

$$\text{Equation (3.33)} \Rrightarrow \Theta_{\text{H(ads)}} = K_2\text{p}^{1/2}_{\text{H}_2\text{(g)}}\Theta_{\text{rad}} \tag{3.38}$$

$$\text{Equation (3.34)} \Rrightarrow \Theta_{\text{CuCl(ads)(neighbouring H)}}\Theta_{\text{H(ads)(neighbouring CuCl)}} = \frac{K_4}{\Theta_{\text{CuClH(ads)(TS)}}} \tag{3.39}$$

$$\text{Equation (3.35)} \Rrightarrow \Theta_{\text{CuClH(ads)(TS)}} = \frac{\Theta_{\text{CuClH(ads)}}\,\Theta_{\text{rad}}}{K_5} \tag{3.40}$$

$$\text{Equation (3.36)} \Rightarrow \Theta_{CuClH(ads)} = \frac{p_{HCl(g)}\Theta_{rad}}{K_6} \tag{3.41}$$

$$\text{Equations (3.40 and 3.41)} \Rightarrow \Theta_{CuClH(ads)(TS)} = \frac{p_{HCl(g)}\,\Theta^2_{rad}}{K_5K_6} \tag{3.42}$$

$$\text{Equations (3.39 and 3.42)} \Rightarrow \Theta_{CuCl(ads)(neighbouring\ H)}\Theta_{H(ads)(neighbouring\ CuCl)} = \frac{K_4K_5K_6}{p_{HCl(g)}\,\Theta^2_{rad}} \tag{3.43}$$

$$\text{Equations (3.31, 3.32, 3.33 and 3.43)} \Rightarrow \nu_{tot} = k_3\frac{K_4K_5K_6}{K_1K_2\Theta^4_{rad}p_{HCl(g)}p^{1/2}_{H_2(g)}p_{CuCl(g)}} \tag{3.44}$$

$$\Theta_{tot} = \Theta_{2CuCl(ads)(neighbours)} + \Theta_{2CuCl(ads)(separate)} + \Theta_{2CuClH(ads)(separate)} + \Theta_{2CuClH(ads)(neighbours)} + \Theta_{rad} = 1 \tag{3.45}$$

Let us also assume here that we work under a very low pressure; $\Theta_{rad} \approx 1$. The final expression for the total growth rate is then:

$$\nu_{tot} = k_3\frac{K_4K_5K_6}{K_1K_2p_{HCl(g)}p^{1/2}_{H_2(g)}p_{CuCl(g)}} \tag{3.46}$$

As a conclusion, under the assumption of an Eley–Rideal type of mechanism, the final expression for the overall CVD growth rate, as a function of reactant concentrations (i.e., partial pressures in the case of gas phase species), can be expressed as equation (3.46).

The question is now, how to differentiate between these two CVD mechanisms? One plausible experimental way is to vary the concentration of one reactant, while keeping the other constant, and studying the effect on the overall growth rate. (This may work within a specific range of partial pressure variations). Let us start with the assumption of a Langmuir–Hinschelwood type of mechanism and use equation (3.30):

$$\nu_{tot} = k_2\frac{p^2_{HCl(g)}}{K_3K_4\,K_5^2K_1^2p^2_{CuCl(g)}p_{H_2(g)}}$$

We will now take two steps; (1) $p_{CuCl(g)}$ = constant, and(2) $p_{H_2(g)}$ = constant. We shall then for each of these steps estimate ν_{tot} for a very large partial pressure of the second species (i.e. not the one with constant pressure). We shall start here by assuming that p_{HCl} is kept constant in the reactor.

Step 1 ⇒ ν_{tot} = constant $p_{H_2(g)}^{-1}$, for $p_{CuCl(g)}$ = constant.

Step 2 ⇒ ν_{tot} = constant $p_{CuCl(g)}^{-2}$, for $p_{H_2(g)}$ = constant.

Let us now assume that we have an Eley–Rideal start type of mechanism and use equation (3.46):

$$\nu_{tot} = k_3 \frac{K_4 K_5 K_6}{K_1 K_2 p_{HCl(g)} p_{H_2(g)}^{1/2} p_{CuCl(g)}}$$

We do the same procedure, and get the following results:

Step 1 ⇒ ν_{tot} = constant $p_{H_2(g)}^{-1/2}$, for $p_{CuCl(g)}$ = constant.

Step 2 ⇒ ν_{tot} = constant $p_{CuCl(g)}^{-1}$, for $p_{H_2(g)}$ = constant.

These results do not help very much since they in principle show the same qualitative results. That is, the growth rate is decreasing more rapidly with constant partial pressure of HCl, than it does for a constant partial pressure of H. However, let us take the partial pressure of the product HCl(g) into consideration. It is constantly transferred out from the reactor by pumping. This means that we can assume that $p_{HCl(g)} \ll p_{CuCl(g)}, p_{H_2(g)}$. The following must, therefore, hold;

L–H reaction ⇒ ν_{tot} is initially extremely low and reaches zero very fast.

E–R reaction ⇒ ν_{tot} is initially extremely high and decreases from this point and onwards.

References

[1] Carlsson J-O and Martin P M 2010 *Handbook of Deposition Technologies for Films and Coatings: Science, Applications and Technology* 3rd edn (Norwich, NY: William Andrew) pp 314–361

[2] Choy K 2003 Chemical vapour deposition of coatings *Prog. Mater Sci.* **48** 55–170

[3] Li X, He L, Li Y and Yang Q 2020 Diamond deposition on iron and steel substrates: a review *Micromachines.* **11** 719–39

[4] Nandi S, De Luna P, Daff T D, Rother J, Liu M, Buchanan W, Hawari A I, Woo T K and Vaidhyanathan R 2015 A single-ligand ultra-microporous MOF for precombustion CO_2 capture and hydrogen purification *Sci. Adv.* **1** e1500421

[5] Raja P M V and Barron A R 2019 *Physical Methods in Chemistry and Nanoscience* (Rice University) https://hdl.handle.net/1911/112293

Chapter 4

Theoretical methods and methodologies

Theoretical modeling and simulations have become particularly important tools within the modern area of materials science. In fact, the area of theoretical calculations is the fastest growing field within the materials science of today. It has been shown to be unbelievably valuable to support and explain experimentally obtained results. Also, it has been frequently used for the prediction of experimental parameters, and the results thereof. It has also been found to be indispensable for those situations where experiments are practically impossible. As such, theoretical modeling and simulations are nowadays used in disciplines like chemistry, physics, engineering, biology, medicine, etc. Materials with different dimensions are then being studied (e.g. smaller molecules, large clusters, thin solid films, and bulk material).

4.1 General

In order to understand the geometrical structures and properties of, e.g. thin solid films, one needs to know how the electrons interact with their surrounding forces. It was earlier believed that the motions of the electrons could be expressed by classical mechanics.

According to classical mechanics, an electron is a particle and is, thereby, distinctly different to a wave. However, at the end of the 19th century, experiments showed that particles, as well as electromagnetic radiation, have the characteristics of both a particle and a wave–particle duality. As a result of this discovery, a new type of mechanics, called quantum mechanics, was developed (figure 4.1). Since 1920, many different quantum mechanical methods, with a variety of computational technologies, have been developed. These strives have continued during the last decades, which have resulted in a dramatic increase in both computational efficiency and accuracy. This is especially the situation for the density functional theory (DFT) method, which uses the electrons in the systems to describe a vast number of materials properties. DFT is nowadays one of the most used quantum mechanical

doi:10.1088/978-0-7503-3107-4ch4

Figure 4.1. The Schrödinger equation was developed by the Austrian physicist *Erwin Schrödinger* (1887–1961) (https://en.wikipedia.org/wiki/Erwin_Schr%C3%B6dinger).

Figure 4.2. Three examples of material systems.

methods in the materials science area, for use as both a research tool and as an important analysis instrument within the industrial production line. It is thereby easily understood that DFT has become one of the most accurate and versatile calculation techniques of today. As will be described in section 4.3, electron density distributions are used in the DFT calculations when simulating and analysing the materials with their properties. But before going into depth with the DFT calculations, let us start with the basics of quantum mechanical calculations.

4.2 The Schrödinger equation

4.2.1 Introduction

Many different types of materials systems are frequently being investigated by using quantum mechanical calculations. Some examples are shown in figure 4.2. In a solid-state system, like a CVD deposited thin film, a vast number of particles with different masses and charges interacts with each other. For us to be able to calculate the geometrical structure and physical properties of this type of thin film system, it must be treated as a complex many-body problem. The Schrödinger equation is the basic tool for solving such a manybody problem and is generally regarded to be the fundamental equation within quantum mechanics [1]. The time-dependent Schrödinger equation is presented in equation (4.1).

$$i\hbar\frac{\partial}{\partial t}\Psi = \hat{H}\Psi \tag{4.1}$$

A simplified version of the Schrödinger equation can be obtained by using fixed positions of the atoms in the system (see equation (4.2)). This is called the time-independent Schrödinger equation, by which it is possible to find good approximations of the total energy of the system.

$$\hat{H}\Psi = E\Psi \tag{4.2}$$

where $\hat{H}$ is the Hamiltonian operator, which acts on a many-electron wavefunction (Ψ), resulting in the total energy (E) of the system. The energy € is more strictly defined as the eigenvalue to the Hamiltonian operator ($\hat{H}$). For a system that consists of N electrons and M nuclei, the Schrödinger equation can be written as:

$$\begin{aligned} &\hat{H}\Psi(\vec{r}_1 s_1,\ \vec{r}_2 s_2,\ \ldots,\ \vec{r}_N s_N,\ \vec{R}_1,\ \vec{R}_2,\ \ldots,\ \vec{R}_M) \\ &\quad = E\Psi(\vec{r}_1 s_1,\ \vec{r}_2 s_2,\ \ldots,\ \vec{r}_N s_N,\ \vec{R}_1,\ \vec{R}_2,\ \ldots,\ \vec{R}_M) \end{aligned} \tag{4.3}$$

where E is the total energy of the system, $\hat{H}$ is the Hamiltonian operator describing all interactions in the system, $\vec{r}_i$ is the position of the ith electron (where $\vec{r}_i = x_i,\ y_i,\ z_i$), s_i is the spin of the ith electron, R_i is the position of the ith nucleus (where $\vec{R}_i = X_i,\ Y_i,\ Z_i$), and Ψ is the total wavefunction of the system. This is a material-specific wavefunction that contains all information about the nuclei and electrons in the system. For this type of many-body system, the Hamiltonian operator can be expressed as:

$$\begin{aligned} \hat{H} &= -\frac{\hbar^2}{2}\sum_{k=1}^{M}\frac{\vec{\nabla}_k^2}{M_k} - \frac{\hbar^2}{2}\sum_{i=1}^{N}\frac{\vec{\nabla}_i^2}{m_e} + \frac{1}{2}\sum_{k\neq l}^{M}\frac{Z_k Z_l e^2}{|\vec{R}_k - \vec{R}_l|} + \frac{1}{2}\sum_{i\neq j}^{N}\frac{e^2}{|\vec{r}_i - \vec{r}_j|} - \sum_{i,k}^{M,N}\frac{Z_k e^2}{|\vec{r}_i - \vec{R}_k|} \\ &= T^N + T^e + V^{NN} + V^{ee} + V^{Ne} \end{aligned} \tag{4.4}$$

where $\hbar$ is the Planck constant (h) divided with 2π.

For the ith nuclei, M_k is the mass, R_k is the position, and Z_k is the atomic charge. Also, for the ith electron, m_e is the mass, r_i is the position, and e is the electronic charge. Moreover, the Hamiltonian is composed of three different terms. The first term represents the kinetic energies of the electrons (T^e). The other three terms represent potential energies of the nuclei–nuclei repulsions (V^{NN}), electron–electron repulsions (V^{ee}), and attractive interactions between the electrons and the nuclei (V^{Ne}). The last term is described as an external potential for the electrons.

However, it is only possible to find an exact solution to equation (4.4) for two different systems with only one electron in the atomic/molecular structure: H and He_2^+. The reason is that it is not possible to exactly calculate the electron–electron correlations for systems that contain more than one electron. Thus, for these latter types of systems, approximations of the electron–electron correlations must be included in equation (4.4). Consequently, the expression of the Hamiltonian operator must be even more simplified in solving the Schrödinger equation. As a first step, the Born–Oppenheimer approximation must be used (see section 4.2.2). This approximation is based on the fact that an electron is much lighter and faster than a nucleus. As a direct consequence, the atomic nucleus can be regarded as fixed in relation to the movement of the electrons.

4.2.2 Born–Oppenheimer approximation

The Schrödinger equation can be further simplified by assuming that the Born–Oppenheimer approximation is valid for the system under investigation. The heavy nuclei are then assumed to be 'frozen' at certain positions, while the light-weighted electrons are moving in the field of the positively charged nuclei [1]. Hence, the nuclei will act on the electrons with an external potential, V^{ext} (also expressed as V^{Ne}). With this approximation, the total wavefunction will be divided into two parts (Ψ_e and Ψ_N) describing the behaviour of the electrons and nuclei, respectively. The new expression of the total wavefunction is now expressed as:

$$\Psi = \Psi_e \Psi_N \tag{4.5}$$

In addition, the total Hamilton operator (see equation (4.4)) can also be divided into two parts, describing the electrons and nuclei separately:

$$\hat{H}_e = -\frac{\hbar^2}{2}\sum_{i=1}^{N}\frac{\vec{\nabla}_i^2}{m_e} + \frac{1}{2}\sum_{i\neq j}^{N}\frac{e^2}{|\vec{r}_i - \vec{r}_j|} + \sum_{i,k}^{M,N}\frac{Z_k e^2}{|\vec{r}_i - \vec{R}_k|} = T^e + V^{ee} + V^{ext} \tag{4.6}$$

$$\hat{H}_N = -\frac{\hbar^2}{2}\sum_{k=1}^{M}\frac{\vec{\nabla}_k^2}{M_k} = T^N \tag{4.7}$$

where $\hbar$ is the Planck constant divided by 2π, i is the number of the electrons, k is the number of the nuclei, m_e is the mass of an electron, e is the charge of an electron, M_k is the mass of nuclei k, Z_k is the charge of nuclei k, and $\vec{r}_i$ is the position of electron i. Also, T^e, V^{ee}, and V^{ext} describe the total electron kinetic energy, electron–electron potential energy, and nucleus–electron potential energy, respectively. In addition, T^N describes the kinetic energy of the nuclei. As a consequence, the complete Schrödinger equation will also be split into an electron part and a nuclei part. This is also the situation for the total energy:

$$\left(-\frac{\hbar^2}{2}\sum_{i=1}^{N}\frac{\vec{\nabla}_i^2}{m_e} + \frac{1}{2}\sum_{i\neq j}^{N}\frac{e^2}{|\vec{r}_i - \vec{r}_j|} + \sum_{i,k}^{M,N}\frac{Z_k e^2}{|\vec{r}_i - \vec{R}_k|}\right)\Psi_e = E_e\Psi_e \tag{4.8}$$

$$\left(-\frac{\hbar^2}{2}\sum_{k=1}^{M}\frac{\vec{\nabla}_k^2}{M_k}\right)\Psi_n = E_n\Psi_n \tag{4.9}$$

$$E_{tot} = E_e + E_n \tag{4.10}$$

The part of the total energy that belongs to the electrons in the system is thus dependent on the atomic coordinates, $\vec{r}$. The contribution to the total energy by the nuclei in the system should, however, not be neglected. Even if the kinetic energies of the nuclei are assumed to be negligible, their potential energies are not. The nucleus–nucleus interactions form the nuclear potential energy, which, in the Born–Oppenheimer approximation, is merely a constant (i.e. E_n = constant). Hence, the total Schrödinger equation can be simplified to include the wavefunction of the electrons alone (i.e. being identical to equation (4.8)). The constant E_n value will be added to E_e after this Schrödinger equation has been solved.

4.2.3 Hartree–Fock methods

As stated above, it is not possible to solve the Schrödinger equations exactly. The exceptions are H and He_2^+ with their single electrons. This is the situation for all systems that can be formed from the elements in the periodic table. To improve the solutions of the Schrödinger equation, one must therefore consider an approximation of the electron–electron interaction term in equation (4.8) (also called electron correlations). The single particle model, as developed by Hartree [3], will then be of great help. Hartree replaced the total electron wavefunction, Ψ_e, with a product of independent one-electron wavefunctions (i.e. one for each electron in the system under investigation):

$$\Psi_e(\vec{r}_1 s_1,\ \vec{r}_2 s_2,\ \ldots,\ \vec{r}_N s_N) = \varphi_1(\vec{r}_1 s_1)\varphi_2(\vec{r}_2 s_2)\ldots.\varphi_N(\vec{r}_N s_N) \tag{4.11}$$

where $\varphi_N(\vec{r}_N s_N)$ is a one-electron wavefunction for the Nth electron, with its position at $(\vec{r}_N = x_N, y_N, z_N)$ and spin (s_N). Moreover, as shown in equation (4.8), the electron Hamiltonian, $\hat{H}_e$, can also be described as a sum of single electron Hamiltonians:

$$\hat{H}_e = \hat{h}_1 + \hat{h}_2 + \ldots.. + \hat{h}_N \tag{4.12}$$

On the basis of equations (4.11) and (4.12), the total electronic energy, E_e, can now be written as a sum of single electron energies, e_i:

$$E_e = e_1 + e_2 + \ldots + e_N \tag{4.13}$$

The independent particle Schrödinger equation (for electron i) can, thereby, be written as:

$$\hat{h}_{\text{eff},\,i}\varphi_i(\vec{r}_i s_i) = e_i\varphi_i(\vec{r}_i s_i) \tag{4.14a}$$

or

$$\left(-\frac{\hbar^2}{2}\frac{\vec{\nabla}_i^2}{m_e} - \frac{e^2}{4\pi\varepsilon_0}\left\{\sum_{k=1}^{M}\frac{Z_k}{|\vec{r}_i - \vec{R}_k|} + \sum_{j\neq i}^{N}\int\frac{|\varphi_j(\vec{r}_j)|^2}{|\vec{r}_i - \vec{r}_j|}d\vec{r}_j\right\}\right)\varphi_i(\vec{r}_i s_i) = e_i\varphi_i(\vec{r}_i s_i) \tag{4.14b}$$

where $\hat{h}_{\text{eff},\,i}$ is called the effective Hamilton operator of Hartree type. As can be seen in equation (4.14b), the instantaneous electron-interactions (i.e. electron correlations) are approximated here by using an averaged value of the repulsive one-electron integral,

$$\int \frac{|\varphi_j(\vec{r}_j)|^2}{|\vec{r}_i - \vec{r}_j|} d\vec{r}_j$$

Except for the electron correlation, another factor that must be approximated in the all-electron Schrödinger equation (see equation (4.8)) is the electron wavefunction, $\Psi_e(\vec{r}_1 s_1, \vec{r}_2 s_2, \dots, \vec{r}_N s_N)$. This wavefunction should be able to simulate the situation where two electrons in the system exchange positions. First of all, the electron density must become identical = $(\Psi^2_{e,\,after} = \Psi^2_{e,\,before})$, and there are only two solutions to this identity; $\Psi_{e,\,after} = \Psi_{e,\,before}$ and $\Psi_{e,\,after} = -\Psi_{e,\,before}$. (These wavefunctions $(\Psi_{e,\,after})$ and $(\Psi_{e,\,before})$ are antisymmetric). To incorporate these demands on the all-electron wavefunction, Fock proposed to replace the wavefunction with a Slater determinant [4]. Equations (4.15) and (4.16) show the wavefunction for systems with unpaired and paired electrons, respectively. For example, with a two-electron system (e.g. H_2 with paired electrons) the new wavefunction looks like:

$$\Psi_e(\vec{r}_1 s_1, \vec{r}_2 s_2, \dots, \vec{r}_N s_N) = 1/\sqrt{N!}\begin{bmatrix} \varphi_1(\vec{r}_1 s_1)\varphi_2(\vec{r}_1 s_1) & \cdots & \varphi_N(\vec{r}_1 s_1) \\ \vdots & \ddots & \vdots \\ \varphi_1(\vec{r}_N s_N)\varphi_2(\vec{r}_N s_N) & \cdots & \varphi_N(\vec{r}_N s_N) \end{bmatrix} \tag{4.15}$$

$$\Psi_e(\vec{r}_1, \vec{r}_2, \dots, \vec{r}_{N/2}) = 1/\sqrt{\left(\frac{N}{2}\right)!}\begin{bmatrix} \varphi_1(\vec{r}_1)\varphi_2(\vec{r}_1) & \cdots & \varphi_N(\vec{r}_1) \\ \vdots & \ddots & \vdots \\ \varphi_1(\vec{r}_{N/2})\,\varphi_2(\vec{r}_{N/2}) & \cdots & \varphi_N(\vec{r}_{N/2}) \end{bmatrix} \tag{4.16}$$

$$\Psi_e(\vec{r}_1, \vec{r}_2) = 1/\sqrt{2}(\varphi_1(\vec{r}_1)\varphi_2(\vec{r}_2) - \varphi_1(\vec{r}_2)\varphi_2(\vec{r}_1)) \tag{4.17}$$

For a manybody system, the Slater-type wavefunction is, hence, a linear combination of many terms, where each term is a product of all single electron wavefunctions (with various x, y, z, and s values). The combination of these two methods is called the Hartree–Fock (HF) method [4]. Furthermore, the effective Hamilton operator (for electron i) will now also look different, and is, hereby, called the Fock operator:

$$\hat{h}_{\mathrm{eff,\,i}} = -\frac{\hbar^2}{2}\frac{\nabla_i^2}{m_e} - \frac{e^2}{4\pi\varepsilon_0}\left\{\sum_{k=1}^{M}\frac{Z_k}{|\vec{r}_i - \vec{R}_k|} - \sum_{j\neq i}^{N}\int\frac{|\varphi_j(\vec{r}_j)|^2}{|\vec{r}_i - \vec{r}_j|}d\vec{r}_j - \sum_{i\neq j}^{N}\int\frac{|\varphi_j^*(\vec{r}_j)\varphi_j(\vec{r}_i)|}{|\vec{r}_i - \vec{r}_j|}P_{ij}d\vec{r}_i\right\} \tag{4.18}$$

where P_{ij} is called the exchange potential. (The index i replaces the index j within the function $\varphi_i(\vec{r}_i)$ for which the effective operator $\hat{h}_{\mathrm{eff,\,i}}$ operates.) Figure 4.3 shows the

Steps to take when starting a Hartree-Fock calculation

Step 1: Identify the atomic positions and cell parameters (and use as input to the program)

Step 2: Atomic information (incl. electron configurations, and nuclei charges) are stored in the program libraries.

Step 3: Other information, like ionic charges, occurrence of unpaired electrons (i.e., spins), and symmetries within the system, will also be set as input to the program.

Step 4: The effective Fock operator is now constructed by the program.

Figure 4.3. Computer steps in the determination of a Fock operator.

different steps in the construction of the Fock operator (when using a quantum mechanical program).

4.2.4 Basis set specifications

It is not only the effective Hamilton operator that needs to be specified, but also the single electron wavefunctions. The most common approximation of $\varphi_i(\vec{r}_i)$ is a linear combination of analytical functions, and it is generally called the MO-LCAO (molecular orbitals as linear combinations of atomic orbitals) approximation. The set of AOs is normally referred to as the basis set:

$$\varphi_i(\vec{r}_i) = \sum_k c_{ki}\theta_{ki}(\vec{r}_i) \tag{4.19}$$

It should be stressed here that an exceedingly large number of high-quality basis functions is needed for the description of high-quality electron wavefunctions. The larger the number of terms in the linear combination of basis functions, the closer the wavefunction to the true ground-state wavefunction (i.e. to the eigenfunction to the Hamiltonian operator). This most often leads to practical problems, such as the need of excessive disk space and time-consuming calculations. Test-calculations are usually performed with the purpose to find the most optimal number and type of AO terms in the $\varphi_i(\vec{r}_i)$ expression.

There are many different types of analytical basis functions to chose from, and the simplest (and smallest) is the basis set where every filled atomic orbital (s,p,d,…) is described with only one analytical function for every atom in the system. For example, the simplest basis set for a fullerene molecule, C_{60}, consists of $60 \times 5 = 300$ basis functions (since C contains five atomic orbitals that are completely, or partially, filled with electrons). Hence, the larger basis sets will involve a much larger set of basis functions (than 300) for the small C_{60} system! A variety of basis sets are usually stored within the calculation programs.

With these base sets to start from, it should be possible to find more or less good solutions to the electronic Schrödinger equation. A new technique has, therefore, been invented through which it is possible to systematically approach the ground-state energy—the variation principle.

4.2.5 The variation principle

The next step is to decide on the coefficients, c_{ki}, in the single electron wavefunctions, $\varphi_i(\vec{r}_i)$. A new technique has, therefore, been invented by which it is possible to systematically approach the ground-state energy—the variation principle.

The energy of any approximate single-electron wavefunction, $\varphi_i(\vec{r}_i)$, is greater than, or equal to, the exact ground state energy; $e_i' > e_i$. Hence, there is an identity for the eigenvalue only. In solving the effective Schrödinger equation (see equation (4.14)) the similarity between the resulting e_i' and the ground state energy, e_i, depends to a large extent on the coefficients in the electron wavefunctions, $\varphi_i(\vec{r}_i)$. Hence, for a given set of basis functions, their coefficients can be adjusted until the total energy, e_i', becomes minimized. In other words, the best solution we can find is when we come as close as possible to the groundstate e_i value (see figure 4.4). This method is an approximation method that is usually called the variation principle (or the self-consistent field method).

The mathematical equations to be used when iteratively searching for the best coefficients, c_{ki}, are:

$$\hat{h}_{\mathrm{eff},\,i}\,\varphi_i(\vec{r}_i s_i)' = e_i'\varphi_i(\vec{r}_i s_i)' \tag{4.20}$$

By multiplying both sides of equation (4.20) with the complex conjugate of $\varphi_i(\vec{r}_i s_i)'$ we get:

$$\varphi_i(\vec{r}_i s_i)'^{*}\hat{h}_{\mathrm{eff},\,i}\,\varphi_i(\vec{r}_i s_i)' = \varphi_i(\vec{r}_i s_i)'^{*}e_i'\varphi_i(\vec{r}_i s_i)' \tag{4.21}$$

Since e_i' is a constant, we can write equation (4.21) as

$$\varphi_i(\vec{r}_i s_i)'^{*}\hat{h}_{\mathrm{eff},\,i}\,\varphi_i(\vec{r}_i s_i)' = e_i'\varphi_i(\vec{r}_i s_i)'^{*}\varphi_i(\vec{r}_i s_i)' \tag{4.22}$$

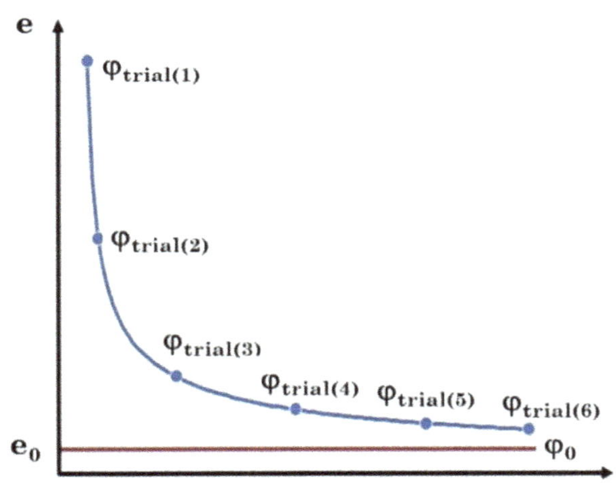

Figure 4.4. According to the variational principle, the energy calculated from an approximate wavefunction (φ_{trial}) will always be larger than the true energy (e_0). Also, the better the wavefunction, the lower the energy.

Moreover, by dividing both sides of equation (4.22) with the product of the wavefunction and its complex conjugate, we get the following expressions for the energy, e_i':

$$e_i' = \frac{\varphi_i(\vec{r}_i s_i)'^* \hat{h}_{eff,\,i} \varphi_i(\vec{r}_i s_i)'}{\varphi_i(\vec{r}_i s_i)'^* \varphi_i(\vec{r}_i s_i)} \tag{4.23}$$

$$e_i' = \frac{\int \varphi_i(\vec{r}_i s_i)'^* \hat{h}_{eff,\,i} \varphi_i(\vec{r}_i s_i)' d\vec{r}_i}{\int \varphi_i(\vec{r}_i s_i)'^* \varphi_i(\vec{r}_i s_i)' d\vec{r}_i} \tag{4.24}$$

where e_i' is the calculated single-electron energy when using a trial wavefunction, $\varphi(r_i,s_i)'$, in solving the one-electron Schrödinger equation. Equation (4.24) is simply a re-writing of equation (4.23), which in turn is a re-writing of equation (4.22). Furthermore, by using a MO-LCAO expression of the $\varphi_i(\vec{r}_i s_i)$ functions, equation (4.25) will be derived from equation (4.24):

$$e_i' = \frac{\sum_{k,l} c_{ki} c_{li} \int \theta_{ki}(\vec{r}_i s_i)'^* \hat{h}_{eff,\,i} \theta_{li}(\vec{r}_i s_i)' d\vec{r}_i}{\sum_{k,l} c_{ki} c_{li} \int \theta_{ki}(\vec{r}_i s_i)'^* \theta_{li}(\vec{r}_i s_i)' d\vec{r}_i} \tag{4.25}$$

Equation (4.25) is used for the determination of those coefficients, c_{ki}, that gives the smallest e_i'value. To achieve this e_i' value, the first derivative of e_i', with respect to c_{ki}, is then being looked for:

$$\frac{\partial e_i'}{\partial c_{ki}} = 0 \quad k = 1,\ 2 \tag{4.26}$$

(number of atomic orbitals in the basis set).

The results from these calculations are called the Secular equations:

$$\sum_l (H_{kl} - e_i' S_{kl}) c_l = 0 \quad k, l = 1,\ 2 \tag{4.27}$$

The S_{kl} term can be expressed as

$$S_{kl} = \int \theta_{ki}(\vec{r}_i s_i)'^* \theta_{li}(\vec{r}_i s_i)' d\vec{r}_i \tag{4.28}$$

and the H_{kl} term can be expressed as

$$\begin{aligned} H_{kl} = \int \theta_{ki}(\vec{r}_i)_i(\vec{r}_i s_i)'^* \hat{h}_{eff,\,i} \theta_{li}(\vec{r}_i s_i)' d\vec{r}_i = \\ \int \theta_{ki}(\vec{r}_i s_i)'^* \hat{h}_i \theta_{li}(\vec{r}_i s_i)' d\vec{r}_i + \frac{e^2}{4\pi\varepsilon_0} \sum_{j \neq i}^{N} \int \frac{\theta_{ki}(\vec{r}_i s_i)'^* \varphi_{pj}(\vec{r}_j s_j)'^* \varphi_{qj}(\vec{r}_j s_j) \theta_{li}(\vec{r}_i s_i)'}{|\vec{r}_i - \vec{r}_j|} d\vec{r}_i d\vec{r}_j \\ - \frac{e^2}{8\pi\varepsilon_0} \sum_{j \neq i}^{N} \int \frac{\theta_{ki}(\vec{r}_i s_i)'^* \varphi_{pj}(\vec{r}_j s_j)'^* \varphi_{qi}(\vec{r}_i s_i) \theta_{lj} \varphi_j(\vec{r}_j)}{|\vec{r}_i - \vec{r}_j|} d\vec{r}_i d\vec{r}_j \end{aligned} \tag{4.29}$$

where

$$\hat{h}_i = -\frac{\hbar^2}{2}\frac{\nabla_i^2}{m_e} - \frac{e^2}{4\pi\varepsilon_0}\sum_{k=1}^{M}\frac{Z_k}{|\vec{r}_i - \vec{R}_k|} \tag{4.30}$$

and

$$h_{kl} = \int \theta_{ki}(\vec{r}_i s_i)'^{*}\hat{h}_i\theta_{li}(\vec{r}_i s_i)' d\vec{r}_i \tag{4.31}$$

The H_{kl} term can now be written as:

$$H_{kl} = h_{kl} + 2\sum_{j,p,q} c_{pj}c_{qj}(kl|pq) - \sum_{j,p,q} c_{pj}c_{qj}(kq|pl) \tag{4.32}$$

where

$$(kl,|,pq) = \frac{e^2}{4\pi\varepsilon_0}\int \frac{\theta_{ki}(\vec{r}_i s_i)'^{*}\theta_{li}(\vec{r}_i s_i)\theta_{pj}(\vec{r}_j s_j)'^{*}\theta_{qj}(\vec{r}_j s_j)}{|\vec{r}_i - \vec{r}_j|} d\vec{r}_i d\vec{r}_j \tag{4.33}$$

and

$$(kq|pl) = \frac{e^2}{4\pi\varepsilon_0}\int \frac{\theta_{ki}(\vec{r}_i s_i)'^{*}\theta_{qi}(\vec{r}_i s_i)'^{*}\theta_{pj}(\vec{r}_j s_j)\theta_{lj}(\vec{r}_j s_j)}{|\vec{r}_i - \vec{r}_j|} d\vec{r}_i d\vec{r}_j \tag{4.34}$$

To simplify equation (4.29), the charge and bond order matrix, P_{kl}, can now be used:

$$P_{kl} = \sum_{i=1}^{N} c_{ki}c_{li} \tag{4.35}$$

By using equation (4.34), the H_{kl} term (equation (4.32)) can be expressed as

$$H_{kl} = h_{kl} + \sum_{p,q} P_{kl}\left[(kl,|,pq) - \frac{1}{2}(kq|pl)\right] \tag{4.36}$$

and the final Hartree–Fock–Roothan equations can be formed:

$$\sum_{l=1}^{M}(H_{kl} - e_i S_{kl})c_{li} = 0 \quad k = 1, 2, \ldots M \tag{4.37}$$

with its corresponding determinant and its non-trivial solutions:

$$\begin{bmatrix} h_{11} - e_i S_{11} h_{12} - e_i S_{12} & \cdots & h_{1M} - e_i S_{1M} \\ \vdots & \ddots & \vdots \\ h_{M1} - e_i S_{M1} h_{M2} - e_i S_{M2} & \cdots & h_{MM} - e_i S_{MM} \end{bmatrix} = 0 \tag{4.38}$$

As shown in the equations above, there is a large number of one- and two-electron integrals that must be solved. This number will increase with an improved basis set, and with a larger model system. There are two choices to be made when performing these calculations, they can either be stored on disk or re-calculated when needed. The former choice puts large demands on disk space, and the latter prolongs the computational time.

The procedure in doing the SCF calculations is the following. One has to start with an arbitrary trial wavefunction, $\varphi_i(\vec{r}_i,s_i)'$, that is as close to the real one as possible, and the coefficients, c_{ik}, in this wavefunctions will be used as input to the first step in the iteration process (step 1). A first energy value, e_i', is, thereafter, obtained by solving equation (4.37). This energy value will be used as input to equation (4.38), with new values of the coefficients, c_{ik}, as the resulting output (and, hence, a new version of the single-electron wavefunction, $\varphi_i(\vec{r}_i,s_i)'$ (step 2). If this new e_i' value is close enough to the e_i' value obtained in the earlier iteration (i.e. if $\Delta e_i'$ is smaller than a pre-set threshold value), the calculations will be stopped (steps 3 and 4). The best $\varphi_i(\vec{r}_i,s_i)'$ and e_i' have then been obtained. However, if $\Delta e_i'$ is larger than (or equal to) the pre-set threshold value, the calculations will start over again (with step 1). (It is only for the very first iteration that there is no possibility to compare with a previous e_i' value. One instead has to go back to step 1 directly). These iterative SCF calculations are demonstrated in figure 4.5.

Based on the knowledge about the iterative steps in the SCF procedure, it is now time to make a similar presentation (as figure 4.5) for the calculations of the total energy, E, of a specific system. As described in figure 4.6, the building of a model is the first step to take in these calculations. The atomic positions and cell parameters are then needed as input to the program (step 1). All detailed atomic information, such as electron configuration and nuclei charges, is stored in the program libraries (step 2). Other types of information (e.g. ionic charges, occurrence of unpaired electrons (i.e. spins), and symmetries within the system) will also be used as input to the program (step 3). The next step is to decide on the basis set (step 4). When these inputs have been made to the program, the calculation of the various one- and two-electron integrals starts taking place (step 5). These integrals can then either be stored on disk or be re-calculated when needed in a later stage of the calculations. The next step is the SCF iterative calculations (step 7), which start with an initial guess of the coefficients, c_i (step 6). The total energy, E, is finally obtained by using equation (4.13).

Steps to take in the SCF calculations

Step 1: Solve the secular determinant → Gives an e_i' value

Step 2: Solve the secular equations with e_i' as input → Gives c_k values (i.e., and a new $\varphi(r_i,s_i)'$)

Step 3: Is e_i' within a pre-set threshold value (i.e., has it converged)? → No (go back to step 1).

↓ Yes

Step 4: The effective Fock operator is now constructed by the program.

Figure 4.5. Computer steps in iterative SCF calculations.

Steps to take when calculating the total energy, E, for a specific model system

Step 1: Atomic positions and cell parameters (as input to the program).

Step 2: Atomic information (incl. electron configurations, and nuclei charges) are stored in the program libraries.

Step 3: Other information, like ionic charges, occurrence of unpaired electrons (i.e., spins), and symmetries within the system, will also be set as input to the program.

Step 4: Specification of basis set (as input to the program).

Step 5: Calculations of one- and two-electron integrals.

Step 6: Calculations of intitial guesses of coefficients, c_{1i}.

Step 7: The self-consistent field (SCF) calculations are being solved.

Step 8: The total energy, E, is calculated as the sum of the electronic energies, e_i.

Figure 4.6. Computer steps in the calculations of the total energy, *E*, of a system.

These steps are the basis of all modern quantum mechanical calculations. However, what are the advantages and disadvantages with these ones, and, thereby, the Hartree–Fock method? The main drawback with these calculations is that the instantaneous electron–electron correlations are neglected. Also, equations (4.15) and (4.16) do not obey the Paulis' exclusion principle, which states that two electrons cannot occupy the same $\vec{r}_i = x, y, z, s$.

These weaknesses are quite severe. In fact, they most often lead to large deviations from experimental results. It is, thus, necessary to also include approximations of the electron–electron correlations in the quantum mechanical calculations. This can be done by using a technique that is generally referred to as post-Hartree–Fock calculations [5]. There are also other disadvantages with the Hartree–Fock method, in addition to advantages, which are all summarized below:

4.2.5.1 Advantages/disadvantages with the Hatree–Fock method

- Commonly used for calculations on molecules and small clusters.
- Frequently used for inorganic and 'main-group' inorganic compounds, in addition to ionic crystals.
- A very time-consuming method. It is only of practical use for calculations on models of sizes up to about 100 atoms.
- Cannot describe the transition metals, which is due to an inadequite description of the electron interactions in the unfilled d-orbitals.
- Can be used for calculations of all bond types (even van der Waals' interactions).

- This method is superior for ground-state structures, excited structures (like transition state structures), energies (e.g. binding energies), vibrations, and properties (like electronic, optical, magnetic, and dynamic properties) since all electrons can be included in the calculations.
- Gives systematic improvement towards more exact results.

4.2.6 Post-Hartree–Fock methods

One of the drawbacks with the Hartree–Fock (HF) method is that it cannot correctly describe the instantaneous electron–electron correlations. More precisely, the movement of an electron is instead assumed to be totally independent of the motions of the other electrons in the system. However, the contribution of the electron–electron correlations to the total energy is quite extensive, with an approximative value of 1%:

$$E_{\mathrm{exact}} - E_{\mathrm{HF}} = E_{\mathrm{corr}}(E_{\mathrm{corr}} = 0.01 \times E_{\mathrm{exact}}) \tag{4.39}$$

The major part of E_{corr} comes from the electron pairs within the chemical bonds. It is, thus, especially important to calculate the electron–electron correlations for chemical processes where bonds are formed, or broken:

$$M + A \rightarrow \mathrm{M} - \mathrm{A} \tag{4.40}$$

In calculating the electron–electron correlations one uses something called post-Hartree–Fock (post-HF) calculations. There are several methods to choose from, and the choice is often based on practical considerations. The most common post-HF methods are the Möller-Plesset perturbation theory [6] and the limited configuration interaction [7].

There are also other severe practical drawbacks with the Hartree–Fock method, and those are connected with the very large number of integrals in the calculations; N^4 (N is the number of electrons in the system, and each of these electrons has four unknown variables (x, y, z, and s)). Hence, the calculation time is proportional to N^4, which can be extremely long for large model systems. As an example, a benzene molecule contains 6 C and 6 H. Hence, the total number of electrons, N, is 6×8 (C) + 6×1 (H) = 54. This number of electrons results in 8503056 integrals in the Hartree–Fock equations. Within the field of materials science, many scientific problems involve a large number of atoms, and, thus, even more electrons. The wavefunctions reach practically impossible sizes for these very large systems. Hence, it has been found necessary to find another quantum mechanical method with a reduced number of integrals, and, thereby, with a shorter calculation time for the same size of system.

The question is now, is it absolutely necessary to use the wavefunction, ψ, in the calculation of the systems groundstate energy, or in the calculations of other properties of interest? The answer is no, it has become possible to use the electron density, ρ, instead:

$$p(\vec{r}_i) = \psi^2(\vec{r}_i) \tag{4.41}$$

In contrast to ψ, the electron density, ψ^2, is a measurable quantity. For any system under investigation:

$$\int \rho(\vec{r}_i)d(\vec{r}_i) = \int \psi^2(\vec{r}_i)d(\vec{r}_i) = N \tag{4.42a}$$

where N is the total number of electrons in the system. However, there is also a very large similarity, and that is the fact that both ψ and ψ^2 contain all necessary information about the system that is being studied. One quantum mechanical method that uses the electron density instead of the wavefunction is the DFT method. The advantage with this method is that it can be used for larger systems (as compared with the Hartree–Fock method), and that it can be applied for extended solid materials without the need of empirical data [8]. The DFT method is a relatively low-cost method, where the number of integrals is just N^3 (the spin variable is here omitted). This approach has therefore been referred to as the main computational method in the present book.

4.3 The density functional theory method

4.3.1 Introduction

Another approach for the Schrödinger equation is to use the electron density, ρ, instead of the wavefunction, ψ (see equation (4.40)). (The relation between ψ and ψ^2 is shown in figure 4.7.) This DFT method describes the properties of a many-electron system by using a functional of the electron positions. (By definition, a functional is a function of a function, and it is in this case a function of the electron density, which in turn is a function of the electron positions.) It was first used to calculate molecular properties but later became very popular for extended material systems as well [9].

The total energy of the system can now be expressed as:

$$E[\rho] = T[\rho] + E_{eN}[\rho] + E_{ee}[\rho] + E_{xc}[\rho] \tag{4.42b}$$

where T, E_{eN}, E_{ee}, and E_{xc} are the energies that correspond to the electron kinetics, nuclear–electron attractions, electron–electron repulsions, and exchange-correlations, respectively, in the system. By expressing these terms as a function of electron density, DFT can now also be used for larger systems and even be extended to liquids and solids without the need to use more simple empirical data as input to the calculations [10].

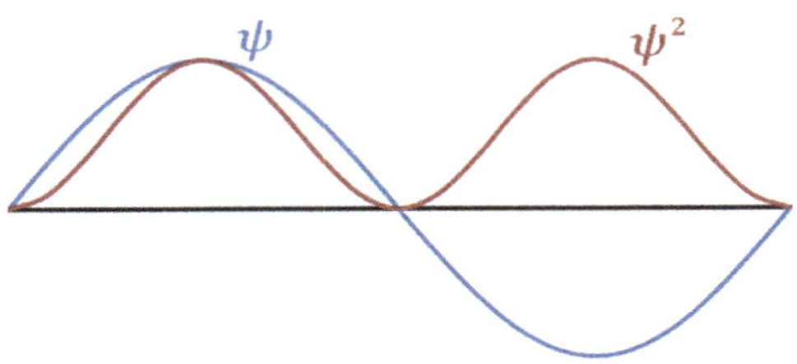

Figure 4.7. The wavefunction (ψ) and its density (ψ^2).

But let us start from the beginning and present the single particle model as developed by Hohenberg–Kohn [11] and Kohn–Sham [9].

4.3.2 The Hohenberg–Kohn theorems

The core theorem of DFT is based on two fundamental Hohenberg–Kohn theorems (figure 4.8):

- Theorem I: *For any system of interacting particles in an external potential* $V_{ext}(\vec{r})$*, the potential* $V_{ext}(\vec{r})$ *is determined uniquely, up to a constant, by the ground state density* $\rho_0(\vec{r})$.
- Theorem II: *A universal functional describing the energy* $E[\rho]$ *in terms of* $\rho(\vec{r})$ *can be defined for any external potential* $V_{ext}(\vec{r})$*. For any particular* $V_{ext}(\vec{r})$*, the ground state energy of the system is the global minimum of the energy functional and the density* $\rho(\vec{r})$ *that minimizes the functional is the exact ground state density* $\rho_0(\vec{r})$.

Hence, theorem I sets the ground for reducing the many-body problem of N electrons with $4N$ variables (i.e. three spatial coordinates (x, y, and z) plus the spin quantum number) to only three spatial coordinates (x, y, and z). As a consequence, the calculation time will now become proportional to N^3, which is considerably shorter than N^4 (i.e. in comparison to the Hartree–Fock method).

According to the Hohenberg–Kohn theorems, the total energy functional can now be expressed as:

$$E[\rho(\vec{r})] = F_{\mathrm{HK}}[\rho(\vec{r})] + \int V_{\mathrm{ext}}(\vec{r})\rho(\vec{r})\, d\vec{r} \tag{4.44}$$

where $F_{\mathrm{HK}}[\rho(\vec{r})]$ describes the electronic kinetic energy and all electron–electron interaction terms. It is a universal functional that looks the same for any N-electron system. However, the external interactions, $V_{\mathrm{ext}}(\vec{r})$, are unique and depend on the specific system. Furthermore, the energy functional expression (equation (4.42b)) is similar to the one obtained from the Schrödinger equation (as presented in section 4.2). The Born–Oppenheimer approximation can also be applied here.

Figure 4.8. Creators of density functional theory. Walter Kohn (left, in 1962) and his two postdoctoral fellows, Pierre Hohenberg (middle, in 1965) and Lu Sham (right, undated), produced their theory in 1964 and 1965. (Photographs courtesy of Walter Kohn and the John Simon Guggenheim Memorial Foundation, Pierre Hohenberg, and Lu Sham.)

4.3.3 Kohn–Sham equations

As was the situation with the Hartree–Fock approximations (see section 4.2.3), the Kohn–Sham ansatz transfers the total wavefunction for N electrons into one-electron wavefunctions [9], with a corresponding independent one-electron Schrödinger equation:

$$\hat{h}_{\text{eff}, i}\varphi_i(\vec{r}_i) = e_i\varphi_i(\vec{r}_i) \tag{4.45}$$

It is thereby assumed that the electrons are moving independently with no correlation in between. Thus, the complete ground-state system can be described by non-interacting electrons in the field of one effective potential. In other words, each independent electron experiences an effective potential, $V_{\text{eff}}(\vec{r})$:

$$V_{\text{eff}}(\vec{r}_i) = V_{\text{ext}}(\vec{r}_i) + \frac{1}{2}\sum_{i \neq j}\frac{e^2}{|\mathrm{r_i} - \mathrm{r_j}|} + \frac{\delta E_{\text{xc}}[\rho(\vec{r}_i)]}{\delta\rho(\vec{r}_i)} \tag{4.46}$$

where $V_{\text{ext}}(\vec{r}_i)$ is the external potential describing the interaction between electron i and its surrounding atomic nuclei. Moreover, the second term describes the electron–electron interactions between electron i and each of the other electrons in the system. Finally, the third term describes the exchange-correlations for electron i in the field of the other electrons. Thus, this term includes all many-body effects of exchange and correlation.

Another similarity with the Hartree–Fock method is that the possibilities for an electron exchange are also included here by the introduction of a Slater determinant. In comparison with equations (4.15) and (4.16), the Kohn–Sham wavefunction does now look like:

$$\Psi_e(\vec{r}_1, \vec{r}_2, \ldots, \vec{r}_N) = 1/\sqrt{\left(\frac{\mathrm{N}}{2}\right)!}\begin{bmatrix} \varphi_1(\vec{r}_1)\varphi_2(\vec{r}_1) & \cdots & \varphi_{\mathrm{N}}(\vec{r}_1) \\ \vdots & \ddots & \vdots \\ \varphi_1(\vec{r}_{N/2})\varphi_2(\vec{r}_{N/2}) & \cdots & \varphi_{\mathrm{N}}(\vec{r}_{N/2}) \end{bmatrix} \tag{4.47}$$

The expressions of the one-electron, e_i, and total energy, E, are then:

$$e_i = H_i + \sum_{j=1}^{N}(J_{ij} - K_{ij}) \tag{4.48}$$

$$E = \sum_{i=1}^{N}H_i - \frac{1}{2}\sum_{i,j=1}^{N}(J_{ij} - K_{ij}) \tag{4.49}$$

For the corresponding Hartree–Fock approach the expression for E looks like:

$$E = \sum_{i=1}^{N}H_i + \frac{1}{2}\sum_{i,j=1}^{N}(J_{ij} - K_{ij}) \tag{4.50}$$

where

$$H_i = -\frac{\hbar^2}{2m_e}\frac{\int \varphi_i^* \nabla_i^2 \varphi_i d\vec{r}_i}{\int \varphi_i^* \varphi_i d\vec{r}_i} \tag{4.51}$$

$$J_{ij} = \frac{e^2}{2\pi\varepsilon_0}\left(\sum_{j\neq i}^{N}\int \frac{/\varphi_j(\vec{r}_j)/^2}{|\vec{r}_i - \vec{r}_j|}d\vec{r}_j + \sum_{i\neq j}^{N}\int \frac{/\varphi_j^*(\vec{r}_j)\varphi_j(\vec{r}_i)/}{|\vec{r}_i - \vec{r}_j|}P_{ij}\right)d\vec{r}_i \tag{4.52}$$

$$Kij = \frac{e^2}{2\pi\varepsilon_0}\sum_{k}^{M}\frac{Z_k}{|\vec{r}_i - \vec{R}_k|} \tag{4.53}$$

As was the case with the solutions of the Hartree–Fock equations, the Kohn–Sham equations can now be solved self-consistently by using a MO-LCAO expression for the electron wavefunctions. Self-consistent electron minimization will be further discussed in section 4.3.5.

Let us now go back to the expression for the total energy of the system (equation (4.43)), for which it is not possible to exactly solve the exchange-correlation part ($E_{xc}[\rho]$). Hence, approximations must be introduced in trying to specifically express the instantaneous electron–electron correlations. It will result in a certain error that accounts for the difference between the many-body approach and the single-particle approach [12].

The exchange-correlation interactions in a system can be split into two parts: the electron exchange and electron correlations. The two most used approximations for the exchange-correlation functional will be further discussed in the next section (i.e. section 4.3.4).

4.3.4 Exchange-correlation functionals

As stated above, the exchange-correlation energy term, $E_{xc}[\rho]$, cannot be solved exactly for systems containing more than 1 electron (e.g. H). The main challenge with DFT is, hence, to find a good approximation of the exchange-correlation energy. There are, though, many different types of approximations that have been developed during the last decades. The simplest approximation is the local density approximation (LDA), where the interactions of the electrons are described by a hypothetical uniform electron gas. Gradient-corrected functionals were later developed by which it is possible to more accurately study the inhomogeneties in the electron gas—which is the situation in materials based on molecular systems. This type of approximation has received the common name generalized gradient approximation (GGA) (figure 4.9).

4.3.4.1 Local density approximations (LDAs)

In the LDA theory, the one-electron exchange-correlation energy is being considered in a local uniform electron gas where the electron density, ρ, is assumed to be

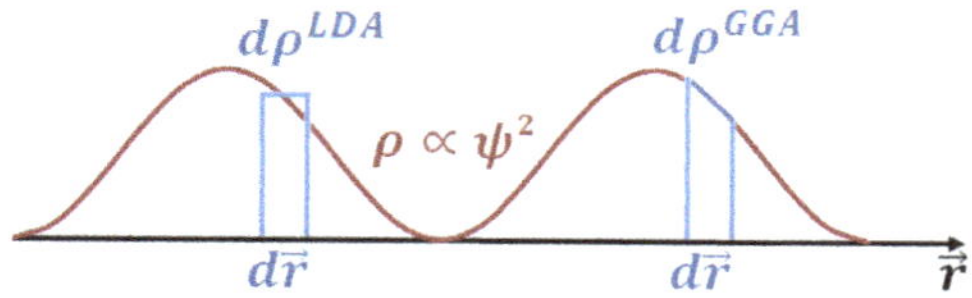

Figure 4.9. Schematic representation of the electron density (ρ) variation in a volume element ($d\vec{r}_i$) for two different electron correlation approximations; local density approximation (LDA) and generalized gradient approximation (GGA).

constant [13]. The total exchange-correlation energy can then be estimated by using the following simple analytical definition:

$$E_{xc}^{LDA}[\rho] = \int \rho(\vec{r}_i)\varepsilon_{xc}[\rho]d\vec{r} \tag{4.54}$$

where $\varepsilon_{xc}[\rho]$ is the one-electron exchange-correlation energy at position $\vec{r}_i$ in the uniform electron gas. Even though equation (4.54) has been developed under the assumption of a constant effective potential, $V_{eff}(\vec{r}_i)$, the usage of LDA gives reasonable results for also systems with a slowly varying electron density. It is, thereby, a commonly used exchange-correlation approximation for most of the metals in the periodic table [14].

However, it is not advisable to use the LDA method for covalent materials (such as semiconductor materials). The main drawback is that it tends to overbind atoms, and that it thereby overestimates the cohesive energy in the material under investigation [15]. An improved description of the inhomogeneity of the electron gas has been developed that includes gradient-corrected, 'non-local' functionals that depend on the gradient of the density at each point in space.

4.3.4.2 Generalized gradient approximations (GGAs)

One commonly used gradient-corrected functional is the generalized gradient approximation (GGA), which treats the gradients of the charge density in the material system (see equation (4.55)).

$$E_{xc}^{GGA}[\rho] = \int \rho(\vec{r}_i)\varepsilon_{xc}[\rho, \nabla\rho]d\vec{r} \tag{4.55}$$

where $\varepsilon_{xc}[\rho, \nabla\rho]$ is the one-electron exchange-correlation energy that depends on both the density, ρ, and the gradient of the density, $\nabla\rho$, at position $\vec{r}_i$. Thus, the gradient of the density is now replacing the local density expression in the LDA equation (see equation (4.55)). As expected, the GGA approximation of the exchange-correlation functional has proven to work quite well. In fact, the accuracy in the calculations of bond strengths and geometries is considerable increased when using GGA for more complex systems (such as semic-conductor materials) [16, 17]. This is especially the situation for systems containing hydrogen.

However, there is not only one single GGA functional, but instead several that depend on the type of density gradient expansion. PBE (Perdew, Burke and Ernzerhof) [17] and PW1 (Perdew and Wang) [18] are the two most commonly used GGA functionals. Another commonly used one is BLYP (Becke, Lee, Yang, and Parr) [19, 20]. These three functionals belong to a group of GGA functionals that have been developed to describe covalent materials [21–24]. In addition, hybrid functionals, including a Hartree–Fock exchange, have also been developed. Examples of hybrid functionals are B3LYP (Becke-3-Parameter, Lee, Yang, and Parr) and PBE0 (Perdew, Burke and Ernzerhof hybrid) [25, 26]. However, it has been found difficult to handle the exchange-correlation functionals for more complex interactions, such as long-range dispersion and strong Columbic interactions within the d and/or f orbitals. Empirical corrections have, therefore, been developed for these types of specific systems (e.g. TS corrections and Hubberd U) [27–29].

4.3.5 Self-consistent electronic minimization

As was the situation with the Hartree–Fock equations, in order to find the ground state solution to the Schrödinger equation one has to solve the Kohn–Sham equations in a self-consistent way (i.e. by using the variation principle) (see equations (4.20)–(4.38)). This method is, in the field of DFT calculations, called self-consistent electronic minimization. In the more traditional version of this method, the total energy is minimized by varying the coefficients in the electron wavefunction. As implemented in many third-party softwares, this type of energy minimization can use an all-band scheme that works on all wavefunctions at the same time, instead of using the variation principle for each individual wavefunction (as is the case in the Hartree–Fock method).

Another type of self-consistent electronic minimization method is based on density mixing, where the sum of individual electron energies is minimized at a fixed external potential [30]. The new electron density is, for each step in this minimization procedure, mixed with the old electron density (i.e. that was used as input to the present minimization step). This procedure is then repeated until convergence is reached. The mixing of new and old densities can, however, take place in different ways:

1. Linear mixing scheme: This is the simplest mixing scheme, where the new density is a linear combination of the input and output densities from the last iteration:

$$\rho_{new} = (1 - \alpha)\rho_{new} + \alpha\rho_{out} \tag{4.56}$$

   where α is the mixing fraction. Other efficient mixing scheme frequently used are the Pulay mixing [31] scheme and the Broyden's method [32].
2. Kerker mixing: The new density is expressed as:

$$\rho_{new}(G) = \rho_{in}(G) + A\frac{G^2}{G^2 + G_{max}^2}\left[\rho_{out}(G) - \rho_{in}(G)\right] \tag{4.57}$$

   where G is the cutoff wave-vector, and A is the mixing amplitude [33].
3. Pulay mixing scheme: This is the most efficient scheme. The new input density is a linear combination of densities from all previous steps [34].

4. Broyden's method: This method is similar to the Pulay's scheme. The difference is that it uses a quasi-Newton relaxation instead of a DIIS minimization of the residuals [32].

4.3.6 Supercell models

The construction of a model is one of the most important tasks in the preparation of quantum mechanical calculations. The model must be as realistic as possible in relation to the material system that is being theoretically examined. This can, hence, be a very difficult task to solve, which will be further discussed in chapter 5. However, some general considerations are in place to also present here.

In modeling a solid material system (e.g. bulk materials, catalytic solid surfaces), it is necessary to use a 2D, or 3D, model. DFT calculations on these types of models are, with a few exceptions, performed under periodic boundary conditions. This means that the models are built in something called supercells (see figure 4.10), which are translated in the *x*-, *y*-, and *x*-directions in the course of the calculations. For a bulk supercell (i.e. for a completely filled supercell), this three-dimensional translation results in an infinite bulk material (see figure 4.10(a)). On the other hand, the supercell is partially filled when constructing a catalytic solid surface. The transitions of this supercell in the *x*-, *y*-, and *z*-directions result in 2D infinitie slabs that are repeatably formed along the *z*-axis (with a vacuum layer in between) (see figure 4.10(b)). However, when constructing these surfaces, one has to fix the positions of the lowest atomic layer, and most often terminate those atoms with suitable atoms (e.g. H), insimulating a continuation into bulk material. In addition, the vacuum layer must be large enough to eliminate any interactions between neighbouring slabs. Hence, the *z*-axis of the supercell must be large enough. When

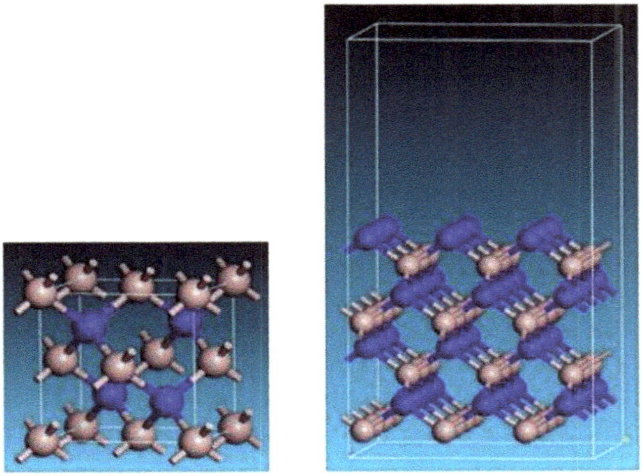

a) b)

Figure 4.10. Supercells demonstrating (a) a bulk material and (b) a solid-state surface by using c-BN and a (100) surface. The pink balls demonstrate the B atoms, whilst the blue balls demonstrate the N atoms.

studying a certain process on the surface in the supercell (e.g. the adsorption of a species to the surface), one must also consider the lengths of the x- and y-axes. The reason is that the adsorption site must be sufficiently separated from the ones in the periodic images to eliminate any adsorbate-adsorbate interactions. One exception is, though, the situation when these interactions are intentionally modelled. However, there are upper limits for both the size of the supercell and number of atoms (with their electrons) in these cells since the computational time is proportional to N^3 (where N is the number of electrons in the supercell model). Except for the demands on computational time, a very large number of electrons (and thereby a very large number of integrals in the Hamiltonian operator) puts huge demands on also disk capacities in the computer. The construction of supercell models will be further discussed in chapter 5.

4.3.7 Planewave basis sets

Some specific expressions for the electron wavefunctions must be used in solving the Kohn–Sham equations. The most common expression for a periodic model system (i.e. for periodic boundary conditions) is based on planewaves. This specific set of planewaves is, like the situation with the Hartree–Fock equations, called a basis set. The electron wavefunction is, therefore, described by a linear combination of planewaves:

$$\varphi_i^k(r) = \sum_G a_{i,\,k+G} e^{(i(k+G)\cdot r)} \tag{4.57a}$$

where k, r, and G are other notations for $\vec{k}$, $\vec{r}$, and$\vec{G}$. k is a wave vector (also called a k-point) and G is a reciprocal lattice vector. Thus, there are k planewaves and, thereby, k terms in the wavefunction description for electron i. Moreover, the k-value is a discrete value in the range $-\pi/2 \leqslant k < \pi/2$, and $a_{i,\,k+G}$ is a mixing coefficient that must be sought (either analytically or numerically). The Kohn–Sham equation will, thereby, take the following form in the reciprocal space:

$$\begin{aligned}\sum\nolimits_G \bigg\{ \frac{\hbar^2}{2m} \mid k + G \mid \delta_{GG'} + V_{\text{ion}}(G - G') + V_{\text{elec}}(G - G') \\ + V_{\text{xc}}\Big(G - G'\Big)\bigg\} a_{i,\,k+G'} = \varepsilon_i a_{i,\,k+G'}\end{aligned} \tag{4.58}$$

where V_{ion}, V_{elec}, and V_{xc} are the electron–nuclei, electron–electron, and exchange-correlation functionals, respectively.

According to Bloch's theorem, the wave function, $\varphi_i^k(r)$, can now be expanded by using a discrete plane wave basis set at every k-point (which is determined by G). In principle, it needs an infinite number of planewaves for this expansion, but one can use a finite number of plane waves with $G \leqslant G_{\max}$ by setting the cutoff energy to a certain value;

$$E_{\text{cut}} = \frac{\hbar^2 G_{\max}^2}{2m_{\text{e}}} \tag{4.59}$$

The planewave have a certain wavelength (λ) that is related to its inherent energy ($E_{planewave} = hc/\lambda$). Consequently, the larger the k-value in the electron wavefunction, the smaller the wavelength of this specific planewave, and the larger the energy of this specific term (see equation (4.57)). The cut-off energy simply puts a limit on the energy value of these terms, and, thereby, determines the number of terms in the wavefunction description. Since a large number of terms is needed for a wave-function description to be close to the true ground-state wavefunction, an increased cutoff energy must be used. A reduced cutoff energy does, otherwise, result in a total energy that diverges from the true ground-state energy value. However, an increased cutoff energy results in an increased computational time. Thus, the cutoff energy must be set up to a sufficiently large value to make sure that the total energy can converge within the required tolerance. However, the drawback with an increased cutoff energy is the increased computational time. Thus, one must use a cutoff energy that gives a total energy within a pre-set tolerance range. This is a procedure that will be further discussed in section 7.2.

4.3.8 Pseudopotentials

There is a problem with the description of an electron wavefunction by using a linear combination of planewaves, and that is the large number of planewaves that is required for the description of the core electrons (i.e. the electrons that are positioned under the valence electrons in the atoms). Another circumstance is that these inner electrons do not take part in the formation of chemical bonds (i.e. they are chemically 'inert'). To solve this problem, a pseudopotential (also called an effective potential) can be used as an approximation for a simplified description of the complex core electrons. A pseudopotential, V_{pseudo} will then replace the effective potential term, $V_{eff}(\vec{r}_i)$, in the Schrödinger equation. However, $V_{eff}(\vec{r}_i)$ can still be used for the valence electrons further away from the nucleus. The energies obtained by using either a peudopotential, or an all-electron potential (i.e. $V_{eff}(\vec{r}_i)$ for all electrons), will, hence, be identical outside a specifically chosen cut-off radius, r_c (see figure 4.11).

The situation with the wavefunctions is similar. The core electrons are assumed to be Ψ, which has the same shape as the all-electron wavefunction outside the cut-off border, but with fewer nodes inside. Since the wavefunctions are expressed by using planewaves, this means that only planewaves with longer wavelengths (i.e, lower energies) will be used for the wavefunction descriptions. In other words, the number of waveplanes will decrease (i.e. the basis set will become smaller, and the calculation rates will increase). The material systems are in this approach described with chemically active valence electrons in the field of rigid non-polarizable ion cores.

The two most common types of pseudopotentials are the norm-conserving pseudopotentials [35] and the ultrasoft pseudopotentials [36]. Both types of methods are based on a basis-set with a significantly low cut-off energy. Consequently, their computational demands are quite reasonable. A norm-conserving pseudopotential was proposed by Hamann, Schlüter, and Chiang (HSC) in 1979 [37]. The operation

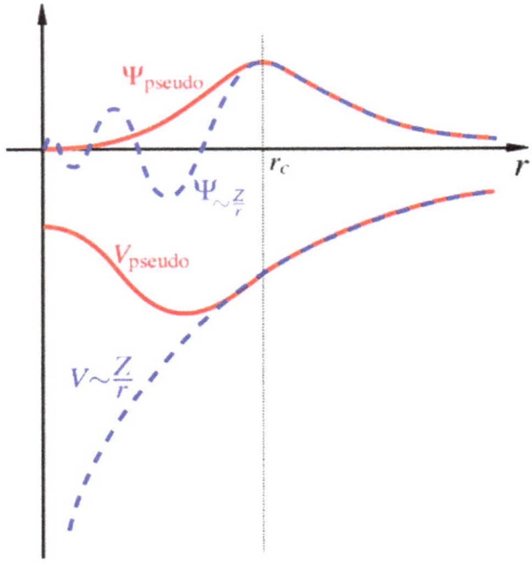

Figure 4.11. A schematic illustration of all-electron (solid lines) and pseudo- (dashed lines) potentials, and their corresponding wavefunctions within, and outside, the cut-off radius r_c (https://en.wikipedia.org/wiki/Pseudopotential).

of this specific pseudopotential on the electron wavefunction depends on the angular momentum state. It is, therefore, declared to be non-local. However, there are two conditions that must be fulfilled before it can be called a norm-conserving pseudopotential:

(1) Within a cut-off radius of $r_c \rightarrow$ the electron densities formed by the norm of the pseudo-wavefunctions must be identical with the corresponding densities formed by the all-electron wavefunctions.
(2) Outside a cut-off radius of $r_c \rightarrow$ the densities formed by the pseudo-wavefunctions must be identical to the corresponding densities formed by the all-electron wavefunctions [38].

For ultrasoft pseudopotentials, the norm-conservation constraints have been relaxed, which results in even softer pseudopotentials and longer cut-off radii [36]. This means that the wavelengths of the corresponding pseudo-wavefunctions are longer and their amplitudes lower. In fact, these pseudo-wavefunctions are set to be as soft as possible within the cut-off radius r_c, and, thus, the cutoff energy is largely reduced. Hence, the ultrasoft pseudopotentials have the capacity to reduce the computational costs even more.

4.3.9 Corrections for van der Waals interactions

The DFT method can be used for almost all types of chemical bonds: covalent, ionic and metallic bonds. However, the van der Waals (vdW) interactions (also called dispersion corrections) cannot be accurately described by using standard DFT methods. Semiempirical dispersion-corrections have been added to the standard DFT equations to make it possible to accurately simulate these types of very weak bonds [27, 40, 41]. Examples of vdW correction methods are the Tkatchenko–Scheffler van der Waals (TS–vdW) scheme [42] and the OBS (Ortmann, Bechstedt, and Schmidt) scheme [41]. These vdW correction schemes have been found to be very successful in improving non-covalent bonding interactions [41, 42]. For instance, the TS–vdW scheme has earlier been used to improve the description of weakly bonded systems [43, 44]. Moreover, the OBS scheme can be used for systems independent of their characteristic bonding [41].

The vdW-corrected total energy, E_{total}, has the following expression:

$$E_{\text{total}} = E_{\text{DFT}} + s_i \sum_{i=1}^{N} \sum_{j>i}^{N} f(S_R R_{ij}^0, R_{ij}) C_{6,ij} R_{ij}^{-6} \tag{4.60a}$$

where E_{DFT} is the standard DFT total energy. The other terms describe the vdW-correction energy. All N atoms in the system are included in the E_{total} expression. For a long-range order, the dispersion potential for the interactions between atoms i and j is given by $C_{6,ij}R_{ij}^{-6}$. The $C_{6,ij}$ parameter is the dispersion coefficient which depends specifically on the material type. For the short- to medium-range order, the dispersion functional is modified by using the parameters s_i and S_R. Moreover, the bridge between the short-range to long-range order is taken into consideration by using the damping function $f(R_{ij}^0, R_{ij})$. Also, the R_{ij}^0 distance and $C_{6,ij}$ factor are obtained from DFT ground-state electron density calculations (with reference values for the free atoms).

4.3.10 Advantages and disadvantages with the DFT method

As with most of the quantum mechanical methods, there are both advantages and disadvantages with the DFT method and the most important are listed below.

- Can be used for clusters (i.e. larger molecules), as well as for solid surfaces and bulk materials.
- Useful for in principle all elements in the periodic table.
- Can be used for calculations of all bond types (the van der Waals interactions needs a specific correction).
- A less time-consuming method. Hence, calculations of quite large models are possible.
- This method is developed for especially metals and is therefore useful for transition metals with unfilled d-orbitals.
- This method is originally developed for ground state systems only.
- Since all electrons can be included in the calculations, the DFT method is especially valuable for calculations of ground-state structures, energies (e.g. binding energies), vibrations, and for calculations of electronic, optical, mechanical, and dynamical properties.
- Does not give a systematic improvement in results.

4.4 Geometry optimizations

4.4.1 Introduction

When constructing a model, its atoms are most probably not positioned in their equilibrium sites (i.e. the most energetically preferred ones). It must be stressed here that the choice of initial positions is crucial for the length of the calculations (as will be demonstrated below). However, despite this problem, there is a variety of geometry optimization methods to choose from. Before presenting these methods, with their pros and cons, the potential energy curve will be presented (see figure 4.12). (With 'potential energy' we here mean the total energy, E, at constant positions of the atoms in the model system.) The position of the initial choice of coordinates is especially marked in figure 4.12. In addition, the final goal with the geometry optimization is also shown in this figure, which is the energy minimum for another set of atomic coordinates. This energy minimum is also shown in figure 4.13, together with analytical expressions for the energy slope and energy minimum. The

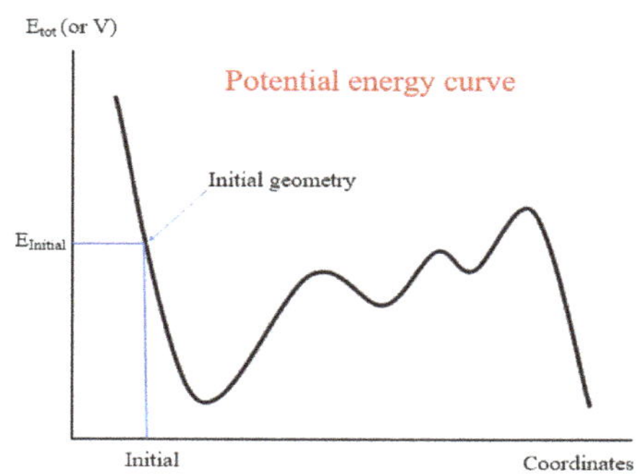

Figure 4.12. A schematic illustration of the potential energy curve as a function of atomic coordinates. The position of the initial choice of coordinates, with its calculated total energy, is especially marked in the figure.

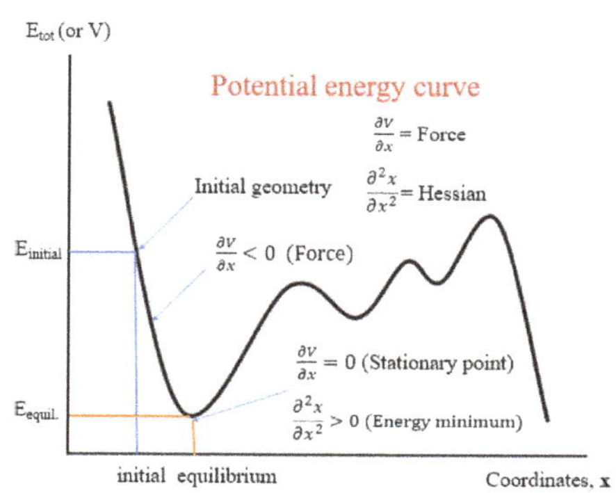

Figure 4.13. A schematic illustration of the potential energy curve as a function of atomic coordinates. The positions of both the initial and final coordinates, with their corresponding total energies, are marked in blue and red, respectively. Analytical expressions for the energy slope and energy minimum are also included in the figure.

first derivative of E_{tot} at the initial position of atom $m, dE_{\text{tot}}/d\vec{r}$, is identical to the force, $\vec{f_m}$, that is acting on this atom. (These forces are vectors in the system model). The sum of these atomic forces (or vectors) gives a total force for the complete system, $\vec{F}$, which is presented in figure 4.13. This force is zero when the geometry optimization reaches an energy minimum or maximum, and the calculation will be finished. To differentiate between these two types of stationary states, the second derivative of the total energy, $d^2E_{\text{tot}}/d\vec{r}^2$, should be calculated (as presented in figure 4.13). These second order derivatives are called the Hessians, and are, hence, calculated to make sure that the energy minimum in figure 4.13 is really a minimum. It is thus calculated after the geometry optimization, and for the resulting atomic coordinates.

If the Hessians are >0, then there is an energy minimum. Understandably, if the Hessians are <0, then there is an energy maximum. The steps to follow when performing these geometry optimization calculations are demonstrated in figure 4.14.

As stated above, there are various methods to choose from when performing these geometry optimization calculations. What is common to all of them is that they are using a general trend to find the stationary points (i.e. where the atomic forces are zero). Moreover, the energy function may be of the force field type or obtained from solving the electronic Schrödinger equation. There are otherwise quite large differences between the various optimization methods. One type of methods uses a simple-minded approach to take one step at a time. However, what is most common

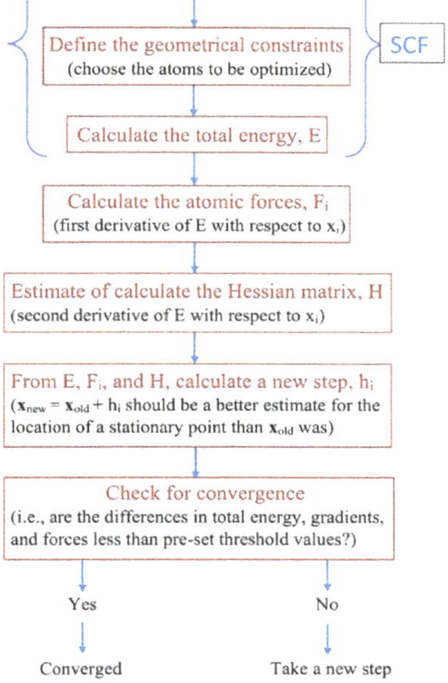

Figure 4.14. General steps in the geometry optimization calculations.

amongst the optimization methods is that they all use forces, $\vec{f}_m$ (i.e. first order analytical derivatives of the total energies). Some methods also use the Hessians (i.e. the second order derivative of the total energies). Here follows an overall description of different types of geometry optimization methods, which will be followed by the optimization methods that are most commonly used for DFT calculations.

4.4.2 Steepest descent

Geometry optimization methods exist that can locate both energy minima and maxima, but the steepest descent method can only locate the former type of transition state (i.e. the energy minima) [45]. A series of steps are in this method taken along the negative energy derivative, $\boldsymbol{dE_{tot}/d\vec{r}}$ (also called the gradient). This movement, along a specific search direction, will guarantee an approach to an energy minimum (see figure 4.15). However, the drawback with this method is that the rate of convergence slows down as the minimum is approached.

4.4.3 Conjugate gradient

When using the conjugate gradient method, each search line is 'conjugate' to the previous search direction(s) [46]. The first step is equivalent to the steepest descent step, but subsequent searches are performed along a line that is a mixture of the current negative gradient and the previous search direction (see figure 4.15):

$$\Delta\vec{r}_i = -\frac{\boldsymbol{dE_{tot}}}{\boldsymbol{d\vec{r}_i}} + \beta_1 \Delta\vec{r}_{i-1} \tag{4.60}$$

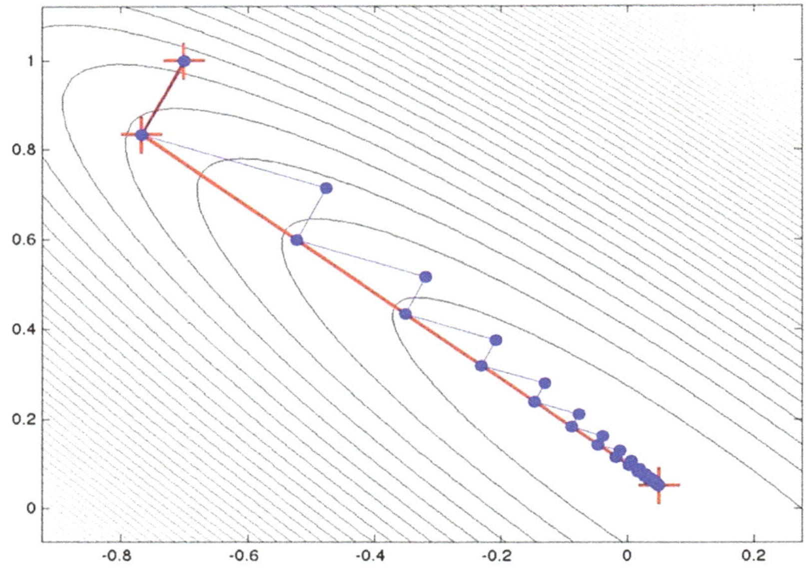

Figure 4.15. Schematic illustrations of common geometrical optimization methods; steepest descent (red) and conjugate gradient (blue).

As was the situation with the steepest descent method, the conjugate gradient method can only locate energy minima. The difference, though, is that it has a much better convergence characteristic. However, this benefit comes along with a requirement of slightly more data storage.

4.4.4 Newton–Raphson

The Newton–Raphson method expands the potential energy curve to second order around the current coordinates, $\vec{r}_i$ [47]. The new step, $\Delta\vec{r}_i$, has the following form:

$$\Delta\vec{r}_i = -\frac{f_i}{\varepsilon_i} \tag{4.61}$$

where f_i is the projection of the gradient, $dE_{\text{tot}}/d\vec{r}$, along the Hessian eigenvector, $d^2E_{\text{tot}}/d\vec{r}^{\,2}$, with eigenvalue ε_i. The advantage with this method is that it converges very quickly when $\vec{r}_i$ is close to a stationary point (either to an energy minimum or an energy maximum).

Let us now see what happens when $\vec{r}_i$ is close to an energy minimum (see figure 4.16(a)). The sign of $dE_{\text{tot}}/d\vec{r}$ is then negative, whilst the sign of $d^2E_{\text{tot}}/d\vec{r}^{\,2}$ is positive. Hence, the direction of the next step, $\Delta\vec{r}_i$, is positive. This means that the step is taken towards the energy minimum. Let us now make the same exercise for the situation with $\vec{r}_i$ close to an energy maximum (see figure 4.16(b)). The sign of $dE_{\text{tot}}/d\vec{r}$ is now positive, and the sign of $d^2E_{\text{tot}}/d\vec{r}^{\,2}$ is negative. The direction of the next step, $\Delta\vec{r}_i$, is therefore also positive. As expected, this means that this step is taken towards the energy maximum. The question is now, what will happen if $\vec{r}_i$ is somewhere in between the energy minimum and maximum, i.e. in the middle of the slope? This is the main disadvantage with this method since infinite steps will be created for these atomic positions. Thus, the system will 'explode'. It is therefore very important that the initial guess of the atomic positions is close to the desired stationary state. This stationary state must be an energy minimum when looking for the most stable state of the whole model system (preferably a global one).

The Newton–Raphson method does also require a full second energy derivative matrix, which makes the calculations computationally very demanding.

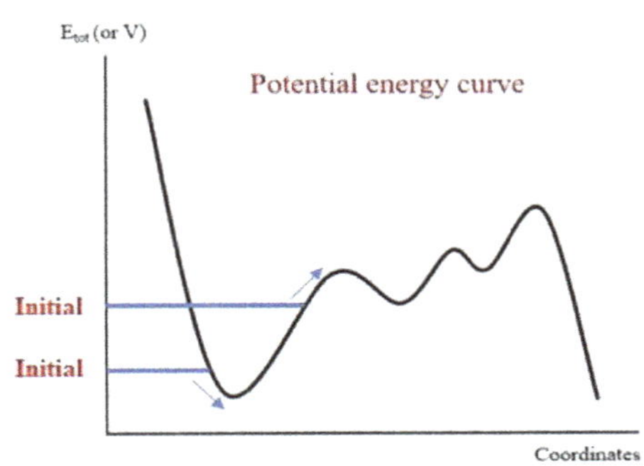

Figure 4.16. Demonstration of the usage of the Newton–Raphson method.

4.4.5 Pseudo Newton–Raphson

Due to the disadvantages with the Newton–Raphson method, a type of geometry optimization method has been developed that starts with an approximation of the second energy derivative matrix (also called the Hessian) [48]. A unit matrix has also been used for this purpose. The Hessian matrix will then be updated during the optimization by using gradients ($d\boldsymbol{E}_{tot}/d\vec{\boldsymbol{r}}$) from earlier iterations. The step lengths and directions will thus be varied. The drawback with this pseudo Newton–Raphson method is that the optimizations do not converge as rapidly as they do when using a true Newton–Raphson method. Examples of pseudo Newton–Raphson methods are DFP (Davidson–Fletcher–Powell) [49] and BFGS (Broyden–Fletcher–Goldfarb–Shanno) [50].

There are also other types of geometry optimization methods that use a combination of those presented above. For instance, the combination of steepest descent and pseudo Newton–Raphson methods is often used (see below). It is then possible to rapidly move closer to an energy minimum in the steepest descent stage, while achieving a very quick convergence in the pseudo Newton–Raphson stage.

4.4.6 Common optimization methods for DFT calculations

For all of the above presented geometry optimization methods, it is, in principle, possible to use any theoretical method in calculating the potential energy curve (with its gradients and Hessians). However, the most common optimization methods for DFT calculations of larger model systems will be presented here. One of these methods is the BFGS method. The main advantage of the BFGS minimizer is the ability to perform cell optimization, including optimization at fixed external stress [50]. The BFGS scheme uses a starting Hessian that is recursively updated during optimization. The CASTEP software (from BIOVIA, Inc.) involves a Hessian in the mixed space of internal and cell degrees of freedom, so that both lattice parameters and atomic coordinates can be optimized.

The low-memory BFGS (L-BFGS) is a variant of the BFGS method that uses a limited amount of computer memory and has thereby proven to be very valuable in the optimization of very large model systems. This specific method has become very popular in the field of machine learning [50, 51]. While the BFGS method stores all $N \times N$ elements in the Hessian (where N is the number of electrons in the model system), L-BFGS uses only a limited number of inverse Hessian updates to construct a new Hessian. When using the L-BFGS method, the past m updates of the position, $\vec{r}_i$, and Hessian, $d^2E_{tot}/d\vec{r}^2$, will be kept (where m can be as small as $m<10$). This scheme has, thus, been shown to scale linearly in memory (i.e. it has a linear memory requirement), while BFGS scales as a square of the system size [52]. This results in a linear memory requirement. The LBFGS algorithm also implements a universal sparse preconditioner that accelerates geometry optimization by a factor of two for small systems and a factor of ten for very large ones [53].

There are situations where it is necessary to fix some specific atoms while performing the geometry optimization [54]. In other words, one must use constraints on the system. Examples of situations where this is necessary are the scanning of potential energy surfaces, identification of reaction pathways, etc. The simplest type of constraint is to fix the atom positions. This can be done for all *x*-, *y*-, and *z*-coordinates, and for any combination thereof. In the CASTEP software, this means fixing the fractional coordinates of the atoms. It is also possible to fix lattice parameters. However, it is only necessary to do this for parameters that are not fixed by symmetry. This type of constraint is often useful in phase stability studies when, for example, you want to fix the angles to 90° [55].

More generally, there are different ways to proceed when using constrained geometry optimization. The coordinates that are held constant can either be removed from the space of variables, or the constraints can be applied by using penalty functions, projection methods, or Lagrangian multipliers. The penalty functions are used for simple constraints. In this method, the constrained optimization problem is replaced by a series of unconstrained problems whose solutions ideally converge to the solution of the original constrained problem [56].

It is also possible to impose more general linear constraints on atomic coordinates. These constraints are specified using a matrix that transforms the Cartesian coordinates of all atoms to the subspace of unconstrained coordinates. This facility is intended for fixing, for example, the *z*-coordinates of atoms in slab calculations of surface processes. Furthermore, non-linear constraints refer to constraints on interatomic distances (bonds), angles, and torsions. Such constraints can be imposed using the delocalized internals optimizer. The projection methods are most effective methods for solving large-scale linear and non-linear equations with constraints [57].

However, the preferred method to use for general constraints is the Lagrangian multipliers. The reason is that it converges much faster than it does with, e.g. the penalty function method, and that the constraints are satisfied exactly [58]. The strategy behind the Lagrangian multiplier is to find the local maxima and minima of a function subject to equality constraints (i.e. subject to the condition that one or more equations have to be satisfied exactly by the chosen values of the variables). The great advantage of this method is that it allows the optimization to be solved without explicit parametrization in terms of the constraints. As a result, the method of Lagrange multipliers is widely used to solve challenging constrained optimization problems.

4.5 Transition state search

4.5.1 Introduction

A transition state is a stationary point that is an energy maximum in one direction (the direction of the reaction coordinate) and an energy minimum in all other directions. During the course of a chemical reaction, the total energy naturally changes. Starting from the reactants, the energy increases to a maximum and then decreases to the energy of the products. The maximum energy along the reaction pathway is known as the activation energy (most often called the energy barrier). The structure corresponding to this energy is called the transition state. In available third party software (like CASTEP from BIOVIA, Inc.) a so-called TS Search task is particularly useful for predicting the energy barriers to chemical reactions and to determine the reaction pathways. (Similar TS Search tasks can also be found in other commercial software.) It can be also used to find diffusion barriers for either solid state diffusion or surface diffusion. The reactants and products can either be optimized separately before running a TS Search or as part of this task.

4.5.2 Synchronous transit methods

In the TS Search task, transition state searches are performed using synchronous transit methods. These work best when reasonable structures for the reactants and products exist, but the location of the transition state is unknown. Starting from reactants and products, the synchronous transit methods interpolate a reaction pathway to find a transition state. The linear synchronous transit (LST) method performs a single interpolation to a maximum energy. The quadratic synchronous transit (QST) method alternates searches for an energy maximum with constrained minimizations in order to refine the transition state to a high degree. The options available include:

LST Maximum: Performs a single LST maximization, bracketing the maximum between the reactants and product. This is the quickest but least accurate of the options. The TS structure determined by this method generally requires further refinement.

LST/Optimization: Performs an LST maximization, followed by an energy minimization in directions conjugate to the reaction pathway. This yields a structure lower in energy and closer to the true TS than a simple LST. Minimization steps continue until an energy minimum is reached or the number of conjugate directions is exhausted.

Halgren–Lipscomb: Is a kind of limited LST/optimization and is designed to reproduce the algorithm popularized by Halgren and Lipscomb. After determining the LST maximum, this method performs a conjugate gradient minimization, but only in a single direction.

Complete LST/QST: Begins by performing an LST/optimization calculation. The TS approximation obtained in that way is used to perform a QST maximization. From that point, another conjugate gradient minimization is performed. The cycle is repeated until a stationary point is located or the number of allowed QST steps is exhausted.

A complete description of these algorithms is presented in [59].

4.6 Process energies

4.6.1 Introduction

As a result of the geometry optimization calculations, let us now assume that the model system has reached a stationary state with a specific geometrical structure (see figure 4.15). It has, for instance, ended up in its most stable form (either locally or globally). In addition to information about the most stable geometrical structure, these quantum mechanical calculations have resulted in the total energy for the system in addition to the corresponding ground state wavefunction. These energies can later on be used in the calculation of reaction energies (such as adsorption energies), construction of potential energy curves, calculation of stabilization energies, etc.

4.6.2 Reaction energies

The adsorption of an atom, H, on a solid surface, S, is an example of a chemical reaction:

$$S+H \rightarrow S-H \tag{4.62}$$

In calculating the adsorption energy (i.e., the reaction energy), ΔE_{ads}, the following mathematical expression will be used; where $E_{S\text{-}H}$ is the total energy for the surface with an adsorbed H species, E_S is the total energy for the surface prior to adsorption, and E_H is the total energy an H atom. Hence, one needs three different energy calculations to obtain the adsorption energy. With just a few exceptions, each of these calculations must undergo geometry optimization. It is also of outmost importance that each of these three energies are calculated by using identical theoretical parameters and supercell size. Systematic errors will, otherwise, be introduced in the final result. The values of the adsorption energies are negative for exothermic adsorption processes, and positive for endothermic ones.

$$\Delta E_{ads} = E_{S-H} - (E_S - E_H) \tag{4.63}$$

Another chemical surface reaction process is the gas phase abstraction process. As an example, a gas phase species, H, reacts with an H adsorbate on the surface, S, leaving a gas phase H_2 species and a radical site on the surface, S^*:

$$S - H + H \rightarrow S^* + H_2 \tag{4.64}$$

The abstraction energy, ΔE_{abs}, can then be calculated by using equation (4.65):

$$\Delta E_{abs} = (E_s + E_{H_2}) - (E_{S-H} + E_H) \tag{4.65}$$

As was the situation with the expression for the adsorption energy, each energy calculation should undergo geometry optimization and all calculation-related parameters must be identical.

4.6.3 Potential energy curves

Another example of process energy calculations is the construction of a potential energy curve. Let us again take the abstraction of an adsorbed H species as an example (see figure 4.17(a)). For the locations of the energy maxima (i.e., transition states) in the potential energy curve, one can either perform an 'automatic' transition state search (see section 4.5.1) or use the following manual method. For the present example, if a TS search method is chosen, one must, however, be certain that there is no interaction between the gaseous H species and the surface in the initial geometry optimizations of the reactants.

For some specific model systems, and sizes, this type of transition state searches may be practically very difficult to perform (which is sometimes due to computer

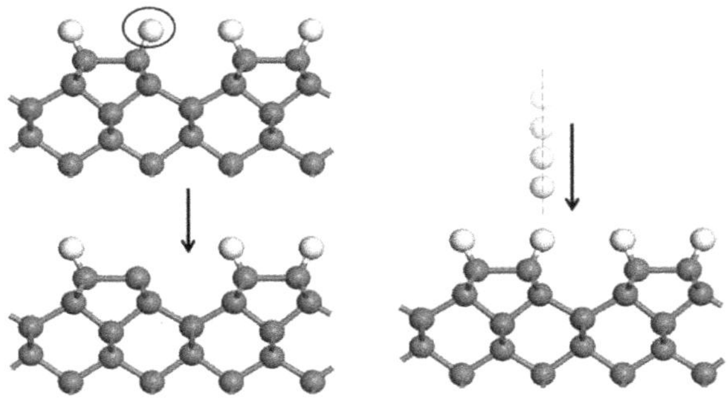

Figure 4.17. An H-terminated diamond (100) surface (a) before and after hydrogen abstraction, and (b) with an approaching H atom towards the H-adsorbate.

limitations). The following manual method is, therefore, a plausible alternative to use. In this type of theoretical simulation, one starts by positioning H at a certain distance, z Å, above the solid surface, S. The hydrogen atom is thereafter allowed to move towards the surface in distinct and small steps (see figure 4.17(b)). For each of these steps, geometrical optimization calculations are performed and finished by the calculation of the total energy of the system. Depending on the chosen methodology, the approaching H atom can be kept fixed for each step in the adsorption process or become relaxed. The energy barrier is, thereafter, expressed as:

$$\Delta E = E_{\text{maximum}} - E_{\text{furthest away}} \tag{4.66}$$

where E_{maximum} is the total energy at the transition state (i.e. the energy maximum), and where $E_{\text{furthest away}}$ is the total energy of the system when the gaseous H atom is furthest away from the surface.

4.6.4 Stabilization energies

The calculated stabilization energy of a specific model system gives information about the bond strengths in the material. As such, the stabilization energy is directly related to the stability of the system (i.e. the most stable system has the large stabilization energy). The stabilization energy is usually calculated as the difference in energy between the complete system and the sum of energies of the individual atoms. Take bulk Al_2O_3 an example. The stabilization energy of this material can be expressed as:

$$\Delta E_{\text{stab},Al_2O_3} = E_{Al_2O_3} - (2E_{Al} + 3E_O) \tag{4.67}$$

where $E_{Al_2O_3}$, E_{Al}, and E_O are the total energies for a supercell filled with bulk Al_2O_3, one Al atom, and one O atom, respectively. As was the situation with the calculations of adsorption and abstraction energies, all theoretical parameters must be kept identical for all terms in equation (4.67). It must here also be stressed that equation (4.67) shall not be mixed up with the expression for the formation energy, for which the difference in energy is between the complete bulk system and the sum of energies of solid Al and gaseous O_2. (Al(s) and O_2(g) are the normal states of Al and O at 25 °C and 1 atm.) Hence, the formation energy of Al_2O_3 ($\Delta E_{\text{form},Al_2O_3}$) is strictly valid at 25 °C and 1 atm.

4.7 Property analysis methods

4.7.1 Introduction

As a result of the geometry optimization calculations, not only the total energy, but also the total wavefunction, Ψ (or density, Ψ^2), will be obtained for a specific model system. As a matter of fact, information about the electronic structure, as well as of all materials properties, are included in this specific wavefunction (e.g. chemical, mechanical, electrical, optical, and magnetic properties). However, it is mainly the chemical properties that are of major importance to study in a CVD growth process. There are, in particular, four theoretical methodologies to use in studying these properties; electronic density calculations, density-of-state calculations, Mulliken charge calculations, and Fukui function calculations.

4.7.2 Electron deformation density calculations

The calculated electron densities are very useful in studying both the various bond strengths and electron distributions in the systems. However, it is most often quite difficult to study minor differences in bond strengths by using these types of total electron densities. Instead, one better use something called electron deformation density calculations. By calculating the difference between the electron density for the complete system and the electron densities for each of its atoms, the outcome is a picture over the including bond electron densities:

$$\Delta\rho_{\text{bonds}} = \rho_{\text{total}} - \sum\rho_{\text{atoms}} \tag{4.68}$$

As an example, figure 4.18 shows this type of electron deformation density for an approaching H atom to an N-doped H-terminated diamond surface. The yellow areas show the excess of electron density as a result of bond formation (i.e. bonding interactions between two atoms). It is clear from figure 4.18 that there are covalent

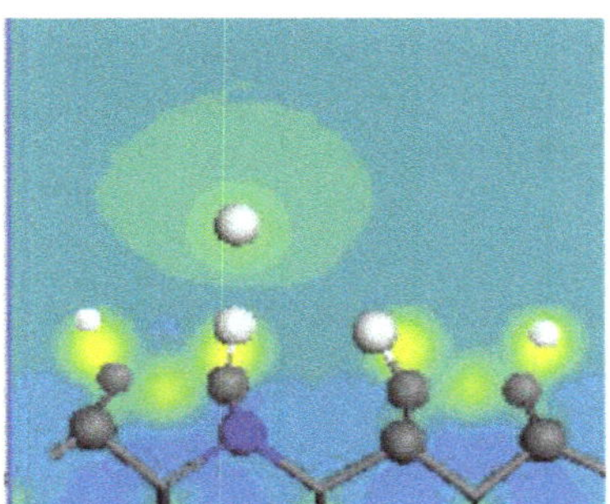

Figure 4.18. Electron deformation densities in a plane that cuts through the C(2), N and H(approach) atoms (where C(2) is neighbouring the N dopant in the 2nd C atomic layer) [61].

bonds between the surface C atoms and the terminating H atoms. In addition, there are pairwise covalent bond formations between surface C atoms. The reason to the absence of electron densities for the other covalent bonds in the model can be explained by the fact that we are looking at a density plane that is cut through the 3D density cloud of the approaching H atom. More specifically, this plane is cut through the C(2)–N–H(approach) formation, where C(2) is the neighbouring C atom (in the 2nd C layer) to the N dopant in the diamond surface. As can be seen in figure 4.18, it is apparent that there is a strong interaction between the surface and the approaching H atom, which has resulted in a certain electron density transfer from the surface to the H atom. It is not mandatory to use atomic electron densities when calculating the electron deformation densities. As a matter of fact, one can instead use any parts of the model system (for instance, the approaching H atom + the rest of the system). However, it is only possible to get information about the bond formation between the approaching H atom and the surface by using the electron deformation densities.

For the situation with one (or several) radical atoms in the model systems, it is necessary to use a spin polarized method in performing something called spin density calculations. The α (+1/2) and β (−1/2) electron spins are described in this method by different wave functions (see equation (4.15)), and spin densities can, thereby, be obtained. It is then possible to plot the alpha spin density, beta spin density, or a difference between these two spin densities. The spin densities can be used in, e.g. explaining the effect of a dopant on neighbouring bond strengths. Take substitutionally N-doped diamond as an example and let us study its H-terminated surface. Nitrogen has an extra valence electron compared to carbon, and it is at first necessary to get information about the geometrical location of this 'extra' electron density. As can be seen in figure 4.19(a), spin density calculations (of a–b type) show that the density from this extra electron is more globally distributed amongst the bonds in the vicinity to N in the diamond surface. In comparison, spin density calculations have also been performed for the situation with a radical NH group chemisorbed to the surface (see figure 4.19(b)). Moreover, the calculated electron deformation density for the situation with an N dopant in the 2nd atomic C layer (figure 4.19(c)) is observed to strongly correlate with the spin density formation for the same dopant situation (figure 4.19(a)). As can be seen in figure 4.19(c), the electron densities in the N–C bonds are smaller, which is a strong indication of weaker bonds. As an explanation, the extra electron (from N) is partially filling the anti-binding molecular orbitals between N and the C atoms, which will cause this

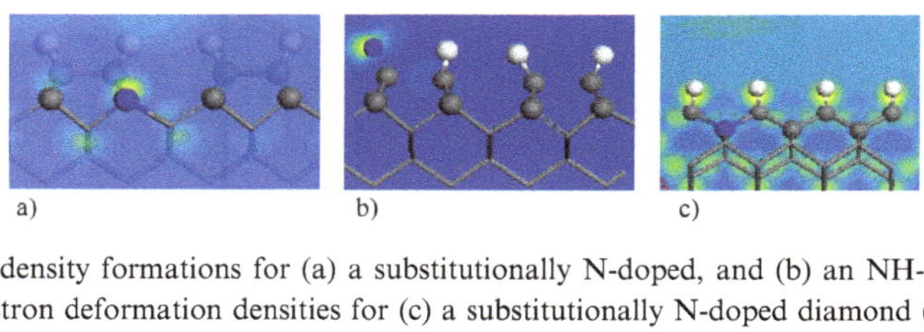

Figure 4.19. Spin density formations for (a) a substitutionally N-doped, and (b) an NH-adsorbed diamond (100) surface. Electron deformation densities for (c) a substitutionally N-doped diamond (100) surface [61].

bond weakening. Something similar is not observed for the NH-adsorbed diamond surface (figure 4.19(b)).

4.7.3 Density of state (DOS) calculations

Density of state (DOS) calculation is an effective mathematical tool for electronic structure analysis. It is strongly connected to the electronic band structure for a specific system. An example of an electronic band structure, with its corresponding DOS, are shown in figure 4.20. The electronic bandstructure (most commonly called the bandstructure) describes the range of electronic energy levels with their specific electron wavefunctions. Both filled and unfilled energy levels are shown in the bandstructure. The differences between the electronic energy levels are, for a solid state system, so small that they instead form bands of energy levels (e.g. sp-bands, d-bands). (This is due to the large number of interacting atoms in a solid material.) The bandstructure is a two-dimensional plot where the x-axis contains a range of k-values of the wavefunctions ($-\pi/2 \leqslant k<\pi/2$) and the y-axis shows the energies for these electronic bands. The corresponding DOS spectrum can be calculated by using the following expression:

$$N_n(E) = \int \frac{1}{4\pi^3}\delta(E - E_n(k))dk \tag{4.69}$$

where $N_n(E)$ is the DOS for the electron band n. Very detailed descriptions of how to create bandstructures and DOS's for different types of periodic structures are given in [60]. As described in [60], longer atomic distances (i.e. weaker interatomic interactions) give sharper DOS peaks, and shorter distances (i.e. stronger interactions) give broader peaks. It is, thereby, possible to evaluate the possibilities for, e.g. an H adsorption by simply studying the surface–H interaction peak in the DOS spectrum. Furthermore, there is a very small slope in energy for the k-values in the central part of the x-axis. Hence, there is a large number of electronic wavefunctions

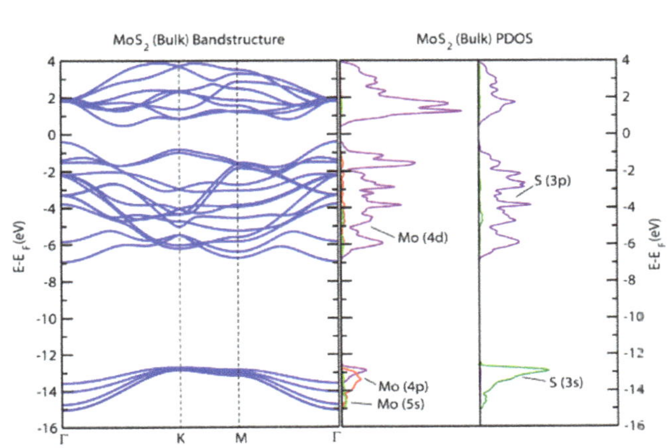

Figure 4.20. An (a) electronic bulk band structure for MoS_2 and the (b) pDOS's for both Mo and S [62].

in a very small energy range in this region with a corresponding maximum in the DOS peak. In other words, the DOS peaks show the energy distribution for the various wavefunctions in the bands. By, in turn, integrating over the whole volume, one gets information about the number of electrons that belong to this peak.

It is very common to calculate the partial DOS (pDOS), which means that the total DOS can be separated into DOSs for s, p, d, and f electrons, and for the various elements in the structure. This is demonstrated in figure 4.20, which shows the bulk band structure for MoS_2 and the pDOS's for both Mo and S. The x-axis represents the k-point mesh along the Γ–K–M–Γ path in the irreducible Brillouin zone.

4.7.4 Mulliken population analysis

The assignment of atomic charges and/or electron bond populations is widely used in analysing the material properties of a model system. Electron bond populations are simply the electron density between two atoms. Due to the delocalized nature of the planewave basis set, the electron bond population analysis is often made by projecting the planewave-related densities onto a localized basis set (by using the methods by Sanchez–Portal *et al*) [63]. This projection of states is, thereafter, analyzed by using the Mulliken formalism [64]. This latter type of method uses a linear combination of atomic orbitals (LCAO) in describing the electron wave-functions of the systems. As such, the electrons will be partitioned to each atom in the system. Hence, the LCAO basis set provides a natural way of specifying both atomic charges and electron bond populations.

Both the atomic charges and bond populations are very useful in analysing processes and properties of, e.g. a solid surface. For instance, the atomic charges make it possible to study the process of charge transfer within a bond, or over an interface. Moreover, a bond population analysis can also identify bond weakening effects. Let us go back and have a look at figure 4.18. The calculations of the bond population between the incoming H atom and the H adsorbate give a value of 0.18, which is an indication of an electron orbital overlap. This positive electron density value means that there is an interaction between these two species at this specific distance. Thus, the bond population calculations do strongly support the results obtained from the electron deformation density calculations. Both types of calculations reveal an increased interaction between the incoming species and the surface.

Another example of the usefulness of the bond population calculations is the analysis of the C–H bond populations in figure 4.19(c). These two bonds are the ones that are positioned closest to the N dopant. Their bond populations are both 0.72, which is to be compared with the bond populations for C–H bonds situated further away from the dopant (0.90). These differences are very small and not visible to the eye. Thus, it is extremely important to use a combination of analysis methods when analyzing material structures and properties.

4.7.5 Fukuji function calculations

Fukui functions, being introduced by Fukui [65, 66] and later on modified by Parr [67] and Yang [68], have been found to be reliable indicators when analysing surface reactivities. These functions describe the change in electron density in a frontier orbital. (The frontier orbitals are the electronic states (i.e. wavefunctions) just below, and above, the highest occupied electron wavefunction. This latter type of wavefunction has several names: valence band edge (VBE) and highest occupied molecular orbital (HOMO). However, the common name is VBE for the solid state.) There are three different types of Fukui functions, being defined as f^+, f^-, and f^0. By adding a small negative charge to the frontier orbitals, the susceptibility for a nucleophilic attack to the surface is being calculated, f^+ (equation (4.70)).

$$f^+ = \frac{1}{\Delta N}(\rho_{N+\Delta}(\vec{r}) - \rho_N(\vec{r})) \tag{4.70}$$

$$f^- = \frac{1}{\Delta N}(\rho_N(\vec{r}) - \rho_{N-\Delta}(\vec{r})) \tag{4.71}$$

$$f^0 = \frac{1}{2}(\rho_{N+\Delta}(\vec{r}) - \rho_{N-\Delta}(\vec{r})) \tag{4.72}$$

This function is visualized as an electron density on the surface, where some parts have a larger value of f^+ and some have lower values (see figure 4.21(a)). To explain this function more in detail, when an extra electron density is added to a surface it prefers to go to positions where f^+ is large. The reason is that the added electron density will be most stabilized at these surface sites. (This is to be compared with the electronegativities of elements.) On the contrary, by removing a small negative charge from the frontier orbitals, the susceptibility for an electrophilic attack can be calculated, f^- (equation (4.71)). Similar to the f^+ function, a solid-state surface is susceptible to an electrophilic attack at the locations where f^- shows a large value. With the same comparison to eletronegativity values, these are the surface locations where the surface becomes stabilized when reacting with an electrophilic species. Moreover, the Fukui function for a radical attack, f^0, is simply the average of f^+ and f^-. For instance, to indicate the surface reactivity towards an H radical attack, the radical Fuikui function,

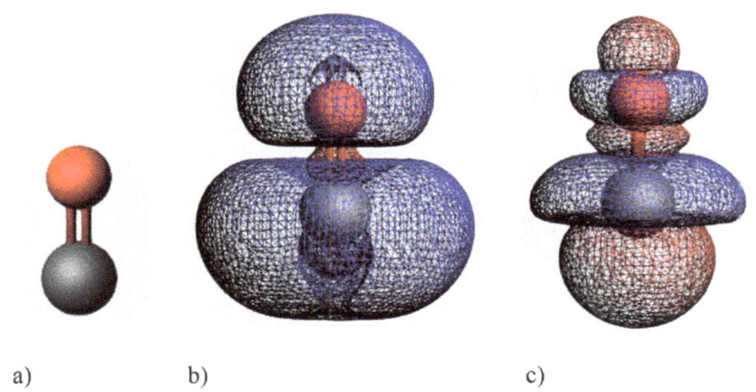

Figure 4.21. (a) Model of a CO molecule, and electron density maps showing the (b) susceptibility for an electrophilic attach (f^+) and (c) susceptibility for both an electrophilic and nucleophilic attach (f^+–f^-). Blue: f^+; red: f^-.

f^0 (equation (4.72)), can be used. Examples of calculated f^+ and f^0 density maps can be seen in figures 4.21(b) and (c), where N is the number of electrons in the system and $\rho_N(\vec{r})$ is the electron density at a position, $\vec{r}$, above the surface. Furthermore, the electron densities $\rho_{N+\Delta}(\vec{r})$ and $\rho_{N-\Delta}(\vec{r})$ are the electron densities when adding, or removing, a small charge to/from the neutral surface, respectively.

In short, the Fukui functions over a specific surface area are generally used in visualizing the surface reactivity. More precisely, they can be used to indicate solid surface reactivities toward different kinds of chemical reactions with the environment (like adsorption or abstraction). For instance, the Fukui functions are very important indicators for electron-transfer controlled reactions. The practical usage of the Fukui functions will be further demonstrated in section 6.7.

References

[1] Schrödinger E 1926 An undulatory theory of the mechanics of atoms and molecules *Phys. Rev.* **28** 1049–70

[2] Born M and Oppenheimer R 1927 Quantum theory of molecules *Ann. Phys.* **84** 0457–84

[3] Hartree D R 1928 The wave mechanics of an atom with a non-Coulomb central field. Part I: theory and methods *Proc. Cambridge Philos. Soc.* **24** 89–110

[4] Fock V 1930 Approximation method for the solution of the quantum mechanical multibody problems *Z. Angew. Phys.* **61** 126–34

[5] Cramer C J 2004 *Essentials of Computational Chemistry: Theories and Models* 2nd edn (Hoboken, NJ: Wiley)

[6] Feyereisen M, Fitzgerald G and Komornicki A 1993 Use of approximate integrals in *ab initio* theory. An application in MP2 energy calculations *Chem. Phys. Lett.* **208** 359–63

[7] Boustani I 1995 A comparative study of *ab initio* SSF-CI and DFT. Example of small boron clusters *Chem. Phys. Lett.* **233** 273–8

[8] Koch W and Holthausen M C 2000 *A Chemist's Guide to Density Functional Theory* (New York: Wiley-VCH)

[9] Kohn W and Sham L J 1965 Self-consistent equations including exchange and correlation effects *Phys. Rev.* **140** 1133–8

[10] Yi S S, Reichert E L, Holthausen M C, Koch W and Weisshaar J C 2000 Crossed-beam study of Co+(F-3(4))+propane: experiment and density functional theory *Chem. Eur. J.* **6** 2232–45

[11] Hohenberg P and Kohn W 1964 Inhomogeneous electron gas *Phys. Rev.* **136** 864–71

[12] Martin R M 2004 *Electronic Structure: Basic Theory and Practical Methods* (Cambridge: Cambridge University Press)

[13] Ceperley D M and Alder B J 1980 Ground state of the electron gas by a stochastic method *Phys. Rev. Lett.* **45** 566–9

[14] Delley B 1990 An all-electron numerical method for solving the local density functional for polyatomic molecules *J. Chem. Phys.* **92** 508–17

[15] Staroverov V N and Scuseria G 2004 Tests of a ladder of density functionals for bulk solids and surfaces *Phys. Rev.* B **69** 075102

[16] Perdew J P, Burke K and Ernzerhof M 1996 Generalized gradient approximation made simple *Phys. Rev. Lett.* **77** 3865–8

[17] Perdew J P, Burke K and Wang Y 1996 Generalized gradient approximation for the exchange-correlation hole of a many-electron system *Phys. Rev.* B **54** L6533–9

[18] Perdew J P, Chevary J A, Vosko S H, Jackson K A, Pederson M R, Singh D J and Fiolhais C 1992 Atoms, molecules, solids, and surfaces: applications of the generalizedgradient approximation for exchange and correlation *Phys. Rev.* B **46** 6671

[19] Becke A D 1988 Density-functional exchange-energy approximation with correct asymptotic behavior *Phys. Rev.* A **38** 3098–100

[20] Lee C, Yang W and Parr R G 1988 Development of the Colle–Salvetti correlation-energy formula into a functional of the electron density *Phys. Rev.* B **37** 785–9

[21] Ziegler T 1991 Approximate density functional theory as a practical tool in molecular energetics and dynamics *Chem. Rev.* **91** 651–67

[22] Labanowski J K and Andzelm J W 1991 *Density Functional Methods in Chemistry* (New York: Springer)

[23] Johnson B G, Gill P M W and Pople J A 1993 The performance of a family of density functional methods *J. Chem. Phys.* **98** 5612–26

[24] Handy N C, Murray C W and Amos R D 1993 Study of CH_4, C_2H_2, C_2H_4, and C_6H_6 using Kohn–Sham theory *J. Phys. Chem.* **97** 4392–6

[25] Yanai T, Tew D P and Handy N C 2004 A new hybrid exchange-correlation functional using the coulomb-attenuating method (CAM-B3LYP) *Chem. Phys. Lett.* **393** 51–7

[26] Adamo C and Barone V 1999 Toward reliable density functional methods without adjustable parameters: the Pbe0 model *J. Chem. Phys.* **110** 6158–70

[27] Grimme S 2006 Semiempirical GGA-type density functional constructed with a long-range dispersion correction *J. Comput. Chem.* **27** 1787–99

[28] Anisimov V I, Zaanen J and Andersen O K 1991 Band theory and Mott insulators: Hubbard *U* instead of Stoner I *Phys. Rev.* B **44** 943–54

[29] Cococcioni M and de Gironcoli S 2005 Linear response approach to the calculation of the effective interaction parameters in the LDA+U method *Phys. Rev.* B **71** 035105

[30] Kresse G and Furthmüller J 1996 Efficient iterative schemes for *ab initio* total-energy calculations using a plane-wave basis set *Phys. Rev.* B **54** 11169

[31] Pulay P 1980 Convergence acceleration of iterative sequences. the case of SCF iteration *Chem. Phys.Lett.* **73** 393–8

[32] Broyden C G 1965 A class of methods for solving nonlinear simultaneous equations *Math. Comput* **19** 577–93

[33] Kerker G P 1981 Efficient iteration scheme for self-consistent pseudopotential calculations *Phys. Rev.* B **23** 3082–4

[34] Pulay P 1977 *Modern Theoretical Chemistry* vol 4 (New York: Plenum Press)

[35] Troullier N and Martins J L 1991 Efficient pseudopotentials for plane-wave calculations *Phys. Rev.* B **43** 1993

[36] Vanderbilt D 1990 Soft self-consistent pseudopotentials in a generalized eigenvalue formalism *Phys. Rev.* B **41** 7892–5

[37] Hamann D R, Schlüter M and Chiang C 1979 Norm-conserving pseudopotentials *Phys. Rev. Lett.* **43** 1494–7

[38] Bachelet G B, Hamann D R and Schlüter M 1982 Pseudopotentials that work: from H to Pu *Phys. Rev.* B **26** 4199–228

[39] Payne M C, Teter M P, Allan D C, Arias T A and Joannopoulos J D 1992 Iterative minimization techniques for *ab initio* total-energy calculations: molecular dynamics and conjugate gradients Rev. Mod. Phys. **64** 1045–97

[40] Jurečka P, Černý J, Hobza P and Salahub D R 2007 Density functional theory augmented with an empirical dispersion term. interaction energies and geometries of 80 noncovalent complexes compared with *ab initio* quantum mechanics calculations *J. Comput. Chem.* **28** 555–69
[41] Ortmann F, Bechstedt F and Schmidt W G 2006 Semiempirical van der Waals correction to the density functional description of solids and molecular structures *Phys. Rev.* B **73** 205101
[42] Tkatchenko A and Scheffler M 2009 Accurate molecular van der Waals interactions from ground-state electron density and free-atom reference data *Phys. Rev. Lett.* **102** 073005
[43] Marom N, Moussa J E, Ren X, Tkatchenko A and Chelikowsky J R 2011 Electronic structure of dye-sensitized TiO_2 clusters from many-body perturbation theory *Phys. Rev.* B **84** 245115
[44] Marom N, Bernstein J, Garel J, Tkatchenko A, Joselevich E, Kronik L and Hod O 2010 Stacking and registry effects in layered materials: the case of hexagonal boron nitride *Phys. Rev. Lett.* **105** 046801
[45] Meza J C 2010 Steepest descent *Wiley Interdiscip. Rev. Comput. Stat.* **2** 719–22
[46] Nocedal J and Wright S J 2006 Conjugate gradient methods *Numerical Optimization* (New York: Springer)
[47] Kao R 1974 A comparison of Newton–Raphson methods and incremental procedures for geometrically nonlinear analysis *Comput. Struct.* **4** 1094–7
[48] Mamen R and Mayne D Q 1972 A pseudo Newton–Raphson method for function minimization *J. Optim. Theory Appl.* **10** 263–76
[49] Oi K, Sayama H and Takamatsu T 1973 Computational schemes of the Davidon–Fletcher–Powell method in infinite-dimensional space *J. Optim. Theory Appl.* **12** 447–58
[50] Pfrommer B G, Cote M, Louie S G and Cohen M L 1997 Relaxation of crystals with the quasi-Newton method *J. Comput. Phys.* **131** 233–40
[51] Malouf R 2002 A comparison of algorithms for maximum entropy parameter estimation *Proc. 6th Conf. on Natural Language Learning (CoNLL-2002)* pp 49–55
[52] Andrew G and Gao J 2007 Scalable training of L1-regularized log-linear models *Proc. 24th Int. Conf. on Machine Learning*
[53] Aarons J 2011 *A New CASTEP and ONETEP Geometry Optimiser* http://www.hector.ac.uk/cse/distributedcse/reports/castep-geom/castep-geom.pdf
[54] Packwood D, Kermode J, Mones L, Bernstein N, Woolley J, Gould N, Ortner C and Csányi G 2016 A universal preconditioner for simulating condensed phase materials *J. Chem. Phys.* **144** 164109
[55] Schlegel H B 2011 Geometry optimization *Adv. Rev.* **1** 790–809
[56] Winkler B and Milman V 1997 Polymorphism of CsI *J. Phys. Condens. Matter* **9** 9811–7
[57] Smith A E and Coit D W 1996 *Penalty Functions. Handbook of Evolutionary Computation* (Oxford: Oxford University Press)
[58] Galántai A 2004 *Projection Methods in Optimization Projectors and Projection Methods* (Boston, MA: Springer) pp 181–213
[59] Govind N, Petersen M, Fitzgerald G, King-Smith D and Andzelm J 2003 A generalized synchronous transit method for transition state location *Comput. Mater. Sci.* **28** 250–8
[60] Hoffmann R 1988 *Solids and Surfaces* (New York: Wiley-VCH)
[61] Yiming Z, Larsson F and Larsson K 2014 Effect of CVD diamond growth by doping with nitrogen *Theor. Chem Acc.* **133** 1432

[62] Ahmad S and Mukherjee S 2014 A comparative study of electronic properties of bulk MoS_2 and its monolayer using DFT technique: application of mechanical strain on MoS_2 monolayer *Graphene* **3** 52–9
[63] Sanchez-Portal D, Artacho E and Soler J M 1995 Projection of plane-wave calculations into atomic orbitals *Sol. Stat. Commun.* **95** 685–90
[64] Mulliken R S 1955 Electronic population analysis on LCAO–MO molecular wave functions. I *J. Chem. Phys.* **23** 1833–40
[65] Fukui K, Yonezawa T and Shingu H 1952 A molecular orbital theory of reactivity in aromatic hydrocarbons *J. Chem. Phys.* **20** 722–5
[66] Fukui K 1982 Role of Frontier orbitals in chemical-reactions *Science* **218** 747–54
[67] Parr R G and Yang W T 1984 Density functional-approach to the Frontier-electron theory of chemical-reactivity *J. Am. Chem. Soc.* **106** 4049–50
[68] Yang J, Parr R G and Pucci R 1984 Electron-density, Kohn–Sham Frontier orbitals, and Fukui functions *J. Chem. Phys.* **81** 2862–3

Chapter 5

Construction of solid surface models

In addition to the choice of a theoretical method and theoretical parameters, the construction of a surface model is a particularly important task in theoretical modeling and simulation of CVD processes. The model must be as realistic as possible, while at the same time be compatible with the choice of theoretical parameters. Hence, the size of the model should not entail practical problems with, e.g. disk space and computational times. Within the present chapter, guidelines will be given considering how to construct optimal solid surfaces for various types of CVD process simulations. However, before that, a more general description of solid surfaces, and their behaviour, will be presented and discussed.

5.1 Surfaces within materials science of today

A deeper knowledge about materials surfaces and their properties have become of a major importance within the material science of today. Furthermore, many applications have completely different demands regarding the reactivity and properties of a specific surface. For instance, materials and surfaces that are used for cutting tools and drills must be very hard and tough (i.e. not brittle). They must also be able to withstand high temperatures and harsh environments. The demands on materials surfaces used for bio applications are, though, completely different. Examples of bio applications include implant construction and bone regeneration. These materials must be bio compatible and show the optimal reactivity towards the specific bio system. They must also be chemically inert in order to be resistant towards, e.g. oxidation. For the situation with a hip joint implant, the materials surface must also be low-frictional. Other examples of applications are car catalysts, where carbon monoxide will undergo oxidation (to carbon dioxide) and nitrogen oxide be reduced to nitrogen on the solid surfaces. For instance, the surfaces must tolerate high temperatures and show the right chemical reactivity towards the specific reactions that must take place. Moreover, the surfaces must also be efficient for the oxidation and reduction reactions also

doi:10.1088/978-0-7503-3107-4ch5

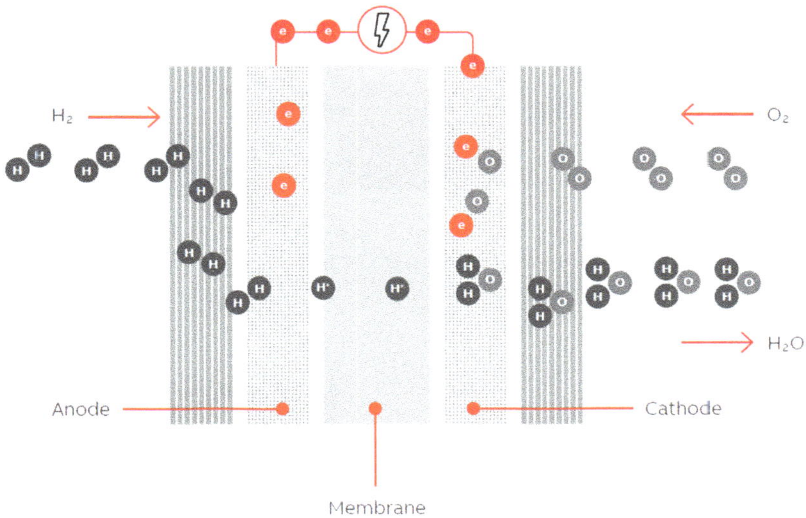

Figure 5.1. Illustration of a fuel cell where H_2 will undergo oxidation and O_2 reduction, forming H_2O and electricity.

at lower temperatures (i.e. for initial cold starts of the engine). A third example of applications is the fuel cell, where hydrogen and oxygen molecules shall undergo reactions on a surface and form water molecules and electricity. These materials surfaces must, thus, be especially reactive towards hydrogen oxidation and oxygen reduction. In addition, there must be a material that is permeable for the resulting hydrogen ions so that H^+ and O_2 can meet and form H_2O (see figure 5.1). Hence, the two electrode areas onto which the reduction and oxidation takes place must be electron conducting and also be able to withstand harsh environments.

5.2 Surface reactivities

The reactivity and properties of a solid material surface is different when compared with the bulk material. In addition to the importance of the material itself, there are several factors that control the surface reactivity, out of which one is particularly important—the number of binding neigbouring atoms. The number of binding atoms to a surface atom is always smaller than the binding situation for a bulk atom (see figure 5.2). Thus, a surface atom has the driving force to become more bulk-like (i.e. the most preferred surrounding electron density is the one for a bulk atom). This is one explanation to the larger surface reactivity (relative bulk). It can also be expressed as the following—each surface atom wants a bulk-like electron density environment. In summary, the lower the electron density, the larger the surface reactivity of that specific surface atom is. Other major parameters that control the solid surface reactivity are:

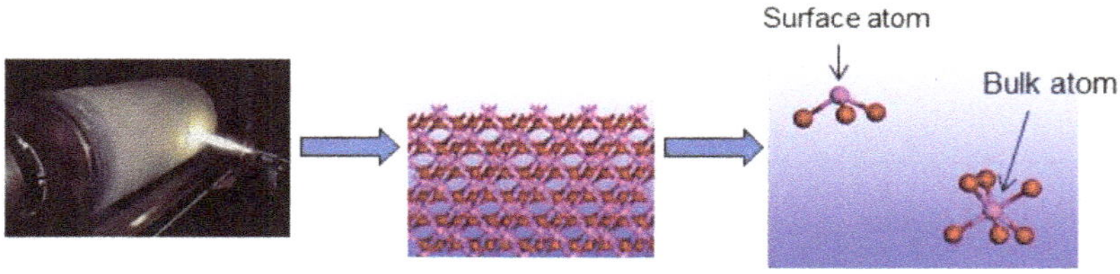

Figure 5.2. Illustration of the (a) fabrication of a solid binary material, with its (b) surface atoms and (c) difference in geometrical surrounding for a surface atom and its bulk equivalent.

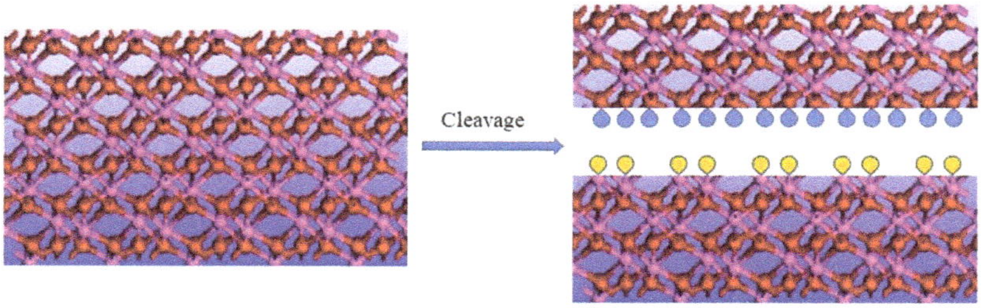

Figure 5.3. Illustration of the cleavage of a bulk (mainly covalent) binary material, forming two surfaces. The blue and yellow drops illustrate the resulting non-paired electrons (prior to any relaxation, reconstruction, or reactions with the environment).

- Surface morphology,
- Surface relaxation,
- Surface reconstruction,
- Adsorption,
- Segregation, and
- Doping.

It must be stressed here that the solid surface reactivity is not identical to the surface energy. When creating a surface by, for example, breaking the bonds along a specific crystallographic orientation in the bulk. One example is the breakage of a covalent binary material into smaller pieces by exerting a force onto it. A specific number of bonds (n) must then be broken in the formation of two surfaces (see figure 5.3). Let us now assume that the bond density between these two surfaces were, prior to the breakage, ρ_n bonds/cm^2.

Hence, the energy (per cm^2) needed to crack this Si sample to two pieces is $\rho_n \times E_{\text{bond}}$. The resulting surface energy (for each of the two surfaces) is thereby: $\gamma = (\rho_n \times E_{\text{bond}})/2$. The Si–Si bond energy, E_{bond}, is a direct measure of the surface reactivity. Hence, the surface energy considers both the reactivity and density of surface atoms, whilst the surface reactivity only takes into account the atomic reactivity.

5.3 Surface planes

Each crystalline solid-state material can form different types of surface planes (also called terraces). Each of these planes have a specific structural geometry, as well as a specific chemical composition. Take cubic boron nitride (c-BN) as an example. It has a face centred cubic (fcc) structure (see figure 5.4(a)). As such, each B (or N) atom in the bulk binds to four N (or B) atoms. When growing c-BN by using a high-temperature high-pressure vapour phase method, either of three different surface planes will be formed; (111), (110), or (100). These three surface planes are shown in figures 5.4(a)–(c), respectively. The (100) and (111) planes are similar in that they are formed by alternate layers of B and N. In contrary, the (110) surface includes a stochiometric number of B and N atoms. Moreover, the (100) and (110) surfaces consist of surface atoms that bind to two neighbouring atoms each, whilst the surface atoms on c-BN(111) bind to three atoms. Hence, the three surface planes (111), (100) and (110) differ with respect to both chemical composition and number of binding neighbouring atoms. As a conclusion, the surface reactivities must also be different for these specific surface types.

Other examples of materials surfaces are those constructed from cubic and hexagonal close packed metals. Let us compare the surface reactivities of the surface planes (100), (110) and (111) for a cubic close packed (ccp) metal (see figure 5.5). As can be seen in figure 5.5(a), a surface atom on the (100) plane has four binding neighbours in the same plane (marked in red). In addition, it binds to four atoms in the atomic layer just beneath the surface plane (marked in blue). Thus, a surface atom on the (100) plane has in total eight binding atoms, which is to be compared with 12 binding neighbours for a bulk ccp metal atom. Figure 5.5(b) shows the situation for the (110) surface. A surface atom has two binding atoms in the (110) plane (marked with red), whilst it has four binding atoms in the second atomic layer (marked with blue). In addition, it has one binding atom in the third layer (which is right beneath the specific surface atom and marked in green). In total, a surface atom on the (110) plane has seven binding neighbouring atoms. Finally, a surface atom on the (111) plane (figure 5.5(c)) has six binding atoms in the (111) plane (marked in red), and three binding atoms in the second atomic layer (marked in blue). Hence, there is a total of nine binding neighbouring atoms to the surface atom in the (111)

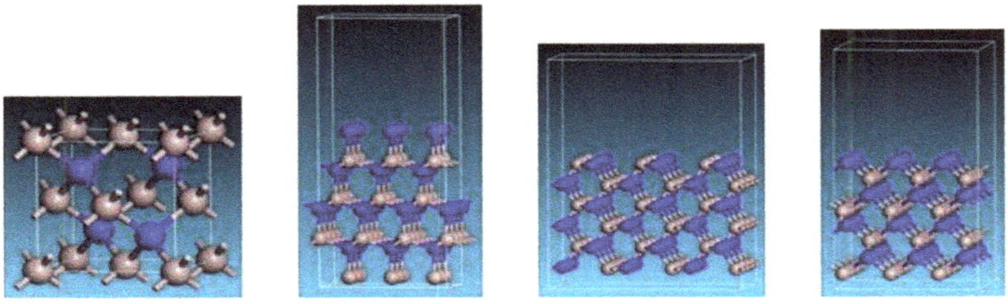

a) b) c) d)

Figure 5.4. Illustrations of the (a) unit cell of c-BN and of the (b) (111), (c) (110), and (d) (100) surface planes. The pink balls demonstrate the B atoms, whilst the blue balls demonstrate the N atoms.

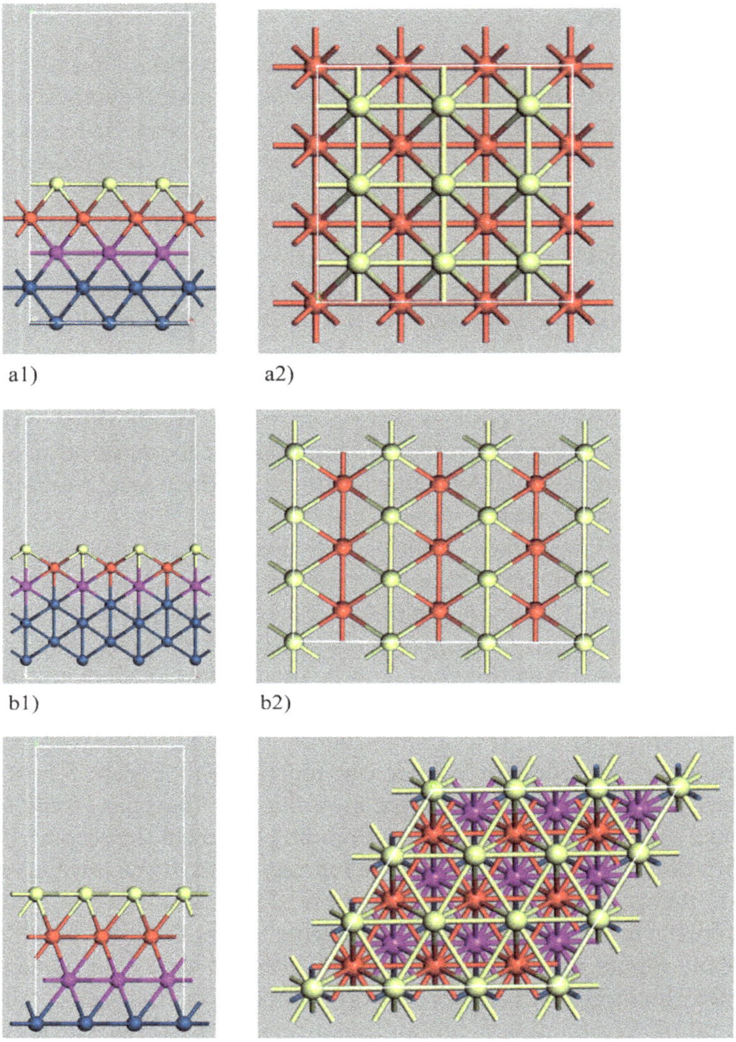

Figure 5.5. Illustration of low-index surface planes for a cubic close packed (ccp) metal; (a) (100), (b) (110), and (c) (111). The growth direction is (111) with ABC packing (as illustrated by A = light green, B = red, and C = purple). Side view: a1, b1, c1; top view: a2, b2, and c2.

plane. Based on the number of binding neighbouring atoms, the order of surface reactivities for the (111), (100) and (110) ccp metal planes is: (111) < (100) < (110). Hence, a surface atom in the (110) plane is the most reactive since it has only seven binding neighbouring atoms, which is much fewer than for the bulk situation (with 12 binding atoms). On the other hand, a surface atom in the (111) plane has nine binding neighbouring atoms, which makes it not so reactive since it is closer to the bulk situation. Let us now make some exercises.

Exercise 1: Iron surfaces of different crystallographic orientations give different reaction rates for, e.g. catalytic reactions. What is the reason for these differences?

Answer: The surface atoms in the different surface planes have different numbers of binding neighbouring atoms and, thereby, different surface reactivities.

Exercise 2: Let us compare the surface energies of Si(100), Si(110) and Si(111). Which surface has the highest surface energy? Which has the lowest?

Answer: Silicon is a covalent compound with a cubic fcc lattice. A surface atom in the surface planes Si(100), Si(110), and Si(111) has 2, 2, and 3 binding Si neighbouring atoms, respectively (see figures 5.6(b)–(d)). Let us assume that the edge length of the cubic Si unit cell is a Å. (It is important to stress that a covalent material atom has fewer binding atoms, as compared with a metal atom (see figure 5.5 with figure 5.6).) As presented above, the surface energy is expressed by $\gamma = (\rho_n \times E_{\text{bond}})/2$, and E_{bond} is a

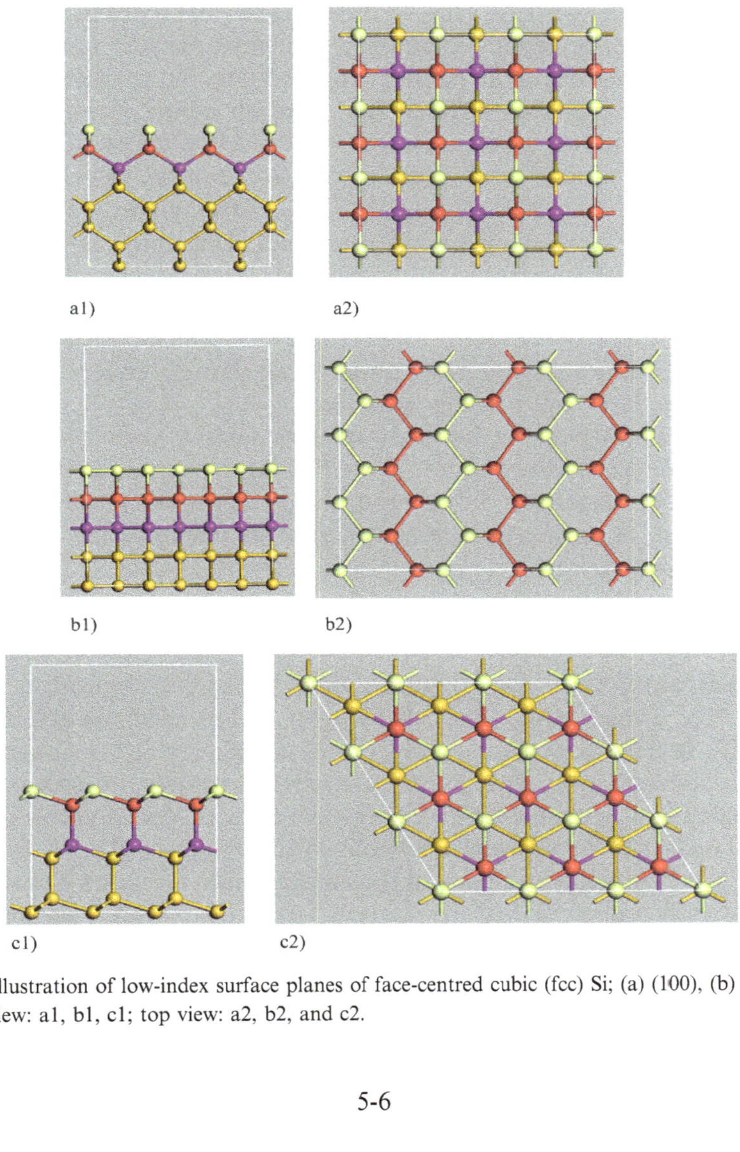

a1) a2)

b1) b2)

c1) c2)

Figure 5.6. Illustration of low-index surface planes of face-centred cubic (fcc) Si; (a) (100), (b) (110), and (c) (111). Side view: a1, b1, c1; top view: a2, b2, and c2.

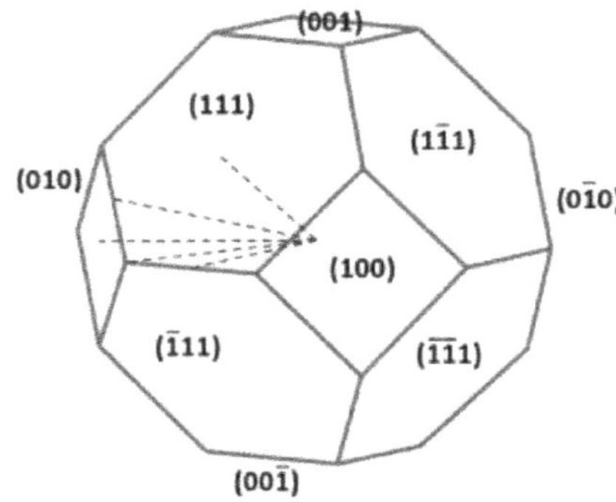

Figure 5.7. A sketch of a metal ccp particle that has been grown under strict thermodynamic conditions.

measure of the surface reactivity. Thus, in order to solve the problem, we must find an expression for the atomic densities on the three different surfaces, in addition to their relative surface reactivities. As can be seen in figure 5.6(b), there are two Si atoms in an area of a^2 Å^2 on the Si(100) surface. This gives an atomic density of $2/a^2$ Å^{-2}. Furthermore, there are two atoms in an area of $\sqrt{2}\,a^2$Å^2 on the Si(110) surface, which gives an atomic density of $\sqrt{2}/a^2$ Å^{-2} (see figure 5.6(c)). Finally, there are two atoms in an area of $\sqrt{3}\,a^2/2$ Å^2 for Si(111) (see figure 5.6(d)), which gives an atomic density of $4/\sqrt{3}\,a^2$Å^{-2}. When comparing the surface reactivities of the three Si planes, we get: Si (111) < Si(100) ~ Si(110) (since a surface atom in Si(111) is binding to one more atom, as compared with the other two surface planes). As a conclusion, the order of surface energy is Si(111) < Si(100) ~ Si(110).

Exercise 3: Assume that you grow a metal ccp particle under strict thermodynamic conditions. How will it finally look like?

Answer: The (111) plane is the least reactive surface plane and is, therefore, dominating on the particle surface (see figure 5.7). Moreover, the (110) plane is most reactive and will be overgrown. For the same reason, the (100) plane will not exist to any larger extent. Also, the particle is spherical to minimize the total area and, thereby, the surface energy.

5.4 Surface morphologies

Real material surfaces are never perfectly planar. There is, instead, a mixture of surface sites like terraces, edges (also called ledges), and corners (also called kinks) (see figure 5.8(a)). In fact, there are various levels of complexity of the surface morphologies. The combination of these types of surface structures are commonly called surface roughness. The main difference between a terrace, edge, and corner site is the number of bonds that can be formed between an adsorbate and the site. For instance, the adsorbate X forms only one bond to the surface, whilst it binds with two and three bonds to the edge and corner site, respectively (see figure 5.8(b)). This means that X binds the strongest to the corner site, whilst it binds the weakest to the terrace. Hence, a very rough surface containing a lot of corners is also a very reactive surface, whilst an almost perfectly planar surface shows a very weak surface reactivity.

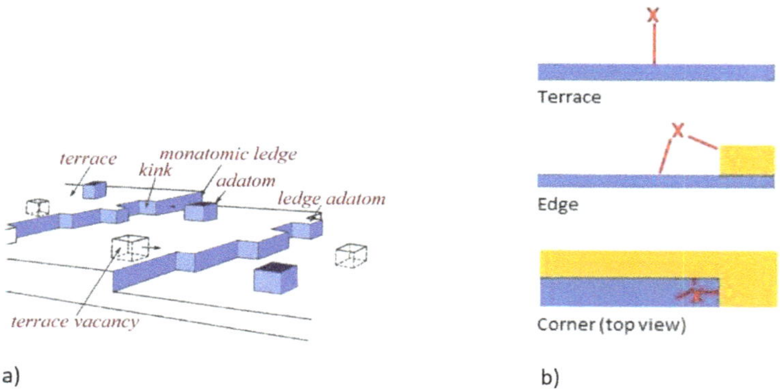

Figure 5.8. (a) Illustration of surface roughness with various types of surface sites (e.g. terrace, edge, and corner). (b) Adsorbate binding to a terrace (top), edge (middle), and corner (lowest).

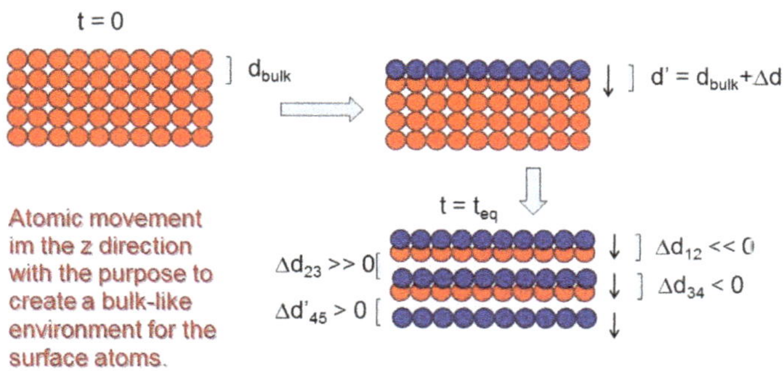

Figure 5.9. Illustration of surface relaxation. The Δd value is the change in distance between two atomic layers.

Let us now do another exercise:

Exercise 4: How does a surface change with respect to steps, corners, etc, when the temperature is increased?

Answer: The surface roughness will decrease when increasing the temperature. This is because the most reactive surface sites (like corners, and thereafter edges) will disappear, which is facilitated by temperature-induced atomic migration.

5.5 Surface relaxation

The surface relaxation is a term that describes the movement of atoms in, predominantly, the z-direction towards a local energy minimum (i.e. towards more stabilized atomic positions in the z-direction). The reason is the strive by the atoms to achieve a bulk-like environmental electron density (see section 5.3). Figures 5.9 and 5.10 illustrate the surface relaxation more in detail. The surface structure directly after its formation (by, e.g. cleavage of a bulk material) (at $t = 0$) and at a certain time later (at $t = t_{eq}$) (at equilibrium) can be seen in figure 5.9. The first atomic layer experiences a very low electron density at $t = 0$. Since these atoms

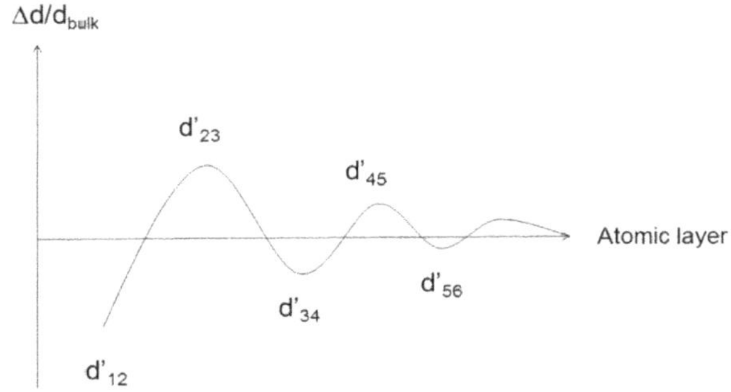

Figure 5.10. Change in $\Delta d/d_{\text{bulk}}$ values as a function of surface atomic layers. Δd is the distance between two relaxed atomic layers, and d_{bulk} is the corresponding bulk distance.

strive to become bulk-like, they will move down towards the second-layer atoms. As a result, the electron density for the second layer atoms becomes too high, which forces them to repel the third layer atoms. The fourth layer atoms will thereby experience a much too high electron density, which also makes these atoms repel the atomic layer 5, and so on. The change in interlayer distance, Δd, will thereby be negative for the distance between atomic layers 1 and 2, and 3 and 4. On the other hand, the change in interlayer distance, Δd, will be positive for the distance between atomic layers 2 and 3, and 4 and 5. These variations in interlayer distances, as a function of atomic layers, are further demonstrated in figure 5.10. The y-axis is the quotient between the change in atomic layer distance and the corresponding bulk distance. Figure 5.10 is, thereby, completely related to figure 5.9. Furthermore, it is clear from this sketch that the relaxation is a process that converges down into the bulk region, and the number of atomic layers being involved in the surface relaxation is completely dependent on the material and surface plane. Let us now continue with another exercise:

Exercise 5: Cu has a tight-binding ccp structure. After the formation of a Cu (110) surface, a change in distance of approximately 10% was observed for the two upper Cu layers. The corresponding change in distance for a Cu (100) surface was 0%. What is the reason for this difference between the two Cu surface planes?

Answer: A Cu (110) surface is more reactive since a surface atom has six binding neighbours. A corresponding surface atom on the Cu (100) surface has eight binding neighbours.

5.6 Surface reconstruction

A material surface will not only undergo surface relaxation. In addition, surface reconstructions will also occur, which is something that takes place predominantly in the x- and y-directions. The underlying reason here is to also decrease the surface reactivity (as well as the surface energy). Materials in general can be divided into three major groups depending on the type of binding between its inherent atoms; metals, ionic materials (e.g. salts), and covalent material (e.g. semiconductors). The degree of surface reconstruction depends on these types of bonding.

Metals have metallic bonds that are non-directed. Due to this circumstance, there is no prominent surface reconstruction in metals. Instead, there is in principle only surface relaxation for these types of materials. The same goes for the situation with nonpolar ionic compounds, which also have non-directed bonds. The situation is, however, completely different for polar ionic compounds. This type of material has a large strive to undergo something called charge compensation. This means, in practice, that any (or a combination) of the following will take place;

- Adhesion of ions with a charge (positive or negative) that is in minority in the lattice. It can, in principle, be of any element type.
- Desorption of ions with a charge that is in majority in the lattice.
- Redox reactions in the lattice (e.g. $Fe^{2+} \rightarrow Fe^{3+} + e^{-}$).
- Creation of facets (i.e. surface reconstruction) (see figure 5.11).

The creation of surface facets means that the positive and negative ions will reposition within the surface region in such a way that an overall charge

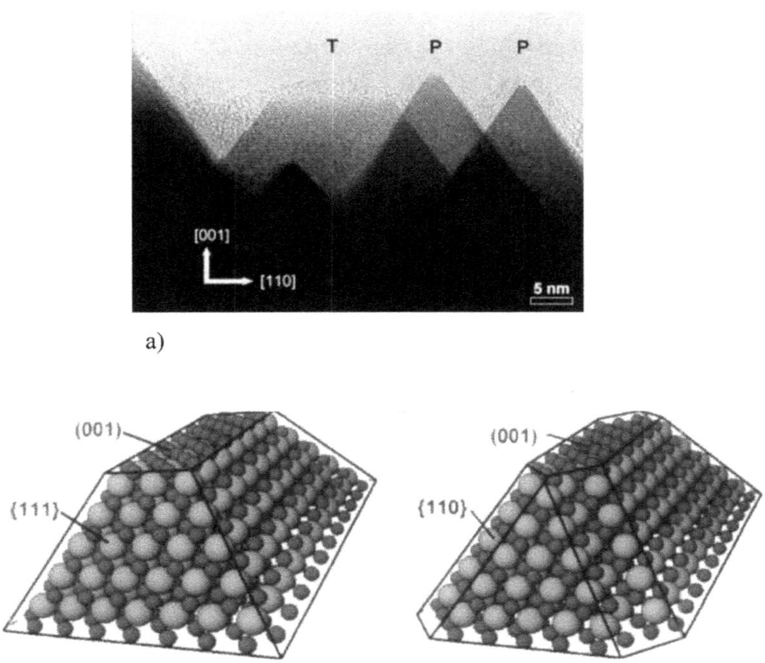

Figure 5.11. (a) High resolution micrograph of cerium dioxide crystals from adjacent columns to the surface, viewed along [1:10] of the cubic lattice. The pointed crystals (P) are close to atomically sharp at the tip. The trapezoidal crystal (T) shows a smoothly graded contrast towards the top. Crystallographic directions are indicated. (b) Schematic of the fluorite structure of CeO_2 as if inserted into the surface topography of the thin film. Note that only {111}-type surfaces are exposed. Large spheres represent cerium and small, dark spheres represent oxygen. The surfaces are shown here as terminated by oxygen. [1], reproduced with permission © Elsevier.

compensation will take place. These rearrangements will then take the form of more, or less, sharp facets.

On the other hand, covalent materials have completely directed bonds. As an example, imagine that you crack a wafer of a mainly covalent material. The new surfaces that are created instantaneously have a bulk-like binding environment, but with dangling bonds (i.e. unpaired electrons) (see figure 5.3). These two surfaces are thereby very reactive, with very high surface energies. In order to decrease the surface reactivity, and thereby lowering the surface energy, new bonds are created in the *x*- and *y*-directions. The surface atoms will then move predominantly in the *x*- and *y*-directions, and to a minor extent in the *z*-direction. This type of surface reconstruction is demonstrated in figures 5.12(c) and (d) for the diamond (100) and (111) surface planes, respectively.

Different types of surface reconstruction generally occur depending on the experimental conditions (e.g. at different temperatures, pressures, and environments). One common reconstruction for the diamond (100) surface is the 2 × 1-reconstruction. As can be seen in figure 5.12(c), the surface C atoms are pairwise forming new covalent bonds. This process also leads to a change in the surface periodicity. The new period of the surface structure is doubled in one direction, giving rise to the so-called (2 × 1) reconstruction. A similar reconstruction is obtained for the diamond (111) surface, which is called the 'Pandey-chain' reconstruction (see figure 5.12(d)). There are, however, also possibilities for other types of surface reconstruction. When a (111) surface of Si(s) is heated to a sufficiently high temperature under ultra-high vacuum (UHV) conditions, the surface atoms will become quite mobile. As such, it will be possible for the surface atoms to rearrange into a more energetically stable configuration. Thus, instead of obtaining the very simple pattern shown in figure 5.12(b), the new arrangement involves a much larger unit cell with repositioning of the atomic positions in the top three atomic layers. A model of this new 7 × 7 surface structure is shown in figure 5.13.

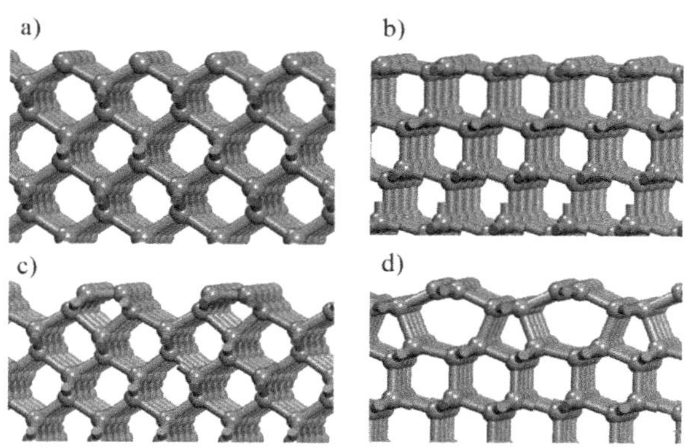

Figure 5.12. Surface structures of (100) and (111) diamond surfaces: (a), (b) before and (c), (d) after surface reconstruction.

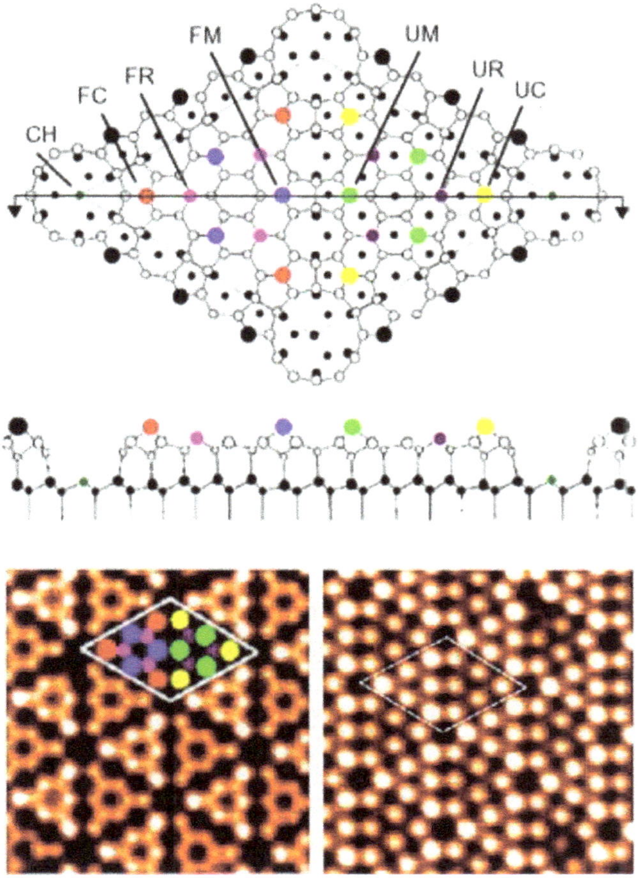

Figure 5.13. 7 × 7-reconstruction of Si (111). [2], reproduced with permission © Elsevier.

Colour-coding indicates different types of Si atoms in the top three atomic layers. The most prominent features are the yellow adatoms (i.e. adsorbates). Corner-holes are also shown in the corners of the 7 × 7 unit cell, where atoms from the top three layers are missing. There has been a focus on the reconstruction of clean surfaces in the present chapter. However, it is worth noting that the reconstruction of a solid surface is frequently induced by the adsorption of molecular or atomic species. This phenomenon is known as adsorbate-induced reconstruction.

This chapter will now end with the following exercise:

Exercise 6: Why do covalent compounds show a more pronounced surface reconstruction (as compared with metals)?

Answer: Covalent compounds have directed bonds that are quite stiff. The only way for a covalent material to change its atomic positions is, therefore, to create new covalent bonds. Everything in between is very unstable. Larger changes in atomic positions will, therefore, take place in predominantly the x- and y-directions. In comparison, the bonds in a metal are non-directed, so the metallic structure can more easily adapt to, e.g. higher temperatures and pressures. More precisely, the metal atoms can more easily move from their stable positions.

Exercise 7: What is the main driving force behind the surface reconstruction of ionic crystalline dipole surfaces?

Answer: The ionic surface strives to undergo charge compensations in trying to reach locally neutral surface areas (i.e. areas that are as stable as possible). Faceting is one of the atomic level processes that the system is using, which is the most common reconstruction process for these types of materials.

Exercise 8: For a covalent material, sketch the fcc(110)–2 × 1 and bcc(100)–2 × 2 reconstructions.

Answer: The unit cells of fcc and bcc are shown in figures 5.14(a) and 5.15(a). Let us now extract the (110) and (100) surfaces from the fcc and bcc unit cells, respectively (see figures 5.14(b) and 5.15(b)). Figures 5.14(c) and 5.15(c) show what a 1 × 2- and 2 × 2-reconstruction of the (110) and (100) surfaces, respectively, will look like.

The following references are strongly recommended for those readers that would like to learn more about surface planes with their relaxation and reconstruction processes [3–5].

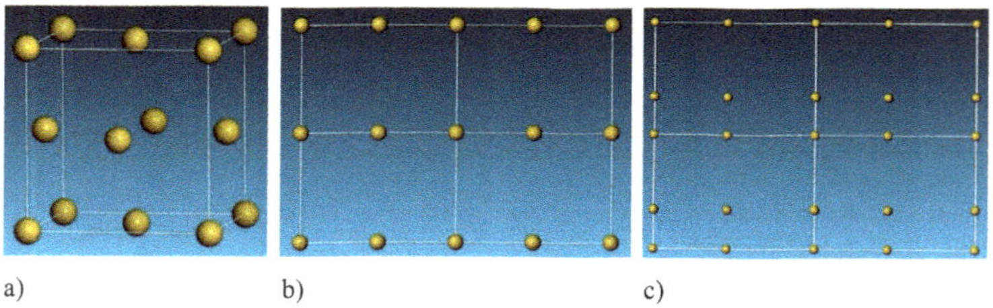

Figure 5.14. Formation of a fcc(110)–2 × 1 surface reconstruction of a covalent material surface: (a) bulk material, (b) (110) surface, and (c) 1 × 2-reconstructed (110) surface. Only the surface atomic layer, in four adjacent unit cells, is shown in (b) and (c). The yellow balls do not specify any specific element in the periodic surface.

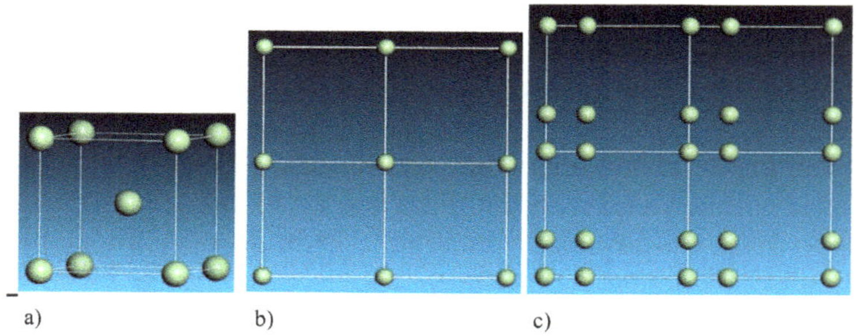

Figure 5.15. Formation of a bcc(100)–2 × 2 surface reconstruction of a covalent material surface: (a) bulk material, (b) (100) surface, and (c) 2 × 2-reconstructed (100) surface. Only the surface atomic layer, in four adjacent unit cells, is shown in (b) and (c). The yellow balls do not specify any particular element in the periodic surface.

5.7 Construction of model surfaces for CVD simulations

5.7.1 Initial plans and preparations

It is utterly important to choose an optimal combination of theoretical parameter qualities and model size in the modeling and simulations of various types of CVD processes. It is tempting to say that if this cannot be done, one must forget about those calculations. The reason is that the results will not be trustworthy, and most probably have some major errors. In the following part of the chapter, there will be a more practical hands-on instruction about how to construct a planar crystalline surface, which could either be the substrate or the thin film in the CVD growth of a specific material (see section 5.8). This instruction is also applicable for the construction of other types of model structures, also amorphous ones.

The first steps to take in the construction of a model surface is the determination of theoretical parameters and eventual surface reconstructions.

5.7.1.1 Choice of theoretical parameters

The best way to determine the most optimal theoretical parameters is to do a geometry optimization of the corresponding bulk material. A bulk supercell is, thereby, constructed where both the atomic positions and the cell parameters will be completely relaxed during the DFT calculations. For most solid-state materials, these data have already been experimentally obtained and published. It is then easy to do some test-calculations (by using DFT), where the theoretical parameters are successively improved until a desired agreement between theoretical and experimental results have been achieved. It is not necessary to go beyond this level of theoretical accuracy since the calculation times would then increase to a large extent.

If there are no experimental crystalline data for the material of interest, it will become a bit tricky. One way is then to look for a similar type of material, for which the necessary structural data is known, and do these test-calculations for that material instead. For a material to be similar, both atomic sizes and electronegativity values should be similar. The obtained crystalline data is then a plausible choice for the unknown structure of the material of interest.

5.7.1.2 Occurrence of surface reconstructions

In constructing a specific crystalline surface, it is very important to have clear information about the bonding types. Is it mainly metallic, ionic, or covalent? Or is there a combination of these bonding types? The reason is, if this material is a metallic compound, or a non-polar ionic compound (i.e. an oxide), then we can exclude the presence of surface reconstructions. However, surface reconstructions are highly expected if there are covalent bonds in the material, or if the material is polar with ionic bonds (see section 5.7). Moreover, when modelling a semiconducting material surface that is built from covalent bonds, it is important to know as much as possible about the experimental CVD conditions (like temperature and pressure). These conditions are very important for the formation of reconstructed surfaces. Since our aim is to construct a surface that is as realistic as possible (i.e. compared to the experimental structure), we have to include the possibility for surface reconstructions in the construction of the model surface.

5.7.2 Choice ofinitial surface structure

As discussed in section 4.4, irrespective of optimization method used in the calculations of the most stable ground state geometry, it is very important to start with atomic positions that are close to final ones. This will speed up the calculations to a large extent and might also avoid optimization to a terrace point (i.e. an energy maximum). However, it is most often extremely difficult to make any prediction about the most stable positions of the atoms in the system. It is only possible to make suggestions based on the chemical common sense. However, the choice of geometry optimization technique will be of great help. It is advisable to start with a steepest descent method (see section 4.4), with which it is possible to rapidly approach an energy minimum (but unfortunately converges very slowly). When getting close to the energy minimum, it is better to go over to, e.g. the pseudo-Newton–Raphson method by which it is possible to quickly converge to the energy minimum atomic position. For example, the DMol3 software (BIOVIA, Inc.) can use several different algorithms for performing the optimization. These include the steepest descent, conjugate gradient, and Newton–Raphson methods. These are not under user control. DMol3 will choose the appropriate method automatically.

In addition to a careful choice of initial atomic positions and geometry optimization procedure, it is also possible to 'play with' the choice of theoretical parameters. For instance, it is for larger systems advisable to start the geometry optimization calculations by using the LDA method. This optimized structure can, thereafter, be used as an input structure in a second round of geometry optimization calculations by using the more accurate GGA method. This will speed up the total calculation time appreciably. With these theoretical tools and strategies, we are now ready to build the most optimal model surfaces.

5.7.3 Construction of a surface model

A supercell is most often used in the construction of a crystalline solid-state surface model (figure 5.16(a)). As can be seen in figure 5.16(b), infinite identical two-dimensional surface slabs will be formed when translating the supercell in the *x*-, *y*-, and *z*-directions. These slabs are separated with something called a vacuum layer, which has to be large enough to diminish the interaction between neighbouring slabs. In addition, the thickness of the supercell slab, as well as the surface area thereof, must be large enough to simulate an optimal model of the problem at hand. This infinite three-dimensional model is generally used in the DFT-calculations of geometrical structure, energy, and materials properties of solid-state materials.

5.7.3.1 Thickness of a surface slab

In the decision of the surface slab thickness, the upper surface atoms should not experience the artificial cutoff of the lattice at the lower surface of the slab. As a guidance, the depth of the combination of surface relaxation and reconstruction

Figure 5.16. (a) Supercell with a solid-state material. (b) Surface slabs formed by translating the supercell in the x-, y-, and z-directions. C=grey; H=white. Bulk diamond and an H-terminated diamond (111) surface are exemplified in (a) and (b), respectively.

should be chosen as a starting thickness. One should, from there on, successively add one atomic layer and, for each of these additions, calculate a specific surface property (e.g. electronic structure, reaction energy (like adsorption energy), etc). The final, and most optimal, surface is the one where the calculated properties for two successive layer additions are within a pre-set threshold value. This procedure is expected to converge very fast when starting from the depth of surface relaxation/reconstruction. However, if for practical reasons it is not possible to start at this depth, a convergence will be more difficult to reach. It will, at least, take longer. One of the main questions is also: which material property to use when doing these test-calculations? As a piece of advice, use the properties that are closest to the process that you finally want to study. We have started here from a very thin diamond slab containing only two C layers. The material system is diamond, which is a covalent material with directed and very strong covalent bonds. Furthermore, the surface plane is an H-terminated non-reconstructed (111) plane. If we leave the surface slab in this condition, the lowest surface C atoms will be radical once. Thus, they have non-paired electrons, which can cause problems with the calculations (like prolonged calculations times). These lower surface C atoms have, therefore, been terminated with H atoms (with electronegativity values close to the C atom). Figures 5.17–5.19 demonstrate the search for a surface slab thickness by studying the pDOS spectra and H adsorption energy for various slab thicknesses. The models of H-terminated diamond (111) surfaces, with thicknesses of 2, 4, 6 and 8 C layers, are demonstrated in figure 5.17, and the corresponding pDOS spectra, and H-adsorption energies, are shown in figures 5.18 and 5.19, respectively.

The pDOS spectra for various thicknesses of an H-terminated (111)–1 × 1 surface of diamond (two, four, six, and eight C layers) are shown in figure 5.18. These spectra include the two upper C layers, in addition to its terminating H layer. Moreover, the

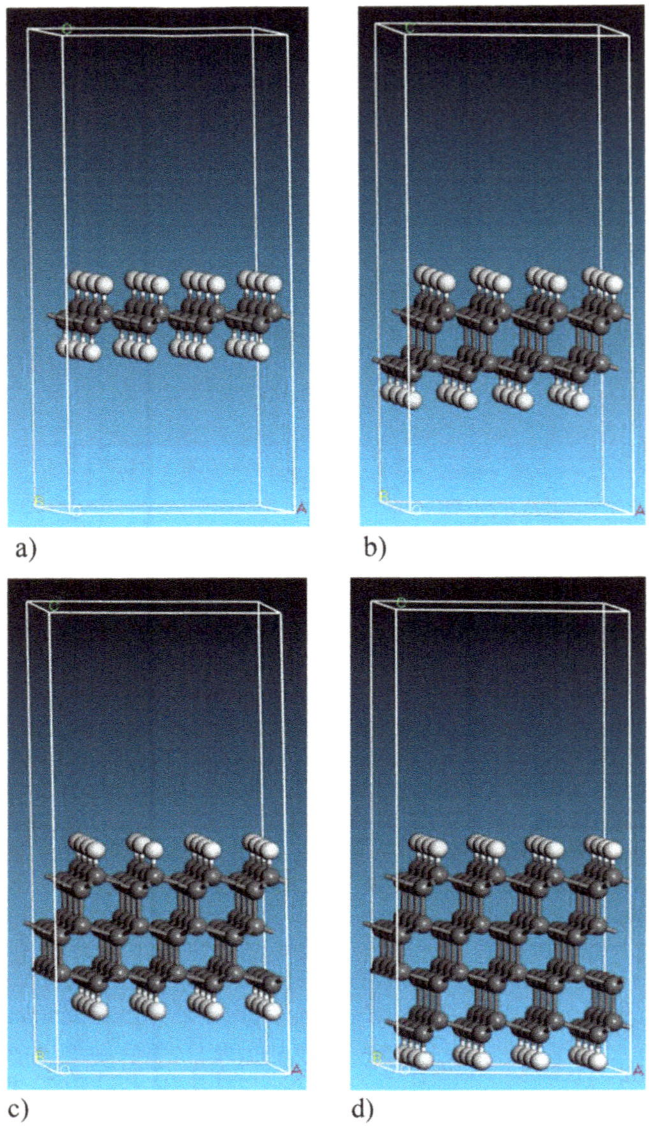

Figure 5.17. Models of H-terminated diamond (111) surfaces with thicknesses (a) two, (b) four, (c) six, and (d) eight C layers. C=grey; H=white.

contributions of the s and p electrons are shown in blue and red, respectively. In addition, the valence band maximum (VBM) is set to 0 eV and the respective conduction band minimum (CBM) is 2.77, 2.56, 2.48, and 2.45 eV for a two, four, six, and eight C-layer thick diamond (111) surface. Thus, there is no convergence in results, which means that one has to continue studying even thicker diamond slabs (until the necessary convergence in VBM values has been obtained). However, the

Figure 5.18. pDOS spectra for various thicknesses of an H-terminated (111)–1 × 1 surface of diamond (see figure 5.17); (a) two C layers, (b) four C layers, (c) six C layers, and (d) eight C-layers. The spectra include the two upper C layers, in addition to its terminating H layer. Blue: s electrons; red: p electrons.

calculated H adsorption energies are not that sensitive. As can be seen in figure 5.19, a convergence in adsorption energies has been achieved for diamond slabs as thin as four C layers. It is then tempting to draw the conclusion that a diamond (111) surface model of this size is optimal to use for further calculations. But this is apparently wrong (as clearly demonstrated by the pDOS spectra in figure 5.18). A friend of order might suggest that we, therefore, need to use several decimals in the adsorption energy values. However, this is not a good idea since we then cross the border to results that are within the error of limit (i.e. the expected accuracy of DFT-calculated adsorption energies is of the order of 1 kJ mol^{-1}). As a conclusion, the pDOS spectra is the best theoretical tool to use when looking for the most optimal size of the material model.

Another aspect of these types of test-calculations is that one should also look for the most optimal number of geometry-optimized atomic layers. Hence, for a given thickness of the surface slab, we should look for the most stable model (i.e. lowest total energy, *E*) by using different numbers of optimized atomic layers. It is, though, not necessary to use more atomic layers than necessary since that would increase the

Figure 5.18. (Continued.)

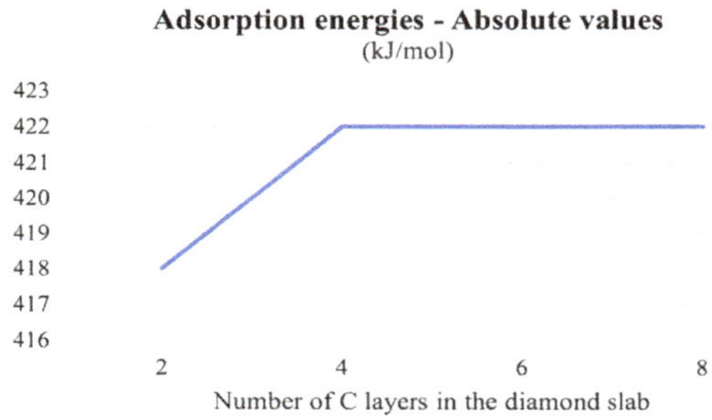

Figure 5.19. Evolution of the H adsorption energy for an increasing number of atomic C layers in an H-terminated diamond (111)–1 × 1 surface.

calculation time to a large extent. The evolution of total energy as a function of geometry optimized atomic layers, is shown in figure 5.20. We have here used the thickest diamond (111) slab in figure 5.17, but in reality we should use the thickness corresponding to a convergence in pDOS spectra.

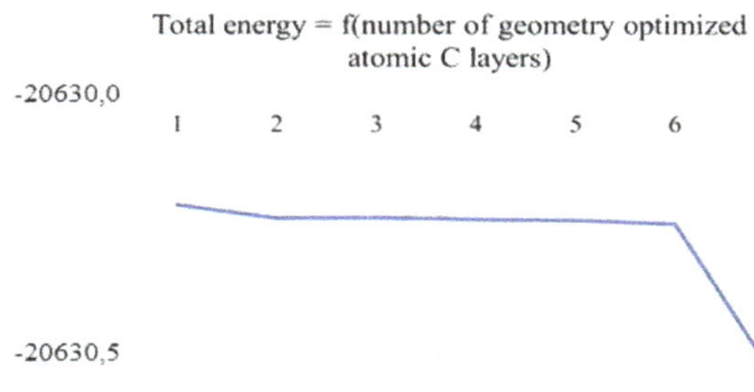

Figure 5.20. Evolution of the total energy (in eV) for an increasing number of geometry-optimized atomic C layers in an H-terminated diamond (111)–1 × 1 surface of thickness 8C layers.

As can be seen in figure 5.20, the total energy, *E*, is slowly declining until we reach an optimization of the upper six layers of C atoms. It, thereafter, takes a major jump. It might also here be tempting to believe that the total energy is slowly decreasing to a convergence if we only consider one to six optimized layers. However, the jump between six and seven layers shows us how wrong this might be. Thus, it is of the largest importance to include a large enough number of optimized layers. The major jump between six and seven can be explained by the sudden release of necessary degrees of freedom (for the optimal optimization to take place). Here, we can draw the conclusion that we need to increase the diamond (111) slab thickness even further. Thus, eight C-layers is a too thin H-terminated diamond (111) surface model.

Here, we have learnt that we must choose even thicker diamond (111) slabs and continue with even more sequential geometry optimizations and pDOS analysis, until convergence have been reached. The surface model thickness has now been carefully considered, and the next step is to study the requirements on the supercell surface area.

5.7.3.2 Size of surface model area

The first demand on the surface area is that it must be large enough to cover the surface reconstruction. One should also consider the fact that all supercells are identical. This means that if we want to study, e.g. the adsorption of F on a diamond surface, there are identical F adsorbates on a cell edge-distance away (both in the *x*- and *y*-directions). Consequently, if we want to calculate the adsorption of F on an otherwise non-terminated diamond surface, we have to do similar test-calculations as we did in the process of finding the optimal surface slab thickness. In other words, the supercell surface area must be big enough for the F adsorbates to not experience the ones in the neighbouring supercells. The strategy is then to start the adsorption energy calculations by relatively small values of *x* and *y*, and then to repeat these calculations for larger values (i.e. for a larger supercell surface area). The optimal *x*- and *y*-values are found when the difference in adsorption energies is smaller than a pre-set threshold value (i.e. when a convergence has been met).

The procedure has, otherwise, to continue with new x and y values. The results from this iterative procedure are shown in figure 5.21 for an (111)–1 × 1 surface of diamond. This is a 3D diagram that demonstrates the calculated adsorption energies for increasing x and y supercell edge lengths.

As can be seen in figure 5.21, it is obvious that the F adsorption energy (in absolute values) is not continuously increasing when increasing the supercell edges in the x- and y-directions. It is instead varying in a more irregular manner. This rather unexpected behaviour is due to the geometry optimization of the model systems. The adsorption energies are calculated by using equation (4.63) (see section 4.5). It is then essential to allow for surface relaxation and reconstruction for each model system in this reaction energy expression. This is being done by performing geometry opimizations in the calculations of the included energy terms. The reason is that relaxations and reconstructions occur at experimental condition during the course of adsorption. This is something that cannot be avoided. Hence, the contribution to the adsorption energy by the new surface–adsorbate bond is only partial. Another main contribution is the relaxation and reconstruction—both of the surface with the adsorbate and of the surface with a radical site (i.e. E_{S-H} and E_S in e.g. (4.63)). In case the adsorbate is not a single atom, but a molecule, one must consider the geometry optimization of this molecule as well. In conclusion, the adsorption energy is not identical to the surface–adsorbate bond strength! And this is the reason why we see these F adsorption energy fluctuations in figure 5.21.

However, it is possible to estimate the surface–adsorbate bond strength by instead omitting the geometry optimizations when calculating the adsorption energy. Thus, it is then advisable to do single-point calculations, and the result from those type of calculations for the present system example can be seen in figure 5.22. As expected, the abs. value of F adsorption energy increases for larger F–F distances. The increase in energy is, however, minor since the electronic repulsions between neighbouring F adsorbates are also minor.

Another way to estimate the surface–adsorbate bond energy is to analyze the ionic part of the bond strength by using the calculated atomic charges of the atoms

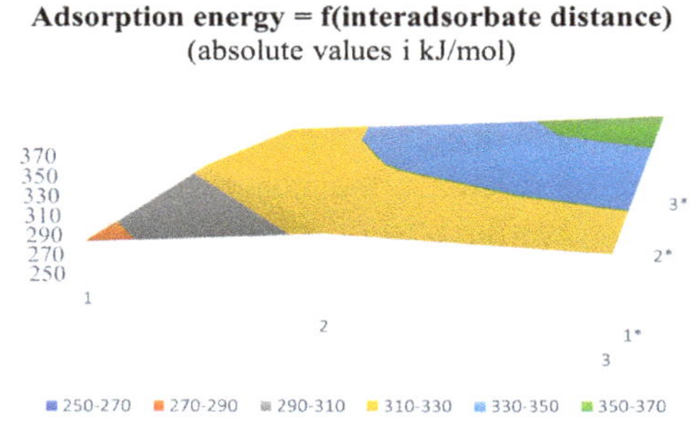

Figure 5.21. Variations in F adsorption energy for increasing x and y supercell cell edges for a (111)–1 × 1 diamond model. The models are geometry optimized. x: 1, 2, and 3 = 2.5, 5.0, and 7.5 Å, respectively. y: 1*, 2*, and 3* = 2.5, 5.0, and 7.5 Å, respectively.

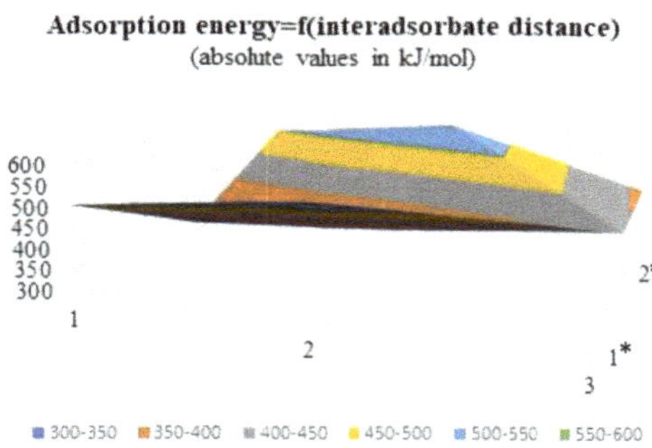

Figure 5.22. Variations in F adsorption energy for increasing x and y supercell cell edges for a (111)–1 × 1 diamond model. The models are not geometry optimized. x: 1, 2, and 3 = 2.5, 5.0, and 7.5 Å, respectively. y: 1*, 2*, and 3* = 2.5, 5.0, and 7.5 Å, respectively.

that are involved in the bond formation. One must then also analyze the covalent part by calculating the electron density of the bond (i.e. by performing a population analyses).

The question is now: how to construct a surface model for the study of an H adsorption at various surface coverages (i.e. at different adsorption concentrations)? When varying the adsorbate concentration, it is tempting to use one adsorbate and change the adsorbate concentration by just varying the supercell x and y values (instead of using a much larger cell size). However, this is not always correct. Let us have a look at this problem by studying an example. Assume that we are interested in studying the adsorption of H to the B-covered surface of c-BN (100)–2 × 1 at a surface coverage of 1/4. Let us also use two different strategies when calculating the adsorption energy of one H to this type of surface.

Strategy 1: Let us start with the adsorption of H to a large supercell area that is free of other H adsorbates and, thereafter, decrease it in specific steps until the surface coverage of 1/4 has been reached. It must be stressed here that we refer to the total solid-state surface when considering the surface coverage, and not only the supercell surface (see figure 5.23). However, the drawback with this strategy is that each H adsorbate is identical to the 'neighbouring' ones (i.e. in the neighbouring supercells). This means that each of these adsorbates have identical geometrical and chemical surroundings. The advantage with this strategy is that we do not need to use a very large supercell size for the adsorption energy calculations. Hence, the calculation time will become decreased (as compared to strategy 2). For the final 1/4 surface coverage, the calculated H adsorption energy for this type of strategy is -181 kJ mol^{-1}.

Strategy 2: Here we start by directly using a large supercell, onto which we can cover the surface with one fourth of H adsorbates. In doing so, we separate the adsorbates as much as possible over the surface, also taking into consideration the H adsorbates in the neighbouring supercells. The reason for this behaviour is that all H adsorbates are positively charged and, thus, experiences repulsive inter-adsorbate forces on the surface.

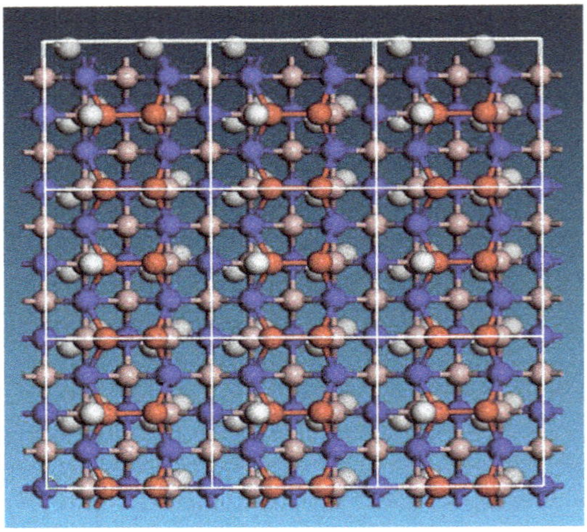

Figure 5.23. A final surface coverage of 1/4 by tuning the supercell size of the model of a B-covered H-terminated c-BN (100)–2 × 1 surface. Nine supercells are shown in this figure. The surface B atoms are coloured red. B, N, and H are otherwise shown in blue, pink, and white, respectively.

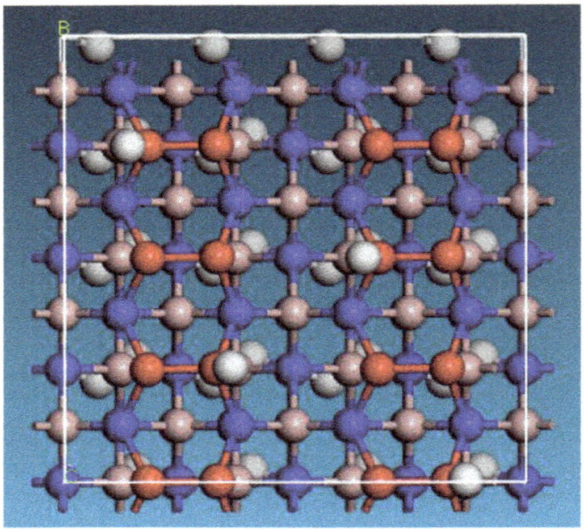

Figure 5.24. Model of a large B-covered H-terminated (100)–2 × 1 diamond supercell surface with a surface coverage of 1/4. The surface B atoms are coloured red. B, N, and H are otherwise shown in blue, pink, and white, respectively.

This model is shown in figure 5.24. The average adsorption energy for the H adsorption here is −203 kJ mol^{-1} for the adsorbates H1, H2, H3, and H4, respectively. The advantage with this strategy is that each adsorbate can be treated as an individual adsorbate, which is most often the situation on a real surface. In other words, they are not identical with each other, experiencing different environments and relaxation/reconstruction conditions. However, the drawback is the longer calculation time (compared to strategy 1).

References

[1] Skårman B, Wallenberg L R, Larsson P-O, Andersson A, Bovin J-O, Jacobsen S N and Helmersson U 1999 Carbon monoxide oxidation on copper oxide thin films supported on corrugated cerium dioxide {111} and {001} surfaces *J. Catal.* **181** 6–15

[2] McNab I R and Polanyi J C 2011 Imprinting atomic and molecular patterns *Frontiers of Nanoscience* vol 2, ed A J Mayne and G Dujardin (Amsterdam: Elsevier) ch 4 pp 79–120

[3] Somorjai G A 1994 *Introduction to Surface Chemistry and Catalysis* (New York: Wiley)

[4] Attard G and Barnes C 1998 *Surfaces* (Oxford: Oxford Science Publications)

[5] McCash E M 2002 *Surface Chemistry* (Oxford: Oxford University Press)

IOP Publishing

Chemical Vapour Deposition
Growth processes on an atomic level

Karin Larsson

Chapter 6

Thermodynamic modelling of CVD growth processes

The chemical vapour deposition (CVD) of a thin film material is a very complex and dynamic process. It involves elementary chemical reactions, such as the adsorption of growth species, surface migrations, nucleation and general chemical reactions on the growing surface, and desorption of residual species. There is a large variety of each of these elementary steps. Take the growth of diamond (111) surface as an example. It has to be terminated with, e.g. H atoms, since it will otherwise collapse to the graphitic phase. However, these terminating species must be abstracted away from the surface to leave room for an incoming growth species (e.g. CH_3). Moreover, in the initial part of the diamond growth process, the abstraction of two surface H species (terminating the surface and one of the H ligands on CH_3) has to take place before any surface migrating will occur, ending with a final incorporation at a step edge.

Theoretical modelling has been found to be extremely powerful in the determination of the most plausible CVD growth mechanism of various thin film solid materials.

6.1 General

As discussed earlier in the present book, one must be very careful when performing theoretical modelling and simulations. One should always, as much as possible, compare with experimental results and tabulated data. In addition, one should also be well aware of any types of errors that might occur. The most important thing to remember is that it is never possible to construct a macroscopic model, press a button, and then just witness the evolution of, e.g. a complete CVD growth process on the computer screen (or any other type of chemical process). This is not possible due to two major contradictory restrictions. One must use quantum mechanical methods for an accurate description of bond formations and ruptures in the course of the growth

doi:10.1088/978-0-7503-3107-4ch6

process, and the model cannot be very large when using this type of theoretical accuracy (i.e. the calculation time is proportional to N^3 when using ordinary density functional theory (DFT) methods). Thus, a mixture of theoretical methods is often required when performing simulations of, e.g. a CVD growth process. The CVD process must be broken down into its components and each of these components should be investigated by using the most appropriate theoretical method for this purpose. The connection of these components will thereafter require a good integration of the results. (It is, in this sense, similar to the experimental characterization methods.) As an example, the following is the strategy to use when calculating the adsorption energy of the growth species when attached to the growing CVD thin film;

- Choose an appropriate theoretical method, with parameters.
- Ask yourself, is this method compatible with the model size?
- If yes → Continue with the planned calculations!
- If not → Choose another theoretical method that is not that computational demanding but still gives trustworthy results (e.g. DFTB [1]). The DFTB method needs parametrizations which can be constructed by using the DFT method [1].

Some third-party software producers offer a range of theoretical calculation methods for the whole range of model size, from the smallest molecular level to the largest macrolevel. As can be seen in figure 6.1, there is a range of five different

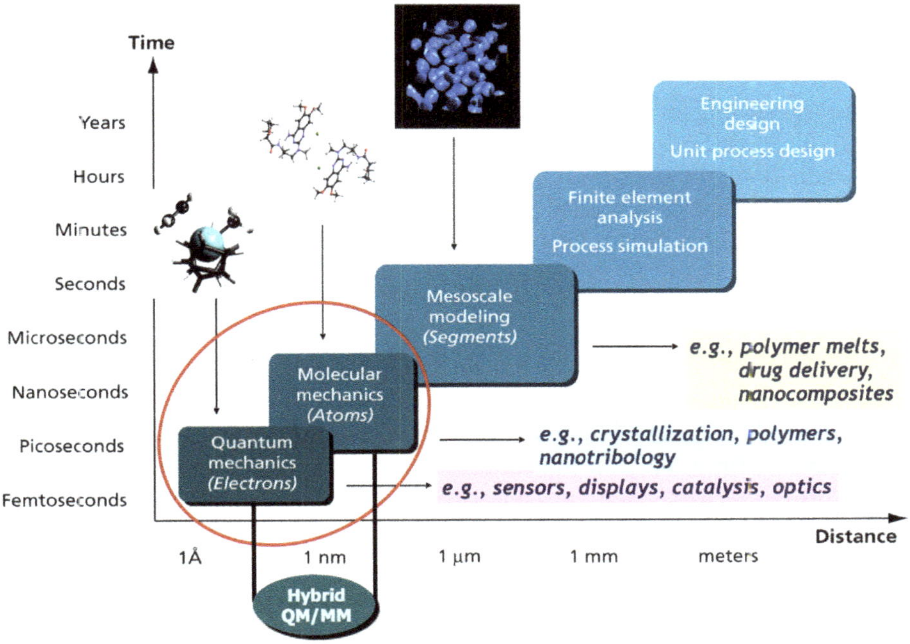

Figure 6.1. Overview of a continuum of theoretical methods for various length and time scales. Examples are also given for the different types of systems that can be studied by using these methods (BIOVIA, Inc.).

groups of theoretical methods that together cover model sizes from Angstrom to meters. In addition, when using these methods in molecular dynamic (MD) simulations, together they cover time scales from femtosecond to years. Thus, each of these length- and time-scales require special modelling techniques. The danger, though, is that each of these types of methods creates its own 'island of' information. It is therefore important to use multiscale modelling schemes, which are demonstrated in figure 6.1 by the pairwise overlap of method types. In this scheme, each lower-level method is validated and fed into the higher-level method. For example, quantum mechanics calculations can be used to construct and validate the computational parameters that are used in molecular mechanics. In short, the theoreticians must seek to understand their problem through a variety of techniques, and to combine the results in building a picture of the system that is being studied.

The method types in figure 6.1 include various theoretical techniques. For instance, the quantum mechanical toolbox includes theoretical methods like DFT, density functional tight-binding (DFTB) methods, and extended Huckel (EH) methods, of which the latter two need parametrization. Each method in the different groups can be used in the modelling scheme presented above.

6.1.1 Finite element analysis and engineering design

Finite element analysis and engineering design include a range of classical methods that model beyond the nanoscale [2].

6.1.2 Mesoscale modelling

Mesoscale modelling of materials or fluids does not include any molecular-level details [2]. This method is, instead, using segments (like monomers) as the smallest unit. As such, it will be possible to effectively model systems on the 10 nm to 100 nm scale. This type of method is particularly useful for studying mechanical properties. Other examples of systems and processes to study are polymer melts, drug delivery, and nanocomposites.

6.1.3 Molecular mechanics

The atoms are treated as distinct balls in the molecular mechanics method. When the smallest unit considered in quantum mechanics is the electron, it is the atom when working with molecular mechanics. Thus, molecular mechanic methods are usually approximations that are based on simplified models of the molecular structure and the forces that govern them [2]. However, this type of method can effectively predict many material properties that are inaccessible to quantum mechanic techniques. By using the DFT method for the construction of the necessary parameters, or using experimentally obtained ones, quite accurate results can be obtained. Examples of systems and processes to study are polymers, nanotribology, and crystallization.

6.1.4 Quantum mechanics

Quantum mechanical methods, such as DFT and Hartree–Fock (HF), mainly require the position and atomic number of each atom as an input to the calculations. They are based on a pre-defined methodology, containing no adjustable parameters, and a solution to the Schrödinger equation is approximated (see chapter 4). These methods work for, in principle, every element in the periodic table, and a vast amount of literature demonstrates their accuracy. Furthermore, the results of these calculations include the positions of all atoms in the system, the forces on them, the electronic structure (i.e. a description of the 'electron cloud'), and the energy of the complete system. Furthermore, other key properties can be derived from these basic results.

Examples of quantum mechanical DFT calculations, with their applications and results, are demonstrated in figures 6.2–6.4. As can be seen in figure 6.2, a bucky ball (i.e. C_{60}) between a STM tip (made of W) and a copper surface will induce an current amplification. Moreover, figure 6.3 demonstrates the binding between a nanotube and various metal surface. It is obvious that the nanctube binds the

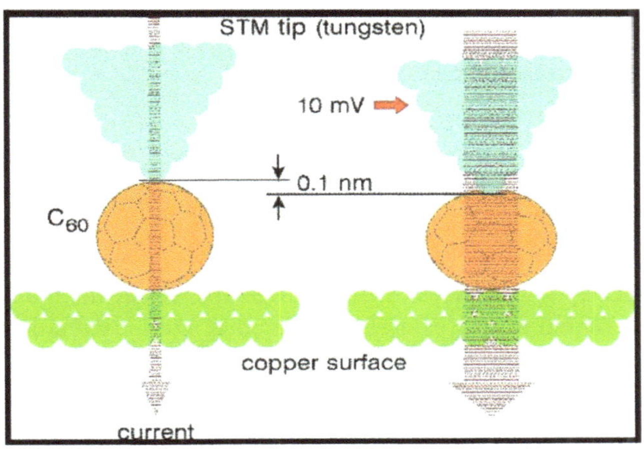

Figure 6.2. Molecular amplifier (BIOVIA, Inc.).

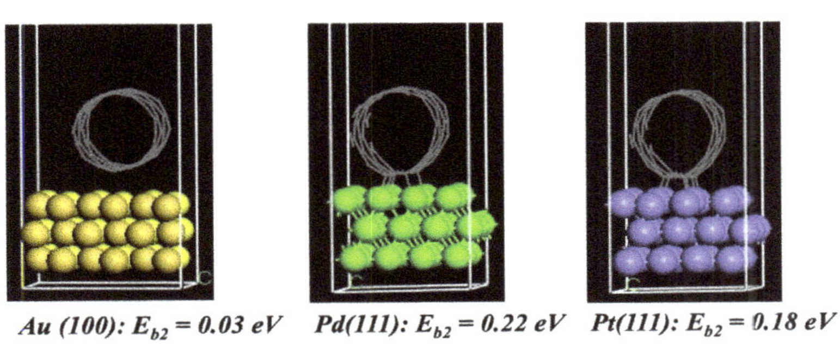

Figure 6.3. Binding of nanotubes to various metal surfaces (BIOVIA, Inc.).

a)

	Ionisation potential (eV)	Position of HOMO (eV)
Free nanotube	6.4	- 5.1
1 H_2O on 1 nanotube	6.3	- 5.0
5H_2O on 1 nanotube	5.8	- 4.5

b)

Figure 6.4. (a) HOMO levels are demonstrated for the nanotube tip with 0, 1, and 5 adsorbed water molecules. (b) Ionization potential and position of the HOMO energy level for the models presented in (a) (BIOVIA, Inc.).

strongest to the Pd(111) surface, and the weakest to the Au(111) surface, with the Pt(111) surface in between. Figure 6.4 shows the effect by adsorbed water molecules (onto a reconstructed nanotube tip) on the electronic properties of this nanotube. The ionization potential is of a special interest to study. It is important for applications that rely on the electron emission properties of the material of interest.

The above presented examples make it clear that DFT calculations play a major role in the field of theoretical modelling and simulations. This type of quantum mechanical method has therefore been chosen as the main method in the present book. Moreover, two different materials have been predominantly used in demonstrating the theoretical modeling of various CVD growth processes; diamond and cubic boron nitride (c-BN). These materials are covalent, and the first one contains only one atomic type (C) while the other contains two (B and N). There are also other dissimilarities. Diamond is very easily grown by using CVD at low temperature and pressure, while c-BN is not. In the present book, the reason for this choice of strongly covalent materials (as demonstration materials) is that they are expected to undergo severe surface reconstructions. Furthermore, the reason for the choice of only two materials is that it is easier to demonstrate the methodology behind the modeling of the CVD growth processes by using a single (or just a few) material types.

6.2 Stability of non-terminated surfaces

As presented in section 5.4, one of the factors that is decisive for the surface reactivity is the number of binding neighbouring atoms for a surface atom. The number of unpaired electrons (i.e. dangling bonds) at a surface atom is, consequently, determined by the number of binding neighbours. Hence, the resulting chemical properties of a surface do mainly depend on the number of dangling bonds on the surface. Let us take diamond as an example. Mainly two facets are formed when growing diamond by using a CVD technique: the (111) and (100) facets. Bulk diamond C atoms have four covalently binding C neighbours in the formation of a sp^3 hybridization. Let us now assume that we have a newly formed diamond (111) or (100) surface plane. The (111) plane has only one dangling bond whilst the (100) atoms have two (see figures 5.13(a) and (b)). Hence, the (100) surface is expected to be more reactive. To lower the surface energy, the diamond (100) surface undergoes a surface reconstruction by forming a 2 × 1 surface structure (figure 5.13(c)). Also, the more ideal (111) surface undergoes a 2 × 1 reconstruction at higher temperatures, thereby forming a Pandey chain structure (see figure 5.13(d)).

The relative total energies for these surface structures have earlier been calculated by using DFT methods [2]. It was shown that the 2 × 1-reconstruction of the bare (i.e. non-terminated) (100) surface results in a stabilization by 1.85 eV per surface C (see figure 6.5). The corresponding 2 × 1-reconstruction of the bare (111) surface (i.e. a Pandey-Chain reconstruction) resulted in an energy stabilization of 0.54 eV per surface C. These results strongly support our prediction that the bare diamond (100) structure is more reactive than the bare diamond (111) surface, and the (100)–2 × 1 surface is, most probably, the most stable one. As can be seen in figure 6.5, this

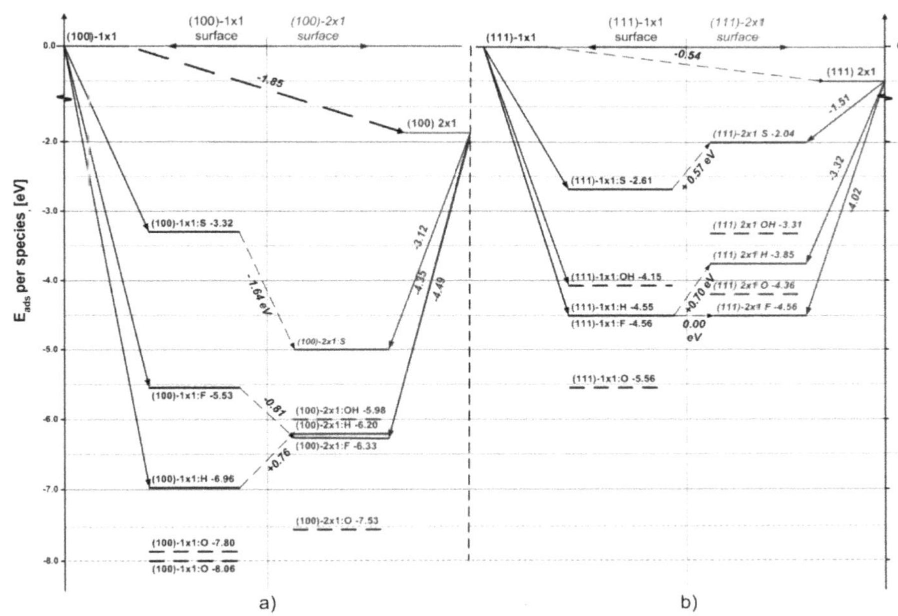

Figure 6.5. Adsorption energies for various terminating species (H, OH, O, F and S) attached to (1 × 1)- and (2 × 1)-reconstructed diamond (a) (100) and (b) (111) surfaces, respectively. For these two surface types, the differences in total energy and adsorption energies are especially highlighted. The adsorption energies are presented as an energy per adsorbate, and the total energies per surface C.

prediction can be verified by the respective adsorption energies of H, F, O, which are all larger for the non-reconstructed (111) surface (as compared with the non-reconstructed (100) surface). It might also be tempting to assume that the (100)–2 × 1 surface is less stable than (111)–2 × 1. However, the respective adsorption energies for these surfaces show that this is not the case. The (100)–2 × 1 surface is still more reactive than (111)–2 × 1.

These stabilization energies do not, however, tell us anything about the kinetics of the transfer from the ideal diamond (111)–1 × 1 structure to diamond (111)–2 × 1. The results from the energy barrier calculations are shown in figure 6.6. (How to calculate these types of barrier energies will be presented in chapter 7.) As can be seen in figure 6.6, a very small energy barrier (also called activation energy) of 0.32 eV was found for the transfer from a bare (111)–1 × 1 surface to the slightly more stable Pandey-chain structure ((111)–2 × 1) [3]. This very small energy barrier is expected to be overcome at experimental CVD conditions (i.e. at higher temperatures). It is also possible to do DFT-based molecular dynamic (MD) calculations (also called qunatum mechanics MD (QMMD)) at an increased temperature [4]. The very small energy barrier was justified by using QMMD (on the diamond (111)–1 × 1 surface) at 500 K [3].

How to achieve information about the energetic stability of bare material surfaces (non-reconstructed and reconstructed) has been exemplified here by using a covalent diamond surface. However, this description can be used for any material type that undergoes reconstructions. The question is now, why is it important to also calculate the energetics of the non-terminated surfaces? Can they really exist in a CVD environment with a specific gas phase pressure? Yes, they can. It is very common

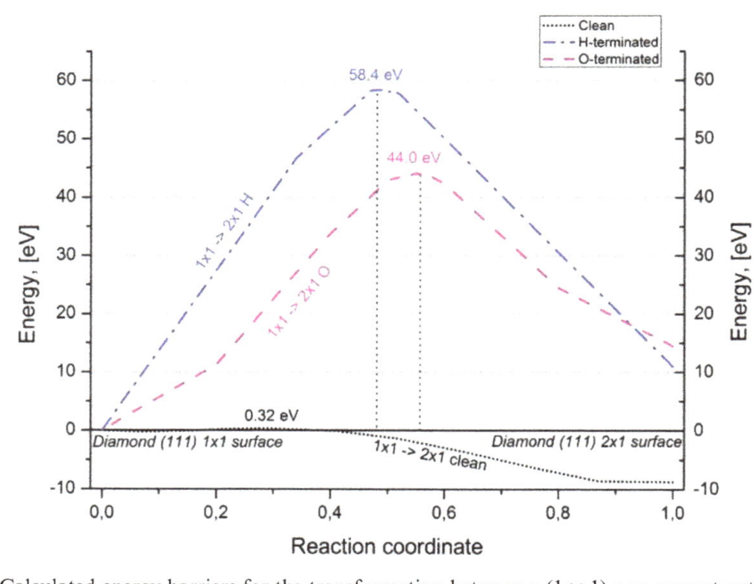

Figure 6.6. Calculated energy barriers for the transformation between a (1 × 1) non-reconstructed and a (2 × 1)-reconstructed diamond (111) surface. The surfaces are either bare, H-terminated, or O-terminated. [3], reproduced with permission © Elsevier.

that gas phase species (i.e. atoms or molecules) do not cover a surface to 100%. Information about the structure and energetics of these bare 'spots' is therefore very important in the understanding of the complete surface, with its detailed geometrical structure and properties.

6.3 Surface termination

6.3.1 Introduction

The termination (i.e. adsorption) of gaseous species to a growing CVD thin film is of the largest importance in the growth process. It has both physical and chemical effects on the material surface and is in practice not possible to completely avoid, not even at ultra-high vacuum (UHV). Let us now estimate how long it takes for a newly formed and bare diamond (100) surface to become completely terminated with H atoms at a temperature of 600 K and a partial H pressure of 40 Torr. (The SI unit for pressure is Pascal (Pa), but the Torr unit is most often used in practice.) Let us use equation (3.2) for these calculations (which is an expression of the collision rate and based on the classical expression for the collision rate). Let us also assume that the sticking coefficient, s, for the chemisorption of H to this specific surface is equal to 0.7, and that each H species adsorbs on top of a surface C atom:

$$\nu_{\mathrm{H,\ adsorption}} = \frac{p_{\mathrm{H}} s_{\mathrm{H}}}{\sqrt{2\pi k T m_{\mathrm{H}}}} = \frac{40*133.322*0.7}{\sqrt{2\pi*1.3806*10^{-23}*600*1*1.66053*10^{-27}}} = \frac{3733.0}{92.9*10^{-30}} = 40.2*10^{30}\ \mathrm{adsorptions/s} \tag{6.1}$$

The diamond (100) unit cell area is $3.556^2*10^{-20} = 12.645*10^{-20}\ \mathrm{m}^2$. The surface density of C atoms is, thereby:

$$\frac{2}{12.645*10^{-20}} = 1.58*10^{19}\mathrm{C\ atom/m^2} \tag{6.2}$$

Thus, by dividing the adsorption rate with the surface density of C atoms, we get the corresponding adsorption rate for the adsorption of a monolayer H per second:

$$\nu_{\mathrm{H,\ adsorption}} = 40.2*10^{25}\ \mathrm{adsorptions/s} = \frac{40.2*10^{30}}{1.58*10^{19}} = 2.5*10^{10}\mathrm{monolayers/s} \tag{6.3}$$

We can now use the following equation to get our final result:

$$2.5*10^{10} = \frac{1}{\mathrm{x}}\ \mathrm{monolayers/s} \tag{6.4}$$

As a conclusion, it takes $4*10^{-9}$ s = 0.000000004 s to completely cover a diamond (100) surface with H adsorbates.

As presented in sections 5.6 and 5.7, a newly prepared solid surface will increase its stability by undergoing surface relaxation and reconstruction. If the surface is subjected to a certain partial pressure, p_i, of gas phase species, i, the adsorption of i will also increase the stability of the surface. This is due to the increase in electron density at the sites of the surface atoms. The surface relaxation, reconstruction, and adsorption are then competing processes with a common goal, to lower the surface energy. Which process dominates depends on the material type and the specific surface plane.

6.3.2 Effect of surface termination on surface structure and properties

As presented above, surface termination is not only non-avoidable, it can also affect the surface structure such as phase and degree of surface relaxation/reconstruction. Taking H-adsorption as an example, the atomic H species are radical atoms that are very efficient as reduction agents in the chemical removal of, e.g. oxygen, O, from oxidized surfaces. They will, thereafter, at a reasonably low temperature, immediately adsorb to the bare surface, forming a more, or less, complete H-terminating atomic layer. It has been found that this type of surface termination is very important for the CVD growth of diamond. But before studying the effect of surface termination on the surface geometrical structure and properties, let us have a look at the most common adsorption sites.

6.3.2.1 Surface adsorption sites

There are different types of surface sites onto which species can adsorb, and those ones differ in adsorption probability (which is closely connected to their respective adsorption energy). The most common surface sites are on top, bridge, and hollow positions, which are demonstrated in figure 6.7. Generally, the number of adsorption bonds to these surface sites are one, two, and three, respectively. Consequently, the trend in increasing adsorption energy follows the same order: $\Delta E_{\text{ads}}(\text{ontop}) < \Delta E_{\text{ads}}$ (bridge) $< \Delta E_{\text{ads}}(\text{hollow})$. Consequently, the trend in increasing adsorption energy follows the same order: $\Delta E_{\text{ads}}(\text{ontop}) < \Delta E_{\text{ads}}(\text{bridge}) < \Delta E_{\text{ads}}(\text{hollow})$. However, the to-adsorbate bond type. Take atomic hydrogen, H, as an example. When it binds covalently to a surface it can only form one bond, and the most possible adsorption site is the on top one. However, it most often not possible to just 'guess' the energetically most prefered adsorption site (even though it might be a very clever guess). As a matter of fact, you must do geometry optimization DFT calculations.

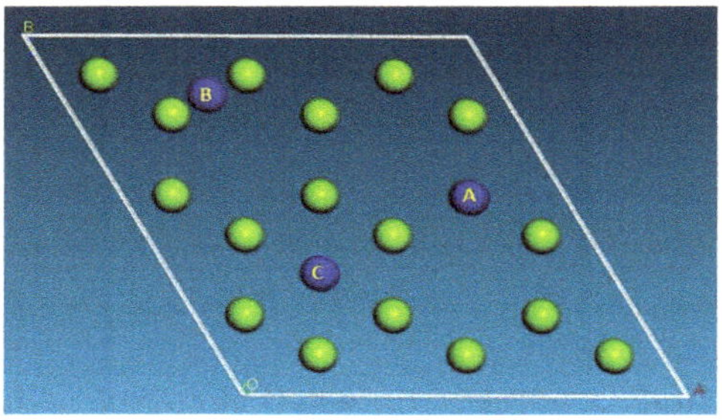

Figure 6.7. Three types of adsorption sites: (a) on top, (b) bridge, and (c) hollow.

6.3.2.2 Structural properties

Let us now continue to study the adsorption of atomic hydrogen, H, to a diamond surface. In addition, let us also assume that the diamond is grown by using a CVD reactor. This is an example of a situation where we want to deposit a specific material phase that is not stable at the temperature and pressure used in the CVD setup. Diamond represents this type of metastable material phase. Diamond, the cubic phase of C(s), is in fact less stable than graphite, the hexagonal phase of C(s), at lower temperatures and pressures (such as room temperature and atmospheric pressure). To avoid growth of the unwanted graphite phase one must terminate the cubic diamond surface with, e.g. H atoms. This is the only way to uphold the sp^3-hybridization of the surface C atoms. They will otherwise collapse to graphitic sp^2-hybridized surface atoms [5]. This is a quite easy way to grow a metastable phase of a material. (In addition, atomic H will also etch away the small amount of graphitic C that eventually is formed.)

In addition to the termination-induced phase stabilization, surface termination will also affect the degree and type of surface reonstruction [6]. This is most often called adsorption-induced surface reconstruction. As can be seen in figure 6.5, the most stable diamond (111) surface structure is the non-reconstructed (1 × 1) when terminating the surface with H, O, OH, F, or S species. All of these species are adsorbed on the on top sites, which is, in fact, the only type of adsorption site on this type of surface plane. (One can try another surface site, but the energetic constraints on the surface will then be too large, which most probably leads to final on top positions as a result of geometry optimization.)

As is also shown in figures 6.5 and 6.6, the total energy of the surface reconstructured diamond (111) surface (i.e. the (2 × 1)-reconstruced Pandey-chain structure) is somewhat lower than the total energy for the non-reconstructed one. One important question to ask is the following. Even though the terminated Pandey-Chain structure is less stable, does it have the possibility to transform to the terminated non-reconstructed counterpart? As can be seen in figure 6.6, the corresponding energy barriers for H- and O-termination are extremely large; 58.4 and 44.0 eV, respectively. Consequently, the probability for this type of transformation is very low (even at larger temperatures).

The situation is a bit different for the diamond (100) surface. Each surface C atom must be saturated with 2 H atoms for the non-reconstructed (1 × 1) surface. However, only one is needed for the (2 × 1)-reconstructed one. The second electron participates in the newly formed C–C bond. As can be seen in figure 6.5, the 2 × 1-reconstruction is more stable for the on top adsorption of F and S, but less stable for the adsorption of H. The latter is also the situation for ontop adsorption of O, as well as for adsorption of O to a bridge site. Moreover, it is not possible to terminate the non-reconstructed (100) surface to 100% with OH species. It will be too crowded. Thus, it is only possible to completely cover the 2 × 1-reconstructed (100) surface with OH adsorbates (in on top positions).

In short, what do figures like figures 6.5 and 6.6 tell us? Figure 6.6 contains information about the various adsorption energies (marked with solid lines). In addition, the dashed lines show the difference in energetic stability for the non-reconstructed and 2 × 1-reconstructed surfacees. Furthermore, figure 6.6 gives both information about the energetical stability of the non-terminated and terminated diamond (111) surfaces, and information about the possibility for a transformation between the non-reconstructed and reconstructed ones. *Hence, it is of an utmost importance to also include barrier energy calculations in the study of specific surface processes.*

6.3.2.3 Chemical properties

Despite the efficiency in maintaining a specific surface reconstruction and phase, different types of surface termination species have been shown to change the chemical and electronic properties of solid surfaces. 'Chemical surface properties' is a very broad term that influences processes like CVD growth, functionalisation, and anchoring to larger molecules (just to mention a few). Different types of terminating species have earlier been shown to change the properties of the diamond surface region (e.g. by inducing surface electronic conductivity and interfacial charge transfer processes). Another example of properties that may be affected by surface termination is the field emission characteristics. The electronegativity of the adsorbed species will thereby influence the emission current and threshold voltage [7]. Moreover, surface termination has also been observed to significantly influence the broad-band infrared reflectivity of diamond [8]. Furthermore, H-terminated diamond surfaces have been found to be hydrophobic [9], and to show unique p-type surface electronic conductivity [10]. On the other hand, O-terminated diamond surfaces do generally show hydrophilic properties [11], and to exhibit positive electron affinity [12]. In addition, F-termination will, compared to H-termination, result in even more strongly hydrophobic diamond surfaces, and have also recently been found to enhance the sp^3 characteristics of the diamond upper C layer (thereby improving the cubic structure of the surface [13]). All of these interesting properties of terminated diamond surfaces make it clear that surface termination is very important for a vast range of applications (e.g. as an electrode material). Similar to diamond, this is also the situation for other solid-state materials.

6.3.3 Calculation of adsorption energies

The process of adsorption will be discussed in more detail in the present chapter. The bond formation between the surface and the adsorbate is then assumed to be of a covalent, or ionic, type. We are, hence, only discussing chemisorption since these bond types are the ones that are involved in the CVD growth mechanism.

The adsorption energy for a gaseous species, X, attached to a surface site, S, can be calculated by using equation (4.63) (for X = H). It is then important that all three terms in the equation are calculated by using identical cell and theoretical parameters, and that geometry optimizations are being performed (where applicable). However, it is also possible to use total energies that have been calculated by other theoretical methods rather than the DFT methods.

Examples are room temperature-corrected energies, energies obtained from MD simulations, and Gibbs free energies from thermodynamic calculations. However, it must here be stressed that the methodology used in studying the atomic level CVD growth processes is the same, regardless of how the energy has been calculated.

It is very important to remember that the calculated adsorption energies strongly depend on the surface coverage (i.e. the concentrations of adsorbates on the surface). It is, therefore, once again stressed that one must be very clear about what type of problem that is to be solve. How does a realistic surface look like? To clarify the importance of these aspects, adsorption energies for various species, being chemisorbed to diamond (111) and (100) surfaces, have here been calculated. The species are H, OH, O_{ontop}, and O_{bridge}, which differ in size and type of adsorption site. (The H and OH species do always bind ontop.)

6.3.3.1 H-terminations

The adsorption energies for chemisorption of H to a surface with various surface coverages, are shown in figure 6.8(a). (The adsorbates have been spread out as evenly as possible on the surfaces. The reason is that they are expected to have, more or less, identical partial charges that may cause inter-adsorbate repulsions.) To be more specific, figure 6.8(a) shows the calculated H adsorption energy for adsorption to a (i) bare surface, (ii) surface with already one adsorbed H species (also called co-adsorbate), (iii) surface with two co-adsorbates, etc. The H atom is the smallest element in the periodic table, so we don't expect any major interactions between these species when adsorbed to the surface. It is, in fact, adsorbate-adsorbate interactions that otherwise (i.e. for other types of adsorbates) cause more or less severe reduction in adsorption energies. However, this is not expected for H adsorption.

The calculated adsorptions energies (equation (4.63)) are negative for an exothermic adsorption process. However, adsorption energies are generally described in absolute terms, i.e. a reduction in adsorption energy is in fact a less negative adsorption energy. This is also the way adsorption energies are presented in the present book (if not otherwise stated). As can be seen in figure 6.8(a), the adsorption energy of H is in principle constant for the various surface coverages. This is the situation for both (111) and (100) surfaces, with their common reconstructions. Hence, the adsorption energies for the first and last H are almost identical for the very small H species.

6.3.3.2 OH-termination

The corresponding situation for OH-termination is shown in figure 6.8(b). The main feature here is that the adsorbate is much larger than H and can have different directions in space. An example of different OH adsorbate directions, for the non-reconstructed diamond (111) surface, is shown in figure 6.9. As can easily be understood, there is a vast number of OH orientation combinations on the surfaces, and each of these OH adsorbate interacts with its neighbours with electrostatic interactions. These interactions are most often repulsions, which will induce a decrease in adsorption energies. In short, each individual combination of OH orientations correspond to a certain degree of intra-adsorbate repulsions, which in turn affects the OH adsorption energies in a very specific way. The question is now, how shall we model our OH-terminated surface? As is always the case, it depends on the problem we want to simulate. But it can nevertheless be very difficult since the real situation depends on so many different factors, like temperature, pressure, etc. One way is, however, to model two extreme situations and perform the calculations for these ones. It is then possible to assume that the results for the real situation is

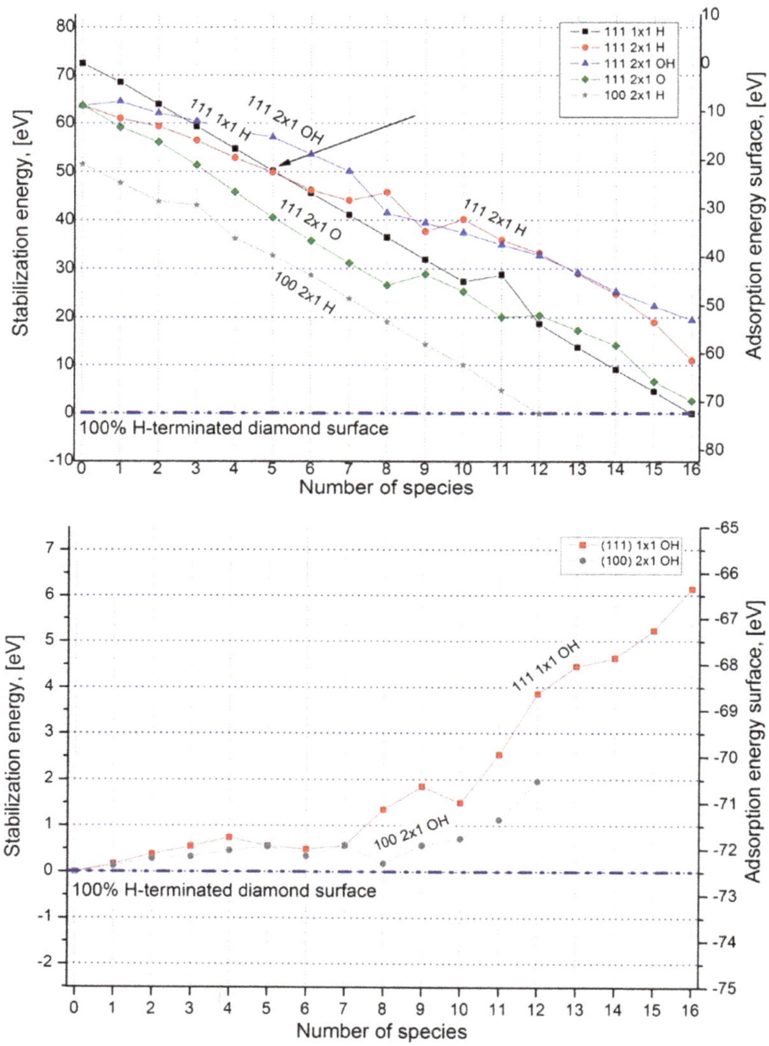

Figure 6.8. Stabilization and adsorption energies of H, OH, O_{ontop}, and O_{bridge} to diamond (111) and (100) surfaces with various degrees of surface coverages. The values are related to the corresponding 100% H-terminated surface. An increase in surface coverage of OH (or O) is obtained by replacing an H adsorbate with OH (or O), respectively. The stabilization and adsorption energies are given per supercell. The (111) supercell contains 16 surface C sites and the (100) supercell 12 surface C sites. [3], reproduced with permission © Elsevier.

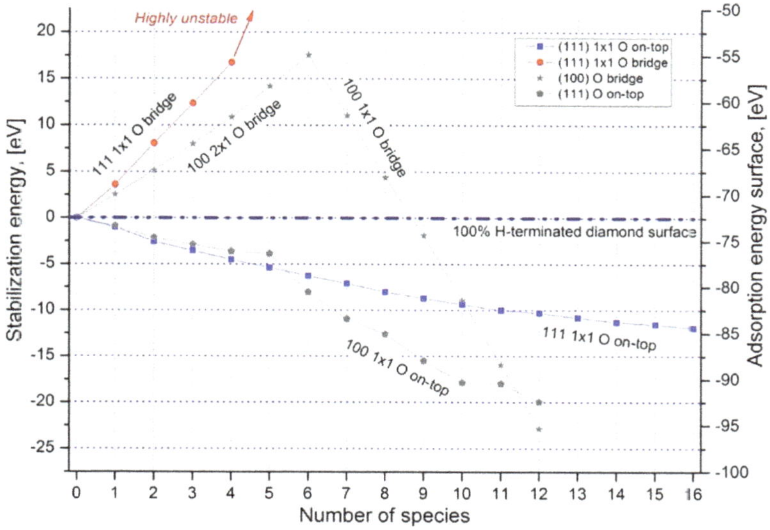

Figure 6.8. (Continued.)

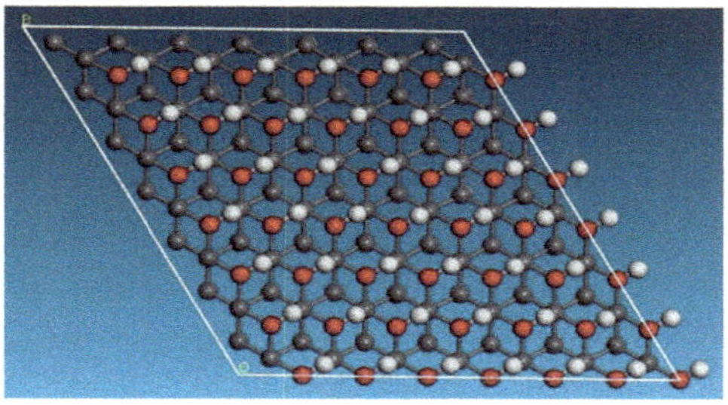

a)

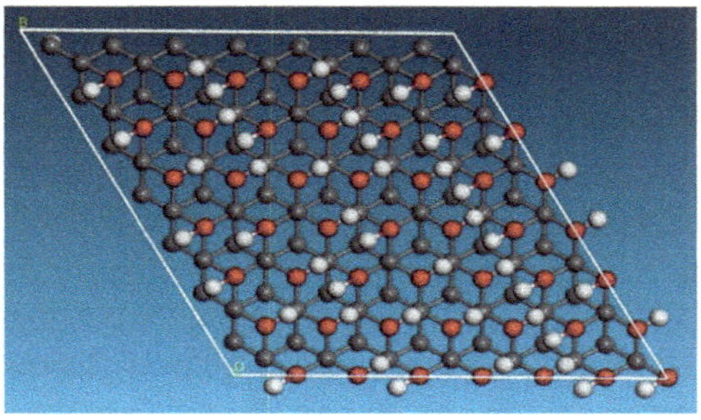

b)

Figure 6.9. Different adsorption structure for a 100% OH-terminated diamond (111) surface; (a) aligned OH adsorbates, and (b) non-aligned OH adsorbates. Grey = C; red = O; white = H.

somewhere in between. This strategy can be of great help in some situations. Otherwise, one has to seek help from experimental analysis of the surface.

The successive adsorption of OH molecules to the non-reconstructed (111) and 2×1-reconstructed (100) diamond surfaces, are shown in figure 6.8(b) for an ideally aligned OH adsorbate structure. As can be seen in figure 6.8(b), the calculated adsorption energies for the adsorption of an OH species to an otherwise H-terminated surface are +1.9 eV [(100)–2 × 1] and −2.0 eV [(111)-1 × 1]. Hence, they are almost identical. When increasing the OH coverage to about 60% [(100)–2 × 1] and 45% [(111)–1 × 1], the adsorption energies for the two surfaces are still very similar but have decreased with about 0.5 eV each. It is clear that there are just weak negative OH–OH interactions up to a surface coverage of 70% [(100)–2 × 1] and 45% [(111)–1 × 1], from where they will be, more or less, dramatically increased (to be further discussed in section 6.3.4). In fact, the last adsorbed OH is binding extremely weakly onto both the (111)–1 × 1 and (100)–2 × 1 surface. The adsorption energies are only +0.03 and +0.01 eV, respectively. Thus, the finally adsorbed OH species does only bind with a weak van der Waals force.

The results for these very ideal situations showed that it is not possible to completely terminate a diamond (111)–1 × 1 or (100–2 × 1) surface with OH species. The adsorption energy for the last OH is, thus, extremely weak for the ideal situation, which is a situation that will not be better for a more realistic surface termination structure. This is a clear example where an extreme situation has been modelled that describes the upper level of the OH adsorption energies.

In this context, a warning is in place for the commonly used methodology in calculating the adsorption energy of OH to an otherwise completely OH-terminated surface. Many researchers simply use an average value:

$$\mathrm{S(bare)} + n\mathrm{X} \rightarrow S - \mathrm{X}(100\%) \tag{6.5a}$$

$$\Delta E_{\mathrm{ads}} = E(S - \mathrm{X}(100\%)) - (E(\mathrm{S(bare)}) + nE(\mathrm{X})) \tag{6.5b}$$

This type of average expression does not give a representative picture of the adsorption energies, nor does it say anything about the possibility to reach a 100% surface coverage. It is only useful for adsorbate types that show a very small surface coverage dependance (like H). Hence, equation (6.1) should be handled with great care.

6.3.3.3 O-termination on surface ontop sites

A third example is the adsorption of O atoms to a diamond surface. As presented above, it can bind to an on top site or a bridge site (see figure 6.7). The reason for these preferences is that the O atom has two unpaired electrons, with which it can bind to either two different atoms, or to one. It will, for the latter situation, either form a double bond or a single bond with one remaining unpaired electron. The successive adsorption of O to either an on top site or to a bridge site. Let us start by looking at the $\mathrm{O_{ontop}}$ adsorption onto diamond (111) and (100)–2 × 1. It is clear that there are small interactions between the O adsorbates on the (111) surface. The adsorption energy of O becomes gradually smaller when going towards larger surface coverages. Moreover, the majority of the O adsorbates are radical ones since they have one non-paired electron each. In addition, some of the O adsorbates form a double bond with the surface C atoms. However, how can this happen since the surface C atoms only have one non-paired electron each? The answer is that the O-binding surface C atom breaks one of its bonds to one of its neighbouring C atoms (in C atomic layer 2). It thereby gets a second non-paired electron by which it can form the second C–O bond. What is left is a radical C atom in the second atomic C layer. So, whatever method is chosen, diamond (111) is left with radical carbon atoms in the upper surface region.

The situation is a bit different for the (100)–2 × 1 surface. The 2 × 1-reconstruction breaks up for the terminated C atoms but remains for the others. This takes place for an oxygen coverage up to approx. 50%. In addition, the electrostatic interactions amongst the O adsorbates are somewhat more severe (as compared with the (111) surface). This is most probably due to a larger fraction of spin-densities on the surface (see below). However, something more dramatic takes place for surface coverages larger than 50%. The surface reconstruction breaks up completely for all surface C atoms in the model, and a non-reconstructed (100) surface is formed. This process is generally called adsorption-induced surface reconstruction. As a result, there are no non-paired electrons that can cause negative repulsive interactions. Hence, the slope of the function $E_{\mathrm{ads}} = f(\text{surface coverage})$ becomes larger. There are, however, some negative interactions at higher surface coverages (above 75%).

In addition to the smaller electrostatic repulsions at higher surface coverages (<50%), the increase in adsorption energy depends on the larger concentration of C–O double bonds. There are double bonds (pairwise) between the reconstructed surface C atoms on the initial bare (100)–2 × 1 surface (see figure 5.12(c)). For a surface coverage up to 50%, when one O atom binds to an on top position on the reconstructed C–C pair, the double bond breaks up leaving C–C and C–O single bonds. The other C atom does now have a non-paired electron (i.e. being a radical). The situation is, however, different for an adsorption coverage larger than 50%. On top adsorptions do now also take place on the radical C atoms. But this is not enough, as the remaining surface C–C single bonds do also break up. This leads to the formation of two non-paired electrons in the C pair that each react with its binding O atom, forming two C = O double bonds (see figure 6.10).

6.3.3.4 O-termination on surface bridge sites

The adsorption of O species to surface bridge sites is different compared to the other adsorption examples. There are two choices for the first adsorption (i.e. the adsorption to a bare surface). As demonstrated in figure 6.11, it can either be adsorbed to a C–C pair or become adsorbed between two of those pairs. The first situation is more likely since those two C atoms are more closely positioned. The successive adsorption energies for O bridge termination are shown in figure 6.8(c). Due to severe structural constraints, this type of termination is not possible for the non-reconstructed diamond (111) surface. As can be seen in figure 6.8(c), the adsorption processes are endothermic for the (111) surface and will, thus, not take

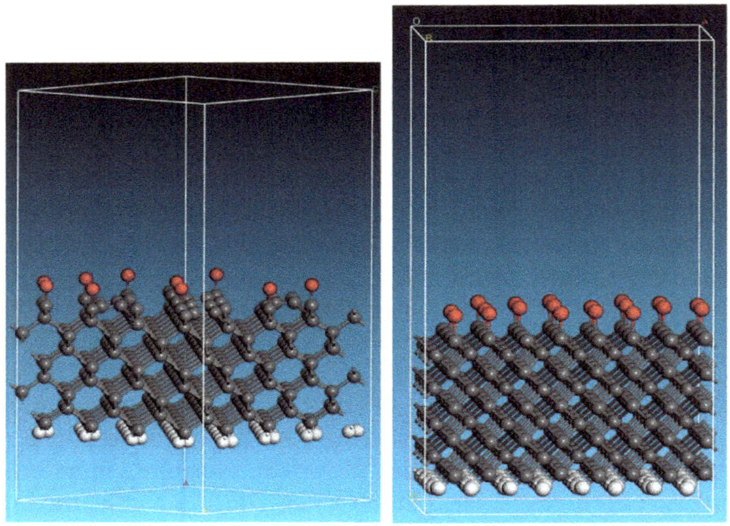

a) b)

Figure 6.10. Adsorption structures for a (a) 30% and (b) 60% O_{ontop}-terminated diamond (100)–2 × 1 surface. Grey = C; white = H; red = O.

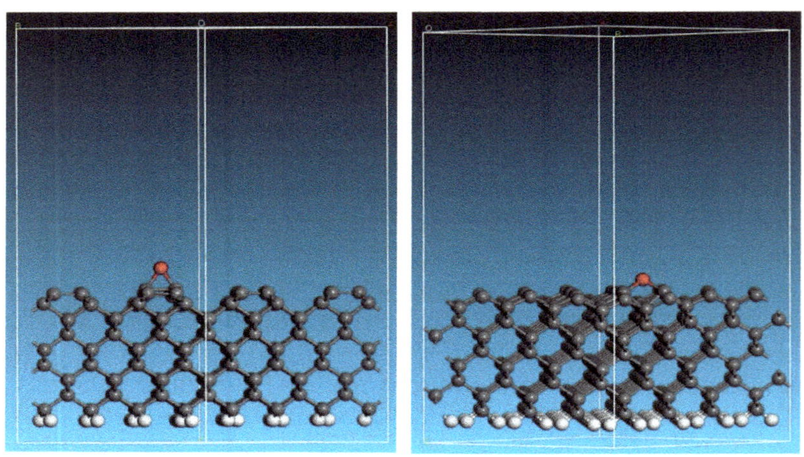

a) b)

Figure 6.11. Adsorption of O onto two different diamond (100)–2 × 1 bridge sites; (a) directly above a C–C pair (i.e. heteroepitaxially), and (b) between two C–C pairs in the direction of the C–C rows (i.e. 90° rotated with respect to the diamond structure). Grey = C; white = H; red = O.

place. The adsorption situation is also a bit tricky for the (100)–2 × 1 surface. As can be seen in figure 6.8(c), the adsorption to the C–C pairs (up to a surface coverage of 50%) is quite strong and seems to not interfere with each other to any larger extent. One of the bonds in the initial C = C double bonds has then broken up (forming C–C bonds) and the resulting unpaired electrons are instead forming bonds with the

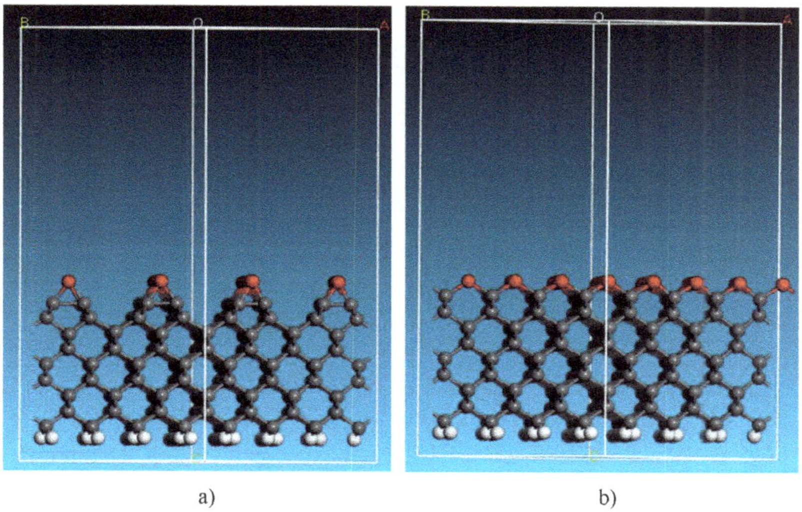

Figure 6.12. Adsorption of O onto diamond (100)–2 × 1 bridge sites for a surface coverage of (a) 50% and (b) 100%. Grey = C; white = H; red = O.

bridging O atom (which has two unpaired electrons) (see figure 6.12(a)). There are, hence, no electron spin densities that can cause pairwise repulsions. However, the problem starts when trying to terminate the surface with coverages larger than 50%. The additional O adsorbates must then form a bridge in between the rows of carbon pairs. The C–C single bond (in the carbon pairs) must also break up in the binding of these additional O atoms. As can be seen in figure 6.12(b), this formation will form a very strained surface structure. The additional O adsorbates are causing a re-reconstruction of the initial (100)–2 × 1 structure, but the C–O–C angles (101.1°) and C–O bond lengths (1.530 Å) are far from the less strained values for a surface coverage below 50% (64.4° and 1.503 Å). As was the situation with the adsorption to a bridge site on the (111) surface, the adsorption processes for coverages larger than 50% are endothermic. Hence, the conclusion is now that a surface coverage of at maximum 50% can be formed by adsorbing O atoms to a diamond (100)–2 × 1 surface (even at low temperatures).

6.3.4 Combination of adsorption and stabilization energies

A lot of information can be gained from the calculated adsorption energies. For the situation with the CVD growth of diamond thin films, when simulating this CVD process you should start with an H-terminated surface (since this is the most commonly observed termination form for as-grown diamond films). This circumstance has been considered in the construction of figures 6.13–6.16. In the replacement of one H adsorbate with another species X, the replacement reaction can be expressed as:

$$\text{Surface} - \text{H} + \text{X} \rightarrow \text{Surface} - \text{X} + \text{H} \tag{6.6}$$

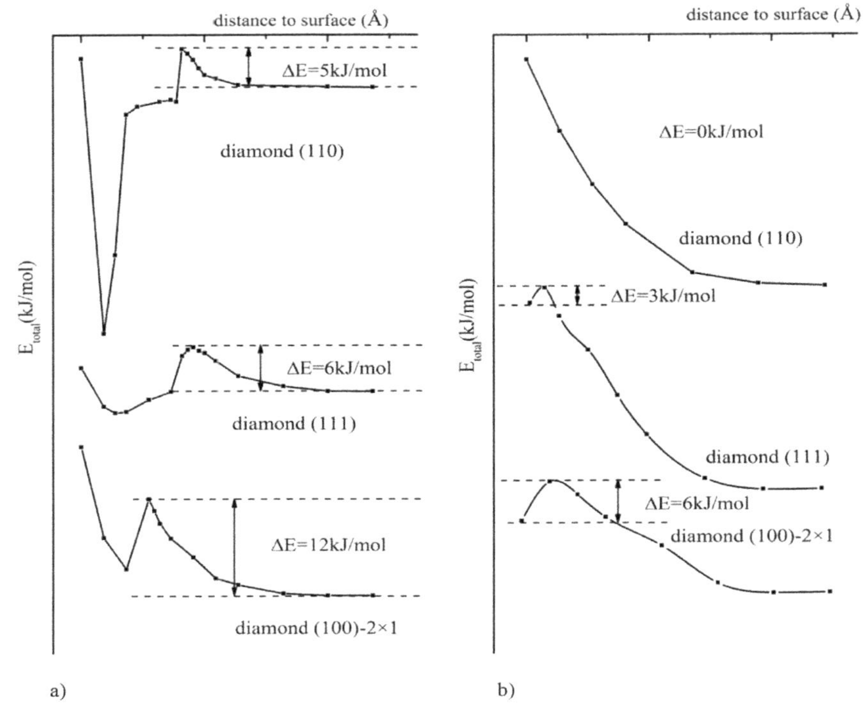

Figure 6.13. The energy evolution curve for (a) an approaching hydrogen radical to a surface terminated hydrogen species, and for (b) the corresponding desorption of H_2 molecule from the surfaces. This is demonstrated for different diamond surface planes; (110), (111) and (100)–2 × 1. [27], reproduced with permission © Elsevier.

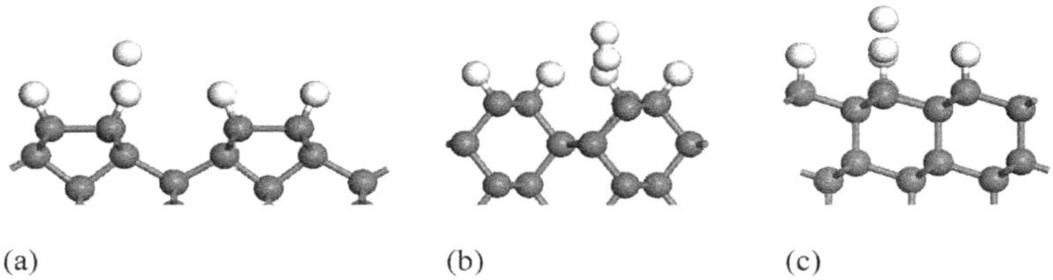

Figure 6.14 Structural geometries at the energy minima for an approaching hydrogen radical to a surface terminating hydrogen species. This is demonstrated for the following diamond planes; (100)–2 × 1, (110) and (111). [27], reproduced with permission © Elsevier.

It can also be written as:

$$\text{Surface} - \text{H} + \text{X} + \text{Surface} \rightarrow \text{Surface} - \text{X} + \text{H} + \text{Surface} \tag{6.7}$$

with the corresponding replacement reaction energy:

$$\Delta E_{\text{replacement}} = \Delta E_{\text{ads, X}} - \Delta E_{\text{ads, H}} \tag{6.8}$$

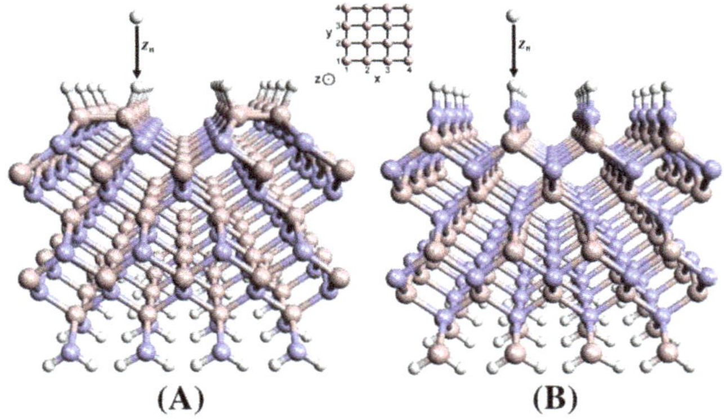

Figure 6.15. Supercell models of 100% H-terminated c-BN (100) surfaces; (a) B-covered 2 × 1-reconstructed, and (b) N-covered non-reconstructed. [28], reproduced with permission © Elsevier.

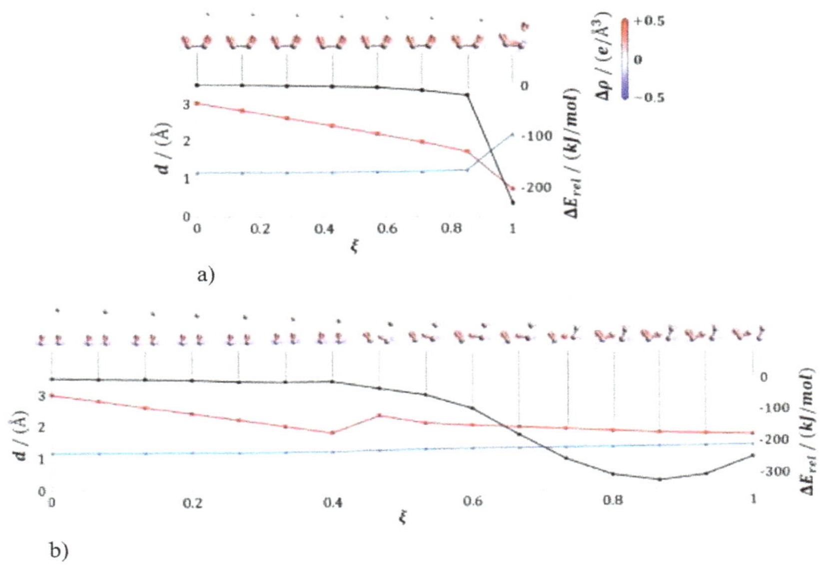

Figure 6.16. ΔE_{rel} (black), $d(H_{ads}–X)$ (red), and $d(B–H_{ads})$ (blue) as a function of ζ for the abstraction of H with gaseous X (X = (a) H or (b) F). The surface is an 100% H-terminated B-covered c-BN(100)–(2 × 1) surface. [28], reproduced with permission © Elsevier.

Hence, the replacement energy is simply the difference in adsorption energy for X and H.

Furthermore, the reaction expressions for the stabilization energies of surface-H and surface-X can be written as:

$$mC + (n + 1)H \rightarrow \text{Surface-H} \tag{6.9a}$$

$$mC + nH + X \rightarrow \text{Surface-X} \tag{6.9b}$$

Moreover, the stabilization energy for a monoradical surface is:

$$mC + nH \rightarrow \text{Surface}^* \quad (6.9c)$$

By continuing with these mathematical exercises, the following results for the relative stabilization energy can be obtained (i.e. when replacing one H adsorbate with one X adsorbate):

$$\Delta E_{\text{rel.stab.}} = \Delta E_{\text{ads, X}} - \Delta E_{\text{ads, H}} + E_X - E_H \quad (6.10a)$$

Hence, the relative stabilization energy is simply the difference in adsorption energy for X and H, plus a constant $E_X - E_H$ value. When comparing equation (6.8) with equation (6.10a) it is clear that equation (6.10a) can be rewritten as:

$$\Delta E_{\text{rel.stab.}} = \Delta E_{\text{replacement}} + E_X - E_H \quad (6.10b)$$

A surface configuration with a negative relative stabilization energy is, thereby, more favourable than the corresponding situation with an H-terminated surface site.

The relative adsorption and stabilization energies are presented in figures 6.14–6.17 for the adsorption of H, OH, and O species over the whole range of surface coverages (0%–100%). Starting from a 100% hydrogen coverage, these plots show the successive

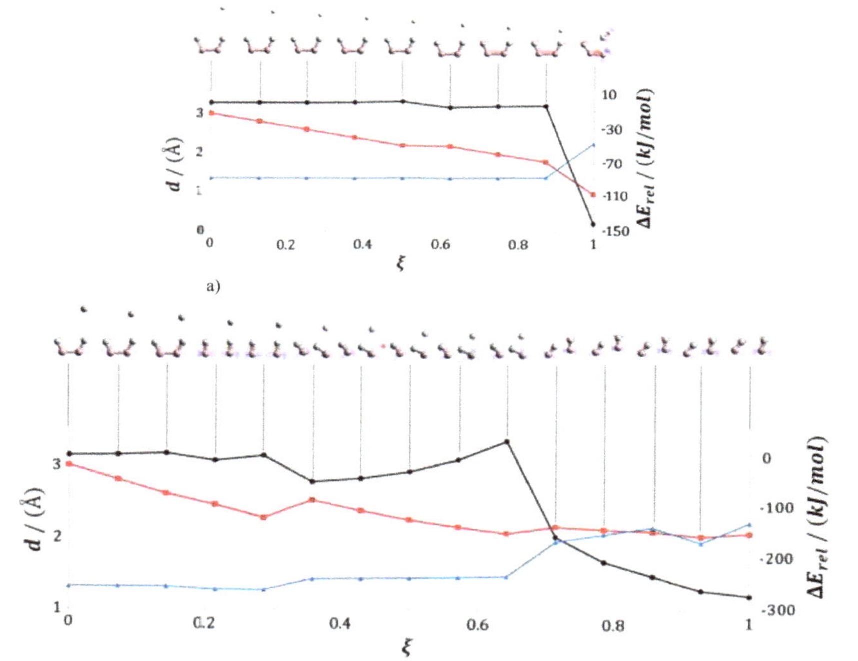

Figure 6.17. ΔE_{rel} (black), $d(\text{F}_{\text{ads}}\text{–X})$ (red), and $d(\text{B–F}_{\text{ads}})$ (blue) as a function of ζ for the abstraction of H with gaseous X (X = (a) H or (b) F). The surface is an 100% F-terminated B-covered c-BN(100)–(2 × 1) surface. [28], reproduced with permission © Elsevier.

relative adsorption and stabilization energies as H is replaced with O, or OH, species. But let us take a step back by studying the formation of the H-terminated surface.

6.3.4.1 Hydrogen termination

A bare diamond surface was the starting configuration for the H adsorption calculations presented in figure 6.8(a). The *x*-axis presents the surface coverages (from 0% to 100%). Furthermore, the left and right *y*-axis present the relative stabilization energy (equation (6.10b)) and replacement energy (equation (6.8)), respectively. As can be seen in figure 6.8(a), the bare (111)–2 × 1 surface (i.e. at zero surface coverage) is energetically more favorable than the corresponding non-reconstructed one (i.e. (111)–1 × 1). This relation continues until a breaking point (for a surface coverage of 5/16) where the H-terminated (111)–1 × 1 and (111)–2 × 1 surfaces show almost identical energies. The (111)–1 × 1 surface is energetically more favourable for an H surface coverage beyond this point (which is slightly above 30%).

As stated earlier in the introduction to this book (chapter 1), it is very important to, as much as possible, compare the theoretical findings with experimental observations. In comparison to the calculated adsorbate-induced surface reconstructions, it has experimentally been found that a small number of adsorbed hydrogen atoms is enough to completely reconstruct the (111)–2 × 1 to the 1 × 1 configuration [15]. A LEED pattern showed that 0.05 Ml would effectively reconstruct the (111)–2 × 1 surface to the 1 × 1 reconstruction. A plausible explanation may be that the delocalized π-bonds in the Pandey-chain formation have become weakened by the partial H adsorption (i.e. the electron configuration that stabilizes the non-terminated Pandey chain structure has been perturbed).

The adsorption energy for adsorption of H onto an otherwise completely H-terminated diamond (100)–1 × 1 surface, has been calculated here to be surprisingly high; −6.96 eV. The numerical value is, though, negative since it is an exothermic process. (It can, therefore, seem confusing to talk about large values when they are in fact negative ones.) This high value of adsorption energy can, though, be explained by the fact that two H adsorbates bind to each surface C site (i.e. it corresponds to the formation of two C–H bonds).

However, this adsorption configuration is most probably not seen experimentally due to the fast and spontaneous 2 × 1 surface reconstruction. In fact, this dihydride configuration has earlier been under much consideration [15–19], and it was assumed to not occur during CVD growth conditions [13]. Furthermore, as can be seen in figure 6.13 at a surface coverage of 0.75, the adsorption energy for H adsorbed onto the (100)–2 × 1 surface has been found to be in excellent agreement with experimental observations [20]; 4.30 vs. 4.58 eV.

In summary, the information in figure 6.8(a) shows that an H-terminated (100) surface only exists in the 2 × 1 reconstruction. Moreover, the non-reconstructed (111) surface is the most energetically stable structure for an H-termination larger than 30%. These results are strongly supported by experimental observations and give us a clue of what to find when H-terminating our specific surface in, e.g. the CVD reactor.

As is the situation for all examples in the present chapter, the diamond material can be substituted with any other crystalline material.

6.3.4.2 Hydroxyl replacement on a hydrogen-terminated surface

Information of the combined adsorption and stabilization energy for OH adsorption can be seen in figure 6.8(b). The axis notations are identical to figure 6.8(a). However, the difference here is that the reference H adsorption energy value, which is constant over the whole coverage range, is shown as a horizontal line. As was discussed in section 6.3.3, there can be severe adsorbate-adsorbate repulsive interactions at larger surface coverages. However, in addition to the repulsive interactions, there are also stabilizing interactions through hydrogen bonding between the OH adsorbates. These types of bonding formations are noticeable at a low OH coverage, whilst the steric repulsions dominate at higher coverages (more for the non-reconstructed (111) surface). The underlying reason for the negative interactions between the OH-groups is the sterical repulsions induced by the geometrical size and form of the adsorbate, and not the mono-valent character of the oxygen atom in the hydroxyl species. The number of electrons (i.e. electron density) that may be withdrawn, and reallocated, is larger for oxygen atoms compared to hydroxyl groups. As can be seen in figure 6.8(b), the breakpoint for the dominance of the OH–OH repulsions occurs at an OH coverage of 45% for the (111)–1 × 1 surface, and 70% for the (100)–2 × 1 surface. Also, only the non-reconstructed (111) and 2 × 1-reconstructed (100) surfaces are included in figure 6.8(b). The reason is that they are more stable than reconstructed (111) and non-reconstructed (100) surfaces, respectively.

As can be seen in figures 6.8(a) and (b), the adsorption energies for OH adsorbed to bare (111)–1 × 1 and (100)–2 × 1 surfaces are almost identical to the corresponding adsorption energies for H. Hence, it is easy to replace only one H adsorbate with a OH species on an otherwise completely H-terminated surface (at least from an energetical point of view). The situation is, however, different when increasing the OH coverage from 0% to 45% for (111)–1 × 1, and from 0 to 70% for (100)–2 × 1. The relative stability of the OH-terminated surfaces is, then, decreasing, with a maximum value of 0.8 eV vs. 0.6 eV.

The tabulated bond energies for a C–O and C–H bonds are on average 3.6 and 4.2 eV, respectively. This value corresponds quite well with the calculated OH and H adsorption energies at zero surface coverage (4.1 vs. 4.6 eV). The decrease in OH adsorption energies for surface coverages up to 45% [(111)–1 × 1] and 70% [(100)–2 × 1] will, hence, make it possible to replace the surface H adsorbates with OH species (even though it will be somewhat difficult). However, the OH-binding becomes much weaker for surface coverages above these limits. Thus, there are very low probabilities for further H adsorbate replacement at coverages larger than 45% [(111)–1 × 1] and 70% [(100)–2 × 1].

In summary, there is a possibility to replace the H adsorbates on the (111)–1 × 1 and (100)–2 × 1 surfaces up to a surface coverage of about 45% and 70%, respectively. Hence, in order to further increase the degree of OH-coverage one has to work with supersaturation levels of OH in the gas phase, and the possibility for success is larger for the 2 × 1-reconstructed (100) surface than for the non-reconstructed (111) surface.

6.3.4.3 Oxygen replacement on a hydrogen-terminated surface

It is most often possible to stabilize a hydrogen-terminated surface with oxygen atoms. In addition, due to the divalent nature of oxygen, and its strong electronegativity value, it can break double bonds on the surface and, thereby, induce surface reconstruction. As presented in figure 6.8(c), oxygen atoms can easily, for O_{ontop} adsorption on a non-reconstructed (111) surface and O_{ontop} adsorption on a non-reconstructed (100) surface, replace the H adsorbates up to a diamond surface coverage of 100%. However, the situation with O_{bridge}-adsorption onto the 2 × 1-reconstructed (100) surface is a bit peculiar. The process depends on the choice of adsorption site. As can be seen in figure 6.8(c), if we start to adsorb the O species at bridge positions between two C–C pairs (i.e. not along a C–C pair row), then the system needs an addition of energy to replace two H adsorbates with one atomic O_{bridge}. This is the situation up to a surface coverage of 50%. However, if we start by adsorbing O to a bridge site formed by the C–C pairs, then they can easily replace two H atoms up to a total O_{bridge} coverage of 50%. Another possibility to test theoretically is the adsorption of O to this bridge site, and the C–C bond will be broken (while keeping the H adsorbates). Also, the results show that it is thermodynamically very difficult to bind O, in bridge formations, to any larger extent on the non-reconstructed (111) surface.

As can be understood from figure 6.8(c), the most favorable position for oxygen on the (100) surface seems to be the bridge position. At least for a surface coverage up to 50%. For the rest of the adsorption sites, on top adsorption has been found to be energetically preferred, which is an observation that is supported by experimental studies [21, 22]. An initial H–C–O–C–H surface configuration was then proposed as a plausible oxidation mechanism [23]. It was also observed that the surface morphology will change and yield more oxygen in on top positions at elevated temperatures. Other experimental studies have also shown that the on top position is more likely to occur [24]. A mixture of on top and bridge positions is, thus, expected at experimental conditions. DFT-based MD calculations are also expected to find a larger degree of on top and bridge formations at elevated temperatures. Especially around the intercept of O_{ontop} and O_{bridge} curves at the surface O coverage of approximately 80%.

In summary, oxygen-termination has experimentally been found to largely stabilize an initial H-terminated surface. This is the situation for an up to 100% coverage of O_{ontop} adsorbates on the (100) and (111) surfaces. On the contrary, it is only possible to cover an (100) surface with O_{bridge} species up to a coverage of 50%. As can be understood from figure 6.8(c), it is expected that a mixture of O_{bridge} and O_{ontop} adsorbates will be found on the (100) surface. At least in the proximity to the crossing points at the surface coverages 0 and 60%–70%. This adsorbate situation can be simulated by performing quantum mechanical MD-simulations at various temperatures. These results were obtained for a diamond material, but any type of crystalline material can be used for these types of calculations.

6.3.4.4 Adsorption-induced surface restructuring

For terminated surfaces, another useful piece of information about energetically preferred structural configurations is given by figures 6.6, 6.8(a) and (c). They give information about adsorption-induced reconstructions, as well as of de-reconstruction of surfaces. The energetical values in figures 6.8(a) and (c) are referred to the corresponding energies for a 100% H-terminated diamond surface. We can also see it in the following way. We are starting with a 100% H-terminated (100) or (111) surface, which is also assumed to be either non-reconstructed or 2 × 1-reconstructed. For instance, the stabilization energies and adsorption energies for an OH-terminated (2 × 1)-reconstructed (111) surface is related to the H-terminated (2 × 1)-reconstructed (111) surface (the (2 × 1)-reconstruction of the diamond (111) surface is also called a Pandey chain structure). The H adsorbates will then sequentially be replaced with other types of adsorbates. In these examples, the relative stabilities (see equation (6.5)) are given for termination with either H_{ontop}, O_{ontop} or O_{bridge} species (at various surface coverages for H_{ontop} and O_{bridge}). Moreover, the H_{ontop}- and O_{ontop}-adsorption takes place on the diamond (111) surface (see figure 6.6), and the O_{bridge}-adsorption on both (111) and (100) (see figures 6.8(a) and (c)). Figure 6.6 shows only the preferred type of surface geometrical structure for a 100% terminated surface. It is clear there that if you start with a non-terminated diamond (111) surface at an experimental temperature, it is most probably 2 × 1-reconstructed. However, a de-reconstruction takes place when you terminate the surfaces with either H_{ontop} and O_{ontop}. Figures 6.8(a) and (c) give more information since they include different degrees of surface terminations (from 0% to 100%). In figure 6.8(a), if we here start with the most stable 2 × 1-reconstructed diamond (111) surface it is clear that the de-reconstruction takes place at a surface coverage of 5/16. In the same way, if we start with a 100% H-terminated non-reconstructed (111) surface and heat the surface to a temperature that causes desorption of the H adsorbates, a Pandey chain reconstruction will take place at 5/16. Furthermore, for the bridge positions of O onto the 2 × 1-reconstructed (100) surface, a de-reconstruction takes place at a surface coverage of 50% (when starting with a non-terminated surface). And a 2 × 1-reconstruction when starting from a 100% O_{bridge}-termination (see figure 6.8(c)). Moreover, the activation barriers have been calculated separately in figure 6.6 for the H_{ontop}- and O_{ontop}-terminated diamond (111) surface, and quite large barrier energies were found for these reconstruction/de-reconstruction processes when trying to alter the surface geometry for a 100% H_{ontop}- or O_{ontop}-terminated diamond (111) surface. The result presented in figure 6.8(a) for the H-terminated diamond (111) surface is, however, more clarifying since it shows that by sequentially varying the adsorption coverage it is possible to induce a barrier-less change in surface geometrical structure. This is, however, is not the case for the O_{bridge}-terminated diamond (100) surface, which is associated with a large barrier energy when going from a non-terminated surface to a 100% terminated one, and vice versa. Thus, the addition of energy is needed for this type of reconstruction/de-reconstruction to occur (e.g. in the form of a higher temperature, or by using a plasma). Similar calculations, as presented in figure 6.8(a) and (c), are, thus, recommended for any terminating species and surface type. This should, for instance, have also been done for the O_{ontop}-terminated non-terminated and 2 × 1-terminated diamond (111) surfaces. In fact, experimental results have shown that a 50% oxygen coverage is expected on the (111)–2 × 1 surface [24].

6.4 Creation of surface reactive sites

6.4.1 Introduction

Before the adsorption of growth species can take place, there must be an available (i.e. empty) surface site for the approaching species. This generally takes place by gas phase abstraction reactions in CVD processes;

$$S{-}X + Y \rightarrow S + X{-}Y \tag{6.11}$$

where S, X, and Y are the surface site, adsorbate, and gas phase species, respectively. Moreover, the X–Y bond must be stronger than the S–X bond for this abstraction reaction to take place. In addition, the gas phase Y species must be both very reactive and small (to avoid sterical repulsions on the surface). Take the H-terminated diamond surface as an example. It has been shown possible to use a radical halogen atom, like F, as an Y species since the H–F bond is much stronger than the C–H bond; 567 versus 414 kJ mol^{-1}. It is, in fact, even stronger than the H–H bond (436 kJ mol^{-1}). These numerical values are tabulated data taken from [25]. Gas phase abstraction by gaseous H radicals is, otherwise, a frequently used species for the removal of H adsorbates from the diamond surface. The question is then, how can gaseous H radical species remove H adsorbates since the H–H bond is weaker than the C–H bond. The answer is the presence of a supersaturation of H radicals in the CVD reactor. That type of environment has also been shown to be efficient [26], although abstraction with F radicals is even more efficient. In fact, by using F species in the gas phase, it has been found possible to decrease the CVD deposition temperature for growth of diamond. Another aspect to consider is that the growth species shall preferentially bind stronger to the surface than the terminating species, X. This circumstance will be further discussed in section 6.4.3.

6.4.2 Gas phase abstraction from the surface

By using gaseous reactive species (Y), an abstraction of a surface-terminating species (X) generally takes place in two steps: (1) adsorption of Y to the surface-terminating X species and (2) desorption of molecular X–H. This process is exemplified here with the most common low-index surface planes of H-terminated diamond: (111), (110), and (100)–2 × 1 (see figures 6.13, 6.14 and 6.16). Another example is the H/F-terminated cubic BN (100)–2 × 1 surfaces (see figures 6.15–6.20). (The (111)–1 × 1 surface is from here on simply denoted as (111).)

6.4.2.1 H-terminated diamond surfaces

It is very important to calculate the energy barrier for each of the gas phase abstraction steps. The gas phase H abstraction reaction of an H adsorbate is in figure 4.16 demonstrated for the diamond (100)–2 × 1 surface. As can be seen in

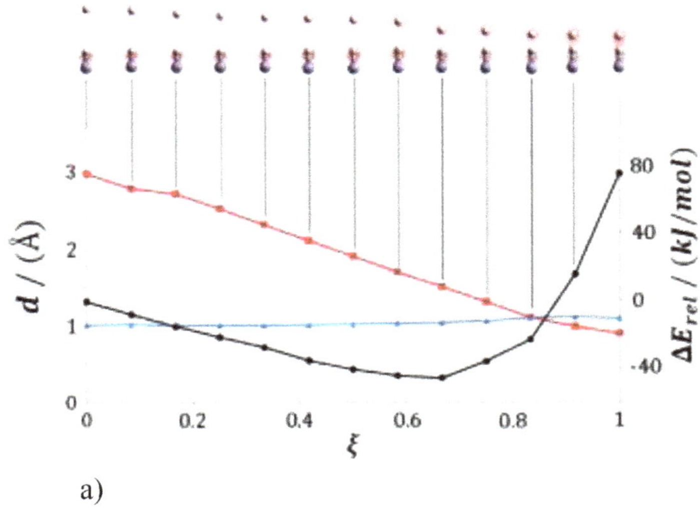

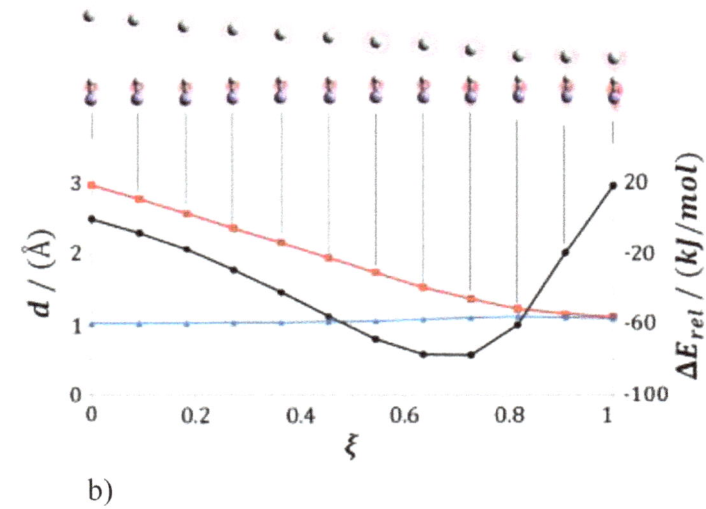

Figure 6.18. ΔE_{rel} (black), $d(H_{ads}–X)$ (red), and $d(B–H_{ads})$ (blue) as a function of ζ for the adsorption step in the abstraction of H with gaseous X (X = (a) H or (b) F). The surface is an 100% H-terminated N-covered c-BN(100)–(2 × 1) surface. [28], reproduced with permission © Elsevier.

figure 6.13, the calculated energy barriers for both step 1 (figure 6.13(a)) and step 2 (figure 6.13(b)) look quite different for the different surface types. Step 1: 12 [(100)–2 × 1], 6 [(110)], and 5 [(111)] kJ mol^{-1}. Step 2: 6 [(100)–2 × 1], 0 [(110)], and 3 [(111)] kJ mol^{-1}. (The methodology for barrier energy calculations (i.e. calculations of potential energy curves) is described in section 4.5.3.) Furthermore, when the gaseous H species continues to approach the diamond surfaces, it will finally end up in an energy minimum at a distance of 1.2–1.4 Å from the surface-binding H species (see figure 6.14). The H radical

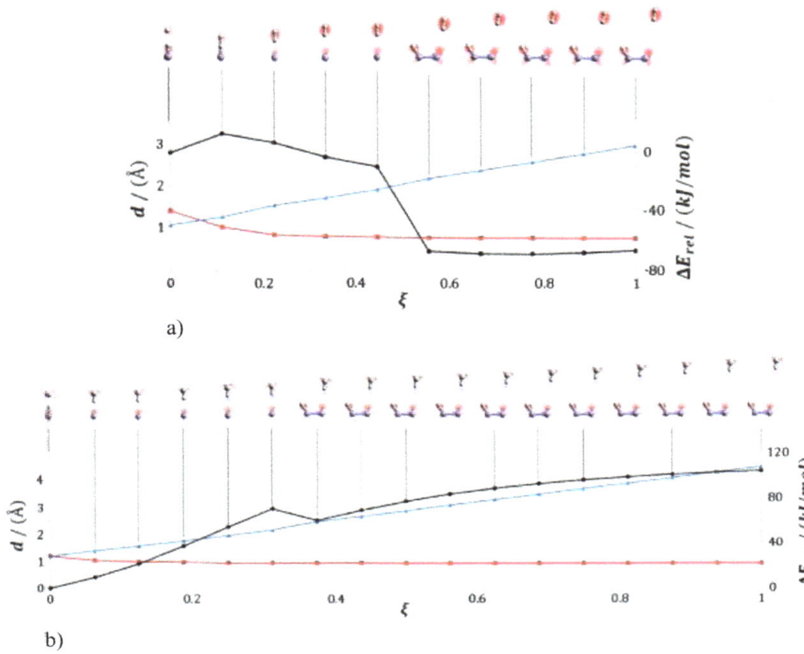

Figure 6.19. $\Delta E_{\rm rel}$ (black), $d(\mathrm{F_{ads}}$–X) (red), and $d(\mathrm{N{-}F_{ads}})$ (blue) as a function of ζ for the desorption step in the abstraction of H with X (X = (a) H or (b) F). The surface is an 100% H-terminated N-covered c-BN(100)–(2 × 1) surface. [28], reproduced with permission © Elsevier.

interacts most strongly with the H adsorbate at this distance, but to various extents depending on the type of surface plane.

As a result of step 1, there is no 'automatic' release of the H_2 molecule from the (111) and (100)–2 × 1 surfaces (see figure 6.13(b)). As will be discussed below, an energy barrier must be overcome for the H_2 molecule to form and leave the diamond (111) or (100)–2 × 1 surface. However, this is not the case for the (110) surface, from which the H_2 molecule will become directly removed (i.e. is energetically feasible).

Moreover, it is apparent from figure 13 that step 2 in the abstraction process takes place with a smaller energy barrier (as compared with step 1 for all surfaces investigated). Hence, it is only the adsorption part of the abstraction process that will be the dominating one in determining the H abstraction reaction rate (which will be further discussed in section 6.6).

In summary, the H abstraction processes on the H-terminated diamond surfaces have not been found to be spontaneous. They are all correlated to a specific barrier energy (either in the incoming stage, or in the outgoing stage). However, it must be stressed that there is only a small barrier energy involved in these reactions, for all surfaces investigated. The reason for these observations is most probably the partial positive charge on the H adsorbates, which originates from the difference between the electronegative values of C and H (2.55 versus 2.20). It is predominantly nucleophilic species that prefer to interact with such surface sites, and, hence, not

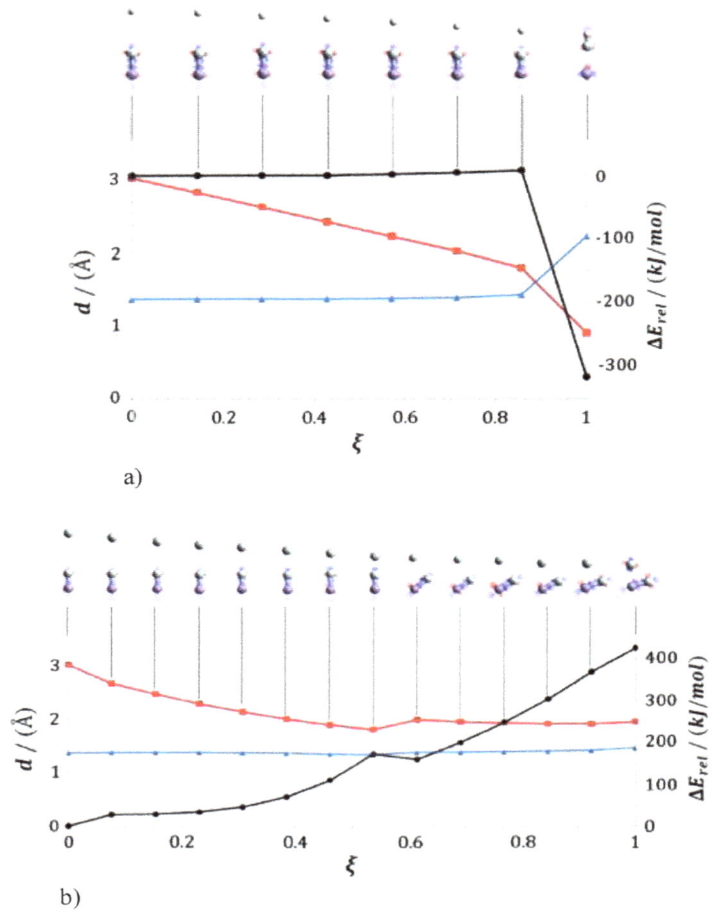

Figure 6.20. ΔE_{rel} (black), $d(F_{ads}–X)$ (red) and $d(N–F_{ads})$ (blue) as a function of ζ for the adsorption process in the abstraction of F with gaseous X (X = (a) H or (b) F). The surface is an 100% F-terminated N-covered c-BN (100)–(2 × 1) surface. [28], reproduced with permission © Elsevier.

electrophilic species like H. (A nucleophile is a species that donates an electron pair to form a chemical bond, while an electrophile is a species that forms by accepting an electron pair.)

6.4.2.2 H-terminated B-covered cubic BN surfaces

Another example of adsorbate abstraction concerns cubic BN (c-BN). This material has a structure that is very similar to diamond. The only difference is that every other atom is replaced by B, or N, atoms. As can be seen in figure 6.21, c-BN in the (100) direction is composed of alternate layers of B and N atoms. Moreover, the B and N atoms have electronegative values that are different, and they are also different from electronegativity value of C: 2.04, 3.04, and 2.55, respectively. Hence, a c-BN surface is built from dipoles, which is not the situation with the diamond

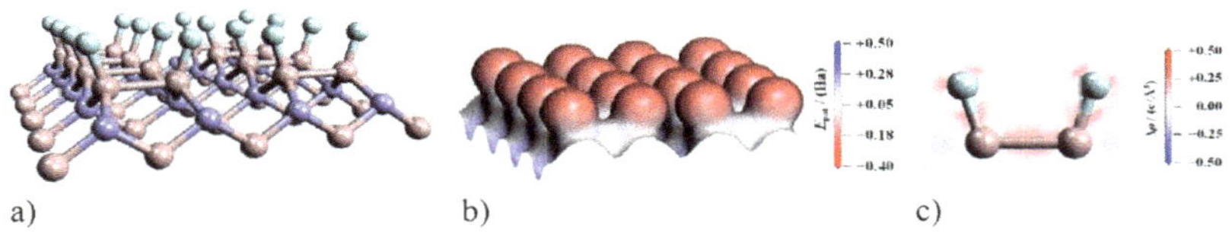

Figure 6.21. (a) 100% F-terminated B-covered c-BN(100)–(2 × 1) surface. (b) Electrostatic potential map for the surface shown in (a). (c) Deformation electron density map of a B–B dimer in a plane perpendicular to the surface shown in (a). [29], reproduced with permission © Elsevier.

surfaces. We will demonstrate here the gas phase abstraction of H or F from an H- or F-terminated c-BN (100) surfaces by using gaseous H and F species. The B-covered (100) surface is 2 × 1-reconstructed, whilst the N-covered one is non-reconstructed (i.e. 1 × 1). The reason for these surface reconstructions will be further discussed in section 8.4.

Let us now follow the energy evolution when a gaseous H or F species is approaching an H adsorbate on a B-covered c-BN (100)–2 × 1 surface (see figure 6.15). We have now improved the content of the information by also including calculations of electron deformation densities in the study. Thus, the following figures (figures 6.16–6.20) are quite complex, showing in addition to the potential energy curves also the distance between the approaching H or F species and the adsorbed H one, as well as the distance between the surface B atom and the adsorbed H species. In addition, an electron deformation density map is added for each position of the approaching H or F species. After geometry optimization, a relative interaction energy, ΔE_{rel}, between the system with an incoming H or F species and the system with an adsorbed H species (step 1) has been calculated according to equation (6.12). The relative interaction energy between the system with a desorbing H_2 (or HF) species and the system with a B radical surface site (step 2) has also been calculated by using equation (6.13).

$$\Delta E_{rel,\ step1} = E\left(BN_{(100\%term.+Y(z_n = 3.0+n(-0.2)))}\right) - E\left(BN_{(100\%term.+Y(z = 3.0))}\right) \qquad (6.12)$$

$$\Delta E_{rel,\ step2} = E\left(BN_{(radical\ site\ +\ Y(z_n = z_{min}+n(0.2)))}\right) - E\left(BN_{(radical\ site+HY(z_0 = z_{min}))}\right) \qquad (6.13)$$

where Y = H or F. Furthermore, $E\left(BN_{(100\%term.+Y(z_n = 3.0+n(-0.2)))}\right)$ and $E\left(BN_{(100\%H+Y(z = 3.0))}\right)$ are the total energies for, e.g. a 100% H-terminated surface, with an abstracting H atom at the z-coordinates 3.0 and $3.0 + n \cdot (-0.2)$ Å, respectively. Moreover, $E\left(BN_{(radical\ site+Y(z_n = z_{min}+n(0.2)))}\right)$ and $E\left(BN_{(radical\ site\ +\ HY(z_0 = z_{min}))}\right)$ are the total energies for the monoradical H-terminated surface with an H_2 (or HF) molecule at the z-coordinates z_{min} and $z_{min} + n{\cdot}0.2$ Å, respectively. Based on equations (6.7) and (6.8), potential energy curves were, thereafter, obtained by plotting ΔE_{rel} as a function of a reaction coordinate, ξ;

$$\xi = n/n_{max} \qquad (6.14)$$

A more elaborate description of the computational methods that have been used for these calculations is presented and discussed in [28].

As can be seen in figure 6.16(a) for the abstraction of H adsorbates with gaseous H species, the relative energy, $\Delta E_{\text{rel, step1}}$, slowly decreases in the region $0 \leqslant \xi \leqslant 0.86$. In addition, the B–H distance becomes somewhat prolonged. These are strong indications for an attractive interaction between the incoming H and adsorbed H (from here on called H_{ads}) atoms. There is a clear breakpoint at $\xi > 0.86$ (corresponding to $d(H_{ads}–H) < 1.72$ Å), after which H_{ads} will be abstracted from the surface by forming an H_2 molecule. This can be understood from the combination of a severely prolonged B–H distance and a strong stabilization of the complete system (i.e. a large decrease in ΔE_{rel}). These results show that there is no energy barrier involved in the abstraction process. It is completely spontaneous. H abstraction from the 100% H-terminated B-covered c-BN(100)–(2 × 1) surface, using gaseous H, is therefore expected to be a highly facile process. Moreover, the calculated deformation electron densities show B–H_{ads} bond electron densities for ξ values up to $\xi = 0.86$, after which the newly formed H–H bond is clearly demonstrated with a high electron density. In addition, the resulting non-paired electron on the B radical (B_{rad}) has now formed a bond with a neighbouring surface B atom (which is an exothermic process). This is a result that strongly support the conclusion that H abstraction takes place without the need for any extra energy.

The energy curve for the corresponding abstraction of H_{ads}, with gaseous F, is shown in figure 6.16(b). It can be seen there that ΔE_{rel} slowly decreases in the region $0 \leqslant \xi \leqslant 0.40$. Furthermore, a significant change in the F–B–H_{ads} angle occurs in region $0.40 < \xi < 0.47$. As can be seen in the corresponding deformation electron density map, it changes from 0 to a value of about 85°. This is a strong indication of repulsions between the approaching F species and the adsorbed H species. The energy will thereafter continue to progressively decrease towards a minimum of -321 kJ mol^{-1} at $\xi = 0.87$. As can be seen in the deformation electron density maps, this decrease in energy is accompanied by a migration of H_{ads} from its binding B atom to the neighbouring one in the B–B pair. In addition, F forms a bond with B, i.e., no bond is formed between F and H_{ads}. As can be seen in figure 6.16(b), there is just slight increase in B–H_{ads} bond length during this process. Also, there is, in addition, just a minor decrease in F–H_{ads} bond length.

In summary, figure 6.16 is an example of a figure that shows a multitude of complementary information that is of great help in the analyses of the abstraction processes. It is, as such, preferable over simple potential energy curves, as those shown in figure 6.13. The main result from figure 6.16 is the spontaneous H abstraction, by using gaseous H, whilst it is not possible to abstract H by using gaseous F. However, this is just an observation and it must be explained. For this to be done, the atomic charges, q, have also been calculated (the theory behind atomic charge calculations is presented in section 4.6.4). As a result, the approaching F species is negatively charged, and the adsorbed H atom is almost neutral. The charge of the F (H) atom changes from −0.45(−0.07)e to −0.27(0.05)e when F is approaching the H adsorbate, while the H-binding B atom change from +0.65e to +0.23e. Thus, as the neutral H_{ads} atom bends away, the negatively charged F atom can directly interact with the positively charged surface B atom. This gives rise to a stabilizing electrostatic contribution that lowers the total energy of the system. Since no bond is formed between H_{ads} and the approaching F species, it can be concluded that a gaseous F radical is most probably not able to abstract a chemisorbed H atom from the 100% H-terminated B-covered c-BN(100)–(2 × 1) surface. There is, thus, a manifold of competing interactions on a solid surface, and it is most often very difficult to find the most probable explanation to the theoretical results. That is why one should calculate as many parameters as possible (i.e. bond lengths, energies, electron densities, atomic charges) to have many supporting results to choose from.

6.4.2.3 F-terminated B-covered cubic BN surfaces

The potential energy curve for abstraction of F adsorbates with gaseous H species, from the 100% F-terminated B-covered c-BN(100)–(2 × 1) surface, is shown in figure 6.17(a). There is a minimal increase in ΔE_{rel} for $0 \leqslant \xi \leqslant 0.50$, which is not observable in the deformation electron deformation maps. This increase in ΔE_{rel} is most probably due to a F_{ads}–H interaction, which in turn causes the relaxation of the upper surface B layer (with 0.14 Å) at $\xi = 0.63$. The result of this relaxation is an increase in the total energy by approximately 5 kJ mol^{-1}. The potential energy curve continues to increase (with a minor slope) in the region $0.63 \leqslant \xi \leqslant 0.88$. For $\xi > 0.88$, corresponding to $d(F_{ads}\text{–}H) = 1.77$ Å, the adsorbed F species will be abstracted from the surface by forming an HF molecule. Thus, this abstraction reaction only involves a minimal energy barrier and is, therefore, expected to be a very facile process in a CVD process. As can be seen in figure 6.17, the potential energy curve is strongly correlated with the $d(F_{ads}\text{–}H)$ and $d(B\text{–}F_{ads})$ distances. The former one is decreasing and the latter one is increasing, in the course of the abstraction process. In addition, the deformation electron density maps show some interesting features. As the H species comes closer to the adsorbed F, there is an induced electron density transfer from the F-binding B species, and its B neighbour, to their binding F species. Also, as was the situation with the corresponding H abstraction, the deformation electron densities show the presence $B\text{–}F_{ads}$ bond electron densities for ξ values up to $\xi = 0.86$, after which the newly formed H–F bond is clearly demonstrated with a high electron density. This density represents a dipole with an excess of electrons on F (marked in blue) and a depletion on H (marked in red). In addition, the resulting non-paired electron on the B radical (B_{rad}) is now involved in the binding with a neighbouring surface B atom.

In comparison to the situation with surface H abstraction using gaseous F species, the abstraction of F shows a completely different behaviour. The potential energy curve for the abstraction of F, with gaseous F, from the B-covered c-BN(100)–(2 × 1) surface, is shown in figure 6.17(b). The energy increases, at first, towards a ΔE_{rel} value of +4.0 kJ mol^{-1} at $\xi = 0.14$. The B–B dimer bond is, thereafter, broken (in the region $0.14 < \xi \leqslant 0.21$) with a following decrease in ΔE_{rel} (to −10 kJ mol^{-1}). The energy is, in region $0.21 < \xi < 0.29$, increasing again (with a total of 10 kJ mol^{-1}). This increase is most probably due to repulsion between the non-bonding electrons on the approaching F species and F_{ads}. As a matter of fact, the calculated Mulliken charges show a negative charge on both F_{ads} (on average −0.13 e) and the gaseous F atom (on average −0.28 e) in the region $0.21 \leqslant \xi \leqslant 0.29$, which points at a repulsive (destabilizing) electrostatic contribution to the F_{ads}–F interaction. Moreover, these repulsive interactions lead to a significant change in the B–B–F angle for $0.29 < \xi \leqslant 0.36$, which was accompanied by a more prominent decrease in energy (with approximately -51 kJ mol^{-1}). The energy was, thereafter, progressively increased towards a maximum of +28 kJ mol^{-1} at $\xi = 0.64$, with an accompanying decrease in B–B–F angle for F_{ads}. This is an indication of an increasing electronic F_{ads}–F repulsion, which finally results in the migration of F_{ads} from its binding surface B atom towards the neighbouring B atom in the dimer pair (for $0.64 < \xi < 0.71$). Thus, the F_{ads} atom has become bonded to the neighbouring B atom at $\xi = 0.71$, and the approaching F has been adsorbed to the initial B atom. The latter migration processes resulted in a large decrease in energy (towards -273 kJ mol^{-1}). As can be seen in figure 6.17(b), the F_{ads}–F and initial B–F_{ads} bond lengths do very well correlate with the changes in ΔE_{rel}. The main conclusion that can be drawn from figure 6.17(b) is that no bond can be formed between the adsorbed F and the approaching F atom. A gaseous F radical is, therefore, not able to abstract a chemisorbed F atom from the 100% F-covered B-terminated surface.

In summary, the gas phase abstraction results for an F-terminated B-covered c-BN surface can be easily understood from the atomic type and charges of F_{ads} and the approaching species. A gaseous H atom is an electrophilic species that prefers to bind to a nucleophilic species, and the adsorbed F atom is a nucleophilic species. In fact, the F adsorbates have an added large degree of electron density since the electronegativity of F is larger than B (3.98 versus 2.04). Hence, the F adsorbates are very susceptible towards an attack by the incoming H radicals. The situation is the opposite for F abstraction by using gaseous F atoms. They are both nucleophiles and are, therefore, not prone to interact with each other. And this is even more the case since the less electronegative B atom is binding to F_{ads}.

6.4.2.4 H-terminated N-covered cubic BN surfaces

To complete the whole picture, we should also have a look at the N-covered c-BN surface. The potential energy curves for abstraction of H with gaseous H, from the 100% H-terminated N-covered c-BN(100)–(1 × 1) surface, is shown in figure 6.18(a). As can be seen in figure 6.18(a), the ΔE_{rel} energies for the abstraction processes decrease progressively towards a minimum (of -46 kJ mol^{-1}) that is situated further away from the surface. The parameter values at this minimum are $\xi = 0.67$, $d(\text{N–H}_{ads}) = 1.04$ Å, and $d(\text{H}_{ads}\text{–H}) = 1.50$ Å. This is a result that is completely different to the corresponding gas phase abstraction from a B-covered surface. The attractive interaction between the approaching H radical and the adsorbed H species is most probably due to the negatively charged H radical ($-0.36\text{e} \geqslant q(\text{H}) \geqslant -0.41\text{e}$) whilst H_{ads} is positively charged ($+0.16\text{e} \leqslant q(\text{H}_{ads}) \leqslant +0.18\text{e}$). (The values in these ranges are obtained from the approaching process $0 \leqslant \xi < 0.67$.) The approaching H species will, thus, receive an electronic density from the surface atoms. Furthermore, the surface N atoms have, due to the difference in electronegativity values between N (3.04) and H (2.20), withdrawn electron densities from the H adsorbate, making them positively charged. This explains the strong positive interaction between the approaching H species and the surface. However, despite this strong interaction, any HF desorption will not take place spontaneously. The main reason is the appearance of negative charge repulsions between the approaching H and the surface N–H species. This is demonstrated by the disappearance of positive charge (marked in blue) between H and H_{ads} at $\xi = 0.74$. The hereby formed H_2 molecule (at $\xi = 0.67$) is thereafter desorbed from the surface, leaving a non-terminated N radical site (see figure 6.18(a)). In this desorption process, the energy will at first increase towards a maximum of +13 kJ mol^{-1} at $\xi = 0.11$. This energy maximum corresponds to the distances $d(N–H_{ads})$ = 1.28 Å and $d(H_{ads}–H)$ = 1.03 Å. The N–H_{ads} bond will, thereafter, break (in the region $0.11 < \xi \leqslant 0.22$), while the H_{ads}–H bond is being formed. This means that the abstraction process must pass over a transition state with a barrier energy of +13 kJ mol^{-1}.

The energy will, thereafter, progressively decrease in the region $0.22 < \xi \leqslant 0.44$. The next-coming step is the formation of an N = N dimer bond, in addition to a gaseous H_2 molecule. This has been found to significantly decrease the energy to −67 kJ mol^{-1} in region $0.44 < \xi \leqslant 0.56$. This ΔE_{rel} energy will, thereafter, remain more or less constant up to $\xi = 1$. It is reasonable, due to the very low barrier energy (ΔE_{barr} = +13 kJ mol^{-1}), to assume that H abstraction (from the 100% H-covered N-terminated c-BN(100)–(1 × 1) surface) will take place under experimental CVD conditions.

The situation is different for the abstraction of H adsorbates by using gaseous F species.

Similar to gas phase abstraction using H radicals, the energy reaches an energy minimum H_{ads} has its origin in the negative charges of F ($-0.54\ e \geqslant q(F) \geqslant -0.63e$) whilst H_{ads} is positively charged ($+0.19e \leqslant q(H_{ads}) \leqslant +0.32e$). (The range go from $\xi = 0$ to 0.73 for these atomic charges.) In relation to the situation with H abstraction using gaseous H species, the larger slope of the potential energy curve can be explained by the more ionic character of the H_{ads}–F interaction. At the energy minima, the charges of H_{ads} and F have become $q(H_{ads}) = +0.32$ e and $q(F) = -0.63$ e, as compared with $q(H_{ads}) = +0.18$ e and $q(H) = -0.41$e for the corresponding H_{ads}–H interaction. Hence, the H_{ads}–F interaction is more electrostatically stable. When moving F further closer to the H adsorbate, there is a repulsion similar to the H_{ads} by H abstraction case. As can be seen in the deformation electron density maps, these repulsions start taking place at $\xi > 0.73$.

In the desorption of H_{ads} by gaseous F, we start from the energy minimum in figure 6.18(b) and continue to step-wise remove the H–F species from the surface. The energy will at first progressively increase towards a local maximum of + 71 kJ mol^{-1} at $\xi = 0.31$. At this local energy maximum (which corresponds to $d(N–H_{ads})$ = 2.18 Å and $d(H_{ads}–F)$ = 0.95 Å), a dimer between two neighbouring N surface atoms is beginning to be formed, which results in a decrease in energy. The energy will thereafter continuously increase up to $\xi = 1$. The most probable explanation to this observation is that there is a positive electrostatic interaction between the desorbing H–F dipole and the bare N surface site. This is clear from the deformation electron density maps in figure 6.19(b). This conclusion is further supported by the atomic charges of H (in H–F) and N that goes from $q(H_{ads}) = +0.38$e and $q(N) = -0.61$e to $q(H_{ads}) = +0.33$e and $q(N) = -0.50$e (for the longest distance from the surface). (The H atom in HF is closest to the surface when desorbing from the surface, HF is perpendicular positioned with respect to the surface.) Hence, the addition of an extra energy is needed to further separate HF from the surface. The desorption reaction is therefore endothermic and cannot take place without the addition of an energy of approximately 108 kJ mol^{-1}.

In summary, depending on which gaseous species to use (H or F), the H abstraction from an N-covered diamond (100) surface is completely different. The first step, which concerns an approaching species to the H adsorbate, is very similar, though. The surface N–H bond is a dipole, with a positive charge on H_{ads} and a negative one on N. When the approaching H (or F) comes close enough (i.e. when it starts to experience an interaction with the adsorbate) there will be a partial electron transfer from the surface to the gaseous species. This is more the situation for the gaseous F species since it has a larger electronegativity value than gaseous H (3.98 versus 2.20). As a consequence, there is a positive electrostatice interaction between the approaching H (or F) and the surface H adsorbates for a distance that is not too short. For even shorter distances, repulsions between H (or F) and N starts to take over. This energy evolution results in energy minima, which is deeper for the H abstraction with F (which can be explained by the larger attraction between H_{ads} and F).

Step 2 is also completely different for the two abstraction situations. When starting from the energy minima presented above, the first part of the desorption process with gaseous H species is to create an N–H–H formation, from where the H–H species will become removed. Since the H_2 molecule is not a dipole, but include a perfect covalent bond, it has a clear electron density 'cloud' which repel to the nucleophilic surface N atom. This explains the decrease in energy when removing H_2 from the surface. However, it is completely the opposite for the HF removal when using gaseous F. This molecule is a dipole, which has its positive part (H) directed to the surface. Hence, there is a positive attraction between H and N for a not too large distance between HF and the surface. This explains the need for an extra energy when abstracting H adsorbates on an N-covered surface by gaseous F species.

6.4.2.5 F-terminated N-covered cubic BN surfaces

The energy curve for the abstraction of F, with gaseous H, from a 100% F-covered N-terminated c-BN(100)–(1 × 1) surface, is shown in figure 6.20(a). In the region $0 \leqslant \xi \leqslant 0.43$, there is just a minor interaction between F_{ads} and the approaching H species. However, for smaller distances ($0.43 < \xi < 0.57$) repulsions start to take place. As can be seen in the deformation electron density maps in figure 6.26(a), there is an electron deficiency in the vicinity of the adsorbed F species, which is directed towards the approaching electrophilic H species. However, this repulsion is very weak, causing just a minimal increase in energy (up to +8 kJ mol^{-1} at $\xi = 0.86$). The corresponding $d(\text{N–F}_{ads})$ and $d(\text{F}_{ads}\text{–H})$ distances are 1.42 Å and 1.79 Å, respectively. For $\xi > 0.86$, and at the distance $d(\text{F}_{ads}\text{–H}) = 1.79$ Å, F_{ads} becomes abstracted from the surface, forming an HF molecule. Thus, this abstraction process is in principle barrier-less (<8 kJ mol^{-1}). The reason for this kinetically favorable abstraction process is the very weak N–F_{ads} bond. In fact, both N and F_{ads} have a negative charge, which results in a repulsive electrostatic contribution that destabilizes the N–F_{ads} bond. (The atomic charges of the surface N atom and its adsorbed F species are ($-0.15e \geqslant q(N) \geqslant -0.20e$) and $-0.17e \geqslant q(F_{ads}) \geqslant -0.19e$ in the region $0 \leqslant \xi \leqslant 0.86$.) F abstraction from an F-terminated N-covered c-BN(100)–(1 × 1) surface, using gaseous H, is therefore expected to be a highly facile process at experimental conditions.

On the contrary, the corresponding abstraction of adsorbed F species, with gaseous F, is completely different (as can be seen in figure 6.20(b)). In fact, the energy increases progressively towards a local maximum of +172 kJ mol^{-1} at $\xi = 0.54$, which corresponds to $d(N\text{–}F_{ads}) = 1.33$ Å and $d(F_{ads}\text{–}F) = 1.78$ Å. This observation is most probably caused by severe electrostatic repulsions between F_{ads} and F. (The calculated charges on F_{ads} and F are $-0.14e \leqslant q(F_{ads}) \leqslant -0.10e$ and $-0.32e \leqslant q(F_{ads}) \leqslant -0.36e$ in the region $0 \leqslant \xi \leqslant 0.54$.) When the approaching F comes even closer to F_{ads} ($0.54 < \xi \leqslant 0.62$), the negative interaction causes a repositioning of the F adsorbate. In fact, it binds down towards the surface, away from the approaching F species. This will, at first, stabilize the surface (by 13 kJ mol^{-1} at $\xi \leqslant 0.62$). The energy will, thereafter, continue to increase (in the region $0.62 < \xi \leqslant 1$), which depends on a continuous increase in electrostatic repulsion as the F species comes even closer to the surface N atom. The calculated atomic charges for this specific region are ($-0.36e \leqslant q(F) \leqslant -0.26e$), ($-0.25e \leqslant q(F_{ads}) \leqslant -0.15e$), and ($-0.36e \leqslant q(N) \leqslant -0.26e$). This is, hence, a very destabilizing situation with negative charges on all involving atoms (N, F_{ads}, and F). This increase in energy will be further accompanied by a decrease in F_{ads}-to-surface angle. Hence, no bond is formed between the approaching F species and F_{ads}, or N. It is, therefore, reasonable to conclude that gaseous F radicals are not suitable for the abstraction of F adsorbates from an F-terminated N-covered c-BN(100)–(1 × 1) surface. These results are similar to the results obtained for the process of F abstraction from the F-terminated B-covered c-BN(100)–(2 × 1) surface (as discussed above).

In summary, the results presented in this chapter indicate that F abstraction from an F-terminated c-BN surface (either B-covered of N-covered) is significantly more favourable if H radicals (instead of F) are used as abstracting species. This is an observation that has been made for halogens in general, and also for other types of materials [23]. The underlying reason is the significantly larger ionic character of the H–F bond, in comparison to the F–F bond. The difference in electronegativity values of H and F is 1.78, which is a difference that gives the H–F bond type notation: 'mostly ionic' [26]. On the contrary, the difference in electronegativity within the F–F bond is 0, which is a value that indicates pure covalency. Because of the difference in electronegativity, the unequal sharing of electrons in the H–F bond makes the F end of the bond negatively charged ($q(F) = -0.35e$) and the H end positively charged ($q(H) = +0.35e$). Consequently, the attraction between these charges will increase the energy needed to break the bond. In comparison, the F–F bond is weakened by strong repulsions between the non-bonding electron pairs in the F atoms. This reasoning is supported by the experimentally obtained bond energies of 567 and 159 kJ mol^{-1} for H–F and F–F, respectively [26]. For the F abstraction from the surface, the formation of H–F, instead of F–F, molecules will, therefore, provide a large energy gain.

6.4.3 Important considerations

The gas phase abstraction from a solid surface has been presented above with several examples. The solid surface was then exemplified with diamond and cubic boron nitride (c-BN). Diamond is a true covalent material, whilst c-BN is built from B–N dipoles. They otherwise have completely identical structures. The only difference is that every other C atom (in diamond) has been replaced with B atoms, and the other C atoms with N atoms. The cubic cell axes is, of course, determined by the types of atoms. When creating the most common surface planes of these materials ((111), (100) and (110)), the diamond planes will, hence, only contain C atoms, whilst this is not the case for all c-BN surface planes. In fact, the (111) and (100) planes are layered with alternate B and N planes, and (110) contains 50% B and 50% N. Hence, the (111) and (100) c-BN planar surfaces will either be covered with a B or N plane. Cubic BN is, therefore, a perfect example of a material where all surface planes look different chemically.

However, all types of material surface also show other types of differences. As presented in section 5.4, the different surface planes show different surface reactivities (which is caused by the deficiency in binding atomic neighbours). This depletion in surface electron density results in a combination of surface relaxation (section 5.6) and surface reconstruction (section 5.7). In addition, it will also increase the surface' reactivity towards reactions with gaseous species (section 6.3). Hence, different materials, with their different types of surface planes, behave completely differently when they are in contact with a gas mixture in, e.g. a CVD reactor. It is then easily understood that surface termination is of an utmost importance to understand (and even to predict) in the CVD growth of a specific material.

The main purpose of the examples above (see section 6.4.2) was to emphasize that it is not good enough to just calculate the reaction energies (e.g. abstraction energies). One must also follow the complete reactions by the construction of potential energy curves (see figures 6.16–6.20). Let us again take gas phase abstraction as an example in the construction of potential energy diagrams. In addition to information about abstraction energies, one gets information about the barrier energies that must be overcome for the abstraction to be realized. However, this is not good enough, especially if you want to learn something from the process. The reason is that these types of constructions are only observations, which is not good enough for the prediction of new and novel gas phase abstraction possibilities. Instead, information about surface geometrical structures and deformation electron density maps (as demonstrated in figures 6.16–6.20) must be added to the information from the potential energy curves. This is information that is highly valuable in the understanding and prediction of CVD growth processes. Furthermore, even more parameters can be added to the list of theoretical analysis methods (e.g. calculations of spin density maps) if other properties of the growing film are of any interest. It is important to understand that the calculations of more parameters will just strengthen and support the final conclusions.

6.4.4 Collapse of surface radical sites

The abstraction of a surface adsorbate results in a non-terminated (i.e. bare) surface site that might partially or completely collapse. It is even worse when neighbouring surface sites have become bare ones, which may have a negative effect on the CVD growth process. It is, therefore, necessary to also study this phenomenon. However, as will be shown below, it is advisable to not only study the structural geometries, but also other parameters as complementary tools. The idea is like above, the more the better (since it will enhance the probability to achieve a deeper understanding of how the material behaves during a CVD growth process). Let us take F-terminated c-BN(100) surfaces as an example.

6.4.4.1 F-terminated B-covered c-BN(100) surface

As a result of geometry optimization calculations, the stabilized F-terminated B-covered c-BN(100) has a (2 × 1) surface reconstruction (see figure 6.21(a)). This restructuring of the surface plane has saturated the dangling bonds (i.e. non-paired electrons) of the surface B atoms, which is the reason for the surface stabilization. Moreover, the surface B atoms exhibit a tetrahedral coordination (i.e, of a large degree of sp^3 hybridization), with an average surface N–B–N bond angle and B–N bond length of 109° and 1.57 Å, respectively. These values are in very good agreement with the corresponding bulk values (109.5° and 1.57 Å). This is evidence of the fact that a 100% F-termination on the B-covered c-BN(100)–(2 × 1) surface is able to uphold a bulk-like geometrical structure of the surface B atoms.

As can be seen in figure 6.22, the removal of an F adsorbate from this surface results in a pronounced, but very local, collapse of the B radical site. This B atom moves downwards with Δz = 0.35 Å and ends up with a more trigonal planar coordination (ν(N–B–N) = 116°). Hence, this collapse is accompanied by a somewhat buckled B–N–N triangle, with the B surface radical in the middle. This change in bonding coordination (for the radical B site) is a strong indication of a change in hybridization, from sp^3 to sp^2. This is most probably caused by a combination of (i) an inductive electron withdrawal from the less electronegative B atom towards the more electronegative second-layer N atoms, and (ii) a hyperconjugation interaction between the unhybridized 2p orbital on B and adjacent σ bonds (as explained in [29]). The electronegativity vales and atomic charges are 2.04 versus +0.31e for the surface radical B atom and 3.04 versus −0.71 e for the second-layer N atoms.

In addition to the geometrical structures presented in figures 6.21(a) and 6.22(a), electrostatic potential maps and deformation electron density maps are shown in figures 6.21(b), (c) and 6.22(b), (c). These various maps give us further information about the non-terminated surface sites. The electrostatic potential maps are shown in figures 6.21(b) and 6.22(b). They are negative electrostatic potentials (shown in

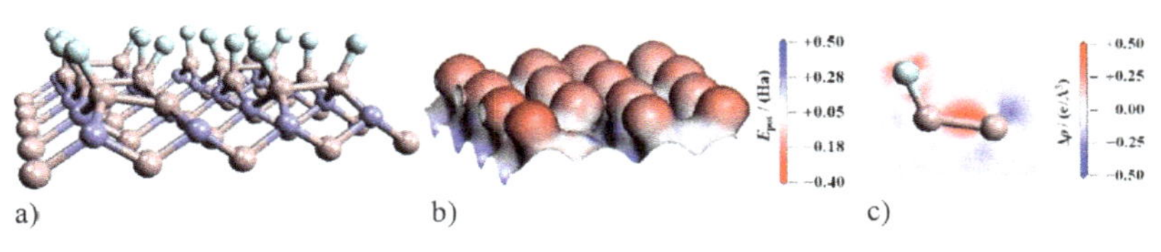

Figure 6.22 (a) 94% (monoradical) F-terminated B-covered c-BN(100)-(2x1) surface. (b) Electrostatic potential map for the surface shown in (a). (c) Deformation electron density map of the B-B dimer in a plane perpendicular to the surface shown in (a). [29], reproduced with permission © Elsevier.

red) since the F adsorbates have a large electronegativity value, and thereby withdraws electrons from the surface B species. The loss of an F adsorbate is clearly demonstrated by a comparison of figure 6.21(b) with figure 6.22(b). What is left after the F abstraction is just a positive electron density (blue) that indicates the presence of a bare B surface site. Furthermore, the abstraction-induced electron redistribution in the surface region is also obvious when comparing figure 6.21(c) with figure 6.22(c). Prior to the abstraction process, the electron densities within the B–B and B–F bonds are easily observed, in addition to the excess of electrons at the more electronegative F adsorbate (see figure 6.21(c)). However, the remaining non-paired electron (at the radical B site) has moved to the B–B bond as a result of the F removal from the surface. This is visualized by a negative electron density (blue) at the bare B site, and a more pronounced positive electron density (red) in the region between the two surface B sites. This B–B bond has, hence, become strengthened.

6.4.4.2 F-terminated N-covered c-BN(100) surface

As can be seen in figures 6.23 and 6.24, the calculations show that an F-terminated N-covered c-BN(100)–(1 × 1) surface has the capacity to maintain its sp^3-configuration, even after the removal of an F adsorbate. More precisely, a non-terminated N surface site will not collapse to its hexagonal form. Instead, a local 2 × 1-reconstruction of the bare N atom, with its neighbouring N atom, will take place.

Electrostatic potential and deformation electron density maps are also, in addition to geometrical structures, presented in figures 6.23 and 6.24. As was the situation with abstraction from an F-terminated B-covered surface, these various maps give us further information about the electronic structure of the non-terminated surface sites. As can be seen in figures 6.23(b) and 6.24(b), there are negative electrostatic potentials on the F adsorbate sites (shown in red) since the F adsorbates have a large electronegativity value. These values are even larger than for

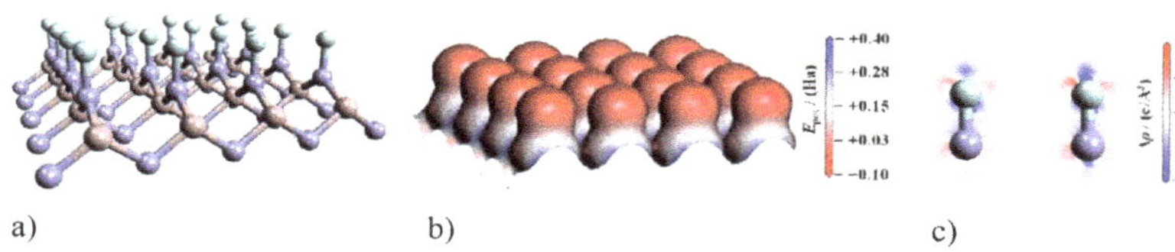

Figure 6.23. (a) 100% F-terminated N-covered c-BN(100)–(1 × 1) surface. (b) Electrostatic potential map for the surface shown in (a). (c) Deformation electron density map of a N–N dimer in a plane perpendicular to the surface shown in (a). [29], reproduced with permission © Elsevier.

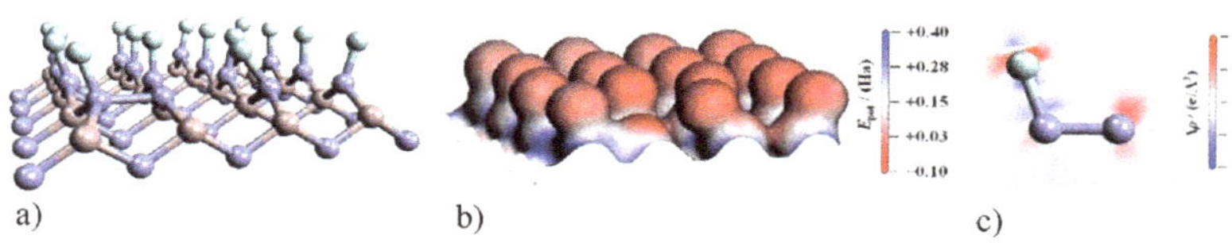

Figure 6.24. (a) 94% (monoradical) F-terminated N-covered c-BN(100)–(2 × 1) surface. (b) Electrostatic potential map for the surface shown in (a). (c) Deformation electron density map of the N–N dimer in a plane perpendicular to the surface shown in (a). [29], reproduced with permission © Elsevier.

N (3.98 vs. 3.04). The loss of an F adsorbate is also clearly demonstrated here by a change in the potential structure. However, in comparison with an F-terminated B-covered surface, the resulting mono-radical N-covered surface shows a negative electrostatic potential (while it became positive (blue) for the mono-radical B-covered surface). Moreover, the electron distributions in the surface regions are also shown in figures 6.23(c) and 6.24(c). Before any abstraction has taken place, there is a clear electron concentration on the nucleophilic N and F atoms, and there is a sign of electrostatic repulsions in between these atoms (marked in blue) (see figure 6.23(c)). However, after the F removal, the remaining non-paired electron (at the radical N site) has been used in the formation of a surface N–N bond (which is demonstrated by a region of electron density).

6.4.4.3 Comparison of F-terminated B- and N-covered c-BN(100) surfaces

The results presented above show that atomic F chemisorbs (i.e. chemically adsorbs) more strongly to the B-covered c-BN(100)–(2 × 1) surface than they do to the N-covered c-BN(100)–(1 × 1) surface. In fact, the calculated adsorption energies, ΔE_{ads}, for the adsorption of F to these surfaces are −471 and −283 kJ mol^{-1}, respectively. The chemisorption of F to the N surface sites is, hence, less exothermic than chemisorption of F to the corresponding B sites. The weaker bond strength for the N–F bond is most probably due to repulsions between the non-bonding electrons on N and F. The reason is the negative charge on both F (q(F) = −0.17 e) and N (q(N) = −0.20 e). This results in a repulsive (destabilizing) electrostatic contribution to the N–F bond energy. It is, however, completely different for chemisorbed F species on the B-covered surface. There is, for this situation, a positive charge on B (q(B) = +0.40 e) and a negative charge on F (q(F) = −0.25 e), which contributes to a strong interaction between these two atoms and a strong B–F bond. The B–F bond will, hence, be stabilized by an attractive electrostatic contribution.

Even though the F atoms on the F-terminated N-covered surface are nucleophilic, the process of F abstraction with gaseous F has been found to be endothermic (ΔE_{abs} = +63 kJ mol^{-1}). However, this abstraction process is less endothermic than the corresponding abstraction process on the B-covered surface (ΔE_{abs} = +252 kJ mol^{-1}). Thus, this observation is in agreement with the lower strength of the N–F bond. As presented above (see section 6.4.2), the F (or H) abstraction is expected to preferentially take place by using gaseous H (or F) species. The reason is that an F–F bond is weakened by strong repulsions between the nonbonding electrons on each of the F atoms. The tabulated bond energies for H–F, H–H, and F–F are 567, 436, and 159 kJ mol^{-1}, respectively [26]. An exchange of the formation of H–H and F–F bonds, with the formation of H–F bonds, will, therefore, provide a large energy gain in the abstraction process of surface H and F adsorbates.

These observations indicate that it is unfavorouble to use only F species as surface stabilizing agents, in the growth of c-BN(100). A better choice is to use only H species. However, the best alternative is to use a mixture of H and F. Such a mixture of gases would make it possible to grow c-BN exclusively through surface reactions, i.e. there is no need for surface damaging ion bombardment (these aspects will be further discussed in section 10.1).

This chapter has dealt with the eventual collapse of a surface radical site (being the result from a surface abstraction reaction). We have, though, only studied the surface restructuring during the formation of a single radical site, and the situation will be even more complex when removing several neighbouring adsorbates. The methodology used for studying this type of bare island formation is, however, identical to the one presented here. As is the situation with all examples in the present book, it is always preferable to add different calculated surface-related properties to the complete picture. For instance, an electrostatic potential map, in addition to a deformation electron density map, has been added here to the calculated surface geometry (as can be seen in figure 6.24). The combination of these pieces of information is highly valuable for the achievement of a deeper knowledge about these surface formations.

Furthermore, it is apparent that we have competing chemical processes in the examples in the present book. Take the radical B surface site, on an otherwise F-terminated c-BN (100)–2 × 1 surface, as an example (see figure 6.20). This is a layered c-BN surface with alternate layers of B and N. Since the N atom is nucleophilic it donates an electron pair to the nucleophilic B atom. At the same time, the N atom is more electronegative than B, and will, therefore, withdraw some electron density from the B site. A dipole will, thereby, be formed. In addition, we have, in fact, a third type of preference. We know from the periodic table that B has three valence electrons and N has five. This is, in fact, a prevalence for these elements in different types of compounds as well. Hence, the B atom prefers a trigonal planar formation (figure 6.21(a)), and the N atoms prefers a trigonal pyramidal formation (figure 6.23(a)). As can be seen in figure 6.21(a), the geometry of the surface B atom is not perfectly trigonal since any (or both) of the other competing factors have been too dominating (especially the tendency of N to donate two electrons to B).

6.5 Adsorption of growth species

As presented in chapter 3, the CVD growth mechanism is very complex and dynamic. It involves elementary reaction steps like adsorption to the growing surface, gas phase abstraction, surface migration, chemical reactions between, e.g. different growth precursors, and, finally, incorporation into a step, corner, etc. Furthermore, it is the slowest elementary reaction that determines the total growth rate. It is like a bottleneck. However, for this to be the situation, the CVD process must take place under strict surface kinetic control (see chapter 3). Hence, it is the heterogeneous surface reactions that govern the overall CVD growth rate. As for chemical reactions in general, it is the rate-determining (i.e. slowest) surface reaction that determines this growth rate. More importantly, this rate-determining step does not need to be in thermodynamic equilibrium. As was also presented in chapter 3, there are chemical reactions in different regions of the CVD chamber. However, in this chapter we will only consider the chemical processes that take place on the growing film (i.e. in the solid/gas interface).

As already emphasized, the CVD growth processes are very complex. Hence, it is therefore a difficult task to simulate them by using theoretical calculations. It is not, as many researchers wish, possible to just make a model, choose some parameters, press a button, and then sit down and watch the complete growth showing up as a movie on the screen. It is, instead, a jigsaw puzzle that must be assembled, with a lot of CVD experience as a helping tool. However, what is common to all of these growth processes is that they must involve the adsorption of a growth species to the growing film. This adsorption (or rather chemisorption) is, thereafter, followed by either a chemical reaction with a gaseous phase species, or a surface migration to another adsorbate on the surface (or to an edge or kink site), by which it undergoes chemical reactions. In addition, these surface processes are facilitated by gas phase abstraction reactions. This is a rather course generalization of the CVD growth processes. There are two general growth mechanisms that govern the growth process, which both involve the presence of two gaseous growth species: Eley–Rideal (E–R) and Langmuir–Hinschelwood (L–H) (see section 3.3.3). According to the E–R mechanism, one growth species will chemisorb to the surface, from where it will react with the other gas phase growth species. On the contrary, both gaseous growth species will first chemisorb to the surface before they react with each other in the L–H mechanism.

Three different types of CVD growth processes will be presented below (see chapters 7 and 8). All of them have different and specific problems that must be solved, and we show here how theoretical modelling can be of great help with these issues.

6.6 Identification of the rate-limiting step in the CVD growth of diamond

6.6.1 Adsorption of plausible growth species to a surface radical site

The following example shows a CVD growth process where only one growth precursor is responsible for the growth. It demonstrates a layer-by-layer growth of a diamond material.

As is the case with CVD growth in general, the CVD of diamond is a very complex and dynamic process [30]. The growing surface must be terminated by, e.g. H atoms since it will otherwise collapse to the graphitic phase. However, these terminating species must, in turn, be abstracted away from the surface to leave room for an approaching growth species (e.g. CH_3). The abstraction of adsorbed H species from a diamond surface has already been dealt with in section 6.4.2. It was observed then that this process is not spontaneous. There is a small barrier energy of +12 kJ mol^{-1} and +5 kJ mol^{-1} for the (100)–2 × 1 and (111) surface, respectively. What we now must do is to identify the growth species for a diamond CVD setup. As expected, these ones differ depending on the type of reactor.

In the very beginning, hot-filament CVD reactors were used for growing artificial diamond. The most common gas mixture was then 1%–4% methane and 96%–99% hydrogen, and this is still a mixture in use. The role of hydrogen is not only to uphold the cubic diamond structure, but to also etch away non-diamond forms of carbon (such as graphite) [31]. Also, if using a larger concentration of methane, a larger excess of non-diamond will be formed. It is, therefore, very important to keep a certain ratio between the carbon source and the H etchant.

In a very early experimental study by Chu *et al*, the growth mechanisms were studied for the growth of diamond onto the (111), (100), and (110) crystal planes of natural diamond [32]. This was a type of competition study where both carbon-13-labeled methane and carbon-12 acetylene were used. For growth of diamond (111) and (100), the carbon-13 mol fraction of the diamond film has indicated that the CH_3 radical is the dominant growth species under hot-filament CVD conditions. However, it is not clear if the methyl radicals are the primary growth species for the deposition of an ideal diamond (110) structure. However, Belton and Harris proposed a detailed deposition mechanism involving the adsorption of C2H2 to a (110) diamond surface [33]. A good agreement between the calculated and observed growth rates was then obtained.

A mixture of CH_3 radicals and C_2H_2 molecules, in the gas phase above the growing diamond surface, has been observed when using a filament-assisted diamond deposition method and a gaseous mixture of methane (CH_4) and hydrogen (H_2). However, the concentrations of each of these two gaseous species must be estimated before any conclusion can be drawn about the dominant growth species. This has been done in different *in situ* experiments [32–34]. The C_2H_2 concentration was estimated to be almost equal to the CH_3 concentration. In the hot filament activation of a CH_4/H_2 mixture, other C-containing molecules will also be formed (such as $CH_{2,}$ CH, and C_2H species). However, kinetic modeling showed that the concentrations of CH_2 and C_2H are much lower than the concentrations of CH_3 and C_2H_2 (by about two and four orders of magnitude, respectively). Thus, characterization of the gas phase right above the growing film surface is very important, if not necessary, especially for the theoretical simulation of the growth species adsorptions. This will be further understood from the results of the calculations shown below.

The models for the chemisorption of CH_3, $CH_{2,}$ CH, C_2H, and C_2H_2 species are shown in figures 6.25 and 6.26. The adsorption of these C-containing molecules is assumed to take place onto a bare surface C site on an otherwise H-terminated diamond (111) (figure 6.25) and (100)–2 × 1 (figure 6.25) surface. (An H-terminated diamond surface is being used since this is the type of termination obtained when using an excess of H_2 in the CVD reactor.) The adsorption energy values of these C-containing species can be seen in table 6.1. The calculated absolute adsorption energies for all species, including that of the competing H species, are in the following order: $C_2H > H \approxeq > CH > CH_2$(triplet) $> CH_3 > C_2H_2$. The CH_2 molecule has two non-binding electrons. It can then either pair these electrons (for the CH_2(singlet)) or keep them unpaired in different orbitals (for the CH_2(triplet)). However, the triplet state has been shown earlier to give stronger adsorption energies, so it has been chosen here as a plausible adsorbate species. Furthermore, this order of energies also shows that the radical C_2H species adsorbs the strongest to the H-terminated diamond (111) surface (even compared to the radical H species). This means that C_2H is a good competitor to H and has a good chance to ‘win this race’, i.e. be the one that chemisorbs to the radical surface site. As can also be seen in table 6.1, the C_2H_2 molecule is binding significantly weaker as compared with the other hydrocarbon species. Moreover, it is obvious that the calculated adsorption energy for the gaseous CH_3 radical is much larger than for the onefold adsorption of

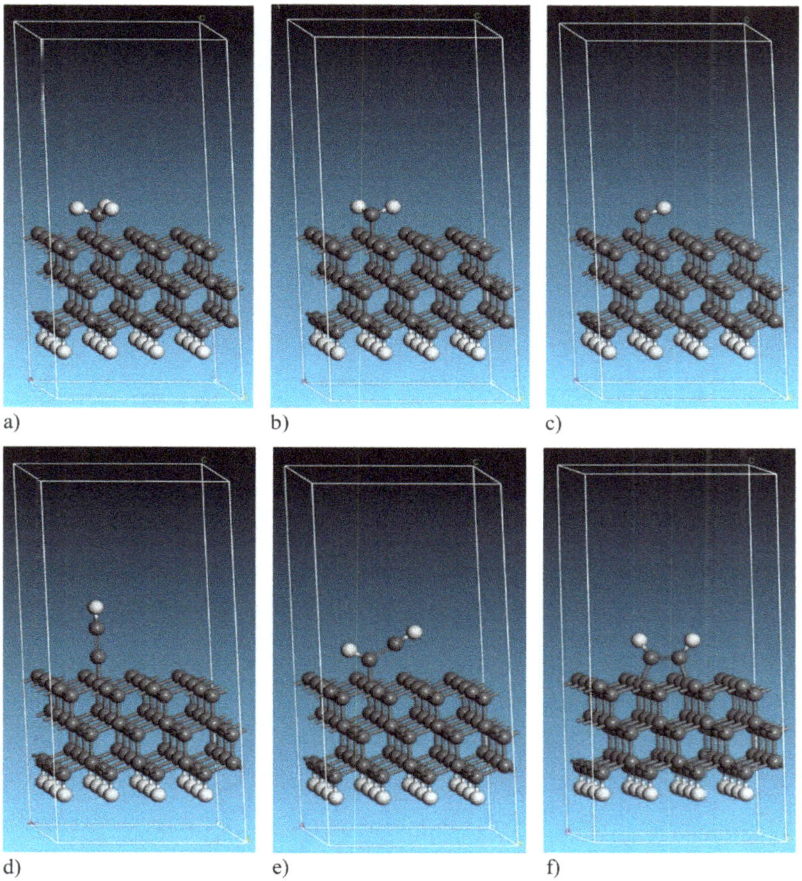

Figure 6.25. Models of the adsorption of (a) CH_3, (b) CH_2, (c) CH, (d) C_2H, and (e) and (f) C_2H_2 onto an otherwise non-terminated diamond (111) surface. The C_2H_2 species can either bind with one bond (e), or two bonds (f).

C_2H_2 (−315 vs. −72 kJ mol^{-1}). There is also the possibility for a two-fold adsorption of C_2H_2, but then we need a surface with two neighbouring surface C radicals (see figures 6.25(f) and 6.26(f)). However, a final two-fold adsorption of C_2H_2 can take place by an initial one-fold site adsorption followed by a combination of gas-phase abstractions and surface migrations of surface H adsorbates (se section 6.6.4). As can also be seen in table 6.1, the calculated adsorption energies of the species C_2H, CH, and CH_2 (triplet) are also larger than the adsorption energy of CH_3. However, one must remember that there is normally a supersaturation of H radicals in the gas mixture over the growing film, in addition to the much smaller concentration of C-containing gas phase species. These relative gas phase concentrations are a direct consequence of the relative concentrations of the inlet gases (appr. 98% H_2 and 2% CH_4) in the CVD reactor. Thus, the effect of these rather high adsorption energies on the diamond growth is counteracted by the low gas-phase concentrations in

Figure 6.26. Models of the adsorption of (a) CH_3, (b) CH_2, (c) CH, (d) C_2H, and (e) and (f) C_2H_2 onto an otherwise H-terminated diamond (100)–2 × 1 surface. The C_2H_2 species can either bind with one bond (e), or two (f).

common CVD setups. On the basis of performed *in situ* gas phase analysis, it is strongly believed that the very low concentration of especially C_zH precludes this species from having a significant role as a growth species during growth of diamond (111). However, the significant difference in adsorption energy for CH_3 and C_2H_2, together with the fact that the gas-phase concentrations of these species are about the same, does strongly support a diamond (111) growth based on CH_3 as the

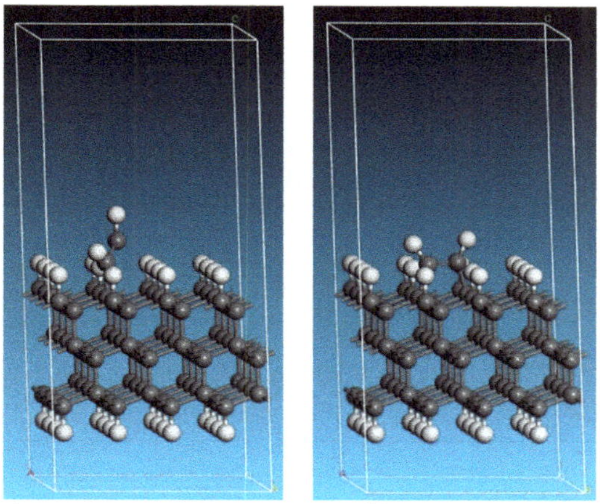

Figure 6.27. Models of the (a) onefold- and (b) twofold-site adsorption of C_2H_2 onto an otherwise H-terminated diamond (111) surface.

Table 6.1. Calculated adsorption energy values for the adsorption of H, and various hydrocarbon molecules, to a radical site on an H-terminated diamond (111) surface.

Adsorbate	Adsorption energy (kJ mol^{-1})
H	−428
CH_3	−315
CH_2	−415 (triplet)
CH	−348
C_2H_2	−72
C_2H	−556

dominant growth species. These theoretical results and conclusions [35–37] are consistent with the experimental results of Chu *et al* [32].

6.6.2 Effect of neighbouring adsorbates

The adsorption of growth species to a growing film is one of the most important steps in the CVD growth mechanism. As such, the influence of the configuration and type of adsorbed species (and vacancies) that surround this specific adsorbate are of the largest importance for the final growth process. This adsorption process is exemplified here with the CVD growth of diamond (111). The importance of the influence of adjacent adsorbates, on the adsorption of H, CH_3, CH_2, C_2H, and C_2H_2 species, is demonstrated below. The neighbouring surface site is then either non-terminated, or terminated with H, CH_3, or CH_2 species. For more detailed information, see [38]. As can be seen in table 6.2, the order of the various adsorption

Table 6.2. Calculated adsorption energy values for the adsorption of H, and various hydrocarbon molecules, next to a radical site, or an H, CH_2 or CH_3 adsorbate, on an H-terminated diamond (111) surface.

Adsorbate	Adsorption energy H (kJ mol^{-1})	Adsorption energy Radical site (kJ mol^{-1})	Adsorption energy CH_2(kJ mol^{-1})	Adsorption energy CH_3 (kJ mol^{-1})
H	−428	−431	−418	−409
CH_3	−315	−317	−253	−199
CH_2	−334	−340	−261	−257
C_2H_2 (onefold)	−72	−75	−101	+97
C_2H_2 (twofold)		−356		
C_2H	−556	−630	−5	−475

energies is $C_2H > H > CH_2 > CH_3 > C_2H_2$ for all types of neighboring environments. The adsorbed species is then either surrounded by only H adsorbates, or one of these H adsorbates has been removed leaving a diradical site or being replaced with CH_2 (or CH_3). However, the order of adsorption energies is different when allowing for a twofold site adsorption of C_2H_2 to a surface site with a neighbouring radical surface site: $C_2H > H > C_2H_2$ (twofold) $> CH_2 > CH_3 > C_2H_2$ (onefold). Since there is an overall large probability for an H abstraction to take place (see section 6.4), this is also the situation for the formation of a radical surface site next to the growth species adsorbate. If we now consider C_2H_2 to be a plausible growth species adsorbate, it becomes highly stabilized via the formation of a neighbouring surface radical site (see figure 6.2). The stabilization of the two neighbouring surface radicals by a C_2H_2 species, hence, produces a stronger adsorption to the surface, as compared with the single bond formation by CH_3 or C_2H_2 (−356, −317, and −13 kJ mol^{-1}, respectively). The question is now if the necessary formation of a neighbouring surface site radical has enough time to be formed. In fact, there are two parallel surface chemical reactions that are competing with each other. The onefold-site binding of C_2H_2 to a monoradical surface sites is rather weak (with an adsorption energy of −75 kJ mol^{-1}) and will, therefore, have a very large probability to become desorbed at normal CVD conditions. Thus, for a strong twofold-site adsorption of C_2H_2 to take place, the abstraction of a neighbouring H adsorbate must be faster than the desorption of the weakly one-fold site binding C_2H_2 adsorbate. So, the question is, which is the fastest surface reaction? The answer to this question will completely change the possibility for C_2H_2 to become the major growth species for the CVD growth of diamond (111).

With just one exception, neighboring CH_3 and CH_2 adsorbates do sterically hinder the adsorption of the different hydrocarbon species, with a lowering in adsorption energy values of about (70–120) and (50–100) kJ mol^{-1}, respectively (see table 6.2). The geometrical structures for these situations are shown in figures 6.28 and 6.29. The exception is the observation that a onefold-site C_2H_2 adsorption cannot take place on a surface site adjacent a C_2H_2 species to an otherwise

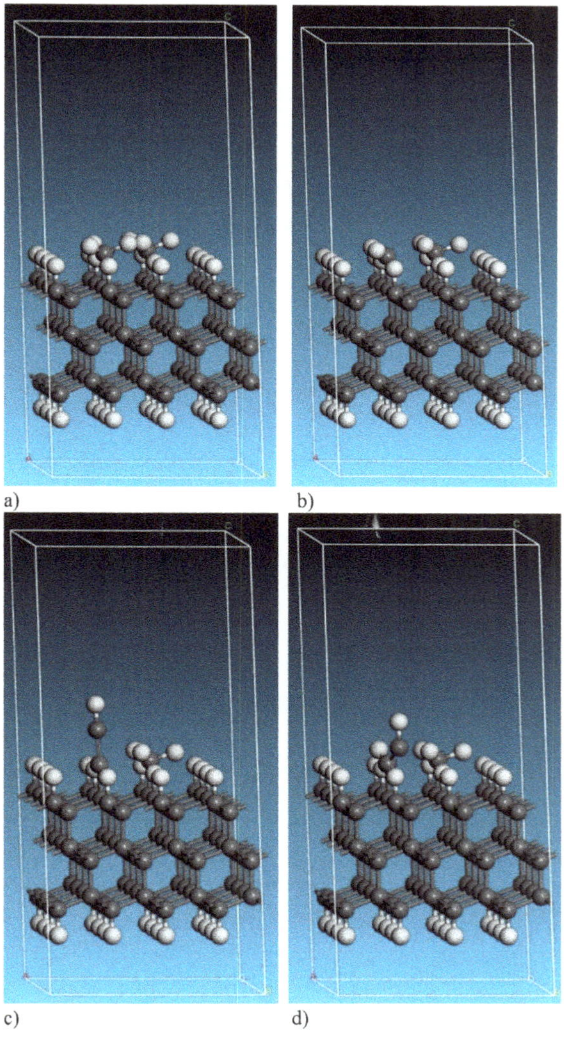

Figure 6.28. Models of the adsorption of (a) CH_3, (b) CH_2, (c) C_2H, and (e) C_2H_2 next to an adjacent CH_3 adsorbate (on an otherwise H-terminated diamond (111) surface).

completely hydrogenated diamond (111) surface. Consequently, the difference in adsorption energies for the C_2H_2 and CH_3 adsorbates will decrease dramatically, by about 170 kJ mol^{-1} (from 319 to 152 kJ mol^{-1}), if these adsorbates end up at a position neighbouring an CH_2 adsorbate. Thus, for the diamond (111) surface, an adjacent CH_2 adsorbate makes the adsorption energy of C_2H_2 (onefold) to approach that of CH_3. This result, together with the still stronger bonding of a twofold-site adsorption of C_2H_2, makes it impossible to preclude the C_2H_2 species from any role as a growth species by using only calculated adsorption energies. In fact, the situation with an adjacent CH_2 adsorbate is very easily accomplished by hydrogen

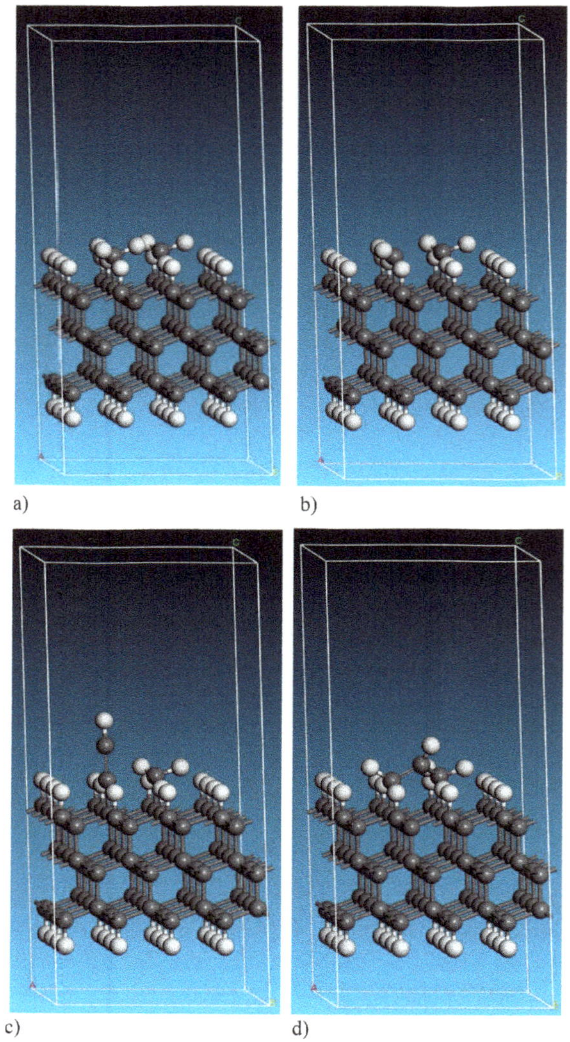

Figure 6.29. Models of the adsorption of (a) CH_3, (b) CH_2, (c) C_2H, and (d) C_2H_2 next to an adjacent CH_2 adsorbate (on an otherwise H-terminated diamond (111) surface).

abstraction of an already adsorbed CH_3 species (which will be further discussed below).

In short, it has been demonstrated here that although a plausible growth species binds strongly to the growing surface, it does not necessarily do so for all types of surface sites. One must be very careful in modeling the most realistic surface under, e.g. CVD conditions. As mentioned elsewhere in the present book, the modeling and simulation of CVD growth process, as well as of other types of surface chemistry processes, is like doing a jigsaw puzzle with your earlier acquired experience/knowledge and experimental evidence as the only input to an CH_3 adsorbate. The sterical hindrances become too severe. Furthermore, the onefold-site adsorption of C_2H_2 next to the CH_2 adsorbate (see figure 6.35(d)) creates a new C–C bond between these two adsorbed hydrocarbon species, which is most probably responsible for the increase in adsorption energy by about 100 kJ mol^{-1} (as compared with the adsorption of important steps in the CVD growth mechanism). However, a gas phase abstraction of surface-terminating species must first take place before any migration of growth species can start to take place. In fact, each following migration step must be initiated by an abstraction from the surface (under the assumption that the surface is completely terminated). Take the CVD growth of diamond as an example. In the initial part of the diamond growth process, the abstraction of two surface H species (terminating the surface and one of the H ligands on, e.g., CH_3) must take place before surface migration can start to take place. The reason is the following. A gaseous CH_3 species is generally assumed to be the dominating growth species in ordinary CVD growth of diamond. However, the growth rate of the CH_3 species is very slow, but the migration rate of CH_2 is much faster. The situation now is that CH_3 adsorbates surface migration (i.e. an instantaneous jumping).

6.6.3 H abstraction from adsorbed growth species

As discussed in chapter 3, surface migration of the growth species is one of the most of a CH_2 species between two surface radical sites that only one H neighbouring adsorbate needs to abstract for each jump. Hence, the process of H abstraction is a very important reaction step in the overall diamond growth mechanism [39]. Since the adsorption energy calculations (as presented above) showed that both CH_3 and C_2H_2 are plausible growth species for HF-CVD growth of diamond (111), we will continue with these types of adsorbates in demonstrating the calculations of H abstraction from already adsorbed species. As was the situation with an H-terminated surface plane, there is also a high possibility for H abstraction to take place from already adsorbed CH_3 and C_2H_2 species. The reason is the super-saturation of H radicals in the CVD chamber. As can be seen in table 6.3, the gas phase abstraction of H from both CH_3 and C_2H_2 adsorbates is weakly exothermic, with the abstraction energies −24 and −28 kJ mol^{-1}, respectively. Hence, these reactions are thermodynamically feasible. However, the abstraction of surface

Table 6.3. Values of adsorption energies, H abstraction energies, and migration barrier energies for various adsorbates onto an otherwise H-terminated diamond (111) surface. The adsorbates include H, CH_2, CH_3, C_2H, and C_2H_2 (onefold chemisorbed).

Adsorbate	Adsorption energy (kJ mol^{-1})	Abstraction energy (kJ mol^{-1})	Migration barrier energy (kJ mol^{-1})
H	−428	+12	+248
CH_3	−315	−24	+353
CH_2	−334		+52
C_2H_2	−72	−28	+186
C_2H	−556		+350

terminating H species is weakly endothermic, which means that an increased temperature, or a supersaturation of gaseous H species, is necessary for the reaction to take place.

6.6.4 Surface migration processes

For the development of new CVD setups, as well as the improvement of already used ones, it is of an outmost importance to obtain a deeper knowledge at the molecular level of, e.g. the surface processes during CVD growth. One important factor is the energetics during the growth. As stated above, adsorption energies, as well as barrier energies for migration, are then of a special interest to study. The barrier energies cannot, unfortunately, be determined experimentally. Instead, one must rely on high-quality theoretical calculations (like DFT). One important aspect to study is whether surface migration should be taken into account, or not, as a rate-limiting step in the complete CVD growth mechanism. We will here exemplify the importance of surface migrations by calculating the barrier energies for some plausible growth species in the HF-CVD growth of diamond (111).

A surface migration process is generally assumed to involve two neighboring-surface radical sites. The migration of an adsorbate is then assumed to occur as a single jump between these two surface sites. For this specific example (i.e. diamond (111) growth), the migration will be of an isotropic character since every surface carbon has six symmetrically equivalent carbon neighbors on the surface (see figures 6.30 and 6.31). As discussed above, gaseous CH_3 and C_2H_2 species are found to be the most plausible growth species in a HF-CVD reactor containing a gas mixture of 1%–2% CH_4 and 98%–99% H_2. In addition, the growing diamond surface is expected to be completely (100%) H-terminated. As a result of this type of supersaturation of H radicals in the CVD chamber, gas phase abstraction of H is expected to take place (both from the surface and from the adsorbed CH_3 or C_2H_2 species).

Let us start with the abstraction of H from the growing diamond (111) surface (see figures 6.32 and 6.33). As is also highlighted in these figures, the abstraction energy is somewhat endothermic (by +12 kJ mol^{-1}). The CH_3 (figure 6.32) and C_2H_2 (figure 6.33) species will, thereafter, become chemisorbed to these surface sites (i.e. become chemically adsorbed to the surfaces). As shown in table 6.1, the calculated adsorption energies are −315 and −72 kJ mol^{-1}, respectively, for these processes. The adsorbed CH_3 and C_2H_2 species will thereafter undergo H abstraction, which results in CH_2 and C_2H adsorbates. However, there is another way for the CH_3 and C_2H_2 adsorbates to be transferred to adsorbed CH_2 and C_2H species. DFT-based molecular dynamic simulations (at 1200 K) have shown that an H atom in the CH_3 or C_2H_2 adsorbates will quickly migrate (within 1–2 fs) to a neighboring surface radical position, leaving a CH_2 or C_2H adsorbate, respectively. However, for this to occur, there must be a vacant neighbouring surface site. Hence, there must also be a gas phase abstraction process for this H adsorbate.

The migration processes of the CH_2 and C_2H adsorbates are, thereafter, assumed to be a simple jump between the radical sites. Furthermore, it is advisable to not only

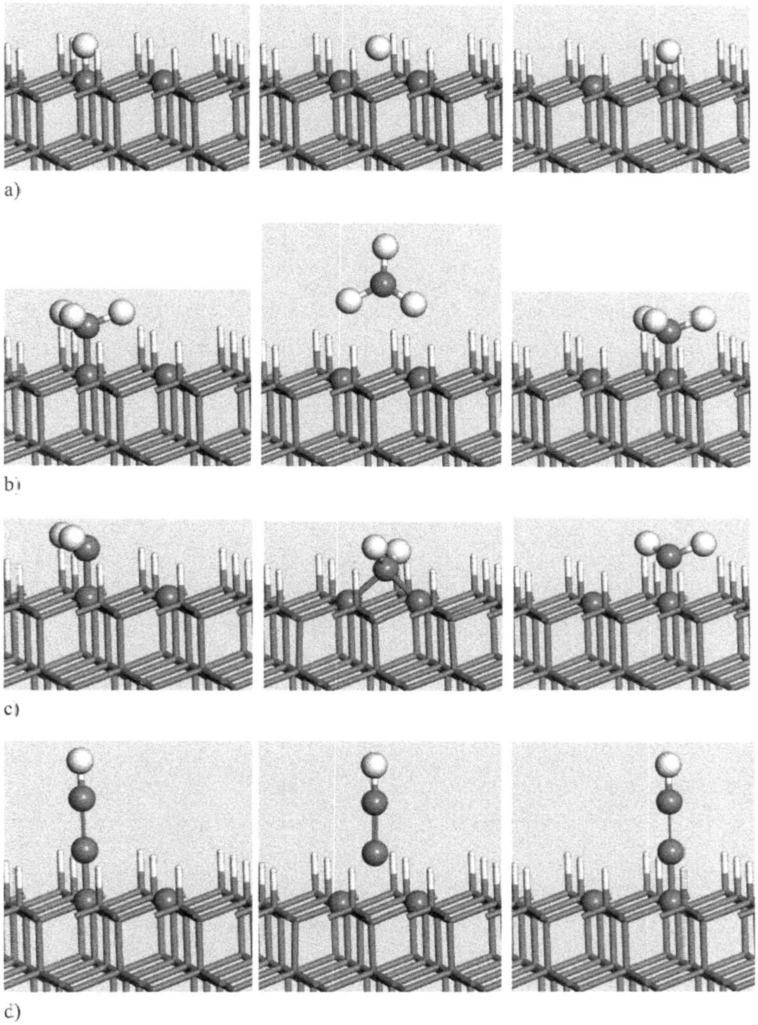

Figure 6.30. Geometrical structures of the surface-adsorbate region; (left column) before the migrational jump, (middle column) transition state at the barrier position, and (right column) after the migrational jump. The adsorbates are (a) H, (b) CH_3, (c) CH_2, and (d) C_2H.

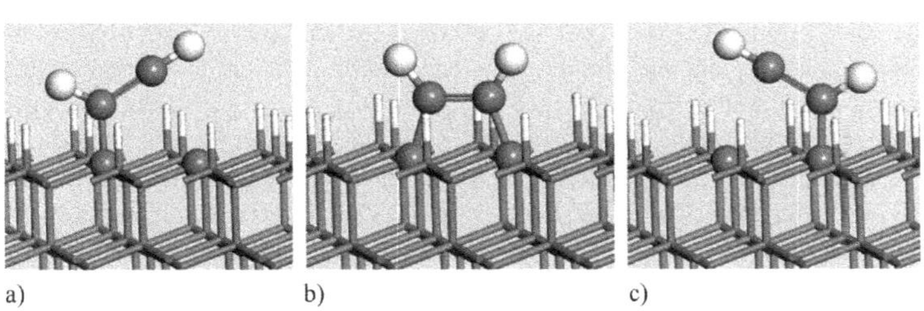

Figure 6.31 : Geometrical structures of the surface-adsorbate region; (a) before the migration, (b) transition state at the barrier position, and (c) after the migration of C_2H_2.

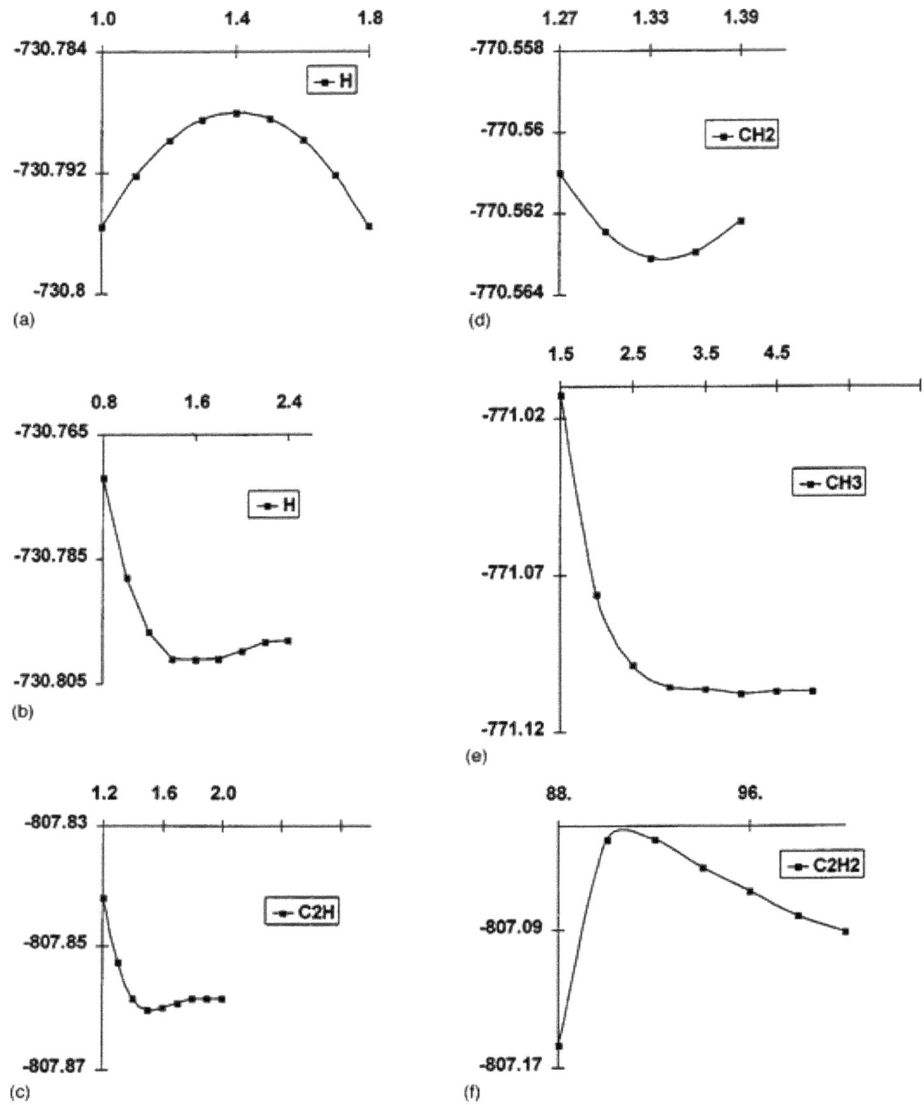

Figure 6.32. Potential energy paths used for the determination of transition-state geometries for (a) and (b) H, (c) C_2H, (d) CH_2, (e) CH_3, and (f) C_2H_2. (a) The obtained potential maximum when moving H in parallel with the surface, and in a direction from one surface radical to the other one (due to symmetrical reasons exactly in between the two surface diradical sites). (b)–(e) The potential path obtained when moving the species in a direction perpendicular to the surface. The y axes show the calculated total energies of the systems. The x-axis in (a) shows the distance from one of the radical surface sites, and (b)–(e) represent the distances between the migrating species and a plane intersecting the surface C atoms. The x-axis in (f) represents the angle formed by the surface carbon and the two C atoms in C_2H_2. The energy unit is au, and that of the distance is Å.

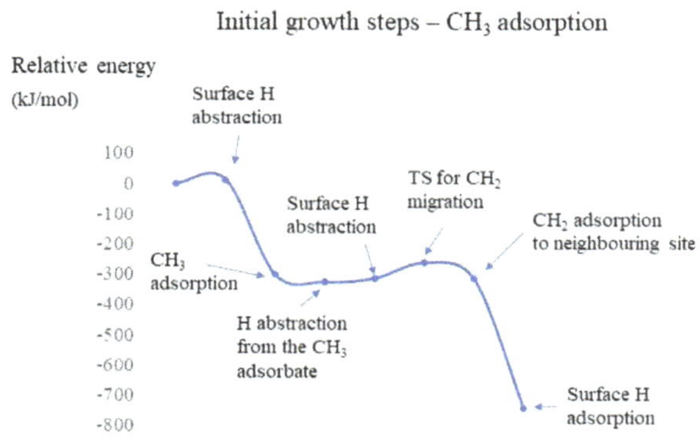

Figure 6.33. Initial diamond (111) growth steps for the use of a CH_3 growth species and starting from a 100% H-terminated surface (with the assumption that the total energy for this system is 0 kJ mol^{-1}). The next-coming data points illustrates the various reaction energies when going from left to right.

simulate these migration processes, but also do the same for the initial CH_3 and C_2H_2 adsorbates. This is also the situation with the H adsorbates. The reason for this latter type of adsorbate is the following. It is not important that the gas phase abstraction of the neighbouring H adsorbate takes place if the surface migration of H on the surface is fast. An abstraction of an H adsorbate further away would then create the necessary radical site, simply a fast H migration (i.e. the neighbouring H atom will migrate away, leaving this radical site). The migration process of CH_3 and H will then be, as was the case for CH_2, simple jumps. The idea of a single jump process between two radical surface sites was then tested by performing MD simulations at 1200 K, which is a temperature close to the diamond deposition temperature. The jumping process of the CH_2 adsorbate was then observed to take place within 50 fs. Hence, the jumping migration of the CH_2 adsorbate was confirmed, and it was verified to easily take place at ordinary CVD conditions.

However, the situation is different for C_2H_2. As shown in table 6.3, a onefold adsorption of a C_2H_2 species (to this type of radical site) is much weaker than a corresponding twofold site adsorption to a diradical surface site (−72 vs −636 kJ mol^{-1}). Due to this observation, the C_2H_2 species is assumed to migrate by an alternating one and twofold site adsorption onto the diamond (111) surface (see figure 6.31).

Due to the difficulties to locate the transition states, the migration barriers are normally more difficult to calculate than adsorption energies. This is an extremely demanding computational task for large models, which is, however, absolutely necessary for simulations of CVD surface processes. Alternative methods for transition state location (i.e. for calculating the barrier of energies) are presented in section 4.5.3. Since all surface sites on a diamond (111) surface have identical surroundings, there is a more time-efficient way to locate this transition state. The transition state will for this type of symmetry be located exactly between any two of surface sites. For the jump-wise type of migrations (for H, CH_3, CH_2, and C_2H) the transitions-state geometries are presented in the middle column in figure 6.30.

Furthermore, the transition state for the special type of migration of C_2H_2 is shown in figure 6.31(b).

The potential energy curve describing the energy path when going from the chemisorbed state to a transition state (during the surface migration) has been determined here as follows. The species was initially positioned at adsorption positions on one of the surface radical sites, and thereafter moved step-wise towards the other adsorption position on the neighbouring surface radical site. The main element in the migrating species (e.g. C in CH_2, and one of the C atoms in C_2H) had, for each step, to be fixed in its position, while the rest of the uppermost atoms (in the model) were geometry optimized. Other underlayer atoms were kept fixed and terminated (here with H atoms) to simulate the continuation into bulk. By using this methodology, potential energy curves along a direction in parallel with the surface, and between the two surface radical sites, could be obtained. As demonstrated in figure 6.32(a) for H, the obtained potential energy curves show a maximum between the approaching start- and end-point of the migration path. Of symmetrical reasons, it is positioned strictly at the midpoint between the two radical sites. However, the stationary point can be a minimum for the situation with a stable bond formation with the surface.

An improved barrier of migration can, thereafter, be obtained by starting at the firstly obtained energy maximum (as described above) and then search for an energy maximum in a direction perpendicular to the intersecting line between the two surface radicals. The next step on the search for the TS is, therefore, to move the adsorbate species towards the surface and on a line that crosses the midpoint between the two surface radical sites. An energy barrier can in this way be localized perpendicular to the surface plane, As expected, the stationary point can here also be a minimum for the situation with a stable bond formation with the surface. It is important to note that one must start with a position far away from the surface and continue with this procedure until the approaching species has reached a position very close to the surface. For the present examples, the initial distances for H, CH_3, CH_2, and C_2H were 1.8, 4.6, 1.4, and 2 Å above a plane intersecting the upper surface C atoms. The species was, thereafter, manually, and stepwise, moved towards the surface. (These steps must be as small as positive to give reliable results.) The main element in the approaching species (e.g. C in CH_2, and one of the C atoms in C_2H) here, for each step, also had to be fixed in its position, while the rest of the uppermost atoms (in the model) were geometry optimized. Other underlayer atoms were kept fixed and terminated (here with H atoms) to simulate the continuation into bulk. By using this methodology, the potential energy curve along the z-direction can, thereby, be obtained. As demonstrated in figures 6.32(b)–(e) for H, CH_3, CH_2, and C_2H, the obtained potential energy curves show a maximum (minimum in the case of CH_2) somewhere in between the approaching start- and end-point of the migration path. Thus, an energy barrier can in this way be localized for the jump-wise migration of H, CH_3, CH_2, and C_2H onto the diamond (111) surface. However, as was the situation with the approaching migration, the stationary point can be a minimum for the situation with a stable bond formation with the surface.

The resulting transition states (TS) for H, CH_3, CH_2, C_2H_2, and C_2H can be seen in figure 6.30. (The migration process of C_2H_2 will be discussed further below.) As a result of these calculations, we have now determined the position of the transition states, and can, thereafter, go on with the calculations of the barrier energies. (The procedure for the location of the TSs that was used in the present calculations is rather simplified, and should be followed or eventually replaced by any of those theoretical methods that exist for the location of TSs (see section 4.5.3). However, they might be impractical to use due to the size of the model.) These barrier energies are very easily calculated. They are just calculated as the difference in energy for the TS and the position of the gaseous species furthest away from the surface (see equation (6.7)).

$$E_{\text{barrier}} = E_{\text{TS}} - E_{\text{longest distance}} \tag{6.7}$$

where E_{TS} is the energy at the transition state position, and $E_{\text{longest distance}}$ is the system's total energy for the starting position of the migration process.

For these specific calculations, the positions of the transition states were also re-checked by performing geometry optimization to a transition state (see [40] for more details). As a matter of fact, this was done to also evaluate the specific methodology used for locating the TSs. A discrepancy of ~8% was then obtained. However, if this value is to be regarded as a minor error or not depends completely on the problem at hand. It was, in [40], considered to be a minor one for the purpose of that specific work.

As can be seen in figure 6.30(a), the position of the TS for the migrating H species is, as expected, not situated at the line intersecting the two adsorption sites (i.e. the final and last position in the jump process). It is, instead, positioned about 0.5 Å above the midpoint of that line. This is reflected in the energy barrier, which is appreciable smaller, by about 69 kJ mol^{-1}, compared to the energy reported in [41]. In that work, they just calculated the energy barrier for a TS position that was exactly in the middle of the positions of the starting position and ending position of the H adsorbate. Thus, there were no change in the z-direction during their assumed H migration. This oversimplified picture of the migrations process is expected to give an erroneous result in the energy barrier for the H adsorbate.

However, it might be difficult to find the true TS state with its barrier energy for a migrating species on the surface. The situation is, though, that the true TS position (i.e. a TS position at experimental conditions) gives the lowest barrier energy. This is why you always calculate a barrier energy that is larger than, or equal to, the true one. In other words, the true barrier energy is a lower boundary to your calculated one.

The barrier energies, together with the adsorption energies, for the candidate species involved in diamond growth, are presented in table 6.3. As can be seen in this table, the order of energy barriers is $+52 < +248 < +350 \leqslant +353$ kJ mol^{-1} for CH_2, H, C_2H, and CH_3, respectively. The surface-terminating H species is, thus, more mobile than C_2H and CH_3, but much less mobile compared to CH_2. This is understandable since CH_2 is forming a stabilized two-fold bond with the surface at the TS, while H shows no chemical bond. The corresponding adsorption energies are -334, -428, -556, and -315 kJ mol^{-1}, respectively.

When studying the migration of an adsorbate, what one must consider is the possibility for the adsorbate to desorb from the surface (instead of migrating). Hence, we have two competing surface reactions here! Let us now continue by coarsely assume that what we gain in energy during adsorption is what we must add for the adsorbate to desorb from the surface (i.e. $\Delta E_{des} = -\Delta E_{ads}$). The order of desorption energies will then be $+315 < +334 < +428 < +556$ kJ mol^{-1} for CH_3, CH_2, H, and C_2H, respectively. When comparing the desorption energies with the barrier energies, it is obvious that the barrier energies are all considerably smaller. Thus, since we know that the barrier energies we determine theoretically are an upper limit to the true ones at experimental conditions, we are also certain that this statement is true (i.e. the barrier energies are smaller than the desorption energies). The energy differences are, with one exception, within the 193–364 kJ mol^{-1} range. The exception concerns the methyl, CH_3, species, for which the adsorption and barrier energies are rather similar (+388 vs +353 kJ mol^{-1}). In fact, the barrier of migration for this very species is the largest amongst the species investigated. On the other side of the range, the smallest energy barrier was found for CH_2 (+52 kJ mol^{-1}). An explanation for these results is presented in the transition state structures in figures 6.30(b) and (c). It is obvious that the methyl, CH_3, species experiences a very weak interaction with the surface at the TS (figure 6.30(b)). It has a trigonal planar structure with H atoms in the direction towards the surface. In fact, this structure is identical to the one in the gaseous state. On the other hand, the triplet CH_2 species is a radical (with two unpaired electrons), and can, thereby, strongly interact with the surface at the TS position. This is obvious from its position close to, and with its C atom in the direction towards, the surface (figure 6.30(c)). These results strongly indicate that CH_2 species are much more mobile on the surface compared to CH_3.

In summary, a CH_3 adsorbate can easily become transformed to a CH_2 adsorbate by a gas phase abstraction of an H ligand (on the CH_3 adsorbate). This CH_2 adsorbate will in turn form a very stable transition state during migration on a diamond (111) surface. This is the reason why the removal of two H atoms (one adsorbate and one ligand) is needed for the first migration step of this radical species. However, for the forthcoming migration steps only one H abstraction (per CH_2 jump) is needed.

At first sight, when increasing the temperature for a diamond (111) surface with CH_3 adsorbates, the possibility for a desorption from the surface is almost identical to the possibility for surface migration. Hence, it seems that there is a 50% probability for a CH_3 migration to take place, but for this to occur there must be a neighbouring surface radical site. In an ordinary CVD environment with a super-saturation of gaseous H radicals, the formation of a neighbouring surface radical will most probably take place. However, it will not stop there. There is also a very large possibility for gas phase abstraction of an H species from the CH_3 adsorbate, which results in a CH_2 adsorbate. In short, a CH_3 adsorbate may (i) desorb back to the gas phase, (ii) jump to a neighbouring radical surface site, or (iii) undergo abstraction with gaseous H, forming a CH_2 adsorbate. The most probable scenario depends on the experimental setup. For a not too high temperature, the CH_2

formation will most probably take place. Hence, the surface migration, of one of the generally assumed growth species, CH_3, will most probably take place via CH_2 migration. (It is, of course, possible here to continue the study by abstraction H from the CH_2 adsorbate also.

The surface migration of the C_2H_2 adsorbate must, however, be treated a bit differently. It is in its free state a molecule with a triple C≡C bond, and is, thus, a very stable molecule. It must break one of its C–C bonds to be able to bind covalently to the surface. Hence, a monovalent C_2H_2 adsorbate is a radial adsorbate (containing a C = C double bond). This explains the very low adsorption energy (-72 kJ mol^{-1}) for a monovalent binding to an H-terminated diamond (111) surface. On the other hand, a gaseous CH_3 molecule is a very reactive radical with one unpaired electron. It is, thus, very prone to bind to a radical surface site, thereby forming a covalent bond with paired electrons.

As was the situation with the CH_3 adsorbate, after the mono-valent adsorption of the C_2H_2 adsorbate, various reactions can continue to take place. However, they are a bit more complex than for the CH_3 adsorbate, which is due to the unsaturated nature of this species (i.e. C_2H_2 has a triple bond). First of all, there is a big risk for this weakly bound C_2H_2 adsorbate to desorb from the surface. However, if gas phase abstraction will first take place, it can do so in two different ways; either from the adsorbate, or from a surface site neighbouring the adsorbate. (These first steps are similar to the ones for a CH_3 adsorbate.) If the H abstraction takes place from the adsorbate, a strongly binding C_2H species will be formed (see table 6.3). This species binds with an adsorption energy of -556 kJ mol^{-1}. It is a non-radical linear molecule with a very strong C≡C triple bond. It is, thus, very stable. On the other hand, if the H abstraction takes place from a neighbouring surface site, a two-fold binding of C_2H_2 will most probably take place (see figure 6.31). This adsorbate has only a C–C double bond, but its two-fold binding gives a very high adsorption energy (-636 kJ mol^{-1}; see table 6.2). A competing reaction would be the jump-wise migration of the one-fold binding C_2H_2. However, due to the very stable two-fold site adsorption, this process doesn't seem so probable.

When it comes to migration on the surface, there are now two scenarios to choose from. It will either be the strongly binding C_2H species that simply jumps to a neighbouring radical surface site, or the two-fold binding C_2H_2 adsorbate that moves over the surface. As can be seen in table 6.3, the barrier energy for the C_2H migration is very large (+350 kJ mol^{-1}). It is, in fact, almost identical to the corresponding energy barrier for a CH_3 migration (+353 kJ mol^{-1}). However, the barrier of energy for the C_2H adsorbate has almost half the value of its approximated desorption energy (+628 kJ mol^{-1}), so any desorption of C_2H radicals is not expected to take place. It is instead very likely that the C_2H adsorbate will stay immobile on the surface (except at very high temperatures).

For the migration of a two-fold binding C_2H_2 adsorbate, the following sequence is assumed to take place (see figure 6.31). It will take place by alternating twofold- and onefold-site adsorptions to the diamond (111) surface. This sequence starts with a one-fold site adsorption in figure 6.31(a). The migration is thereafter assumed to take place by the breakage of one of the C≡C bonds (figure 6.31(a)) and a folding down into a twofold-site adsorption (figure 6.31(b)). Finally, the C≡C triple bond will be formed again while the C_2H_2 moves up to an upwards position. A more or less barrier-less rotation around its $C_{surface}$–$C_{adsorbate}$ axis is then assumed to take place. The calculations of the potential energy surface for this process has, therefore, been divided into three parts, out of which the second is just a mirror image of the first one (see figure 6.34(f)).

As can be seen in table 6.2, the adsorption energy for a twofold adsorption of C_2H_2 is −636 kJ mol^{-1}, whilst the adsorption energy for a onefold site adsorption is only −72 kJ mol^{-1}. The reaction energy for step 1 is, hence, strongly exothermic with −636 kJ mol^{-1}. Consequently, the second step in this migration process must be strongly endothermic (+564 kJ mol^{-1}). Any or both of these steps might also be accompanied with a barrier energy. The methodology used in calculating the eventual barrier energy in step 1 is similar to the one used for calculating the migrating jumps. The $C_{surface}$–C–C angle has been step-wise changed and the $C_{adsorbate}$ that is binding to the surface has for each step been kept fixed, whilst the rest of the model was allowed to freely relax (except for the lowest positioned C and H layers). Due to the very high symmetry of the diamond (111) surface, this is a very accurate method to use. At least if the step length is

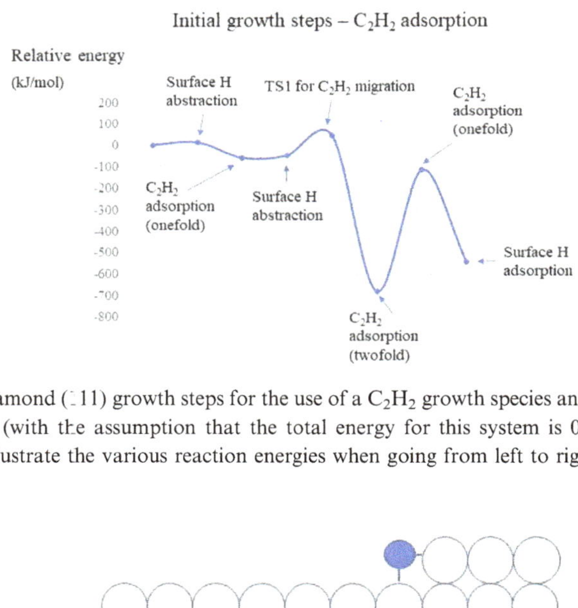

Figure 6.34. Initial diamond (111) growth steps for the use of a C_2H_2 growth species and starting from a 100% H-terminated surface (with the assumption that the total energy for this system is 0 kJ mol^{-1}). The next-coming data points illustrate the various reaction energies when going from left to right.

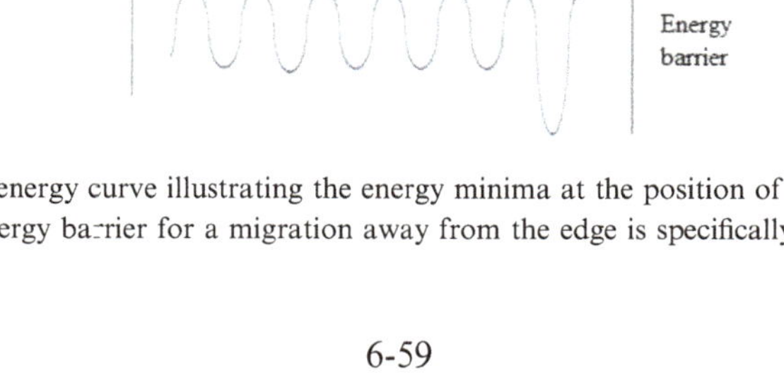

Figure 6.35. Potential energy curve illustrating the energy minima at the position of an adsorbate next to an edge. The increased energy barrier for a migration away from the edge is specifically highlighted.

sufficiently short. The largest barrier is observed for step 2, with a barrier energy of approximately +186 kJ mol^{-1}. However, there is an energy barrier for also step 1 (of approximately +92 kJ mol^{-1}). Thus, the combination of C_2H_2 adsorption barrier energies indicates that step 1 is governed by a barrier of +92 kJ mol^{-1} (even though it is an exothermic process), while step 2 is governed by the large endothermic reaction process with a reaction energy of +564 kJ mol^{-1} (which is much larger than the corresponding barrier energy). In addition, step III is not expected to be governed by an energy barrier larger than +564 kJ mol^{-1}. Thus, the C_2H_2 migration rate is expected to be determined by the endothermic reaction process of step type 2. On the other hand, the C_2H_2 adsorbate can become easily desorbed from the surface when it is in its one-fold adsorbed state (with a desorption energy of +72 kJ mol^{-1}).

In short, it has been shown that neither C_2H_2, nor C_2H, can undergo surface migration. The C_2H species will most probably stay rigid on the surface while C_2H_2 becomes desorbed. On the contrary, CH_3 can do this via the formation of a CH_2 adsorbates. It is, thereby, possible to diminish the role of C_2H_2 as a dominating growth species for diamond (111) deposition. However, a large excess of CH_3 and CH_2 species in the CVD gas mixture is energetically ideal due to their large adsorption energies and high possibilities for surface migration on diamond (111).

Potential energy curves for the initial growth steps using an initial CH_3 adsorption, or C_2H_2 adsorption, are demonstrated in figures 6.33 and 6.34, respectively.

6.6.5 Final incorporation into the lattice

It must be emphasized here that a surface migration on a perfect (i.e. non-defect) surface is a 'random walk' process. However, it has an end, and that is a position with a deeper energy well (see figure 6.35). The most wanted end position during the CVD growth of a material is the incorporation of the migrating growth species at, e.g. a ledge site on the growing surface. However, the complexity of a surface structure is very large, so there are also other types of final growth species positions. What most of these positions have in common is that the growth species adsorbs very strongly to these types of sites. In other words, the corresponding wells are very deep in the potential energy surface. It should, ideally, be so deep that any migration back onto the planar surface is prohibited. The underlying cause for this strong bonding is the possibility to form more than one bond to this specific surface site.

This is demonstrated in figure 5.8 for the incorporation into the lattice at ledge and kink sites. Let us now exemplify the simulations of growth species incorporation by continuing the study of the CVD growth of diamond (111). A stepped morphology on the surface has experimentally been shown to be characteristic for CVD deposited diamond [42].

Furthermore, earlier experiments have indicated that surface diffusion takes place during diamond growth [43]. For such a case, the growth of diamond occurs through propagation of steps over the diamond surface. Two different types of monolayer steps (also called ledges) can be formed on the diamond (111) surface. One type of step is composed of a (111) and (100) surface (step type A), and the other is only composed of two (111) surfaces (step type B) (see figures 6.36 and 6.37). As presented above, the

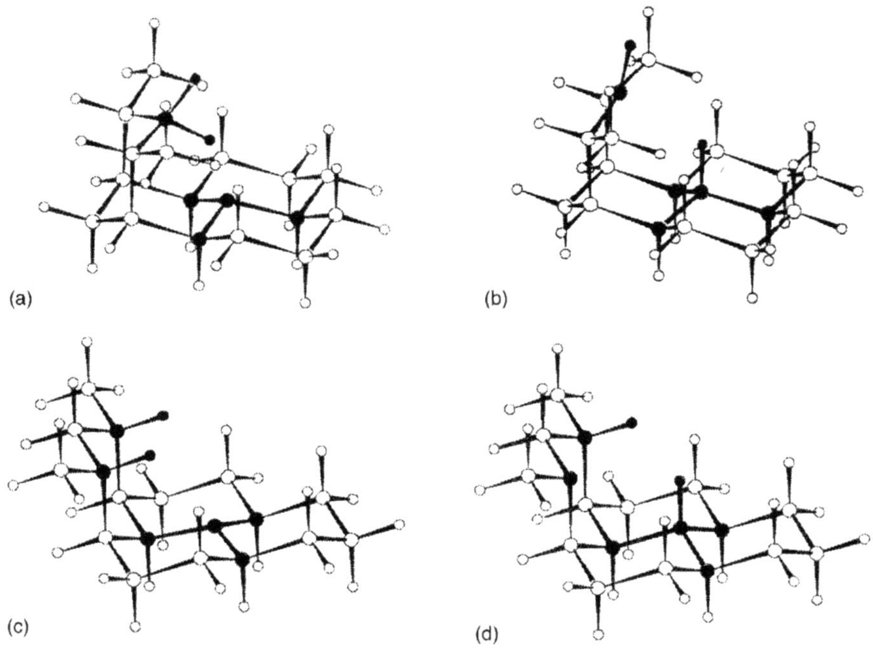

Figure 6.36. Two different monolayer steps on an H-terminated diamond (111) surface. Step type A is composed of A lower (111) surface and a vertical (100) surface (a), (b). Step type B is composed of two (111) surfaces (c), (d). (a) Step type A with a radical site on the lower surface; (b) Step type A with a radical site on the vertical surface; (c) Step type B with a radical site on the lower surface; (d) Step type B with a radical site on the vertical surface. For clarity, the C and H atoms that are closest to the radical sites are marked in black.

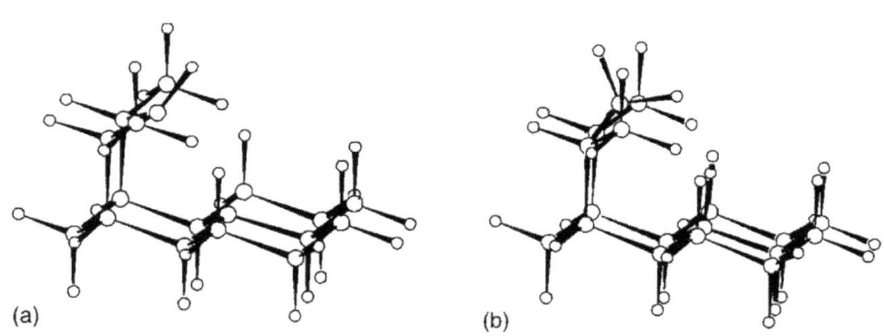

Figure 6.37. Model of a (a) diradical surface site and the corresponding (b) saturated surface at step type A.

CH_2 species has been found to be very mobile on the diamond (111) surface. The C_2H species did also show a certain probability for migration, even though a high temperature is needed for the process to take place. Thus, the processes of adsorption and incorporation, of the CH_2 and C_2H species, into the diamond (111) lattice, will be presented and discussed here. More details are given in [40].

As has been stated above, the ordinary CVD growth of diamond is a very complex and dynamic growth process. Abstraction and adsorption reactions are two of the most frequently occurring elementary surface reactions. A steady-state condition, involving adsorptions and abstractions of surface- and adsorbate-terminating H species, will evolve rather quickly.

Hence, the bonding situation for the CH_2 and C_2H species, at the respective ledge, must be carefully investigated as a function of local H composition. It is important to note here that the vertical plane at the ledges is expected to undergo gas-phase abstraction reactions, leaving a second radical C site (in addition to the one at the lower plane). Furthermore, the growth species is assumed here to chemisorb to the latter type of radical surface site, resulting in CH_3 and C_2H_2 adsorbates. Consequently, it is now possible to identify four distinct situations: (i) adsorption of CH_2 (or C_2H) next to a step edge with an H-terminated vertical plane; (ii) adsorption of CH_2 (or C_2H) next to a step edge with a monoradical vertical plane; (iii) adsorption of CH_3 (or C_2H_2) next to a step edge with an H-terminated vertical plane, and (iv) adsorption of CH_3 (or C_2H_2) next to a step edge with a monoradical vertical plane. The reason for considering the CH_3 and C_2H_2 species, despite the fact that they migrate very poorly, is that a gaseous H species can become chemisorbed to the already adsorbed CH_2 and C_2H species (at the ledge). The results of the calculations of the corresponding adsorption processes are presented in table 6.4 and figures 6.38–6.40. It has been assumed here that the incorporation of C into the diamond lattice is initiated by the binding of ahydrocarbon species to a surface site on the lower (111) surface (i.e. terrace) at a growth step. A lateral growth of diamond has, then, been assumed to take place, which is also referred to as a layer-by-layer growth.

Let us start with step type B. As can be seen in table 6.4, the large sterical hindrances induced by the step formation of type B, with its H-terminating species, is, with one exception, causing a large decrease in adsorption energies for the adsorption to the lower (111) surface; from −428 (H), −315 (CH_3), −334 (CH_2) and −586 (C_2H) kJ mol^{-1} for a planar diamond (111) surface to −379 (H), −14 (CH_3), −279 (CH_2) and −410 (C_2H) kJ mol^{-1} (step B). The exception is adsorption of C_2H to this specific step, which also forms a bond to the vertical step with an accompanying increase in adsorption energy (from −556 (planar) to −563 (step B) kJ mol^{-1}). The most interesting feature here is that one terminating H species at the vertical edge has moved over to the C_2H adsorbate and, thereby, forming a C_2H_2 adsorbate. It is important to stress that this is not an ethene, but instead a $C{=}CH_2$ species. Thus, original triple bond in C_2H has been transferred to a double bond upon adsorption to this specific type of edge. An extra electron on both the edge-binding C atom and on the H-binding C atom was thereby formed, which made it possible for the adsorbate to bind with two bonds to the edge and also to bind with an additional H atom. The geometrical structure of the resulting adsorption formation is shown in figure 6.38 for step type A (with an identical two-fold binding of a resulting $C{=}CH_2$ species).

The situation is, generally, a bit different for the adsorption to the lower diamond (111) surface at step type A. As can be seen in table 6.4, as can be understood from the adsorption energy of H (−379 versus −428 for a planar surface) the situation is now more crowded. However, despite the more crowded surroundings, the

Table 6.4. Adsorption energies, ΔE_{ads}, for attached species at steps A and B on the H-terminated diamond (111) surface (kJ mol^{-1}). Corresponding adsorption energies for a planar surface is also presented as a reference.

Surface step	Adsorption energy H (kJ mol^{-1})	Adsorption energy CH_3 (kJ mol^{-1})	Adsorption energy CH_2 (kJ mol^{-1})	Adsorption energy C_2H_2 (kJ mol^{-1})	Adsorption energy C_2H (kJ mol^{-1})
Planar	−428	−315	−334	−4	−556
A[a]	−379	−14	−28	−260	−563
B[a]	−408	−142	−279	−130	−410
A[b]	−385	−232	−341	−24	−572
B[b]	−395	−217	−295	−49	−440
A[a] (diradical)	−285	−41	−574(two-fold)	−290(two-fold)	−430(two-fold)
B[a] (diradical)	−269	−193	−406(two-fold)	−337(two-fold)	−395(one-fold)
A[b] (diradical)	−293	−151	−574(two-fold)	−290(two-fold)	−430(two-fold)
B[b] (diradical)	−387	−236	−406(two-fold)	−337(two-fold)	−443(one-fold)

[a] Adsorption on the lower (111) surface at the step.
[b] Adsorption on the vertical surface at the step.

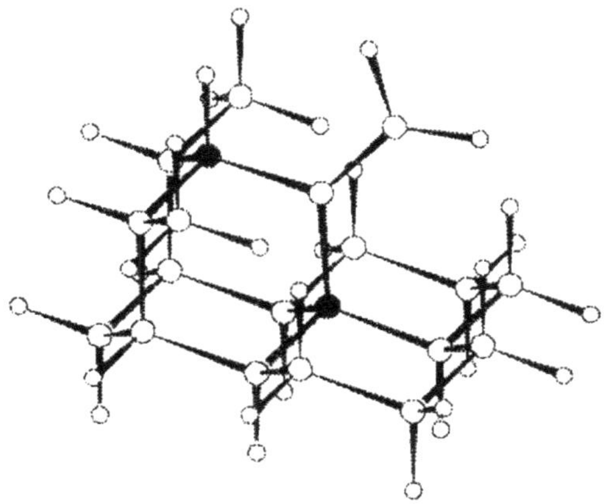

Figure 6.38. Model of the resulting adsorbate geometrical formation for the adsorption of C_2H to a lower surface radical step at step type A.

adsorption energy for both CH_3 and CH_2 will decrease dramatically (to −14 and −28 kJ mol^{-1}, respectively). On the contrary, this is not the situation for C_2H_2 and C_2H. Instead, the non-radical C_2H_2 species binds much stronger to the A step (−260 kJ mol^{-1}), as compared with step type B (−130 kJ mol^{-1}), which is also the situation for C_2H (−563 versus −410 kJ mol^{-1}). It is interesting here to note that the adsorption of C_2H to this specific site is almost identical to the adsorption to a planar surface (−563 versus −556 kJ mol^{-1})!

As was the situation for step type B, the underlying explanation to the strong bonding of C_2H to step type A can be explained by new C–C bond formation that takes place between the C_2H adsorbate and the nearest surface edge, C*, atom during the adsorption process (see figure 6.39(c)). The initial C*-H bond will thereby break, and this terminating H atom will migrate over and bind to the other C atom in C_2H (i.e. the C atom that is not binding to the step edge). Thus, as was the situation for the binding to step type A, a minimum in total energy will be the result for a barrier-less H migration from the edge surface to the non-surface binding C atom in C_2H. The hereby formed adsorbate that binds to step type A with two covalent bonds is also identified here as a $C{=}CH_2$ adsorbate with a C=C double bond, instead of the initial C≡C triple bond. However, the stronger bonding of C_2H_2 to the lower plane of step type A (−260 kJ mol^{-1}), as compared with step type B (−130 kJ mol^{-1}), is more difficult to explain solely on the basis of the resulting adsorbate geometry and bond formation. In fact, the binding of C_2H_2 is much stronger to these edges than it is to a planar surface. There must thus be a type of stabilization at the edge sites.

With the exception of C_2H adsorption, the situation is much different for the adsorption to the vertical edge at the steps. With one exception, the binding of species is somewhat stronger to these vertical surfaces than they are for the lower surfaces. The reason is most probably that one-atomic thick edges have been used for these calculations. Consequently, the edge C atoms do then have the capability to relax more easily during the process of adsorption. However, one-atomic thick

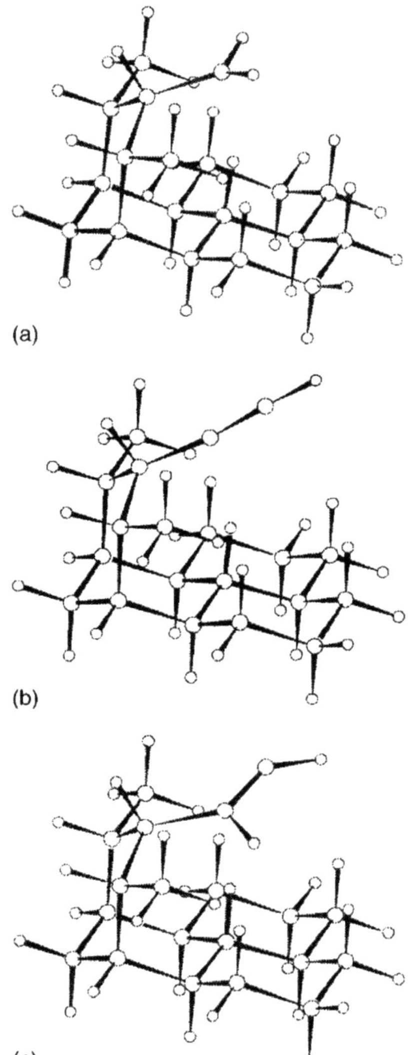

Figure 6.39. Models of the resulting adsorbate geometrical formations for the adsorption of (a) CH_2, (b) C_2H, and (c) C_2H_2 to the vertical plane of step type A.

steps are frequently observed during layer-by-layer growth of solid-state materials. The exception is the non-radical C_2H_2 species, which binds much more weakly to the vertical surfaces than to the corresponding lower surfaces (step A: −24 versus −260 kJ mol^{-1}; B: −49 versus −130 kJ mol^{-1}). These adsorption energies are to be compared with the one for a planar surface (i.e. without any step): −4 kJ mol^{-1}. Hence, the adsorption energies for the attachment of C_2H_2 to the vertical step edges are almost identical to the one for planar surface, which means that the former type of C_2H_2 adsorbates neither experience attraction nor repulsion to the lower surface.

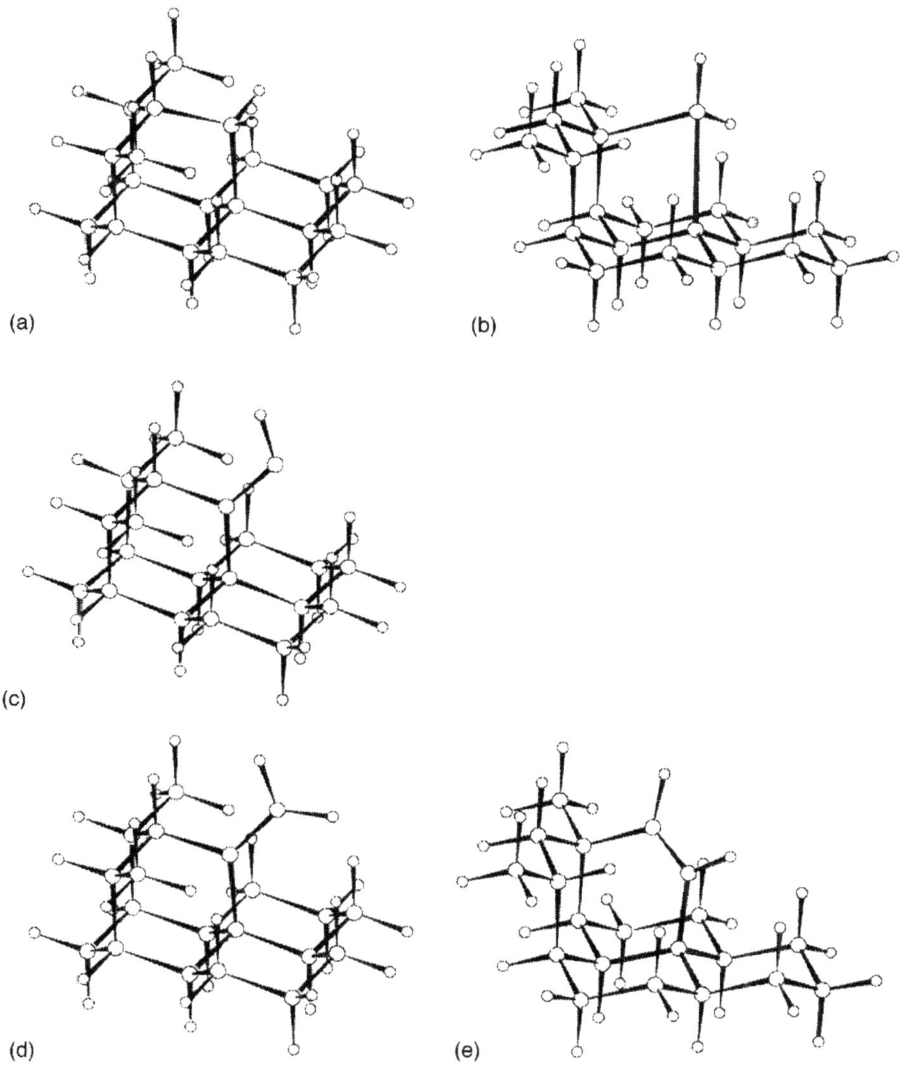

Figure 6.40. Models of the resulting adsorbate geometrical formations for the adsorption of CH_2, C_2H, and C_2H_2 to diradical surface sites at step types A and B; (a) CH_2 to step A, (b) CH_2 to step B, (c) C_2H to step A, (d) C_2H_2 to step A, and C_2H_2 to step B.

Let us now assume that our growth species are binding to either the lower surface or to the vertical surface at the step edges A and B. The question is now, what happens if an H adsorbate on the other plane is being abstracted from the edge (i.e. an H adsorbate on the vertical surface for an initial adsorption of growth species to the lower surface, and vice versa)? The adsorption energies are presented in table 6.4, and the resulting geometrical structures for CH_2, C_2H_2, and C_2H are shown in figure 6.40. As can be seen in figures 6.40(a) and (b), a second C–C bond will be formed between the diamond step and the surface-binding C atom in the CH_2

adsorbate. This is the situation for both step types A and B. The CH_2 adsorbate has, thereby, become well incorporated into the diamond lattice, and the formation of two bonds is reflected in the larger adsorption energy (step type A: -574 kJ mol^{-1}; step type B: -406 kJ mol^{-1}), as compared with the adsorption to either a planar diamond (111) surface (-334 kJ mol^{-1}) or to the lower or vertical surfaces at an otherwise H-terminated step (step type A: -28 vs. -341 kJ mol^{-1}; step type B: -279 vs. -295 kJ mol^{-1}). On the contrary, this type of two-bond formation will only take place for step type A for the C_2H species (-430 kJ mol^{-1}) (see figure 6.46(c)). The original C≡C triple bond in the C_2H adsorbate has then been transferred to a C=C double bond and the resulting C_2H adsorbate is, thereby, a radical (which can in the next step adsorb a gaseous H atom). On the contrary, the adsorption of C_2H to a diradical surface B site is one-fold, with the following adsorption energies; lower surface: -395 kJ mol^{-1}; vertical surface: -443 kJ mol^{-1}. This adsorbate is, thus, a non-radical species with a surface-to-adsorbate bond energy that is very similar to the one for step type A. Hence, the C_2H species cannot be ideally incorporated into a diamond step of type B. Instead, a defect will be formed in the diamond lattice.

Another question is now, what happens if a gaseous H species succeeds in adsorbing to the already chemisorbed CH_2 species to a diradical site? Since the CH_2 adsorbate is two-fold adsorbed to the steps, they are highly stabilized with no tendency to form a two-fold adsorbate CH_3 species. The reason is that a methyl (CH_3) species only has one unpaired electron, by which it binds to either the lower surface or to the vertical surface at the edges. Hence, there is no possibility for this species to also bind to the other side of the edge. Thus, an eventual formation of a CH_3 adsorbate must be accompanied with one-fold adsorption situations with the following adsorption energy; step type A: -41 (lower surface) versus -151 (vertical surface) kJ mol^{-1}; step type B: -193 (lower surface) versus -236 (vertical surface) kJ mol^{-1}. As expected, the CH_2 adsorbate until then goes from a stabilized bonding situation to an appreciably weaker one (for both types of steps), and will, in addition, not be incorporated into the lattice.

As presented above, the adsorption of C_2H species to step type A is two-fold, while it is one-fold adsorbed to step type B. For step type A, the radical C_2H adsorbate can, thereby, easily adsorb a second H atom, and this resulting C_2H_2 adsorbate remains in the same two-fold site adsorption as C_2H (see figure 6.46(d)). As can be seen in table 6.4, this type of two-fold site adsorption is, surprisingly, considerably weaker than the one for C_2H (-290 versus -430 kJ mol^{-1}). For step type B, with a one-fold adsorption of a non-radical C_2H species, the adsorption of an extra H atom will also result in a two-fold site adsorption of C_2H_2 (see figure 6.40(e)). The adsorption energy of C_2H_2 to this type of surface is just somewhat weaker than the corresponding one-fold adsorption of C_2H; -337 versus -395 (lower surface), -443 (vertical surface) kJ mol^{-1}. In addition, it is only somewhat stronger than the binding to step type A (-337 vs. -290 kJ mol^{-1}). Even though the adsorption energies for C_2H_2 are very similar for step type A and B, the geometrical structures look completely different. Both have a C=C double bond, but the binding of C_2H_2 to a diradical site at step type A shows two surface-to-adsorbate bonds from one of its C atoms while the original H ligand (to this very C atom) has barrier-less moved over the other C atom, forming a CH_2 group. On the other hand, the binding of C_2H_2 to a diradical site at step type B show one surface-to-adsorbate bond per C atom. Moreover, the binding of C_2H_2 to step type B has a larger resemblance to a continued cubic growth of c-BN. Especially after that, the adsorbate has been saturated with two more H atoms (forming a C_2H_4 adsorbate with a C–C single bond). The bond strengths of C_2H_2 to these types of steps are reflected by the calculated adsorption energies: −290 (step type A) and −337 (step type B) kJ mol^{-1}.

In summary, an excess of CH_2 adsorbates is energetically ideal due to their large possibility for surface migration and their strong adsorption onto both terraces (i.e. surface planes) and growth steps at the diamond (111) surface. In addition, the CH_2 species has been found to be ideally incorporated into the diamond lattice at both step types A and B. As a prerequisite, a second H abstraction process must first take place at the vertical plane of the steps. But this is easily accomplished in the H supersaturation environment in the CVD chamber.

The overall results can also diminish the role of C_2H_2 as a dominating growth species for any surface process largely based on surface migration on diamond (111) surfaces. At least at more moderate deposition temperatures. The low possibility for surface migration, together with the rather strong chemisorption to step edges (especially for C_2H), may lead to secondary nucleation (i.e. to non-smooth surfaces). However, the theoretical results will not preclude the C_2H_2 species from being one of the dominant growth species during growth governed by the direct adsorption of species to a diradical terrace site, or to a diradical growth step.

Furthermore, it can also be concluded that both types of growth steps (A and B) are beneficical for a defect-free incorporation of CH_2, while this is only the situation for C_2H at step type A. Thus, the corresponding incorporation of C_2H at step type B may result in diamond with higher degree of defect density. The situation for the incorporation of C_2H_2 to the step edges is not that clear. However, it is obvious that it can form a continuation of the c-BN lattice on step type B (after further saturation), but for the formation of a crystalline lattice at step type A, further gas phase reactions and saturation are needed.

In addition, the CH_2 and C_2H_2 species were found to more strongly incorporate to the edge sites as compared to a planar (111) surface (see table 6.4). Hence, the energy minimum at the edge site is deeper than the minima for a terrace site further away from the edge (figure 6.41 demonstrates this observation). It is thereby

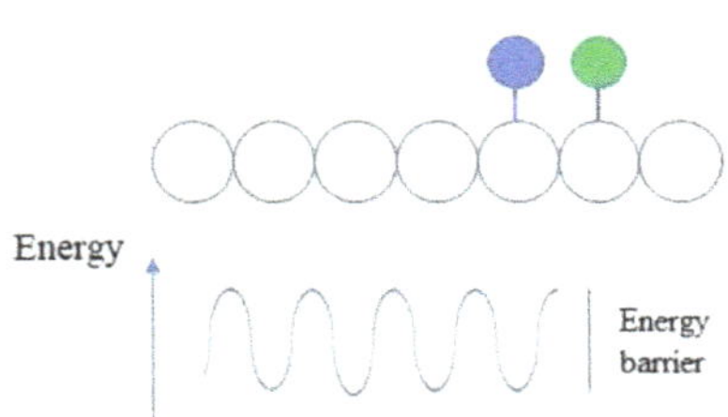

Figure 6.41. Potential energy curve illustrating the energy minima at a position close to a non-interacting co-adsorbate.

expected that both CH_2 and C_2H_2 can be 'trapped' at the edge site, and not continue to migrate out again towards the planar surface. This is, in fact, the desired situation for a controlled layer-by-layer growth of a specific material. However, as stated above, only a direct adsorption of C_2H_2 (from the gas phase) has the energetic possibility to take place at moderately low temperatures. Since the probability for the process is quite low, in comparison to the abondance of CH_2 on the surface (with its large migration possibility), we are left with the CH_2 species as the most probable incorporation species (and most probably formed from adsorbed CH_3 species).

6.6.6 Effect of neighbouring adsorbates on surface migration

As presented above, the surface migration process is a random walk process that ends when the migrating species has reached a very stable adsorption site (with a deep enough energy well in the potential energy curve). There is, during these migrations, a very large possibility for the species to pass another adsorbate, or a radical surface site. This will lead to, at least, three distinctly different situations. As the first type of situation, nothing will happen. The migrating species will experience neither a change in attractive nor repulsive interactions, and will just continue unperturbed. This situation is exemplified with the sketch shown in figure 6.41. The second type of situation is when the migrating species experiences a repulsion towards the passing co-adsorbate. This means that when the migrating species comes closer to the co-adsorbate, the surface-to-adsorbate bond strength will decrease (and thereby the corresponding adsorption energy). This causes lower wells in the potential energy curve (for the migrating species), and thereby smaller migration barriers and larger migration rates. This is like a catalytic reaction where the reaction rate will be increased by the use of a catalyst that will not be destroyed. The catalyst is in this case the co-adsorbate (see figure 6.42). The third example is an attractive interaction between the migrating species and the co-adsorbate. The opposite to the repulsion situation will then take place. The migrating species will bind more strongly (to the surface) at positions close to the co-adsorbate, which causes deeper wells in the migration potential energy curve. Consequently, the migration barriers will increase, which leads to decreased migration rates (see figure 6.43). As the worse scenario, the migrating species might be trapped by the co-adsorb, thereby hindering any further migration. On the other hand, this is maybe the most wanted situation for, e.g. when attempting to increase the nucleation

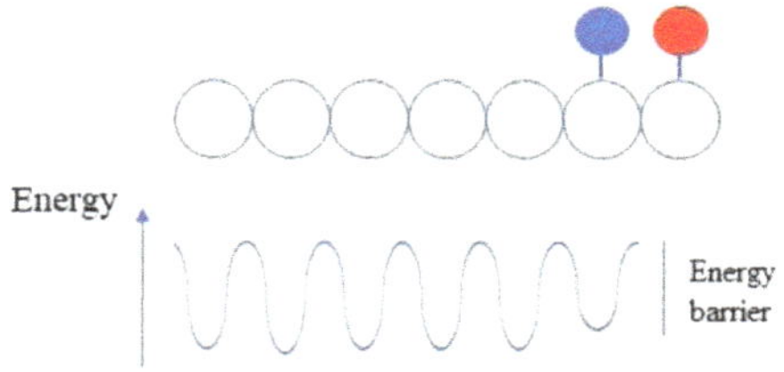

Figure 6.42. Potential energy curve illustrating the energy minima at a position close to a repulsive co-adsorbate.

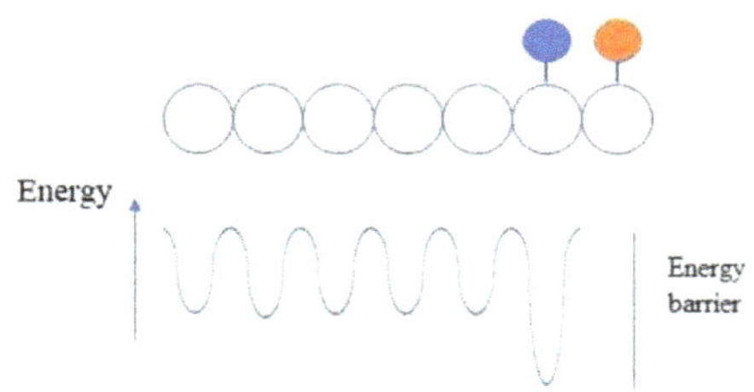

Figure 6.43. Potential energy curve illustrating the energy minima at a position close to a repulsive co-adsorbate.

density. Finally, a fourth situation is chemical reactions with either gaseous species or with co-adsorbates. However, these types of reactions are not considered in the present example.

Let us now go back to the diamond example and see if we can identify any hindering or promoting co-adsorbates to a migrating CH_2 adsorbate. As presented in section 6.6.2, the migrating CH_2 adsorbates will, with just a few exceptions, sterically experience hindering by neighbouring hydrocarbon co-adsorbates (CH_3, CH_2, and C_2H). One exception is when CH_2 passes a one-fold adsorbed C_2H_2 species. A C–C bonding formation will then take place between these two adsorbates, which lowers the total energy of the system by about 100 kJ mol^{-1}. Furthermore, the existence of neighbouring radical surface sites (i.e. non-terminated surface sites) does not have any effect on the migrating CH_2 adsorbate, other than it will offer an alternative migration path. In short, surface radical sites are not expected to change the CH_2 migration rate, while the CH_3, CH_2, and C_2H adsorbates will have an improved effect. This is strongly supported by the calculated adsorption energies presented in table 6.2. In addition, the co-adsorption by C_2H_2 is very special since these adsorbates will effectively stop the migration of CH_2 species, thereby creating the initial stage of a secondary nucleus.

It is also possible to deliberately add another co-adsorbate, with the purpose to, e.g. speed up the migration rate. What then are the general demands on this co-adsorbate? It should be repulsive towards the migrating species. If we want to hinder the migration, we just add a co-adsorbate with an attraction towards the migrating species. It is then very useful to calculate the atomic charges of the adsorbates (for both the migrating one and the neighbouring one), or to do an estimation based on electronegativity values. Two positively charged adsorbates will repel each other, and this is also the situation with two negatively charged adsorbates. In addition, a very large co-adsorbate will induce sterical hindrances and thereby cause inter-adsorbate repulsions. This is due to the overlap (or near overlap) between the electron 'clouds' on the different adsorbates. This was the situation for our diamond example above.

6.6.7 Growth mechanism and determination of the rate-limiting step

As has been shown above, the diamond growth process not only involves reactions between the growth species and the surface, but also H abstraction reactions to a large degree. In fact, after the creation of the first radical surface site, each reaction that involves the growth species is accompanied by an H abstraction, or the migration of an H adsorbate to another surface site. Under the assumption that CH_3 is the dominating growth precursor, summaries of the diamond growth steps discussed above are presented in figures 6.44 and 6.45. In these figures, the steps started by surface H abstraction, followed by a single CH_2 jump (or the movement of the C_2H_2 adsorbate), and ended by an H adsorbate to the surface, are assumed to be repeated many times. Thus, these 'elementary' steps together form the diamond growth mechanism. All of these steps are not truly elementary, but we call them 'elementary' for simplicity reasons. The overall mechanisms are also, to some extent, oversimplified, but the purpose of this example is just show the methodology for

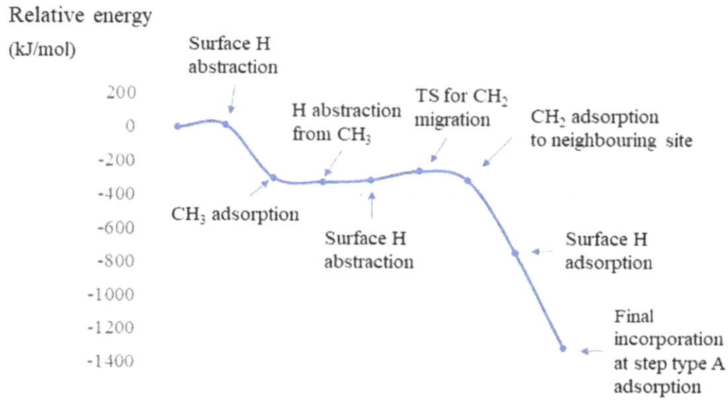

Figure 6.44. Initial diamond (111) growth steps for the use of a CH_3 growth species and starting from a 100% H-terminated surface (with the assumption that the total energy for this system is 0 kJ mol^{-1}). The next-coming data points illustrates the various reaction energies when going from left to right, and ended with the final incorporation of CH_2 at step type A.

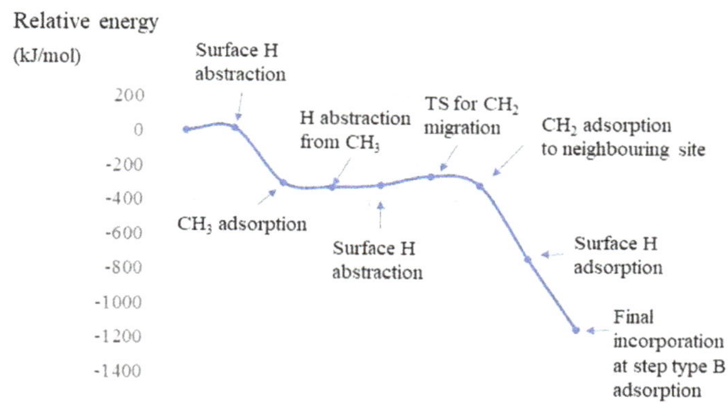

Figure 6.45. Initial diamond (111) growth steps for the use of a CH_3 growth species and starting from a 100% H-terminated surface (with the assumption that the total energy for this system is 0 kJ mol^{-1}). The next-coming data points illustrates the various reaction energies when going from left to right, and ended with the final incorporation of CH_2 at step type B.

simulating a CVD growth process. As stressed in various parts of this book, the CVD process is very complex, and it is not possible to just make a model and press a button when trying to theoretically simulate this type of process. It is more like doing a jigsaw puzzle where all types of earlier acquired knowledge and experiences are necessary as inputs. Not to forget, the comparison with experimental CVD activities is also very important for this task.

A total chemical mechanism can be decomposed into elementary reactions steps, and the slowest elementary reaction, also called the 'bottleneck', is the rate determining reaction. Hence, the rate-determining elementary reaction determines the total CVD growth rate. The CVD process must then work under surface kinetic control (i.e. it is the surface chemical processes that controls the film growth).

It has generally been assumed that the H abstraction process from a diamond surface is the rate-determining step in the ordinary CVD growth of diamond. This assumption can be verified by setting up plausible reaction pathways with energies, as has been done in figures 6.44 and 6.45. The only difference between these figures is the final incorporation into a monolayer step at the diamond surface. This step is either of type A, or of type B.

The first step in figures 6.44 and 6.45 is the H abstraction from an H-terminated diamond (111) surface, and the second step is the adsorption of the commonly accepted CH_3 growth species to this radical surface site. Since this adsorbate cannot migrate, we instead form a highly mobile CH_2 adsorbate by abstracting one H from the CH_3 adsorbate (step 3). Further on, we must abstract a neighbouring H adsorbate so that the CH_2 adsorbate can start jumping (step 4). MD studies at experimental temperatures have shown that this jump really takes place at experimental conditions. After the single jump of the CH_2 adsorbate (steps 5 and 6), the surface radical site that was earlier occupied by the CH_2 adsorbate, is V now assumed to be terminated by a gaseous H species (step 7). The next-coming elementary steps will be successive H abstractions and CH_2 jumps until the CH_2 species has reached a step (of either type A (figure 6.44) or B (figure 6.45)). However, what is missing in figures 6.44 and 6.45 are barrier energies corresponding to the final incorporation at the various steps. This is not included in the present example, but similar calculations, as was done for the migrations, can also be performed for these types of chemical reactions. This is also, of course, the situation for every single reaction step in the initial reaction mechanism assumed here.

As can be seen in figures 6.45 and 6.46, all elementary steps that involve the growth species are exothermic. With one exception, we don't know anything about their kinetics, but they are thermodynamically driven. (We must stress, though, that the entropies have not been estimated in the present example.) The exception is, however, the migration of CH_2, but the calculated energy barrier for this process is extremely low. On the other hand, the H abstraction reactions are found to be endothermic.

Let us now compose the total CVD growth mechanism by using the elementary reactions presented in figure 6.45 (i.e. with the final incorporation at step type A).

1. Surface-H(site 1) + H(g) → Surface(rad)(site 1) + H_2(g)
2. Surface(rad)(site 1) + CH_3(g) → Surface-CH_3(site 1)

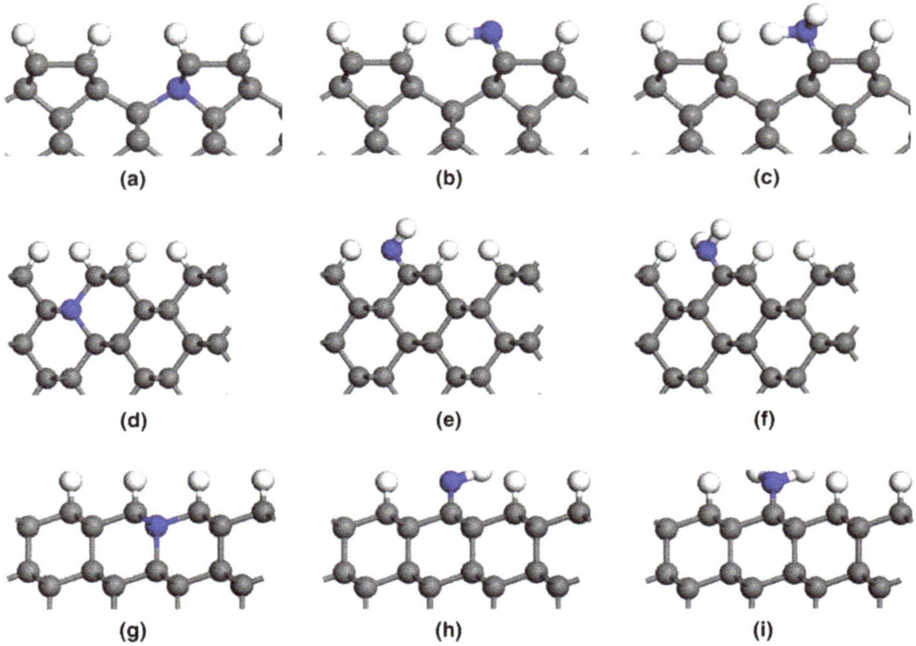

Figure 6.46. Representations of nitrogen doping in different diamond surface structures; (a)–(c) (100)–2 × 1, (d)–(f) (110), and (g)–(i) (111). (a), (d), (g) Substitutional doping into the 2nd C atomic C layer. (b), (e), (h) An adsorbed NH radical. Similarly, (c), (f), and (i) adsorbed NH_2 species. [27], open access © Springer.

3. Surface-CH_3(site 1) + H(g) → Surface-CH_2(site 1) + H_2(g)
4. Surface-H(site 2) + H(g) → Surface(rad; site 2) + H_2(g)
5. Surface-CH_2(site 1) → Surface-CH_2(site 2)
6. Surface(rad)(site 1) +H(g) → Surface-H(site 1)
7. Steps 4, 5, and 6 are repeated until CH_2 reaches step type A (i.e. Step A (CH_2; lower surface))
8. Surface(rad)(site $n-1$) +H(g) → Surface-H(site n-1)
9. Step A-H(vertical surface) + H(g) → Step A(rad; vertical surface) + H_2(g)
10. Step A(rad; vertical surface) + Step A(CH_2; lower surface) → Step A(CH_2; two-fold)

Thus, the final incorporation at step type A in figure 6.45 has here been replaced by (8)–(9). By adding all element to the left in the expressions above, and the ones to the right, we get the following total reaction:

Step A(CH_2; lower surface) + Step A-H(vertical surface) + (4+n) H(g) + CH_3(g) → Step A(CH_2; two-fold) + 4H_2(g)

We must remember here that *Surface-H(site 1), Surface-H(site 2)*, and *Surface-H(site 2)* are identical, which is due to the high symmetry on the diamond (111) surface. This is, therefore, also the situation for *Surface-H(rad; site 1), Surface-H (rad; site 2)*, and *Surface-H(rad; site 2)*.

Reactions 1, 3, 4, 5, 7, 8, and 9 are slightly endothermic, while reactions 2, 6, 8, and 10 are strongly exothermic. As reaction 1 is the first endothermic process, it can be assumed to be the rate determining elementary reaction in the diamond growth mechanism. Hence, the total growth reaction rate can be determined from the rate of step 1 in the diamond growth mechanism;

$$\nu(\text{total reaction}) = \nu(\text{reaction1}) = k_1[\theta_H]p_H \tag{6.16}$$

where k_1 is the reaction rate constant for reaction 1, θ_H is the surface coverage of H adsorbates (on the terrace), and p_H is the partial pressure of H in right above the growing film (i.e. in the gas phase). The same type of reaction mechanism can be constructed for the second reaction mechanism (see figure 6.50). However, the growth rate expression will be identical to the one presented in equation (6.8), since these mechanisms only differ in the last incorporation step (i.e. the rate-determining step is the same).

Table 6.5 gives information about the relative growth rates (experimental and theoretical) of H-terminated diamond surfaces under the assumption that the first H abstraction process is the rate-limiting step. It is, thereby, only the barrier for the approaching H radical, to a 100% H-terminated diamond surface, that has been used for the rate estimations. Table 6.5 not only presents results for the (111) surface of diamond, but also for (100)–2 × 1 and (110) surfaces. Similar calculations, as presented above, have therefore also been performed for the (100)–2 × 1 and (110) surfaces of diamond.

In table 6.5, the Arrhenius relation was used to make a careful estimation of the growth rates:

$$\nu = \nu e^{-\Delta E/RT} \tag{6.17}$$

where T is the temperature during diamond growth, R is the universal gas constant, and ΔE is the energy barrier for the first H abstraction process. Under the assumption that the pre-exponential factor, v_0, is almost identical for the different diamond surface planes, the relative H abstraction rates, and thereby the relative diamond growth rates, can be rewritten as:

$$\nu_1/\nu_2 = e^{-(\Delta E_2 - \Delta E_1/RT)} \tag{6.18}$$

In addition to the calculated growth rates, experimentally determined growth rates are also presented in table 6.5 for the various diamond planes. Chu *et al* investigated

Table 6.5. Values of experimental and theoretically estimated growth rates of diamond (111), (100)–2 × 1, and (110) thin films.

Plane	Growth rate (experimental) (μm/hr)	Relative growth rate (experimental)	Energy barrier (kJ mol^{-1})	Relative growth rate (theoretical)
110	0.24	1.9	4.73	2.4
111	0.19	1.4	6.06	2.0
100	0.13	1.0	11.83	1.0

the growth of different diamond planes, and also presented different kinetic curves for each plane [45]. They used a hot-filament CVD technique, at a substrate temperature of 600–1000 K, for this purpose, in addition to a methane mole fraction of 0.4%. Chu *et al* used a surface kinetic control during their CVD growth of diamond, which means that the diamond growth rate could be determined by the activation energy of the rate determining step. This circumstance makes it ideal to add theoretical modeling and simulations for support and explanations of obtained experimental results. As can be seen in table 6.5, the experimental diamond growth rates at 700 °C were found to be 0.13, 0.19 and 0.24 μm h^{-1} for diamond (100), (111) and (110), respectively. If we now relate these experimental rates to the slowest one, the following relative rates will be obtained: 1.0, 1.4 versus 1.9. Hence, Chu *et al* found that the (100) surface is the slowest growing surface, while the (111) surface grows with a factor larger than 1.5. In addition, the growth rate of diamond (110) is doubled compared to diamond (100).

As can be seen in table 6.5, the experimentally observed trend in diamond growth rate is identical to the order of theoretical H abstraction rates; (100) < (111) < (110). There are, though, some minor differences when comparing the experimental and theoretical relative rates strictly numerically. The experimental relative growth rates of 1.0, 1.4, and 1.9 corresponds to the theoretical values of 1.0, 2.0, and 2.4. However, considering the large number of parameters that may influence the growth rate experimentally, these small differences are acceptable. Hence, the large similarity in relative H abstraction rates with relative experimental growth rates strongly supports the conclusion that the H abstraction is indeed a rate-determining step in the CVD growth mechanism of diamond (111), (100) and (110). The identity in trends, and large similarity in numerical values, will, furthermore, justify the theoretical method and methodology used for the calculations.

6.7 Influence of dopants on the growth process

6.7.1 Introduction

Doping of solid-state material can be performed by introducing impurities into the lattice (substitutional or interstitial) and, thereby, modifying the properties of the material. We shall continue here with diamond as a material example. The most important dopants in CVD-grown diamond are nitrogen, boron, phosphorous, and sulphur, which all will be substitutionally incorporated into the diamond lattice. The dopants will not only change the electronic properties, but it is also clear that these dopant elements can have a great impact on the diamond surface morphology and growth rate. It is generally believed that the enhancement of the diamond growth rate will reduce the production cost. By using the CVD method, it is very convenient and well-controlled to add a specific dopant concentration into the diamond lattice during the growth process. Thus, the effect of dopants on the diamond growth rate will be presented here as an example where theoretical methods will strongly aid the experimental strives in finding new ways to both increase the CVD growth rate of a specific material, and to decrease the corresponding costs.

However, before going into detail concerning the theoretical methodologies and techniques, we have to give some background information.

6.7.2 Diamond doping using nitrogen, phosphorous, sulphur, or boron

6.7.2.1 N-type doping using nitrogen

Although diamond has been proven to be a genuine multifunctional material, thereby useful for quite many applications, it is still quite expensive to grow using vapour phase methods like CVD. However, it has been experimentally proven that, e.g. N doping will enhance the diamond growth rate, and thereby reduce the production costs. The influence of nitrogen on the deposition rate has earlier been reported for both poly- and mono-crystalline diamond [46–49]. The effect by N on the CVD diamond growth rate enhancement has been studied using laser reflection interferometry [50]. In addition, gaseous N concentrations as low as a few ppm were found to strongly boost up the crystal growth rate in the (100) orientation. As an explanation to this observation, Bar-Yam and Moustakas [31] proposed a defect-induced stabilization of diamond. However, this model does not explain the crystal orientation dependence of the increase in growth rates. Frauenheim *et al* [39] suggested an alternative model, in which donor electrons originating from sub-surface N will lead to a lengthening of the (100) surface reconstruction bond. Nitrogen (with its extra electron compared to C) is a well-known n-type dopant, with a very deep donor level in diamond (1.7 eV below the conduction band) [51].

6.7.2.2 N-type doping using phosphorous or sulphur

The effects of substitutional doping by sulfur, or phosphorous, have experimentally been thoroughly studied [52, 53]. It is not only the changes in the electronic properties that have been observed, but it is clear that these dopant elements can have a great impact on the diamond surface morphology and growth rate when using chemical vapor deposition methods [52–54].

Regarding the influence on the diamond growth rate, various results have also been obtained when using an addition of hydrogen sulfide (H_2S) during the CVD process. It has been observed that the doping reagent will cause both a raise in growth rate, as well as a decrease, depending on the substrate temperature that has been used [52]. While it was concluded that hydrogen sulfide will reduce both diamond nucleation and growth of individual crystals onto silicon substrates, the addition of this dopant resulted in an increase in diamond layer growth rate [54]. Similarly, during the addition of phosphine (H_3P) to the CVD reaction chamber, the diamond growth rate was shown to be largely dependant on the substrate temperature. Opposite to the situation with non-doped diamond, a significant increase in diamond growth rate was obtained by using phosphine [54].

The improvement of the diamond growth rate could be one step in the development of a less expensive synthesis method for diamond thin film growth of different dimensions and thicknesses. For example, there is a strong need for a cost-effective method whereby it would be possible to grow large single-crystalline diamond of high quality. This method is to be compared with high-pressure high-temperature (HPHT) synthesis techniques, which will restrict both the size and the quality of the diamond material [55]. A well-controlled growth process using dopants is also of utmost importance for the growth of specific crystalline planes. It is well known that the doping effect is different for the different low-index surface planes; (111), (100), and (110) [56–61].

The very complex and dynamic nature of the basic mechanism for diamond growth has been highlighted earlier in several theoretical investigations [62]. However, only a small number of theoretical studies have been focused on the CVD growth of sulfur- or phosphorus-doped diamond [63, 64]. One report stated that phosphorus would decrease the growth rate due to the high energy associated with the strains caused by phosphorus atoms in the diamond lattice [65]. Another report concluded that sulfur atoms are spontaneously incorporated in a hydrogenated (100) diamond surface, and that they will also enhance the hydrogen desorption from the surface [66].

In choosing a suitable donor (or acceptor) for diamond, however, one has to not only to consider its donor (or acceptor) level but also its solubility and mode of incorporation (e.g. incorporation during growth by in-diffusion or by ion implantation). Impurities have been shown to become introduced into diamond during CVD or HPHT growth [67]. A review article has, in addition, reported on doping by implantation [68]. A substitutional n- (or p-) type doping of diamond during growth has been of a special interest to study in the present example. The resulting substitutional doping, hence, depends not only on the solubility of the impurity, but also on the kinetics of growth. Kinetic trapping may then be possible although the final n- (or p-) type doped product is thermodynamically somewhat unfavourable.

6.7.2.3 P-type doping using boron

The growth process of boron-doped (B-doped) diamond material, with its extraordinary electrochemical properties, has recently been well developed. It has been found that the B dopant will increase the electronic conductivity, as compared with the intrinsic non-doped diamond [69]. For different doping levels, the diamond material can show metallic-type conductivity [70, 71], p-type semi-conductivity [72, 73], or superconductivity [74]. In particular, the superconductivity property makes diamond very attractive for electronic applications. In order to achieve a good quality of B-doped diamond films and to better control the effects of density of boron, the CVD technique has been shown to be a very useful technique. Thus, substitutional doping with boron (using CVD) will not only change the electronic properties, but it will also have a great impact on the diamond surface morphology and growth rate [74–77].

It has been shown that when using a boron precursor (such as trimethylborate, boron trioxide, diborane, or trimethylboron) during the CVD process, the growth rate decreases compared to non-doped diamond. On the other hand, it is also observed that the doping reagent might cause a slight increase in growth rate, which depends on the substrate temperature [78] or pressure [79]. In the study by Hartmann [80] and Miyayta [81], it was found that at very low boron concentration increases the diamond growth rate. It is experimentally very difficult though to explain how B affects the growth process.

6.7.3 Effect of position of the dopant in the diamond lattice

At first, the effect of dopant position within different C atomic layers close to the diamond surface has to be evaluated before any further calculations are to be made.

This will be exemplified here by the substitutional doping with N atoms. More specifically, the effect of N position on the surface H abstraction energy for the diamond (100)–2 × 1, (110) and (111) surfaces should be calculated. This specific surface reaction has been chosen since it is generally assumed to be the rate-determining step in the diamond CVD growth mechanism.

Examples of N positions close to these upper diamond surfaces are shown in figure 6.46. The other N positions that are not included in figure 6.51 involve substitutional doping in the 1st, 3rd, and 4th atomic C layer in the three different diamond surface types.

The calculated energies of H abstraction by gaseous H species can be seen in figure 6.47. For each of the three surface types ((111), (100)–2 × 1, and (110)) there is a large similarity in the abstraction energies when going from a non-doped surface (denoted layer number zero) to a substitutional incorporation within the 1st atomic C-layer. However, the abstraction energy will decrease drastically (i.e. become more negative) when going from the 1st atomic C layer to the 2nd atomic C layer. Thus, the H abstraction becomes more exothermic.

However, the H abstraction energy will increase when replacing a C atom with N further down in the lattice (at least when replacing a C atom in the 3rd and 4th atomic C layer). The situation is identical for substitutional doping with P but differs for the usage of S and B dopants. The situation with B doping is the following. The abstraction energy is just somewhat smaller when going from the 2nd to the 3rd atomic C layer for diamond (111) and (110). On the other hand, it is much more pronounced for diamond (100)–2 × 1. Moreover, substitutional doping with S atoms has only been investigated for the diamond (111) surface, and it was found that the H abstraction energy will decrease when going from the upper surface to the 4th atomic C layer, which is completely different compared to doping with N, P, or B atoms.

As an explanation to the observations made for the N-doped H-terminated diamond (100)–2 × 1 surface, there are peaks in the corresponding adsorption energy

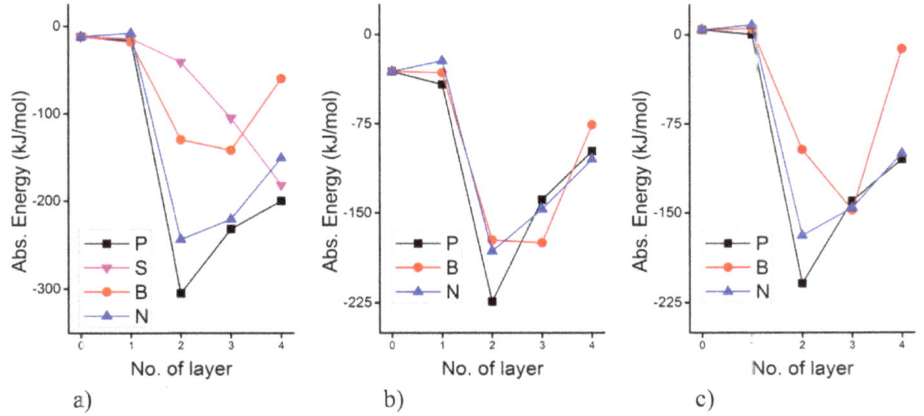

Figure 6.47. Surface H abstraction energy for different dopant positions within the diamond surface lattice: (a) (111), (b) (110), and (c) (100)–2 × 1.

curves for the position of N within C layers 2 and 3 in the upper diamond surface (see figure 6.48). Hence, the H adsorption process becomes less energetic feasible, whilst the H abstraction becomes much more exothermic (see figure 6.45). Two different factors have in the present study been found to be responsible for this effect. When positioning N within the second, or third, atomic C layer in an otherwise completely H-terminated diamond (100)–2 × 1 surface, the extra electron in N occupies the anti-bonding state within one N–C bond. This results in the breakage of the N–C bond, which, however, will re-bind again after the H abstraction from the surface (with the formation of an electron pair at the vacant surface position). This rather severe surface reconstruction is not possible to take place when positioning N further down in the diamond surface (from the fourth atomic layer, and deeper below). This explanation makes it clear that N will improve the abstraction process but show a negative effect on the adsorption process. For the H adsorption process, the values are −458 (non-doped) and −319 (N in 2nd layer) kJ mol^{-1}. The corresponding values for the surface H abstraction process are −38 (non-doped) and −170 (N in 2nd layer) kJ mol^{-1}.

For the diamond (111) surface, the calculated H adsorption energies are: −434 (non-doped) and −349 (N in 2nd layer) kJ mol^{-1} (see figure 6.53). The corresponding values for the surface H abstraction process are 7 (non-doped) to −240 (N in 2nd layer) kJ mol^{-1}. For this specific diamond surface, the only dopant effect is the electron transfer towards the C(surface)–H bond, thereby weakening it. As was also the situation with the (100)–2 × 1 surface, an electron pair is formed on the radical C atom as a result of the H abstraction process. This is a clear indication that this effect willalso become less pronounced with an increased distance from the surface.

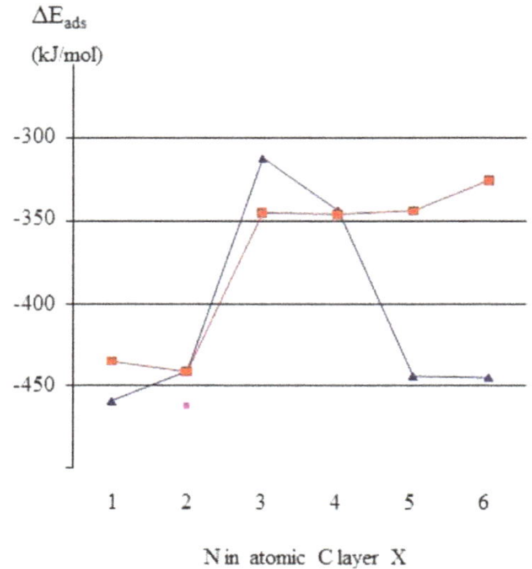

Figure 6.48. Variation in H adsorption energy when positioning N in different atomic C layers. The energy values at X = 0 represent non-doped diamond surfaces. Red: diamond (111); black: diamond (100)–2 × 1. [27], open access © Springer.

Based on all these calculations, it has been proven reasonable to continue the investigation by positioning the N, or P, dopant within the 2nd atomic C layer (for all surface types studied), and it is also quite reasonable to position the B dopant in the 2nd atomic layer for diamond (111) and (110). However, the B dopant should be positioned in the 3rd atomic C layer when using the diamond (100)–2 × 1 surface. The reason for these choices is the larger effect by the dopant on the surface reactivity. However, further studies have to be made to fully understand where the replacement with atomic S should be made.

6.7.4 Effect of substitutional nitrogen doping on the diamond growth mechanism

As presented above, doping with nitrogen has experimentally been shown to increase the diamond CVD growth rate. It is, thus, of the largest interest to study the effect of N doping on the various steps in the diamond CVD mechanism, including H abstraction, adsorption of growth species, migration of growth species, and the final incorporation of a growth species in the diamond lattice. The major question is then —will the N doping change the type of rate-determining step? Or is it still the H abstraction reaction? For the rate-determining step—how will the N doping affect the total growth rate? An answer to each of these questions are given below after that a series of calculations have been made. For each of these calculations, an N atom has been substitutionally positioned within the 2nd atomic C layer in the supercell (see figures 6.46(a), (d), and (g)) for the c-BN (100)–2 × 1, (110), and (111) surface, respectively. As presented above, this is a position that has the largest effect on the H abstraction reaction (and also most probably on other surface reactions).

6.7.4.1 Effect of N-doping on the H abstraction process from the H-terminated diamond surface

As presented above, it has been experimentally proven that N-doping will enhance the diamond growth rate, and thereby reduce the production costs. For instance, Dunst *et al* [50] studied the effect of diamond growth by introducing nitrogen into a microwave plasma-activated CVD reactor in depositing single crystalline diamond (100) at 800 °C. The results of that study state with the introduction of N_2 will improve the diamond growth rate with a factor of approximately 2.5 for a CH_4 concentration of 2%. However, the underlying cause for this growth improvement is not clear since it is experimentally very difficult to determine in which form N will be efficient in this process. To be more specific, it is experimentally very difficult, if not in many cases impossible, to discriminate between the effect of nitrogen in the form of an N-containing species adsorbed on the surface, or as N substitutionally positioned within the upper part of the diamond surface. For this purpose, high-level DFT calculations have proven to be most useful.

In addition to the growth enhancement effect of N doping, it has also been shown possible to steer towards the growth of diamond (100) [82]. An issue that should also be highlighted is the catalytic effect by the N dopant. The situation is that a very low concentration of N dopants is able to give the most pronounced effect on the overall diamond growth rate. However, it has been shown earlier that the surface hydrogens are very mobile on the diamond surface [40]. Hence, the desorption of an H species nearby the N dopant will immediately be replaced by a neighbouring H species on the surface, causing radical surface C sites at longer distances from the N dopant. Thus, the substitutional N dopant will improve both the H abstraction rate and concentration of radical surface C sites, the latter being distributed evenly over the surface.

The main advantage with theoretical DFT calculations is that surface processes, including the breakage and formation of bonds, are most carefully calculated and analyzed. The main aim with the present example is to use DFT-calculations to understand the influence of nitrogen onto the diamond growth rate on an atomic level of theory. The nitrogen dopant is then either substitutionally positioned into the upper diamond lattice (see section 6.6.2) or chemisorbed onto the surface in the form of NH, or NH_2, adsorbates. However, these calculations also put an emphasis on proving that the used method and methodology is appropriate and reliable. As will be shown in the coming sections of this chapter, they will show that the H abstraction from the surface is the rate-determining step in the overall diamond growth rate (i.e. also for the situation with N doping). As such, these calculations can support the experimental observations (regarding the effect of nitrogen on the diamond growth rate) and to further explain these observations by very careful analyses methods on the electronic level.

6.7.4.1.1 N substitutionally positioned into various C atomic layers

As presented in section 6.6.2, the effect of substitutionally positioning N within different C atomic layers close to the surface has, at first, been evaluated. Based on the results presented in figure 6.47, substitutionally positioned N, within the second C layer of the different diamond surface planes, can be chosen since it has been found that this is a position that has a large effect on the H abstraction reaction, and, hence, on the surface reactivity in general. To analyze this effect on the surface reactivity, Fukui function calculations have been performed with the intention to localize the surface sites that are most susceptible towards a radical attack. As can be seen in figure 6.49, there is a major effect by the substitutionally positioned N dopant on the radical susceptibility of the surfaces (100)–2 × 1, (110) and (111). This effect is more delocalized for the (110) surface, whilst the others show a more local but still very strong effect (local means close the N dopant). Based on these Fukui function calculations, the choice of which H adsorbate to abstract (with gaseous H) can be made (which is, thus, determined from figure 6.49).

As presented in section 6.6.2, the most interesting result has been obtained for substitutionally nitrogen doping in the second C atomic layer, for which there is both the most exothermic H abstraction reaction and a process without any energy barrier (for any of the surface planes investigated) (see table 6.6). The question is now to explain this theoretical observation. Nitrogen has an extra valence electron compared to carbon, and it is at first very important to get information about where this extra electron density will be positioned. For the diamond (100)–2 × 1 surface, spin electron density calculations have shown that the density from this extra electron is more globally distributed amongst the bonds in the vicinity to N (see figure 6.50(a)). Electron deformation densities have also been calculated

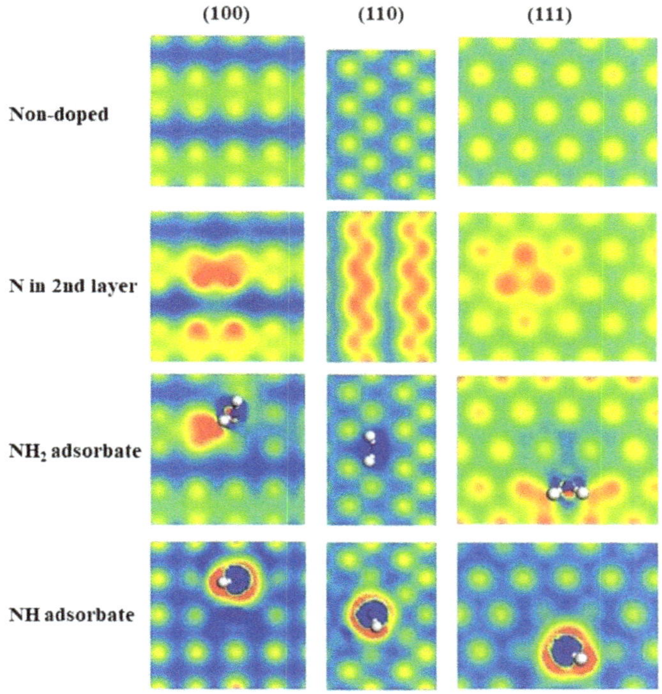

Figure 6.49. Isosurface maps of Fukui functions showing the distribution of radical susceptibility for the diamond surfaces (100), (110), and (100)–2 × 1, and for various dopant situations. The red colour visualizes the most susceptible part on the surface, and the blue colour represents the least susceptible part. [27], open © Springer.

Table 6.6. Effect of N doping on the H abstraction energy, with its energy barrier in a relation to the respective non-doped situation for the diamond surface planes (110), (111) and (100)–2 × 1.

Doping situation	H abstraction energy (kJ mol^{-1})	Energy barrier for the H abstraction (kJ mol^{-1})
(110)		
Non-doped	−38	+5
N in 2nd layer	−170	0
NH_2 adsorbate		+6
(111)		
Non-doped	7	+6
N in 2nd layer	−240	0
NH_2 adsorbate		+6
(100)		
Non-doped	+3	+12
N in 2nd layer	−165	0
NH_2 adsorbate		+8

with the purpose to investigate the effect of N dopants on the individual bonds within the H-terminated diamond surface (i.e. atom–atomic interactions) within the systems. As can be seen in figure 6.50(a), the bond population analysis supports these results, showing that the extra spin will cause a weakening of both the N–C (layer 1) (0.55) and C–H (0.63) bonds. The former one is to be compared with the N–C(layer 3) bonds: 0.60 versus 0.65, and the latter with the C–H bonds for non-doped diamond: 0.88. It is thereby obvious that the extra electron in the N dopant (positioned in the 2nd C atomic layer) weakens the upper part of the diamond (100) surface, and it is assumed that this is also the case for diamond (111) and (110) surfaces. The electron deformation densities have also been calculated for the approaching H radical at a so-called transitions state (TS) distance from the diamond (100)–2 × 1 surface (i.e. identical to the TS distance for non-doped diamond). As can be seen in figure 6.51(b), there is a strong interaction between the incoming H and the surface H. There is an electron density between these two atoms (of 0.18) that indicates an electron orbital overlap. In addition, there has been a partial electron transfer to the approaching H radical, and there is a weakening of the C–H bond (from 0.90 to 0.72). All of these values strongly support the zero-level energy barrier value for the abstraction of H from the three different diamond surfaces.

6.7.4.1.2 N chemisorbed onto the surface in the form NH or NH_2

Since it was not experimentally clear in which form nitrogen will aid the improved diamond growth rate, nitrogen has been attached to the surface in the form of a radical NH group (i.e. chemisorbed onto the diamond surface). As can be seen in

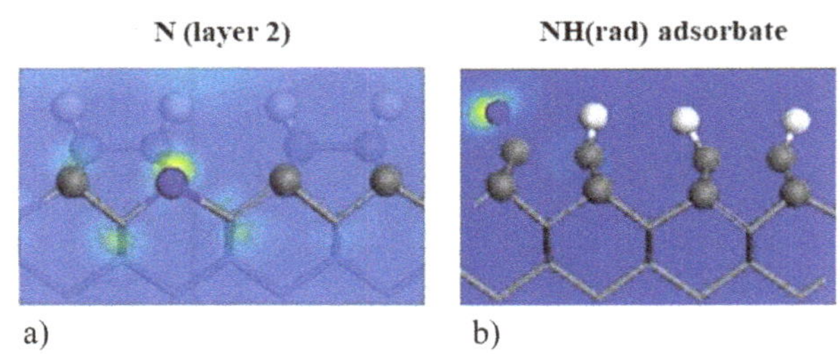

Figure 6.50. Demonstration of the electron spin density for diamond (100)–2 × 1 with N (a) substitutional positioned within the 2nd C atomic layer, or (b) chemisorbed in the form of a radical. [27], open access © Springer.

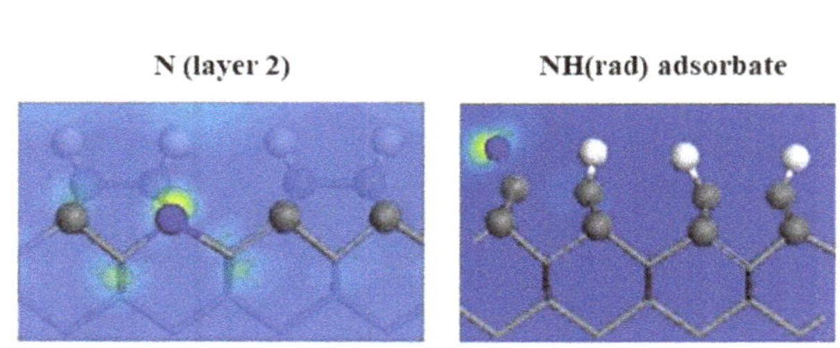

Figure 6.51. Electron density difference maps showing the (a)–(c) dopant-induced lattice weakening, and the (d) interaction between the approaching H species and the surface H adsorbate. [27], open access © Springer.

figure 6.50(b), the non-paired electron stays locally at the adsorbate NH place. This local position of the extra electron is supported further on by the deformation density results in figure 6.51(b). The extra electron in NH does not show any tendency to interact with the surface in weakening bonds by filling anti-bonding orbitals. Due to this reason, the energy barrier for the H abstraction is expected to be very similar as for the non-doped diamond surfaces.

However, when nitrogen is chemisorbed in the form of NH_2 onto the diamond surface, the situation becomes more complex (even though NH_2 is not a radical species when adsorbed to the diamond surface). For the (100)–2 × 1 surface, the reactivity of the surface hydrogen in the vicinity to one of the hydrogen atoms within the NH_2 group is enhanced. The situation with the (110) surface is, though, quite different in that the reactivity close to NH_2 becomes weakened. However, the situation with the (111) plane is similar to the one for the (100)–2 × 1 surface. The reactivity of the surface hydrogen closest to the hydrogen atom in the NH_2 group is reinforced, whilst the one closest to the nitrogen atom decreases in reactivity. However, these changes in surface reactivity are apparently not large enough to affect the H abstraction barrier energy (see figure 6.51 and table 6.6). The usefulness with these Fukui maps is that they are visualizing the most reactive surface H atoms, to which the gaseous H species is being approached. Similar to the situation with NH adsorbates, the nitrogen doping in the form of NH_2 adsorbates does not show any larger effect as compared to the non-doped scenario. To be more specific, the energy barriers for the H abstraction from the diamond (100)–2 × 1 and (111) surface are observed to decrease, whilst the barrier for the (110) plane is increased; NH_2 doping (+8, +6, vs. +6 kJ mol^{-1}) versus non-doping (+12, +6, vs. +5 kJ mol^{-1}) (see table 6.6). However, it must be stressed that these differences are marginal for diamond (111) and (110). There is also a correlation observed between these variations in energy barriers and the effect by NH_2 doping on the surface reactivities. The Fukui function maps for the diamond (100)–2 × 1 and (111) surfaces show a slight improvement in surface reactivity, whilst the corresponding one for the (110) plane show a decrease (see figure 6.49).

6.7.4.2 Effect of N doping on the CH_3 adsorption process

The question is now how N-doping might affect the chemisorption of growth species to the various diamond surfaces. Different positions of N in the first and second C atomic layer have been considered here (see figure 6.52) in order to highlight possible differences in the N dopant effect on the initial diamond growth processes. In addition, both (100) and (111) surfaces have been taken into account to outline the experimental observations that the N effect depends on surface orientation. The initial diamond growth process is then assumed to be dominated by CH_3 adsorption to a radical C site on the diamond surfaces, and gas phase H abstraction from the adsorbed CH_3 species.

The calculated energetic values for CH_3 adsorption are presented in table 6.7 for both diamond (111) and (100) surfaces, and with different positions of the N dopant. Corresponding adsorption energies for the non-doped situation are added for comparison. Similar calculations for also other growth species can easily be done by following the methodology presented here.

6.7.4.3 An N dopant within the 1st carbon atomic layer

A comparison of the calculated CH_3 adsorption energies in the presence of N in the first carbon layer, with the values obtained for the corresponding non-doped situation, does not show any indication of dopant-induced effects (see table 6.7). These results are confirmed by a careful structural and electronic analysis of the surface—both before and after CH_3 adsorption reaction (see below).

6.7.4.3.1 An N dopant within the 2nd carbon layer

As can be seen in table 6.7, the CH_3 adsorption appears to be strongly disfavoured by the presence of N in the second carbon layer. The reason for this effect is most

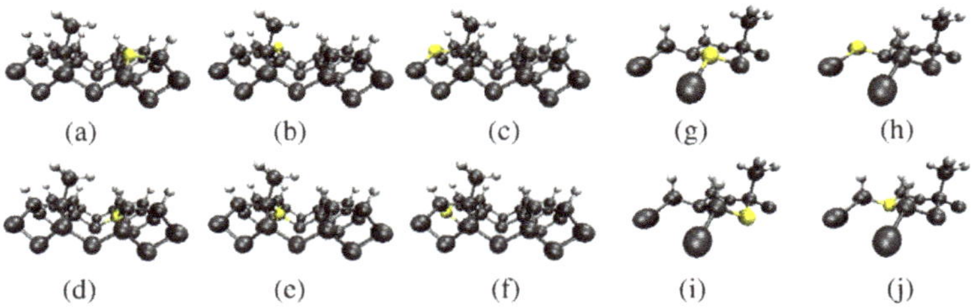

Figure 6.52. For the (100)–2 × 1 surface: N in position (a) 1, (b) 2, and (c) 3 within the 1st C layer, and in position (d) 1, (e) 2, and (f) 3 within the 2nd C layer. For the (111) surface: N in position (g) 1, and (h) 2 within the 1st C layer, and in position (i) 1 and (j) 2 within the 2nd C layer. C: grey; H: white; dopant: yellow. [83], reproduced with permission © Elsevier.

Table 6.7. Calculated CH_3 adsorption energies for N positioned in the first or second atomic layer for both diamond (111) and (100)–2 × 1 surfaces. See figure 6.51 for an explanation of the positions.

	(111)		(100)–2 × 1		
ΔE_{ads} (kJ mol^{-1})	Position 1	Position 2	Position 1	Position 2	Position 3
N in the 1st C layer	−333.5	−319.7	−379.6	−373.6	−363.3
N in the 2nd C layer	−125.5	−176.6	−237.3	−171.6	−212.1
Non-doped	−317.7		−358.5		

Table 6.8. $B_{1C\text{–}C}$, $B_{1C\text{–}N}$ and B_2 bond lengths for CH_3, or CH_2, adsorption to a diamond (100)–2 × 1 surface (with N in position 3). The results are compared with the corresponding non-doped situation. $B_{1C\text{–}C}$: the reconstructed C–C surface bond; $B_{1C\text{–}N}$: the C–N reconstructed surface bond; B_2: the C–C bond between CH_3, or CH_2, and the surface carbon.

	(100)–2 × 1—Position 3		(100)–2 × 1—Non-doped	
Bond length (Å)	CH_3	CH_2	CH_3	CH_2
$B_{1C\text{–}C}$	1.63	1.62		
B_2	1.51	1.46	1.50	1.34
$B_{1C\text{–}N}$			1.60	2.43

probably the transfer of one valence electron from the nitrogen to the surface carbon radical, which, thereby, becomes less reactive towards the adsorption by the formation of an electron lone pair.

6.7.4.3.2 N chemisorbed onto the surface in the form NH or NH_2

Since it was not experimentally clear in what form nitrogen will aid the improved diamond growth rate, nitrogen was attached to the surface in the form of a radical NH group (i.e. chemisorbed onto the diamond surface). As can be seen in figure 6.50(b), the non-paired electron will stay locally at the adsorbate NH place. This local position of the extra electron is supported further on by the deformation density results in figure 6.51(b). The extra electron in NH does not show any tendency to interact with the surface in weakening bonds by filling anti-bonding orbitals. Due to this reason, the energy barrier for the H abstraction is expected to be very similar to 1.54 Å. A comparison between these two C–C bond lengths indicates that $B_{1C–C}$ is constrained with CH_3 adsorbed to the surface. With CH_3 adsorbed to the N doped surface, $B_{1C–N}$ was also found to be longer (1.60 Å) compared to the calculated C–N bond length in the methylamine molecule (1.47 Å). The larger difference observed in the presence of N is a difference in constrained bonds that has been confirmed by performing bond population analysis. The C–N bond ($B_{1C–N}$) presents a significantly lower value compared to the C–C bond ($B_{1C–C}$) when CH_3 is adsorbed on the surface (0.79 versus 0.67), which indicates that this bond is weaker. The observation of a β-scission rearrangement during the geometry stabilization process (in the presence of N) can be explained by the lack of an energy barrier for the CH_3 adsorption process.

6.7.4.4 Effect of N doping on the abstraction process of H from the CH_3 adsorbate

6.7.4.4.1 An N dopant within the 1st carbon layer

The calculated energy values for the abstraction of H from the CH_3 adsorbate are presented in table 6.9 for both diamond (111) and (100–2 × 1) surfaces, and with different positions of the N dopant. This H abstraction is assumed to take place by using gaseous H radicals as reactants, which results in gaseous H_2 molecules. The corresponding abstraction energies for the non-doped situation are added for comparison.

As can be seen in table 6.9, the variation in H abstraction reaction energies, for the different N positions, indicates that this reaction is generally not affected by N for both types of diamond surfaces. Compared to the corresponding non-doped situations, the differences in abstraction energy are, with one exception, not larger

Table 6.9. Calculated H abstraction energies from adsorbed CH_3 species as a function of N position in the 1st or 2nd atomic layer in both diamond (111) and (100).

	(111)		(100)–2 × 1		
ΔE_{abs} (kJ mol^{-1})	Position 1	Position 2	Position 1	Position 2	Position 3
N in the 1st C layer	−10.6	−21.0	−7.9	−23.1	−53.2
N in the 2nd C layer	−262.6	−154.7	−138.0	−180.5	−140.0
Non-doped	−21.4		−15.6		

than 11 kJ mol^{-1}. The exception is the situation with N in position 3 in the (100)–2 × 1 surface, which presents a difference of around 38 kJ mol^{-1}. This relatively important difference, compared to the total abstraction energy (+53.2 kJ mol^{-1}), can be explained by a spontaneous structural rearrangement of the surface after H abstraction. As shown in figure 6.58, the N–C bond (B_1) is broken, and a double bond is formed between the surface carbon and the binding CH_2 species (B_2). This rearrangement is usually called a β-scission. As was the situation with a CH_3 adsorbate, the length of $B_{1C–C}$ is around 1.62 for the resulting CH_2 adsorbate on the non-doped (100)–2 × 1 surface (see table 6.10). Thus, it is quite strained (in comparison to the C–C bond length in the ethane molecule; 1.54 Å).

For convenience, the notation B_1 for the undoped surface will be renamed $B_{1C–C}$, and B_1 for the doped surface with N in position 3 will be renamed $B_{1C–N}$. As can be seen in table 6.8, for the non-doped (100) surface with CH_3 as an adsorbate, the length of $B_{1C–C}$ is around 1.62 Å. (For comparison, the C–C bond in an ethane molecule has been calculated to be equal to N.) The structural and electronic analyses of the surfaces show two different effects, depending on the position of N relative the surface carbon to which CH_2 is binding (C_{CH2}). In the first case, N is not directly binding to $C_{CH2.}$ This is, thus, the situation for N in position 2 in the diamond (111) surface, and N in position 1 and 3 in the diamond (100) surface (see figures 6.52(d), (f) and (j)).

As can be seen in table 6.10, the atomic charge of C in CH_2 is, compared to the non-doped surface, significantly more negative for all these cases. This is the result of a lone pair formation on the C atom (in CH_2), which is more stable compared to a radical. This electron transfer can also be observed by following the geometrical evolution of CH_2 from a planar geometry (being the normal geometry for an non-doped surface) to a tetrahedral geometry on the doped surface. In the second case,

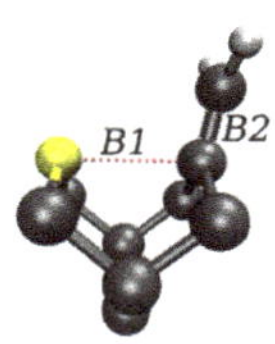

Figure 6.53. Reconstruction of the (100) surface when N substitutionally replaces the surface carbon in position 3 (see figure 6.56). N: yellow; C: black; H: white. [83], reproduced with permission © Elsevier.

Table 6.10. Evolution of the atomic C charge of the surface carbon radical and the adsorbed CH_2 species, for the non-doped surface and in the presence of N in the second carbon layer (position 1 and 2 on the (111) surface, and position 1, 2 and 3 on the (100) surface).

	(111)			(100)–2 × 1			
C atomic charge (e)	Position 1	Position 2	Non-doped	Position 1	Position 2	Position 3	Non-doped
Surface radical	−0.28	−0.39	−0.01	−0.42	−0.31	−0.40	−0.03
CH_2	−0.73	−0.94	−0.53	−0.90	−0.66	−0.89	−0.59

with N being directly bonded to C_{CH} (for position 1 on the (111) surface, and position 2 on the (100) surface (cf. figures 6.52(e) and (i))), a more complex geometrical rearrangement occurs, with an induced surface stabilization. In both of these cases, the N–C_{CH} bond is broken (the bond length is around 2 Å, and the bond population is close to 0.0 e, which is a clear indication of the lack of any kind of interactions), and CH_2 becomes attached to the surface carbon through a double bond (with a bond length around 1.3 Å and the bond population is around 1.3 e, which are all typical values for a C–C double bond).

There is a correlation between the theoretical results presented here and the experimentally observed increase in growth rate and improvement of surface morphology, which are generally obtained by using a small concentration of nitrogen in the gas phase [84]. One possible explanation for these correlations is the larger H abstraction energy in the presence of substitutional N dopants in the lattice, which facilitates the formation of CH_2 adsorbates on the surface. Since this latter type of species is known to be important for surface migration [85], H abstraction from the CH_3 adsorbates will most probably improve the surface migration, which, in turn, will most probably affect the diamond growth rate and surface morphology.

The observed β-scission rearrangement can also be related to the experimental observation that a polycrystalline diamond surface shows a rougher morphology at high N concentrations [86]. This type of surface morphology is known to be induced by a high degree of re-nucleation during growth, which is mushrooming in the presence of sp^2 carbon. This re-nucleation is, then, consistent with the β-scission rearrangements (i.e. the formation of sp^2 carbons) that has been experimentally observed to occur more frequently for larger N dopant concentrations.

6.7.4.5 *Effect of N doping on the CH_2 surface migration process*

As one alternative to the formation of a CH_2 adsorbate from a chemisorbed CH_3 species, by using gas phase abstraction, this can take place by the migration of H (in CH_3) towards a neighboring surface C radical. In fact, this migration is often considered as an alternative mechanism to H abstraction from CH_3 [85, 87–89]. This specific reaction step is, thereby, important for the entire growth process, and the eventual changes induced by the presence of a substitutional N need to be considered.

For the situation with a diamond (100)–2 × 1 surface, the resulting CH_2 adsorbate is, thereafter, assumed to be inserted within the surface carbon dimer. This so-called inserted CH_2 can, thereafter, migrate on the surface and finally reach a surface step, where the carbon will finally be incorporated into the lattice. Moreover, the insertion reaction (within a carbon dimer) is generally described as occurring in two steps: (a) a dimer opening step, which forms an intermediate 'open-ring' structure, and (b) a ring closing step [90–92]. It is, thus, of a large interest to also study the effect by substitutional N dopants on the (i) CH_2 insertion into the C–C, (ii) dimer H migration between an adsorbed CH_2 and a neighboring surface radical, and (iii) migration of inserted CH_2 species.

6.7.4.5.1 *CH_2 insertion into a dimer*

The insertion reaction of an adsorbed CH_2, into the surface carbon dimer, is shown in figure 6.54. The ring closing step of the reaction is assumed here to be the rate

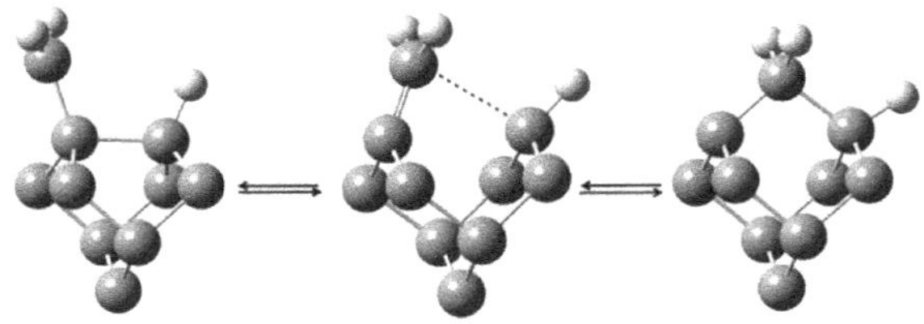

Figure 6.54. Mechanism describing the incorporation of an adsorbed CH_2 into a carbon dimer. R, P, and TS denote the reactant, product, and transition state structure of the reaction, respectively. C: black; H: white. [93], reproduced with permission © American Chemical Society.

determining step since it is expected to have a large energy barrier. It is, thereby, regarded to be the rate-determining step for the complete surface migration process.

6.7.4.5.2 Effect by substitutional N on reaction energy

The calculated CH_2 insertion reaction energy is presented in table 6.11 for N in atomic C layer 2 in three different positions. As can be seen in table 6.11, the presence of substitutional N is not strongly affecting the energy of the carbon dimer insertion reaction for N in position 3. However, the energy barrier for the insertion reaction is much larger for N in position 1 (with a factor of around 2.5 when compared with the non-doped situation). Moreover, the situation is much more severe for N in position 2, for which the insertion reaction becomes endothermic (with a reaction energy of +85 kJ mol^{-1}). In addition, the corresponding energy barrier is very large (with a factor of around 3.5 with respect to the non-doped situation). To explain the energy differences for N in position 2, the total energy of the diamond model with N in position 2 has been compared with the total energy for N in position 1 (both before and after CH_2 insertion). As a result of these comparisons, the observed endothermicity of the CH_2 insertion reaction for N in position 2, can be explained by a structural stabilization of the surface region of the diamond lattice (with about 157 kJ mol^{-1}). This structural stabilization can in turn be explained by the presence of N in a β-position with respect to a surface radical carbon, whereby a structural rearrangement of the surface occurs through a β-scission mechanism. Prior to CH_2 insertion, the radical site is positioned on the CH_2 adsorbate (see figure 6.54). The breakage of the C–N bond (B1 and B3), in favor of the formation of a double C–C bond (B2 and B4), is shown in figure 6.55.

Table 6.11. Reaction energies ($\Delta E_{\text{reaction}}$) and energy barriers ($\Delta E_{\text{barrier}}$) for the incorporation of CH_2 into the carbon dimer on a diamond (100)–2 × 1 surface. Values are presented for both N-doped and non-doped diamond lattices, with N in atomic layer 2 in either of the positioned 1, 2, or 3 (see figure 6.51).

	(100)–2 × 1			
Energy (kJ mol^{-1})	Position 1	Position 2	Position 3	Non-doped
$\Delta E_{\text{reaction}}$	−14	+85	−11	−20
$\Delta E_{\text{barrier}}$	+252	+355	+80	+107

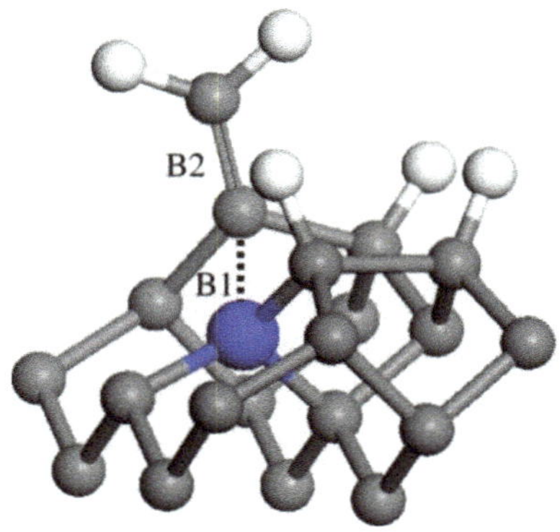

Figure 6.55. Representation of the β-scission reconstruction observed when a surface carbon radical (here the CH_2 adsorbate) is in β-position with respect to N (which is in position 2 in the second carbon layer). [93], reproduced with permission © American Chemical Society.

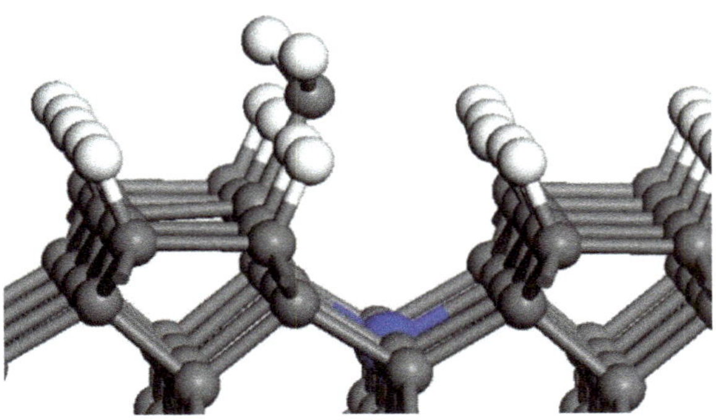

Figure 6.56. sp^3 Configuration of the adsorbed CH_2 in the presence of substitutional N within the third carbon layer. [93], reproduced with permission © American Chemical Society.

For the other situations, with N not being in β-position with respect to the surface radical, the system cannot relax through a β-scission mechanism. These situations involve N in positions 1 and 3 in the second atomic C layer in the diamond (100)–2 × 1 lattice. These systems will instead relax through a transfer of an electron from N to the surface radical carbon, thereby inducing the formation of an electron lone pair. The effect of this electron transfer process is visualized by the geometry of the adsorbed CH_2 species prior to the dimer insertion (see figure 6.56). The geometry of an adsorbed CH_2 species is planar for a non-doped diamond lattice, while this type of geometry, for the N-doped scenario, shows evidence of sp^3 hybridization (being induced by an electron transfer and the formation of a lone pair). This transfer is further confirmed by studying the calculated atomic charges. The C atomic charge within the CH_2 adsorbate has evolved from −0.33 for the non-doped structure to about −0.95 in the presence of N (see table 6.10).

After the CH_2 insertion, the presence of N does not induce any major structural changes. However, the electron transfer is still visible. For the non-doped situation, the threefold coordinated surface carbon has an atomic charge of +0.09, while, in the presence of substitutional N, the charge becomes much more negative (around −0.45). An exception is observed with N in position 2, and with CH_2 inserted within the carbon dimer. N is then directly bonded to the threefold coordinated surface C, which then presents a lower negative atomic charge of −0.25.

In summary, the influence by a substitutionally positioned N dopant (in β-position to a surface radical carbon) on the thermodynamics of the CH_2 insertion reaction, has been identified here. Moreover, the diamond surfaces have been found to be more stable with N directly connected to the surface carbon radical. The CH_2 insertion reaction is, then, disfavoured. For the other N positions, the CH_2 insertion reaction energy was not found to be strongly affected by the presence of N.

6.7.4.5.3 Effect by substitutional N on activation energy

The calculated energy barriers, for the CH_2 insertion reaction into the C–C dimers, are presented in table 6.11. With one exception, the CH_2 insertion reaction is associated with a much larger activation barrier in the presence of substitutional N (as compared with the non-doped situation). There is an increase of 145 kJ mol^{-1} (position 1) and 252 kJ mol^{-1} (position 2). For the situation with N in position 3, the activation barrier is lowered by 27 kJ mol^{-1}.

The large barrier calculated for N in position 2, of +355 kJ mol^{-1}, can be explained by the β-scission rearrangement that strongly stabilizes the surface prior to CH_2 insertion. It is important to emphasize the strong influence on surface reactivity that surface structure stabilization may induce. For N in position 2, the large activation energy that is needed to be overcome clearly shows how much the stabilization of the initial structure will affect the reaction kinetics. For the other N positions considered, where no β-scission rearrangements where observed (i.e. positions 1 and 3), the variations in energy barrier are only related to the stability of the transition structures. These energetic variations can be further investigated by a careful analysis of the geometrical structures and atomic charges of the different carbon atoms involved in the reaction.

As can be seen in figure 6.54, the CH_2 insertion reaction on a non-doped diamond surface passes through a transition state where the C–C dimer bond is broken and the C–CH_2 bond becomes shorter (around 1.35 Å), which is similar to the length of a tabulated C–C double bond: 1.34 Å [94]. In the presence of N, the extra electron that is transferred towards the surface will be localized on the surface C to which the CH_2 will form a new bond. This is represented by a dashed line within the TS structure in figure 6.54. The atomic charge of this specific surface C (of about −0.74 for N in position 1) is much more negative compared to the corresponding surface C in the non-doped situation (−0.22). In addition, the charge of C within the adsorbed CH_2 species is negative and is not much affected by the presence of N (about −0.36 with N vs −0.40 for the non-doped lattice). In the presence of N, the more negative surface carbon will induce a stronger electrostatic repulsion between the two carbon atoms involved in the new bond formation. The presence of this repulsion is confirmed by the longer distance between these two C atoms (2.33 Å without N versus 2.80 with N in position 1). The more intense electrostatic repulsion observed in the presence of N will, thus, induce a destabilization of the TS structure, which explains the larger energy barrier. Furthermore, for N in position 3, where N is directly bonded to the surface C, the electronic charge appears to be less localized on the surface C within the transition structure (this charge is −0.40), thereby reducing the electrostatic repulsion. This tendency is confirmed by a shorter distance between CH_2 and the surface carbon (2.36 Å), and also explains the smaller activation barrier observed for N in that position.

In conclusion, it appears that a large degree of surface stabilization, being induced by the formation of an sp^2 carbon species, causes a significant increase in energy barrier for the CH_2 insertion mechanism. When N is not in β-position, the energy barrier for the CH_2 insertion into a C–C dimer is generally increased, thereby reducing the probability for the insertion reaction to take place. Substitutional N is thereby expected to hinder the growth.

6.7.4.5.4 H transfer reaction between an adsorbed CH_2 and a neighbouring surface carbon

The H transfer process that takes place from a CH_2 adsorbate to a neighbouring C surface site, is shown in figure 6.57. In addition, the calculated reaction energies and energy barriers are presented in table 6.12. The initial reactant structure is considered here to be the adsorbed CH_2 on an otherwise H-terminated surface (R in figure 6.57), and the product is an adsorbed CH_3 species with a neighboring radical carbon (P in figure 6.57). An H is, then, considered to be transferred from a surface carbon toward the adsorbed CH_2 species via a transition state structure (TS in figure 6.57). An adsorbed CH_2 species on the surface is, hence, assumed to undergo two distinct surface reactions: the reaction with a migrating H or the

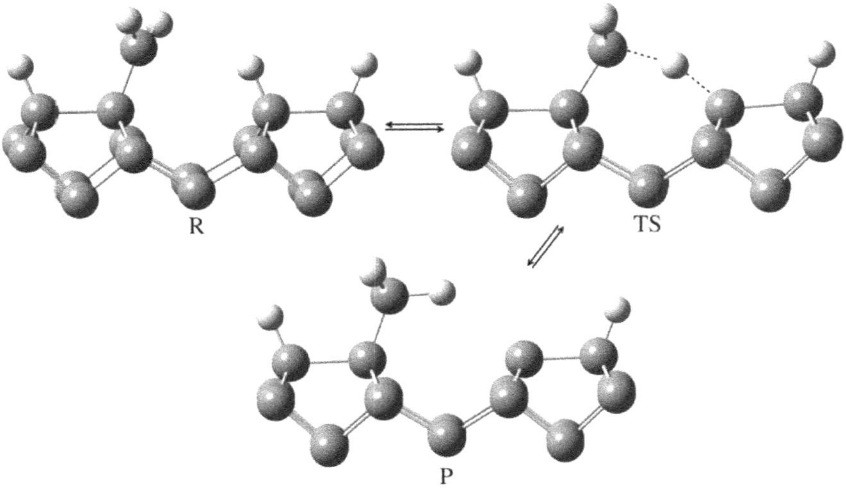

Figure 6.57. Representation of the H transfer reaction, with R, TS, and P as notations of the reactant, transition state, and product structures, respectively. [93], reproduced with permission © American Chemical Society.

Table 6.12. Reaction and barrier energies for the H transfer reaction on a diamond (100)–2 × 1 surface. Values are presented for both N-doped and non-doped diamond lattices, with N in the atomic layer 2 in either of the positioned 1, 2, or 3 (see figure 6.56).

	(100)–2 × 1			
Energy (kJ mol^{-1})	Position 1	Position 2	Position 3	Non-doped
$\Delta E_{reaction}$	−69.4	+161.5	−20.0	+9.9
$\Delta E_{barrier}$	+4.7	+186.5	+77.9	+90.2

insertion into a C–C dimer. It is, therefore, of greatest importance to also study the effect of substitutional N on this H migration reaction.

6.7.4.5.5 Effect of substitutional N on the H migration reaction energy

As can be seen in table 6.12, there are major differences in the H migration reaction energies in the presence of substitutional N in specific positions. As can be seen in table 6.12, reaction energies vary from −69.4 to +161.5 kJ mol^{-1}, which is to be compared with +9.9 kJ mol^{-1} for the non-doped situation. Two major energetic effects are observed with N in the positions 1 and 2. The H migration reaction is, then, favored by −69 kJ mol^{-1} and disfavored by +161 kJ mol^{-1}, respectively. The reaction energy is only slightly affected by the presence of N when it is situated in position 3. The more pronounced exothermic H migration with N in position 1, indicates that the situation with a CH_3 adsorbate and a neighboring surface radical C site is energetically more stable than a situation with CH_2 adsorbed on an otherwise fully H-terminated surface. This observation can be explained by the fact that the substitutional N atom is directly bonded to the surface C radical, thereby stabilizing this specific structure. This result strengthens the conclusion that the situation with N in position 1 is, with N directly bonded to the surface radical C, more stable compared to the situation with N further away. The smaller effects induced by N in position 3 can also be explained by the relative position of N. Moreover, the fact that the H transfer reaction is strongly disfavoured by N in position 2, can be explained by the β-scission rearrangement that strongly stabilizes the surface structure with an adsorbed CH_2.

6.7.4.5.6 Effect by substitutional N on the activation energy

The presence of substitutional N has been shown here to have a major effect on the energy barriers for the H transfer reactions. There is a decrease in energy barrier for the situations with N in positions 1 and 3, while it is increased for N in position 2. The very small energy barrier induced by N in position 1 means that the adsorbed CH_2 species will rapidly react with the neighboring H species, thereby converting to a CH_3 adsorbate. Consequently, it will then hinder the C–C dimer insertion reaction for which an adsorbed CH_2 is a prerequisite. Moreover, for N in position 3, the energy barrier for the H transfer reaction was found to be only slightly lowered compared to the non-doped situation (+78 versus +90 kJ mol^{-1}). However, the inverse reaction (i.e. the H migration from the CH_3 towards the neighboring surface carbon) is also interesting to consider since it is thermodynamically more favourable. For this inverse reaction, the energy barrier is found to be much lower compared to the non-doped surface (+23 versus +87 kJ mol^{-1}). This decrease in energy barrier implies that N in this specific position will hinder the formation of an adsorbed CH_3 by H migration towards an adsorbed CH_2 species. Furthermore, the large energy barrier (+187 kJ mol^{-1}) observed for N in position 2 can be explained by the β-scission mechanism which strongly stabilizes the adsorbed CH_2 and thereby increases the energy barrier.

Substitutional N shows a very strong effect on the migration of H between a CH_2 adsorbate and a nearby positioned surface radical C. The lowering in energy barriers can be explained by the interaction between the CH_2 adsorbate and a neighboring H adsorbate. The CH_2 species has an electron lone pair, which is the result of the electron transfer from the substitutional N dopant towards the CH_2 adsorbate. This CH_2–H interaction is visualized here by the graphical representation of electron density differences in figure 6.58. These representations clearly show that, for N in position 1, the positively charged H adsorbate is attracted by the lone pair in the adsorbed CH_2 carbon (being somewhat localized between the CH_2 carbon and H). Moreover, with N in position 3, the electrons are observed to be more localized to the surface-to-CH_2 bond. The interaction between the CH_2 and neighboring H adsorbates is, thereby, weaker.

As a conclusion, a substitutional N dopant will generally lower the energy barrier for the H transfer between an adsorbed CH_2 species and a nearby positioned radical surface C site. More specifically, the kinetics of this specific surface reaction are, thereby, expected to be energetically favoured for N in positions 1 and 2. However, the direction of the H transfer depends more strongly on the N position, and the overall effect on the atomistic growth process needs further study.

6.7.4.5.7 CH_2 migration

The migration reaction of CH_2 across dimer rows is based on the mechanism proposed by Frenklach *et al* [85] and is considered to occur in two steps (see figure 6.59). The first migration step (R → I) corresponds to the inverse of the CH_2

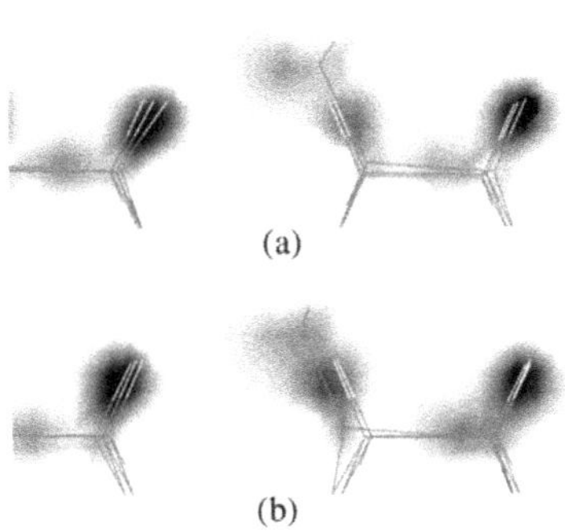

Figure 6.58. Electron density difference maps for CH_2 adsorbed on the diamond (100)–2 × 1 surface with N in position (a) 1 and (b) 3. [93], reproduced with permission © American Chemical Society.

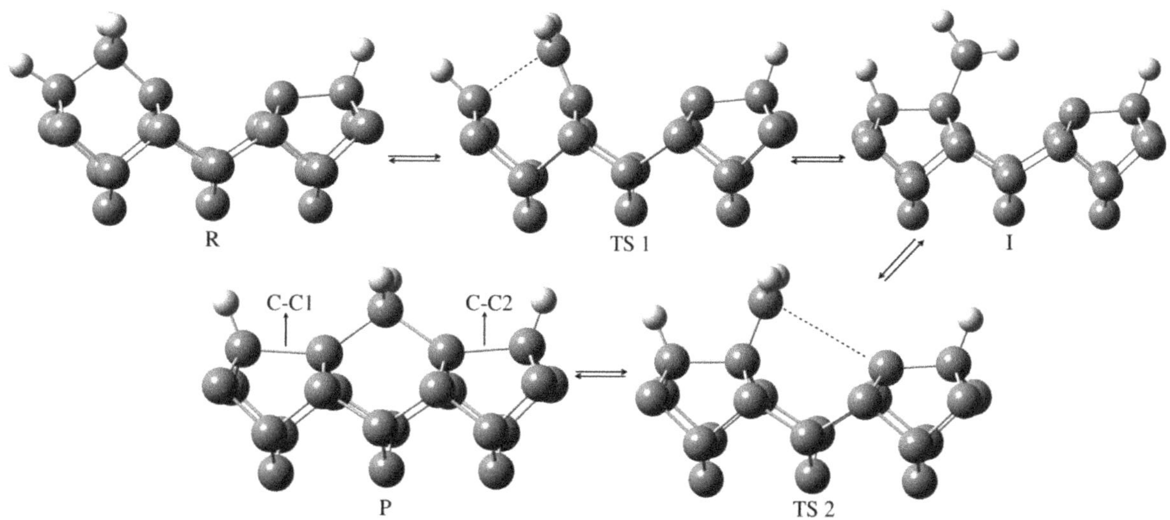

Figure 6.59. CH_2 migration reaction from an inserted position within a carbon dimer towards a bridged position between two C–C dimers. R, I, and P represent the reactant, intermediate, and product structures, respectively. TS1 and TS 2 represent the two transition states. [93], reproduced with permission © American Chemical Society.

Table 6.13. Reaction and barrier energies for the first (R1) and second (R2) reaction steps within the CH2 surface migration process on a diamond (100)–2 × 1 surface. Values are presented for both N-doped and non-doped diamond lattices, with N in the atomic layer 2 in either of the positioned 1, 2, or 3 (see figure 6.56).

	(100)–2 × 1			
Energy (kJ mol^{-1})	Position 1	Position 2	Position 3	Non-doped
$\Delta E_{reaction}$ (R1)	+34.0	−49.4	+32.0	+21.0
$\Delta E_{reaction}$ (R2)	+49.3	+108.8	−41.8	−92.7
$\Delta E_{barrier}$ (R1)	+154.2	+251.1	+125.0	+139.9
$\Delta E_{barrier}$ (R2)	+92.6	+154.2	+11.0	+12.4

insertion reaction. The main difference is that the resulting radical CH_2 adsorbate (I) will be positioned very close to the neighboring surface carbon radical. The proximity of these two radicals induces a rapid formation of a new C–C bond that forms the final bridged structure. This rapid formation constitutes the second step (I → P) in the process. The calculated reaction and barrier energies are presented in table 6.13. For the first migration reaction step (R1), the reaction energies are consistent with the values calculated for the C–C dimer insertion reaction (considering the fact that migration is the forward direction, and the insertion is the inverse direction, of the same reaction type).

6.7.4.5.8 Reaction step 1

As can be seen in table 6.13, the calculated reaction energies for the first step are, with one exception, observed to be quite similar for the N-doped and non-doped situations. The reaction energies are endothermic and have the values +34.0 (position 1), +32.0 (position 3), and +21.0 (non-doped) kJ mol^{-1}. However, the exception is observed for N in position 2 with its exothermic reaction energy value of −49.4 kJ mol^{-1}. This exothermic value can be easily explained by the β-scission reconstruction which stabilizes the surface when N is in β-position with respect to the surface radical.

Furthermore, the energy barriers for reaction step 1 are observed to be only slightly affected by the presence of N in positions 1 and 3; +154.2 (position 1), +125.0 (position 3), and +139.9.0 (non-doped) kJ mol^{-1} (see table 6.13). This is to be compared with a value of +251.1 kJ mol^{-1} for the situation with N in position 2, which is a value that is larger than the non-doped value by a factor of almost 2. With N in position 1, the N atom is directly bonded to the carbon radical within the dimer that is neighboring the reaction site. The analysis of the spin density shows that the extra electron from N is predominantly located on the surface radical carbon that is directly connected to N (demonstrated as a very low spin density on the surface carbon to the right in figure 6.60(a)). The reaction is, thereby, not affected by the presence of the dopant, and the energy barrier is similar to the non-doped situation. This can be confirmed by a structural analysis of the transition structure. The C–C distance, represented with a dashed line in TS1 in figure 6.59, is increasing with an increased electrostatic repulsion between the two carbon atoms. These repulsions

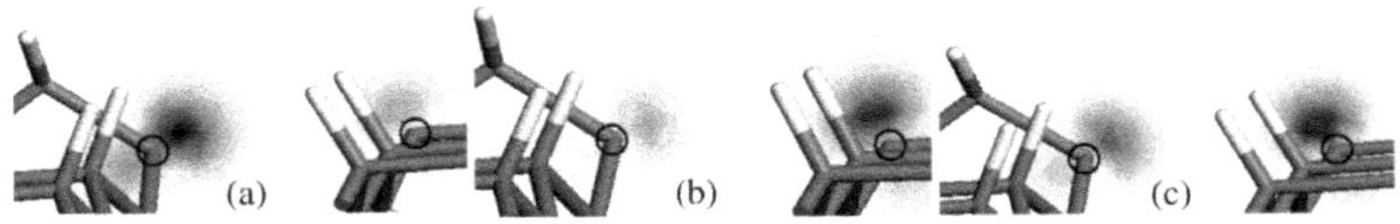

Figure 6.60. Representation of the spin densities on the two surface carbon radicals (identified by black circles) within the initial reactant structure (R in figure 6.64). N is positioned within the second carbon layer in position (a) 1, (b) 2, and (c) 3. [93], reproduced with permission © American Chemical Society.

will also affect the stability of the transition state structure, as discussed above in the description of CH_2 insertion reactions. The distance between these two C atoms is 2.38 Å, which is similar to 2.31 Å for the non-doped situation energy barrier.

The spin density analysis, with N in position 3, indicates that the extra electron from N will be predominantly localized on the surface carbon radical that is participating in the reaction (the spin density is lower on the carbon to the left in figure 6.65(c)). Substitutional doping by N is, thus, affecting the CH_2 migration reaction. As was observed for the CH_2 dimer insertion reaction, N in position 3 tends to slightly reduce the corresponding presence of the dopant, and the energy barrier is similar to the non-doped situation. This can be confirmed by a structural analysis of the transition structure. The C–C distance, represented with a dashed line in TS1 in figure 6.59, is increasing with an increased electrostatic repulsion between the two carbon atoms. These repulsions will also affect the stability of the transition state structure, as discussed above in the description of CH_2 insertion reactions. The distance between these two C atoms is 2.38 Å, which is similar to 2.31 Å for the non-doped situation energy barrier. The spin density analysis, with N in position 3, indicates that the extra electron from N will be predominantly localized on the surface carbon radical that is participating in the reaction (the spin density is lower on the carbon to the left in figure 6.60(c)). Substitutional doping by N is, thus, affecting the CH_2 migration reaction. As was observed for the CH_2 dimer insertion reaction, N in position 3 tends to slightly reduce the corresponding

With N in position 2, the extra N electron is directly being transferred to the surface radical C that is involved in the first step of the migration reaction. The spin density plot clearly shows that the spin density is lower on that specific carbon (i.e. the radical carbon to the left in figure 6.60(b)). As a result for the TS structure, the C–C distance (represented with a dashed line in figure 6.59) is being increased, which is due to electrostatic repulsions between these two C atoms; 2.63 Å for N in position 2 and 2.31 Å for the non-doped surface. The transition state structure is, thereby, destabilized, which explains the large energy barrier.

6.7.4.5.9 Reaction step 2

The reaction energies for the second reaction step (i.e. the formation of a bridged structure) are also presented in table 6.13. Rather large energetic variations can be observed for the presence of N within the surface lattice. This second reaction, with an exothermic energy value of -108 kJ mol^{-1} for the non-doped situation, becomes less energetically favourable with N in position 3 (by about 16 kJ mol^{-1}). On the other hand, N in position 1 and 2 is observed to disfavour the second reaction step with an increase of about 31 and 93 kJ mol^{-1}, respectively. With N in position 2, the larger difference in reaction energy can again be explained by the β-scission mechanism, which takes place prior to the reaction.

An analysis of the final bridged surface structure shows apparent variations in the presence of N (as compared to the non-doped situation). As can be seen in table 6.14, the C–C bridge distances are equal for the neighboring dimers on the non-doped diamond (100)–2 × 1 surface. This symmetry in dimer bond length is, however, not kept for the situation with N in position 1 and 3 in the second atomic C layer. Only one of the C–C dimer bond lengths is elongated in these situations, which has to do with the occupation of anti-bonding orbitals. The N dopant is positioned closer to one of the C–C dimers for both positions 1 and 3, and this specific C–C bond is also the longest one. The underlying cause for this bond elongation is the pre-dominant localization of the extra electron from N within this C–C bond. (It is also worth mentioning that the C–C bonds within the other dimers, to which CH_2 is bonded in a bridge formation, will become shorter as compared with the non-doped situation.)

For N in position 1, the increase in energy barrier can be explained by the presence of the extra N electron on the reactive surface carbon, onto which CH_2 will bind. A resulting electrostatic repulsion has been observed between the negatively charged C (in CH_2) and the surface carbon containing the electron lone pair. With N in position 3, a structural analysis of the surface indicates that the intermediate structure prior to the second reaction step is 'activated' by the presence of the substitutional N dopant. The C–C dimer bond is longer compared to the other N position (1.96 Å (position 3) versus 1.72 Å (position 1)), and the C–CH_2 bond becomes shorter (1.43 Å (position 3) versus 1.48 Å (position 1)). This surface destabilization that takes place before the reaction can, thereby, explain the exothermicity of the reaction and the small energy barrier.

In conclusion, the study of the CH_2 migration reaction shows that the effect of a substitutional N dopant within the second carbon layer is very local. The extra electron from N will only be transferred toward the closest surface carbon radical atom, and a second radical (that is present on the surface) will not be affected. In case the surface carbon radical, to which the electron is being transferred, is not involved in the CH_2 migration reaction, the effect of N on this reaction will be negligible.

The limited size of the model used for study corresponds to a high substitutional N concentration. However, the aim with the present study was not to focus on the effect of N concentration, but instead on the effect of nearby positioned N on the most important growth steps within the diamond CVD synthesis. In the presence of N, the general increase

Table 6.14. Lengths of the carbon dimer (C–C1 and C–C2) bond lengths for the bridged diamond (100)–2 × 1 surface structure shown in figure 6.64 (structure P).

	(100)–2 × 1		
Bond length (Å)	Position 1	Position 3	Non-doped
C–C1	1.82	2.49	1.94
C–C2	2.49	1.80	1.94

in energy barriers, i in addition to the occurrence of β-scission rearrangements (that induces the formation of C–C double bonds), can be an explanation to the experimentally observed surface degradation and decrease in growth rate for very large nitrogen concentration in the gas phase. However, as will be shown in section 6.7.4, a smaller concentration of nitrogen can induce a considerable increase in diamond growth rate.

6.7.4.6 Effect of an N-containing co-adsorbate on the binding of CH_3 to the diamond surface

Even though it has been shown (in the previous part of this chapter) that an N-containing co-adsorbate will not affect the chemical surface reactivity, it might still be the case due to induced physical sterical hindrances. Thus, the N-containing co-adsorbates may still be of importance for the CVD growth of diamond in various stages of the growth mechanism. Firstly, it can be a co-adsorbate for the first chemisorption of the gaseous CH_3 species. Secondly, it can also be a co-adsorbate that is passed during the surface migration of the CH_2 species. Finally, it can also function as a co-adsorbate at the step, or corner, on the growing diamond surface. The adsorption nearby an already chemisorbed NH, or NH_2, species on the otherwise non-doped diamond (100)–2 × 1 surface have, therefore, also been included in the present example. These specific NH_x species are expected to be found on the surface for a gaseous environment containing NH_x species.

In the present example, the effects induced by a neighboring dopant on the CH_3 adsorption are highlighted and compared with the corresponding ones for only neighboring H adsorbates. The model used for the calculations is presented in figure 6.59(a), where the radical surface C site is especially marked. For this adsorption site on the diamond (100)–2 × 1 surface, there are three distinct types of positions for a neighbouring co-adsorbate: position 1, position 2, and position 3 (as presented in figures 6.61(b)–(d), respectively).

6.7.4.6.1 CH_3 adsorption next to NH_2

The calculated energies for CH_3 adsorption to a (100)–2 × 1 diamond surface, with NH_2 dopant species bonded to a neighboring surface carbon site, are presented in table 6.15

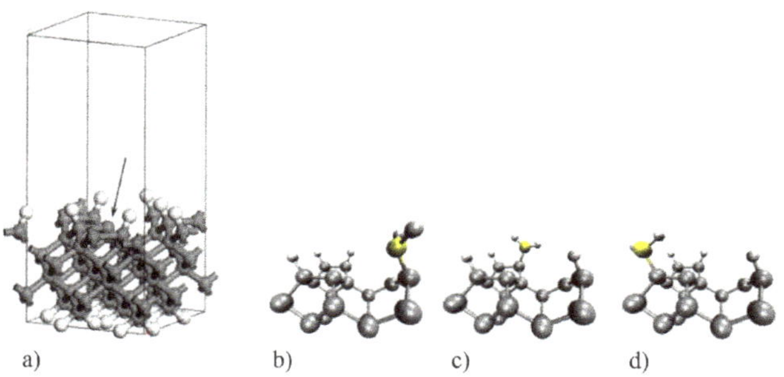

Figure 6.61. (a) Supercell used in modeling the diamond (100)–2 × 1 surface, with an arrow marking the adsorption site for CH_3. Position (b) 1, (c) 2, and (d) 3 for the neighbouring co-adsorbate. [94], reproduced with permission © American Chemical Society.

Table 6.15. CH_3 adsorption energies in the presence of co-adsorbed NH, or NH_2, dopants in positions 1, 2, or 3 (see figure 6.54).

	(100)–2 × 1		
Co-adsorbed species	ΔE_{ads} Position 1 (kJ mol^{-1})	ΔE_{ads} Position 2 (kJ mol^{-1})	ΔE_{ads} Position 3 (kJ mol^{-1})
NH	−355.3	−375.2	−229.1
NH_2	−317.9	−330.0	−349.6
H	−358.5	−358.5	−358.5

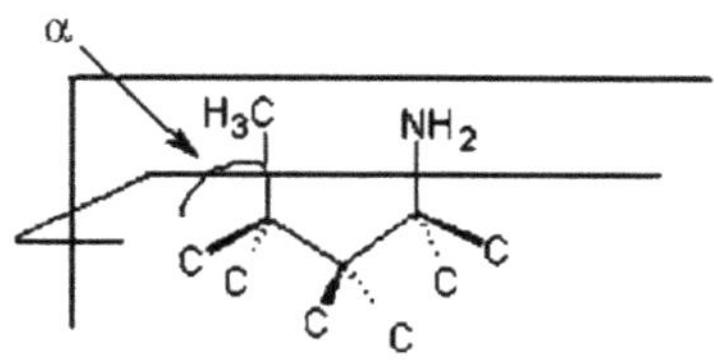

Figure 6.62. Definition of the angle α. [94], reproduced with permission © American Chemical Society.

(the surface is otherwise completely H-terminated). These energies are compared with the corresponding adsorption energies for the situation with no dopant on the surface (i.e. the adsorption site is only surrounded by H adsorbates). In addition, the effects induced by the presence of a co-adsorbed dopant are also compared with interactions generated by a neighboring H, CH_3, or CH_2 adsorbate. The purpose with these latter types of comparison is to highlight the effect of differences in electronegativity values between N and C (or H), which in turn may cause electronic delocalization effects within the surface region of the diamond lattice. In addition, the difference in sterical repulsions will be highlighted for the different co-adsorbates.

With one exception, the adsorption energy values show that the presence of NH_2 on a neighboring surface carbon site will disfavour the CH_3 adsorption. However, when compared to the situation with only H surface neighbours, the decrease in adsorption energy is very small for NH_2 adsorbed to position 3 (with an energy of 8.9 kJ mol^{-1}). (When comparing the adsorption energies, we refer here to the absolute values.) The decrease in energy is larger for NH_2 in positions 1 and 2 (with 40.6 and 28.5 kJ mol^{-1}, respectively). This variation in energy decrease is understandable when studying the geometrical surface structure for the three different neighbouring position sites (see figures 6.61(b)–(d)). it is obvious that the initial distance (before geometry optimization) between the adsorbed CH_3 species and the neighbouring NH_2 adsorbates, at positions 1, 2, and 3, increases in the same order. Hence, the decrease in adsorption energy when going from position 1 to 3 is expected.

The question is now—is the major cause of these energy reductions the sterical repulsions induced by the neighbouring NH_2 adsorbates? As a measure of the degree of CH_3–NH_2 repulsion, an angle R has been defined here by the surface plane (including the first carbon layer) and the $C_{surface}$–CH_3 bond in the plane formed by the two adsorbates (see figure 6.62). For each surface plane and each dopant

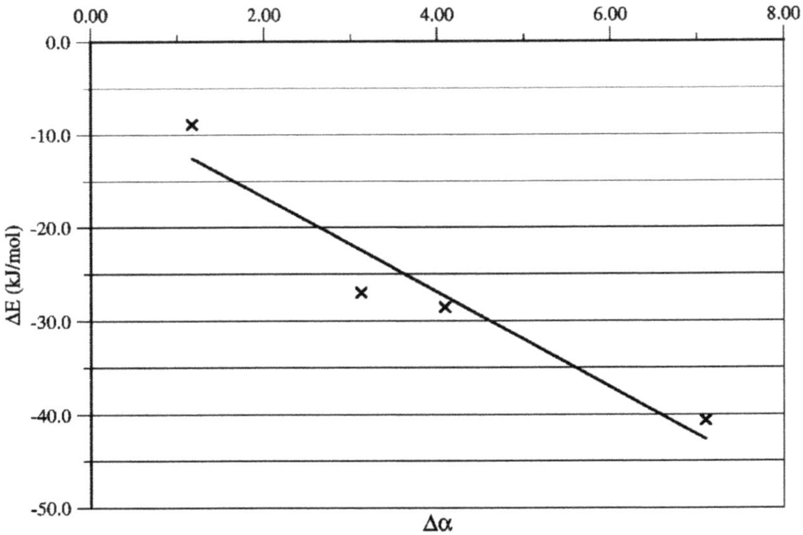

Figure 6.63. Graphical representation of the correlation between ΔE and $\Delta\alpha$. The individual data points are identified as 3.1 for (111) and 7.1, 4.1, and 1.2 for position 1, 2, and 3, respectively, for diamond (100)–2 × 1. [94], reproduced with permission © American Chemical Society.

position considered, ΔR is defined as the difference between the R values for a NH_2 and an H neighbor. As can be seen in figure 6.63, there is an obvious correlation obtained when relating ΔR with the difference in corresponding adsorption energies. The ΔE-to-ΔR correlation gives a good argument for considering the sterical repulsions between the co-adsorbed CH_3 and NH_2 species as the most important effect that is responsible for the small adsorption energy differences induced by the presence of an NH_2 dopant.

6.7.4.6.2 CH_3 adsorption next to NH

The calculated energies for adsorption of CH_3 next to a neighbouring NH adsorbate onto an otherwise H-terminated diamond (100)–2 × 1 surface, are presented in table 6.15. The NH dopant is here forming a radical adsorbate. As was the situation for a neighbouring non-radical NH_2 adsorbate, the calculated CH_3 adsorption energies will be compared with the corresponding ones for the non-dopant situation (i.e. with only H adsorbates as neighbouring species on the diamond surface). As can be seen in table 6.15, there is a very large variation in calculated CH_3 adsorption energies for the three different positions of the neighbouring NH adsorbate. The co-adsorption of NH at position 1 does not affect the CH_3 adsorption significantly (there is a decrease in adsorption energy of only 3.2 kJ mol^{-1}). (We also are referring here to the absolute values.) The decrease in CH_3 adsorption energies is, however, much more severe with NH adsorbed onto position 3 (with a decrease of 129.4 kJ mol^{-1}). Surprisingly, there is an induced increase in CH_3 adsorption energy for NH adsorbed to position 2, although is not significantly large (of 16.7 kJ mol^{-1}). Thus, we cannot observe any trend in change here in CH_3 adsorption energy with the

CH_3-to-NH distance on the surface. Moreover, the NH adsorbate is not that bulky as the NH_2 adsorbate, but it still causes a larger reduction in CH_3 adsorption energy when adsorbed in position 3 (129.4 vs. 8.9 kJ mol^{-1}). Thus, there must be something else other than sterical repulsions that is dominating in causing these differences for the NH_2 and NH co-adsorbates.

When studying the geometrical surface structures prior to the CH_3 adsorption to a surface radical C site, with a neighbouring NH co-adsorbate, it has been found that a second bond has been formed between the NH adsorbate and the reactive surface C site. The interaction between the radical surface carbon and the radical NH adsorbate is expected since these two surface radicals, when they are close enough, tend to react with each other and, thereby, to reduce the surface energy (i.e. increase the surface stabilization). To study the intensity of this interaction, the distances between the NH adsorbate and the two neighboring surface carbons (B_3 and B_4) have been compared (see figure 6.64 and table 6.16).

As can be seen in table 6.16 for positions 1 and 2 on the diamond (100)–2 × 1 surface, the large difference between B_3 and B_4 clearly demonstrates that two neighboring surface radicals (in this case the adsorbed NH species and the surface radical C site) do not always manage to form a bond. The distance between the nitrogen atom (in NH) and the reactive surface carbon is too long to be able to create a B_4 bond without causing severe geometrical constraints. This lack of strong C–NH interaction correlates here to the small energydifference in CH_3 adsorption energy (in relation to the situation with only H adsorbates surrounding the CH_3 adsorbate) for the various NH position sites on the diamond (100)–2 × 1 surface; 3.2 kJ mol^{-1} for position 1 and 16.7 kJ mol^{-1} for position 2. Thus, no B_4 bond needs to be broken so the adsorption of the CH_3 species will only be perturbed by the

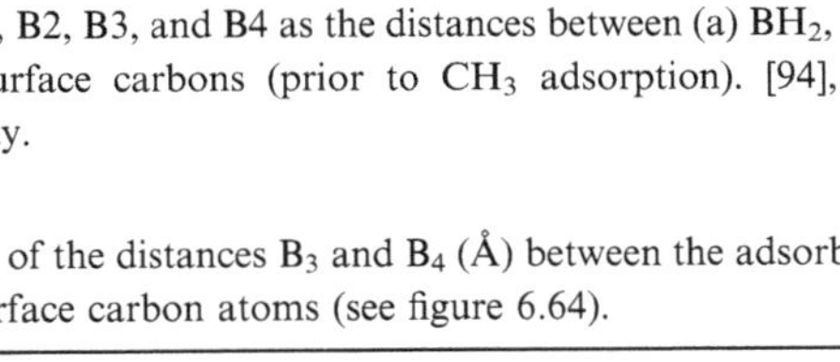

Figure 6.64. Definition of B1, B2, B3, and B4 as the distances between (a) BH_2, or (b) XH (X = N, P, S, or B), and the two neighboring surface carbons (prior to CH_3 adsorption). [94], reproduced with permission © American Chemical Society.

Table 6.16 Values of the distances B_3 and B_4 (Å) between the adsorbed NH dopant and its two binding surface carbon atoms (see figure 6.64).

	(100)–2 × 1		
NH adsorbate	Position 1 (Å)	Position 2 (Å)	Position 3 (Ål)
B_3 distance	1.41	1.40	1.46
B_4 distance	2.66	2.75	1.46

presence of the sterical hindrances induced by the neighbouring NH adsorbate. The situation is completely different for the NH co-adsorbate in position 3 on the surface. The C–NH distance is now much shorter, which means that the B_4 bond can be formed. Consequently, this B_4 bond needs to be broken to make it possible for the adsorption of CH_3 to take place. This explains the large decrease in the CH_3 adsorption energy, of 129.4 kJ mol^{-1}, with respect to the situation with only H adsorbates surrounding the CH_3 adsorbate (see table 6.15).

As discussed above, it is important to also investigate the effect by a co-adsorbed NH, or NH_2, dopant on the migration process for a passing CH_2 species on the diamond surface. This is also the situation for the final incorporation of CH_x into the diamond lattice. However, those examples will not be dealt with here, but the same methodology as above can be used for these purposes.

6.7.4.7 Effect of N doping on the diamond growth rate

As was presented in section 6.6.2, the introduction of nitrogen into the CVD reaction chamber has been found to induce an increased diamond growth rate. For instance, Dunst *et al* [51] were studying the effect of diamond growth by introducing nitrogen into a microwave plasma-activated CVD reactor in depositing single crystalline diamond (100) at 800 °C. The results of that study state that the introduction of N_2 will improve the diamond growth rate with a factor of approximately 2.5 at a CH_4 concentration of 2%.

However, the underlying cause of this growth improvement was not clear: does N have this effect when incorporated into the diamond lattice or was the effect induced by chemisorbed N-containing species? The main advantage with theoretical DFT calculations is that surface processes, including the breakage and formation of bonds, can be most carefully calculated.

The examples presented above in the present chapter have shown that doping with N species (either substitutionally positioned with the upper part of the diamond surface lattice or chemisorbed on the surface) can predominantly affect both reaction energies and energy barriers for the various elementary reaction steps in the diamond CVD growth mechanism. The next step in the overall study of the effect by N dopants on the diamond growth rate is to estimate the effect by N on the slowest elementary reaction steps. This should preferentially be done for the rate-determining step, but since it might be difficult to determine which step that is the bottleneck in the mechanism, it is best to include several very slow elementary steps in this study. However, only one step is chosen in the present example, but the same methodology can also be used for other candidate steps.

The present example deals with the first bottleneck in the CVD growth process of diamond, which has been shown to be the gas phase abstraction of H from the surface, by gaseous H species (see section 6.6). The effect by an N dopant has been studied with N substitutionally positioned into the upper diamond lattice, or with N chemisorbed onto the surface in the form of NH or NH_2 species (the formed being a radical).

6.7.4.8 N substitutionally positioned into the 2nd atomic C atomic layer

Atomic nitrogen, substitutionally positioned within the second C layer of various diamond surface planes, has been chosen here, since it was found that this is a position that has a large effect on the H abstraction reaction, and, thus, on the surface reactivity in general. This latter assumption also confirmed calculated Fukui function maps presented in figure 6.49. It is clear that this specific substitutional N doping has a major effect on the radical susceptibility of all three surfaces; (100)–2 × 1, (110). Furthermore, the increasing electron density between the approaching H species and the surface H adsorbate, in addition to the weakening of the surface-to-H bond, will strongly support the zero-level barrier energy value for abstraction of H from the three different diamond surfaces (see figure 6.56(b)).

6.7.4.9 N substitutionally positioned into the 2nd atomic C atomic layer

Since it was not expill aid for the improved diamond growth rate, nitrogen has also been attached here to the surface in the form of a radical NH group (i.e. chemisorbed onto the diamond surface). As can be seen in figure 6.55(b), the non-paired electron stays locally at the NH adsorbate. This local position is also supported by the deformation density results in figure 6.51(b). Moreover, the extra electron in NH does not show any tendency to interact with the surface in weakening bonds by filling anti-bonding orbitals. Thus, the barrier energy for the H abstraction is, therefore, expected to be very similar to the corresponding non-doped situation. Consequently, calculations of H energy barriers for this specific co-adsorbed dopant are not included in the present study.

However, when the N dopant has been chemisorbed in the form of NH_2 adsorbates onto the diamond surface, the situation becomes more complex (even though NH_2 is not a radical species when adsorbed to the diamond surface). For the (100)–2 × 1 surface, the reactivity of the surface hydrogen in the vicinity to one of the hydrogen atoms within the NH_2 group is enhanced (see figure 6.49). The situation with the (110) surface is, however, quite different in that the reactivity close to NH_2 becomes weakened. Furthermore, the situation for the (111) surface is similar to the situation for (100)–2 × 1. The reactivity of the surface H adsorbate, being positioned closest to H in the adsorbed NH_2 species, will be reinforced whilst the H adsorbate that is positioned closest to the nitrogen atom will decrease in reactivity. However, these changes in surface reactivity are apparently not large enough to affect the H abstraction energy barrier (see table 6.17).

Similar to the situation with the NH adsorbates, the nitrogen doping in the form of NH_2 adsorbates do not show any larger effect on the H abstraction energy barriers, as compared to the non-doped scenario. To be more specific, the H abstraction energy barriers for the diamond (100)–2 × 1 and (111) surface were observed to decrease, whilst the barrier for the (110) surface did increase. As can be seen in table 6.17, the calculated energy barriers for the H abstraction reactions are 8.2, 6.0, versus 5.7 kJ mol^{-1} for doping with a NH_2 adsorbate, and 11.8, 6.2, versus 4.7 kJ mol^{-1} for the corresponding non-doped situation. However, it must be stressed that these differences are marginal for diamond (111) and (110), and quite small for the (100)–2 × 1 surface. There is also a correlation observed between these differences in energy barriers and the effect by NH_2 doping on the surface reactivates. As can be seen in figure 6.49, the Fukui functions for the diamond (100)–2 × 1 and (111) surfaces show a slight improvement in surface reactivity,

Table 6.17. Effect of N on the surface H abstraction energy and relative growth rate (in relation to the respective non-doped surface planes: (110), (111) and (100)–2 × 1).

	Energy barrier (kJ mol^{-1})	Relative growth rate
(110)		
Non-doped	4.7	1.0
N in 2nd C layer	0	1.7
NH_2 adsorbate	5.7	0.9
(111)		
Non-doped	6.1	1.0
N in 2nd C layer	0	1.9
NH_2 adsorbate	6.0	1.1
(100)–2 × 1		
Non-doped	11.8	1.0
N in 2nd C layer	0	3.7
NH_2 adsorbate	8.2	1.5

whilst the corresponding one for the (110) surface show a decrease in reactivity. However, as stated above, these effects are relatively small and there is, therefore, no effort put in trying to understand this correlation in the present example.

6.7.4.10 N substitutionally positioned into the 2nd atomic C atomic layer

As is the situation for non-doped diamond surfaces, the energy barrier values for the various N doping situations can be used in forming ratios of CVD diamond growth for the various diamond surface planes (see table 6.5). In the Arrhenius relation that is used for this purpose (see equation (6.9) in section 6.6.7), the temperature, T, has been set to 1073 K since this is the substrate temperature that was used in the experimental work that the present study is referring to [51]. Moreover, the experimental work did only consider the effect by N doping on the diamond (100)–2 × 1 growth. A comparison with experimental results for the situation with N doping can, hence, only be done for this specific surface. It is also assumed in table 6.5 that the H abstraction step is the rate-determining one in the CVD growth of diamond, which is shown here to be a very realistic conclusion.

As can be seen in table 6.5, the order of the growth rates, for the various non-doped diamond surface planes, are identical for the experimental and theoretical results (even though the numerical results differ to some extent). The diamond (110) has the largest CVD growth rate, followed by (111), and the slowest growing surface is the 2 × 1-reconstructed (100) surface. When comparing these results with the ones for N-doped diamond, it is quite clear that there is a remarkable increase in growth rate when introducing N within the 2nd C atomic layer in the diamond surfaces (see table 6.17). (The growth rates are presented in relation to the corresponding ones for the non-doped scenarios.) This is especially the situation for the diamond (100)–2 × 1 surface, for which a growth enhancement has been calculated with a factor of 3.7 compared to the corresponding non-doped diamond structure. However, the experimental growth rate enhancement for diamond (100)–2 × 1 was found to depend on the combination of N_2 and CH_4 concentration in the MWPA-CVD reaction chamber, with a maximum value of 3.5 [51]. Hence, there is an almost perfect resemblance between the factor calculated here and the experimentally observed one. This result is first of all a justification of the theoretical methods and methodology used in the present study. Secondly, it also strongly supports the general assumption that the H abstraction process from the diamond surface is the rate-limiting step in the complete diamond growth mechanism. Finally, the very good correspondence between theory and experiments confirms the experimental observation and takes a step further in making the assumption that it is most probably N, incorporated in the lattice, that is responsible for the observed growth rate enhancement.

For the situation with N substitutionally positioned in the 2nd atomic C layer, the theoretically obtained growth rate enhancements for the (111) and (110) surfaces are not that pronounced as for the (100)–2 × 1 surface (1.9, 1.7, versus 3.7) (see table 6.17). But it is still considerably large. For the situation with a chemisorbed NH_2 species onto the surface, instead of being positioned within the upper lattice, only a minor effect has been observed for the diamond (100)–2 × 1 surface (of 1.5), and in principle none for the other two surfaces.

As a conclusion, N-doping in the form of NH_2 adsorbates will not affect the CVD growth rate of diamond to any considerable extent. However, the growth rate will be largely improved by substituting with N in the second C atomic layer in the diamond surfaces. This is especially the situation for the diamond (100)–2 × 1 surface, for which the enhanced growth factor of 3.7 is almost identical to the experimentally obtained one (of 3.5).

The present results have proven that (i) the chosen level of theory and methodology are adequate to use for the present example, and (ii) it is substitutional N that will affect the growth rate of diamond (100)–2 × 1.

References

[1] Hourahine B *et al* 2020 DFTB+, a software package for efficient approximate density functional theory based atomistic simulations *J. Chem. Phys.* **152** 124101

[2] Sharma S 2019 *Molecular Dynamics Simulation of Nanocomposites Using BIOVIA Materials Studio, Lammps and Gromacs* (Amsterdam: Elsevier)

[3] Petrini D and Larsson K 2008 Theoretical study of the thermodynamic and kinetic aspects of terminated (111) *J. Phys. Chem.* C **112** 3018–26

[4] Ziegler T 1991 Approximate density functional theory as a practical tool in molecular energetics and dynamics *Chem. Rev.* **91** 651

[5] Spear K E and Dismukes J P 1994 *Synthetic Diamond: Emerging CVD Science and Technology* (New York: Wiley)

[6] Somorjai G A 1994 *Introduction to Surface Chemistry and Catalysis* (New York: Wiley)

[7] May P W, Stone J C, Ashfold M N R, Hallam K R, Wang W N and Fox N A 1998 The effect of diamond surface termination species upon field emission properties *Diam. Relat. Mater.* **7** 671–6

[8] Mackey J, Russell J N Jr, Crowell J E and Butler J E 1995 Effect of surface termination on the electrical conductivity and broad-band internal infrared reflectance of a diamond (110) surface *Phys. Rev.* B **52** R17009(R)
[9] Martin H B, Argoitia A, Landau U, Anderson A B and Angus J C 1996 Hydrogen and oxygen evolution on boron-doped diamond electrodes *J. Electrochem. Soc.* **143** L133–6
[10] Nebel C E, Ertl F, Sauerer C, Stutzmann M, Graeff C F O, Bergonzo P, Williams O A and Jackman R B 2002 Low temperature properties of the p-type surface conductivity of diamond *Diam. Relat. Mater.* **11** 351–4
[11] Maier F, Ristein J and Ley L 2001 Electron affinity of plasma-hydrogenated and chemically oxidized diamond (100) surfaces *Phys. Rev.* B **64** 165411
[12] de Theije F K, Roy O, van der Laag N J and van Enckevort W J P 2000 Oxidative etching of diamond *Diam. Relat. Mater.* **9** 929–34
[13] Touhara H and Okino F 2000 Property control of carbon materials by fluorination *Carbon* **38** 241–67
[14] Yamada T, Chuang T J, Seki H and Mitsuda Y 1992 Chemisorption of fluorine, hydrogen and hydrocarbon species on the diamond C (111) surface *Mol. Phys.* **76** 887–908
[15] Yang S H, Drabold D A, Adams J B and Sachdev A 1993 First-principles local-orbital density-functional study of Al clusters *Phys. Rev.* B **48** 5261
[16] Yang Y L and D'Evelyn M P 1992 *38th National Symposium of the American Vacuum Society (AVS)* (Seattle, Washington)
[17] Zheng X M and Smith P V 1991 The topologies of the clean and hydrogen-terminated C (100) surfaces *Surf. Sci.* **256** 1–8
[18] Kubiak G D, Hamza A V, Stulen R H, Sowa E C, Kolasinski K W and Van Hore M A 1990 Hydrogen desorption and subsequent reconstruction on natural diamond surfaces *Carbon* **28** 751–2
[19] Hamza A V, Kubiak G D and Stulen R H 1990 Hydrogen chemisorption and the structure of the diamond C(100)–(2 × 1) surface *Surf. Sci.* **237** 35–52
[20] Thoms B D and Butler J E 1995 HREELS and LEED of HC (100): the 2 × 1 monohydride dimer row reconstruction *Surf. Sci.* **328** 291–301
[21] Pehrsson P E and Mercer T W 2000 Oxidation of the hydrogenated diamond (100) surface *Surf. Sci.* **460** 74–90
[22] Badziag P and Verwoerd W S 1987 MNDO analysis of the oxidised diamond (100) surface *Surf. Sci.* **183** 469–83
[23] Nakamura J and Ito T 2005 *Oxidization Process of CVD Diamond (1 0 0):H Surfaces (Hamamatsu, Japan)* (Amsterdam: Elsevier)
[24] Zheng X M and Smith P V 1992 The stable configurations for oxygen chemisorption on the diamond (100) and (111) surfaces *Surf. Sci.* **262** 219–34
[25] Kian Ping L, Xie X N, Yang S W and Zheng J C 2002 Oxygen adsorption on (111)-oriented diamond: a study with ultraviolet photoelectron spectroscopy, temperature-programmed desorption, and periodic density functional theory *J. Phys. Chem.* B **106** 5230–40
[26] Larsson K 2017 Surface chemistry of diamond *Novel Aspects of Diamond* (Berlin: Springer) pp 53–83
[27] Yiming Z, Larsson F and Larsson K 2014 *Theor. Chem. Acc.* **133** 1432–44
[28] Karlsson J and Larsson K 2013 Kinetic considerations of gas-phase abstraction of H and F from the c-BN(100) surface *Thin Solid Films* **548** 280–7
[29] Karlsson J and Larsson K 2011 Hydrogen-induced de/reconstruction of the c-BN (100) surface *J. Phys. Chem.* C **115** 22910–6

[30] Kern G, Hafner J and Kresse G 1996 Atomic and electronic structure of diamond (111) surfaces. I. Reconstruction and hydrogen-induced de-reconstruction of the one dangling-bond surface *Surf. Sci.* **352-354** 745–9

[31] Bar-Yam Y and Moustakas T D 1989 Defect-induced stabilization of diamond films *Nature* **342** 786–7

[32] Chu C J 1991 Mechanism of diamond growth by chemical vapor deposition on diamond (100), (111), and (110) surfaces: carbon-13 studies *J. Appl. Phys.* **70** 1695–705

[33] Belton D N and Harris S J 1992 A mechanism for growth on diamond (110) from acetylene *J. Chem. Phys.* **96** 2371–7

[34] Celii F G, Pehrsson P E, Wang H -T and Butler J E 1988 Infrared detection of gaseous species during the filament-assisted growth of diamond *Appl. Phys. Lett.* **52** 2043–5

[35] Harris S J, Weiner A M and Perry T A 1988 Measurement of stable species present during filament-assisted diamond growth *Appl. Phys. Lett.* **53** 1605–7

[36] Harris S J, Belton D N, Weiner A M and Schmieg S J 1989 Diamond formation on platinum *J. Appl. Phys.* **66** 5353–9

[37] Larsson K, Lunell S and Carlsson J-O 1993 Adsorption of hydrocarbons on a diamond (111) surface: an *ab initio* quantum mechanical study. *Phys. Rev.* B **48** 2666–74

[38] Larsson K, Carlsson J -O and Lunell S 1995 Nearest-neighbour influence on hydrocarbon adsorption on diamond (111): studied by *ab initio* calculations *Phys. Rev* B **51** 10003–12

[39] Frauenheim T, Jungnickel G, Sitch P, Kaukonen M, Weich F, Widany J and Porezag D 1998 A molecular dynamics study of N-incorporation into carbon systems doping, diamond growth and nitride formation *Diam. Relat. Mater.* **7** 348–55

[40] Larsson K and Carlsson J-O 1999 Surface migration during diamond growth studied by molecular orbital calculations *Phys. Rev.* B **59** 8315–20

[41] Chang Y, Thompson D L and Raff L M 1994 Hydrogen-atom migration on a diamond (111) surface *J. Chem. Phys.* **100** 1765–6

[42] Enckevort W J P 1994 *Synthetic Diamond: Emerging CVD Science and Technology* ed K E Spear and J P Dismukes (New York: Wiley) pp 307–53

[43] van Enckevort W J P, Janssen G, Vollenberg W, Schermer J J, Giling L J and Seal M 1993 CVD diamond growth mechanisms as identified by surface topography *Diam. Relat. Mater.* **2** 997–1003

[44] Larsson K and Carlsson J-O 2001 Incorporation of C into growth steps of diamond (111) *Phys. Status Solidi* **186** 319–30

[45] Chu C J *et al* 1992 *Novel Forms of Carbon* **270** 341–6

[46] Cao G Z, Schermer J J, van Enckevort W J P, Elst W A L M and Giling L J 1996 Growth of {100} textured diamond films by the addition of nitrogen *J. Appl. Phys.* **79** 1357–64

[47] Müller-Sebert W, Wörner E, Fuchs F, Wild C and Koidl P 1996 Nitrogen induced increase of growth rate in chemical vapor deposition of diamond *Appl. Phys. Lett.* **68** 759–60

[48] Yan C S and Vohra Y K 1999 Multiple twinning and nitrogen defect center in chemical vapor deposited homoepitaxial diamond *Diam. Relat. Mater.* **8** 2022–31

[49] Liu T and Raabe D 2009 Influence of nitrogen doping on growth rate and texture evolution of chemical vapor deposition diamond films *Appl. Phys. Lett.* **94** 211191

[50] Dunst S, Sternschulte H and Schreck M 2009 Growth rate enhancement by nitrogen in diamond chemical vapor deposition—a catalytic effect *Appl. Phys. Lett.* **94** 22410

[51] Bachelet G B, Hamann D R and Scluter M 1982 Pseudopotentials that work: from H to Pu *Phys. Rev.* B **26** 4199

[52] Sternschulte H, Schreck M, Stritzker B, Bergmaier A and Dollinger G 2003 Growth and properties of CVD diamond films grown under H_2S addition *Diam. Relat. Mater.* **12** 318–23

[53] Haubner R and Sommer D 2003 Hot-filament diamond deposition with sulfur addition *Diam. Relat. Mater.* **12** 298–305

[54] Bohr S, Haubner R and Lux B 1995 Influence of phosphorus addition on diamond CVD *Diam. Relat. Mater.* **4** 133–44

[55] Liang Q, Yan C, Meng Y, Lai J, Krasnicki S, Mao H and Hemley R J 2009 Recent advances in high-growth rate single-crystal CVD diamond *Diam. Relat. Mater.* **18** 698–703

[56] Kato H, Barjon J, Habka N, Matsumoto T, Takeuchi D, Okushi H and Yamasaki S 2011 Energy level of compensator states in (001) phosphorus-doped diamond *Diam. Relat. Mater.* **20** 1016–9

[57] Kato H, Makino T, Yamasaki S and Okushi H 2007 n-type diamond growth by phosphorus doping on (0 0 1)-oriented surface *J. Phys. D: Appl. Phys.* **40** 6189–200

[58] Nishitani-Gamo M, Xiao C, Zhang Y, Yasu E, Kikuchi Y, Sakaguchi I, Suzuki T, Sato Y and Ando T 2001 Homoepitaxial diamond growth with sulfur-doping by microwave plasma-assisted chemical vapor deposition *Thin Solid Films* **382** 113–23

[59] Eaton S C, Anderson A B, Angus J C, Evstefeeva Y E and Pleskov Y V 2003 Diamond growth in the presence of boron and sulfur *Diam. Relat. Mater.* **12** 1627–32

[60] Koizumi S, Kamo M, Sato Y, Ozaki H and Inuzuka T 1997 Growth and characterization of phosphorous doped {111} homoepitaxial diamond thin films *Appl. Phys. Lett.* **71** 1065–7

[61] Miyazaki T, Kato H, Okushi H and Yamasaki S 2006 *Ab initio* energetics of phosphorus impurity in subsurface regions of hydrogenated diamond surfaces *Surf. Sci. Nanotechnol.* **4** 124–8

[62] Wada H, Teraji T and Ito T 2005 Growth and characterization of P-doped CVD diamond (111) thin films homoepitaxially grown using trimethylphosphine *Appl. Surf. Sci.* **244** 305–9

[63] Mainwood A 2006 Theoretical modelling of dopants in diamond *J. Mater. Sci. Mater. Electron.* **17** 453–8

[64] Miyazaki T and Okushi H 2002 Theoretical modeling of sulfur–hydrogen complexes in diamond *Diam. Relat. Mater.* **11** 323–7

[65] Gheeraert E, Casanova N, Tajani A, Deneuville A, Bustarret E, Garrido J A, Nebel C E and Stutzmann M 2002 N-type doping of diamond by sulfur and phosphorus *Diam. Relat. Mater.* **11** 289–95

[66] Shah Z M and Mainwood A 2008 A theoretical study of the effect of nitrogen, boron and phosphorus impurities on the growth and morphology of diamond surfaces *Diam. Relat. Mater.* **17** 1307–10

[67] Kalish R 2001 The search for donors in diamond *Diam. Relat. Mater.* **10** 1749–55

[68] Kajihara S A, Antonelli A, Bernholc J and Car R 1991 Nitrogen and potential n-type dopants in diamond *Phys. Rev. Lett.* **66** 2010–3

[69] McCreery R L 2008 Advanced carbon electrode materials for molecular electrochemistry *Chem. Rev.* **108** 2646–87

[70] Pleskov Y V 1987 Photoelectrochemical properties of semiconductor diamond *J. Electroanalyt. Chem* **228** 19–27

[71] Swain G M and Ramesham R 1993 The electrochemical activity of boron-doped polycrystalline diamond thin-film electrodes *Anal. Chem* **65** 345–51

[72] Williams A W, Lightowl E and Collins A T 1970 Impurity conduction in synthetic semiconducting diamond *J. Phys. C: Solid State Phys.* **3** 1727–35

[73] Chrenko R M 1973 Boron, dominant acceptor in semiconducting diamond *Phys. Rev.* B **7** 4560-4567

[74] Li H D 2010 Investigation on crystalline structure, boron distribution, and residual stresses in freestanding boron-doped CVD diamond films *J. Cryst. Growth* **312** 1986–91

[75] Ferreira N G, Abramof E, Corat E J and Trava-Airoldi V J 2003 Residual stresses and crystalline quality of heavily boron-doped diamond films analysed by micro-Raman spectroscopy and x-ray diffraction *Carbon* **41** 1301–8

[76] Miyata K 1993 Morphology of heavily B-doped diamond films *J. Mater. Res.* **8** 2845–57

[77] Cifre J 1994 Trimethylboron doping of CVD diamond thin-films *Diam. Relat. Mater.* **3** 628–31

[78] Haubner R, Bohr S and Lux B 1999 Comparison of P, N and B additions during CVD diamond deposition *Diam. Relat. Mater.* **8** 171–8

[79] Wang L 2015 Effect of pressure on the growth of boron and nitrogen doped HFCVD diamond films on WC–Co substrate *Surf. Interface Analysis* **47** 572–86

[80] Hartmann P *et al* 1998 Diamond growth with boron addition, diamond growth with boron addition *Int. J. Refract. Met. Hard Mater.* **16** 223–32

[81] Tallaire A, Achard J, Silva F, Brinza O and Gicquel A 2013 Growth of large size diamond single crystals by plasma assisted chemical vapour deposition: recent achievements and remaining challenges *Physique* **14** 169–84

[82] Zou Y, Larsson F and Larsson K 2022 Effect of sulphur and phosphorous doping on the growth rate of CVD diamond (111)thin solid films *Diam. Relat. Mater.* submitted

[83] Van Regemorter T and Larsson K 2008 Effect of substitutional N on the diamond CVD growth process: a theoretical approach *Diam. Relat. Mater.* **17** 1076–9

[84] Frenklach M and Skokov S 1997 Surface migration in diamond growth *J. Phys. Chem.* B **101** 3025–36

[85] Netto A and Frenklach M 2005 Kinetic monte carlo simulations of CVD diamond growth—interlay among growth, etching, and migration *Diam. Relat. Mater* **14** 1630–46

[86] Locher R, Wild C, Herres N, Behr D and Koidl P 1994 Nitrogen stabilized ⟨100⟩ texture in chemical vapor deposited diamond films *Appl. Phys. Lett.* **65** 34–6

[87] Skokov S, Weiner B and Frenklach M 1994 Elementary reaction mechanism for growth of diamond (100) surfaces from methyl radicals *J. Phys. Chem.* **98** 7073–82

[88] Gali A, Mészáros A and Deák P 1996 Calculations of migration barriers on hydrogenated diamond surfaces *Diam. Relat. Mater.* **5** 613–6

[89] Garrison B J, Dawnkaski E J, Srivastava D and Brenner D W 1992 Molecular dynamics simulations of dimer opening on a diamond {100}(2 × 1) surface *Science* **255** 835–8

[90] Harris S J and Goodwin D G 1993 Growth of the reconstructed diamond (100) surface *J. Phys. Chem.* **97** 23–8

[91] Cheesman A, Harvey J N and Ashfold M N R 2008 Studies of carbon incorporation on the diamond {100} surface during chemical vapor deposition using density functional theory *J. Phys. Chem.* A **45** 11436–48

[92] Van Regemorter T and Larsson K 2009 Effect of substitutional N on important chemical vapor deposition diamond growth steps *J. Phys. Chem.* A **113** 3274–84

[93] Aylward G and Findlay T 1998 *SI Chemical Data* 4th edn (New York: Wiley)

[94] Van Regemorter T and Larsson K 2008 Effect of coadsorbed dopants on diamond initial growth processes: CH_3 adsorption *J. Phys. Chem.* A **112** 5429–35

Chapter 7

Identification of growth mechanisms for ALD deposition of Cu

Cu films can be deposited by using an ALD technique and CuCl(g) as the growth precursor [1]. This type of ALD process consists of four steps that are repeated until a film of desired thickness has been formed: (1) gaseous CuCl is first introduced into the reactor, and these molecules become adsorbed onto the growing surface (or substrate). (2) This is followed by a purging inert gas, which removes the excess of CuCl species. The purpose of the combination of the first and second step is, thus, to create a monolayer of adsorbed CuCl species. (3) As a third step, H_2(g) is added to remove the chlorine atoms from the CuCl adsorbates. Gaseous HCl will then be formed, which is removed from the deposition chamber by pumping. (4) Finally, as step number four, the reactor is again purged with an inert gas to remove any physisorbed HCl, in addition to eventual remaining hydrogen species. As the CuCl molecule approaches the surface, CuCl–surface interactions will occur and, at a close enough distance, a chemical bond is formed. The CuCl adsorbate can thereafter undergo many different reactions on the surface.

The removal of halogen ligands by the addition of hydrogen does, normally, require a dissociation of H_2 on the metal surface. This step is often regarded to be the rate determining step. It has experimentally been shown that the dissociation of H_2 onto a Cu surface requires an activation energy (i.e. barrier energy) of 40–60 kJ mol^{-1} [1].

Theoretical modelling is perfectly suited to outline the growth mechanism for the ALD deposition of Cu, especially the activation energies for the rate-determining step.

7.1 General

In comparison with the elementary steps in the diamond growth example (see chapter 6), the CVD growth in the present example, which is about metal Cu deposition, does also include reactions between adsorbates on the surface.

doi:10.1088/978-0-7503-3107-4ch7

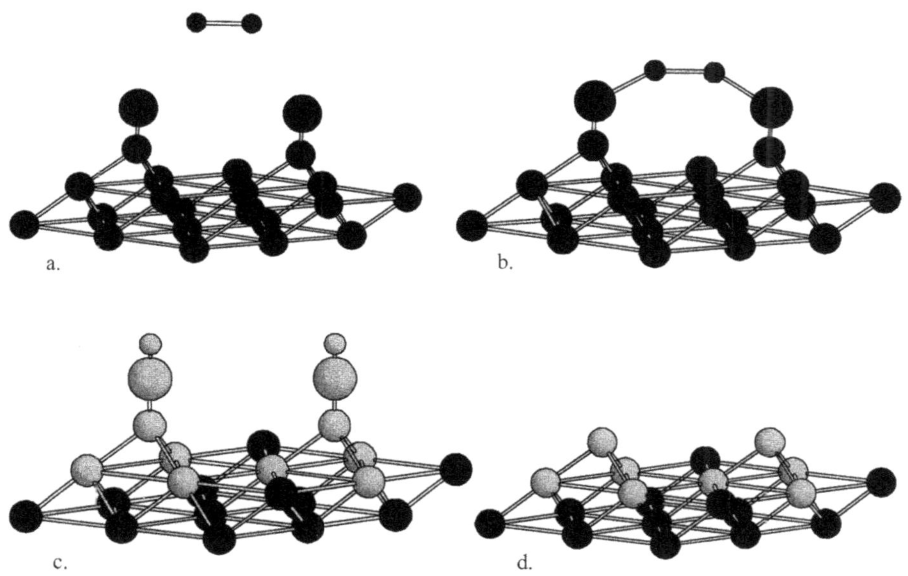

Figure 7.1. Suggestion of the four different elementary steps in the CVD growth of Cu(s). (a) Adsorption of CuCl with Cu binding to the surface. (b) Reaction between two neighbouring CuCl adsorbates and a gaseous H_2 molecules. (c) Breakage of the H–H bond. (d) Desorption of CuCl(g). [9], reproduced with permission © Elsevier.

Cu films were deposited earlier by using an ALD technique and CuCl(g) as the growth precursor [1]. This type of ALD process consists of four steps that are repeated until a film of desired thickness has been formed: (1) gaseous CuCl is first introduced into the reactor, and these molecules become adsorbed onto the growing surface (or substrate). (2) This is followed by a purging inert gas, which removes the excess of CuCl species. The purpose of the combination of the first and second step is, thus, to create a monolayer of adsorbed CuCl species. (3) As a third step, H_2(g) is added to remove the chlorine atoms from the CuCl adsorbates. Gaseous HCl will then be formed, which is removed from the deposition chamber by pumping. (4) Finally, as step number four, the reactor is again purged with an inert gas to remove any physisorbed HCl, in addition to eventual remaining hydrogen species. These steps can be further divided into several chemical elementary steps (one suggestion is shown in figure 7.1). As the CuCl molecule approaches the surface, CuCl–surface interactions will occur and, at a close enough distance, a chemical bond is formed. The CuCl adsorbat can thereafter undergo many different reactions on the surface, e.g. dehalogenation by the reaction with hydrogens, disproportionation, migration, etc. Dehalogenation of halide growth precursors in CVD, or ALD, by means of the addition of hydrogen, is often used in the fabrication of thin films of various materials. Hydrogen has, for instance, been used extensively as a reducing agent in the deposition of many different metals [2]. The removal of halogen ligands by the addition of hydrogen does normally require a dissociation of H_2 on the metal surface. This step is often regarded to be the rate determining step.

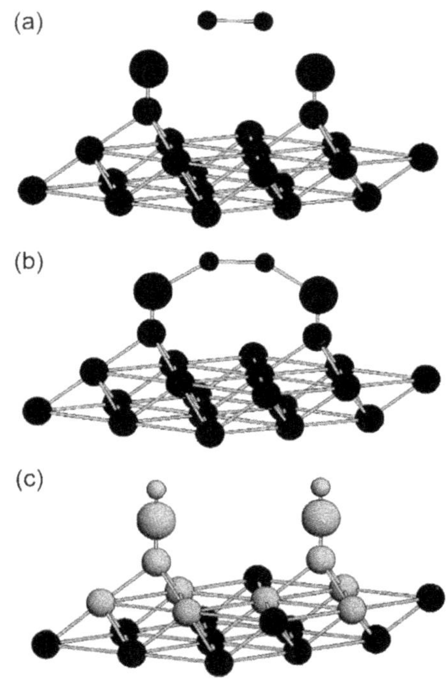

Figure 7.2. A Langmuir–Hinschelwood reaction between chemisorbed CuCl species and gaseous H_2. [10], reproduced with permission © Elsevier.

The adsorption and dissociation of H_2 have been thoroughly investigated earlier for a variety of surfaces. It has experimentally been shown that the dissociation of H_2 onto a Cu surface requires an activation energy (i.e. barrier energy) of 40–60 kJ mol^{-1} [1, 3, 4], while the theoretical results have presented barrier energies in the region of 50–120 kJ mol^{-1} [5–8]. For the ALD deposition of elemental Cu, earlier obtained experimental results have shown that the adsorption of CuCl is very fast and proceeds without, or with, a very small activation energy. On the other hand, the hydrogen reduction step is very slow and limited by an energy barrier of approximately 85 kJ mol^{-1} [1]. CuCl(g) was then used as a growth precursor, and H_2(g) was used as a reducing agent. Thus, the rate limiting step was assumed to be the reaction between hydrogen and CuCl on the copper surface, rather than the adsorption and dissociation of H_2. This conclusion is based on the relatively large difference in activation energy between the results found in [1], and what has earlier been reported in [3–8] for different copper surfaces.

In the present example, the ALD deposition mechanism for the growth of Cu films, from gaseous CuCl and H_2, will be studied here in the greatest detail. There is then a special focus on the reduction reaction between CuCl and H_2. This reaction can take place by any of two reaction mechanisms. There can either be a reaction between the CuCl adsorbate and gaseous H_2, or the CuCl adsorbate will react with adsorbed H species (i.e. after the dissociation of H_2) (see figures 7.2 and 7.3).

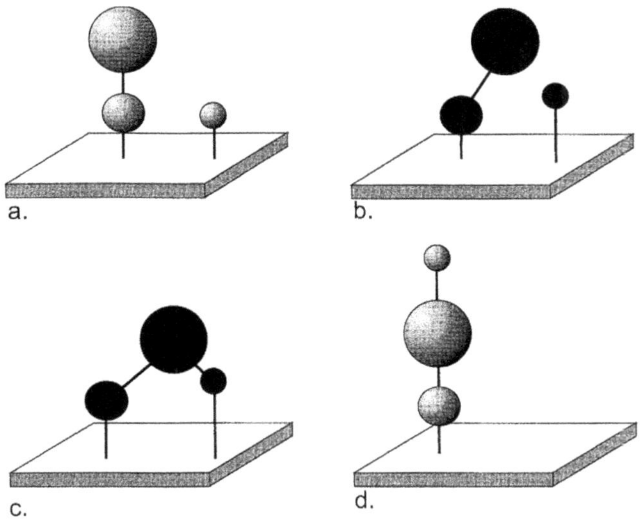

Figure 7.3. An Eley–Rideal reaction between chemisorbed CuCl species and chemisorbed H. [9], reproduced with permission © Elsevier.

The resulting mechanisms, which have been thoroughly discussed in section 3.3.3, are called Langmuir–Hinschelwood (L–H) and Eley–Rideal (E–R), respectively. The main purpose of the present example is to show how it is possible, by using theoretical methods, to identify the specific mechanism that takes place during ALD deposition from CuCl(g) and H_2(g) (i.e. L–H or E–R). As presented in section 3.3.3, this is very useful information for further development and optimization of the overall growth process.

7.2 Test-calculations

When starting with the calculations on a completely new system, it is absolutely necessary to do test-calculations. This is compulsory for the results to be trustworthy. The test calculations for the study of the ALD growth mechanism of Cu (111), from gaseous CuCl and H_2, can take place as follows. This example is taken from [11].

7.2.1 Size of the supercell model

At first, one can start with a surface model including three layers of copper atoms, and with 3 × 3 atoms in each layer. The dimension of the unit cell is then 7.68 × 7.68 × 17.00 Å, with $\alpha = \beta = 90°$ and $\gamma = 60°$. As is always the situation when constructing a surface model, the starting geometry and inter-atomic distances are obtained from the corresponding bulk material. The necessity for using at least three layers of Cu (111) is the fact that one needs to differentiate between the adsorptions to cubic closed packed (ccp) and hexagonal closed packed (hcp) sites. (For description of ccp and hcp, see [12].) Even though it is sufficient to use 2 × 2 atoms in the *a*–*b* plane

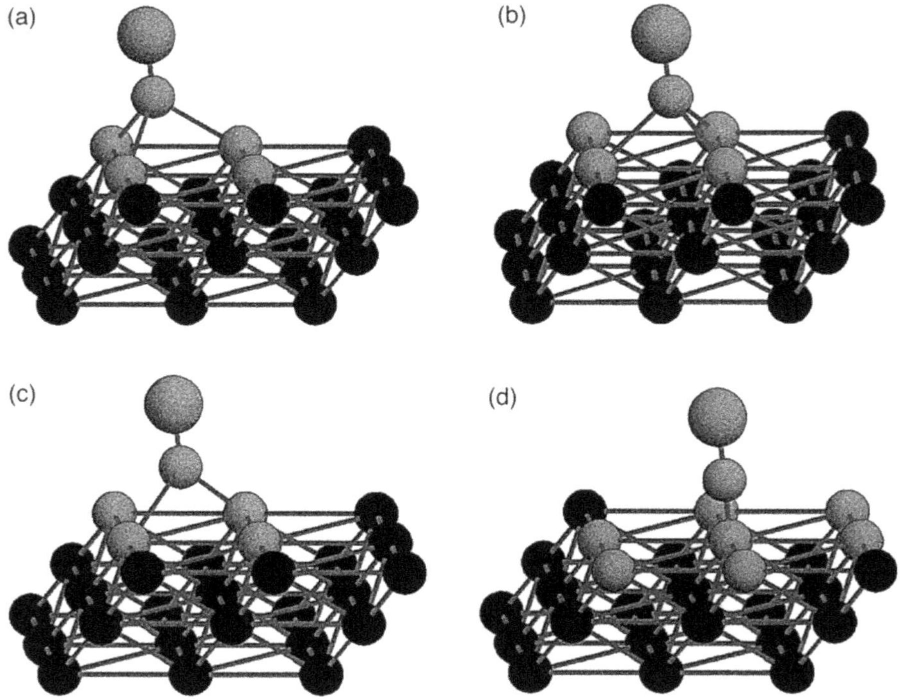

Figure 7.4. Models of three-layer thick Cu model with CuCl adsorbed onto various surface sites. (a) A ccp three-fold site, (b) a hcp three-fold site, (c) two-fold bridge site, and (d) an on top site. The Cl atom is represented by a sphere that is larger than the Cu atoms. The adsorbate, with its neighbouring surface region, is shown in grey, while the other Cu atoms are shown in black. [11], reproduced with permission © Elsevier.

(which includes all different types of adsorption sites), it was decided to use 3 × 3 atoms instead. If we are only using one adsorbate on this supercell surface, this size results in an adsorbate–adsorbate distance of 7.68 Å (i.e. for adsorbates situated in neighbouring supercells). This distance is here considered to be sufficient for avoiding inter-adsorbate interactions. (However, it must be stressed here that it is sometimes necessary to include these types of interactions, and that will be decisive for the choice of supercell size (see section 5.1.3).) The chemisorption of a CuCl adsorbate to various surface sites on this type of surface model is shown in figure 7.4.

As a first test, let us consider the effect of a fourth layer of copper atoms. For the models with three and four Cu layers, all atoms were kept fixed during the DFT calculations. The surface-to-adsorbate bond strength can be calculated in this way, without any contribution from relaxation and/or reconstruction. The Cu–Cl distance was set to 2.094 Å, which is a distance that has been obtained from optimization calculations of a single CuCl molecule. In these so-called single point calculations (i.e. not with geometry optimization), the CuCl molecule was initially positioned at a twofold bridge position (see figure 7.4), with distances of 2, 3, 4, and 5 Å above the surface. For each of these initial positions, the relative energies,

Table 7.1. Calculated results showing the Cu–Cl and surface–CuCl bond distances for a two-layer thick optimized supercell model. Relative total energies, for various Cu–Cl bond distances, are also presented for non-optimized models. (Each of these latter values are related to the corresponding value of the optimized two-layer thick supercell model.)

Surface–Cu_{CuCl} distance (Å)	Relative energy ($kJ\ mol^{-1}$) Three layers	Relative energy ($kJ\ mol^{-1}$) Four layers	Difference between three and four layers (%)
2	444	415	8.5
2 (optimized)	0(Cu–Cl: 2.125 Å) (Surface–Cu: 2.210 Å)	0(Cu–Cl: 2.131 Å) (Surface–Cu: 2.208 Å)	0(Cu–Cl: 0.28) (Surface–Cu: 0.09)
3	1100	1081	1.8
4	1254	1283	2.6
5	1341	1302	3.2

compared to a geometry optimized model with the same number of copper layers, are summarized in table 7.1. The values presented in the second row are those obtained from the geometry optimizations. As can be seen in table 7.1, the final geometries of these optimized structures are almost identical, independent of whether three, or four, Cu layers have been used in the model. The difference in bond length is only 0.28% for the Cu–Cl intra-molecular distance, and 0.09% for the vertical distance from the copper atom in CuCl to the top layer of the surface. Moreover, the differences in relative energies, for the two model thicknesses, are larger, but is still very small. The important results here, however, are that the geometry obtained from the optimizations are very similar, whether we use three, or four, layers in the copper model. Hence, it was found to be sufficient to use three layers in the Cu (111) supercell in the present investigation. This result, together with the fact that the computer efficiencies were quite limited at the time for this investigation, contributed to the decision to continue with a three-layer thick Cu model. Moreover, a similar procedure can be done in testing the length of the *c*-axis (i.e. the inter-slab interactions). As can be seen in [11], a *c*-axis of 17 Å was found to be sufficient for all types of investigations performed in that study.

7.2.2 Number of *k*-points and cut-off energy of the plane waves

The adsorption energies for chemisorption of CuCl, onto the two-fold bridge site of Cu(111), have been compared for different combinations of *k*-points and cut-off energies. The purpose of these calculations was to find the most optimal number of *k*-points and cut-off energy for the plane waves. The results of these test calculations are shown in table 7.2. As can be seen in this table, a cut-off energy larger than, or equal to, 300 eV gives an adsorption energy of approximately −200 $kJ\ mol^{-1}$. On the other hand, the number of *k*-points was not so critical for the final results. Hence, it was concluded that 4 *k*-points, and a cut-off energy of 300 eV, give a sufficient accuracy for the calculations performed in [11].

Table 7.2. Adsorption energies for CuCl adsorbed on the twofold bridge site, calculated with different combinations of *k*-points and cut-off energy for the planewaves.

Cut-off energy (eV)	Adsorption energy ($kJ\ mol^{-1}$) One *k*-point	Adsorption energy ($kJ\ mol^{-1}$) Four *k*-points	Adsorption energy ($kJ\ mol^{-1}$) Eight *k*-points
136	1669	1582	2094
300	211	206	184
500	207		

As a final test, the formation energy and bond length of CuCl, in addition to the dissociation energy for the reaction:

$$(CuCl)_2(g) \rightarrow 2CuCl(g) \qquad (7.1)$$

were calculated and compared with the results obtained by Kölmel and Ahlrichs [13] and Guido *et al* [14]. An excellent agreement was found for molecular CuCl. The values obtained for the formation energy and bond length of CuCl ($-384\ kJ\ mol^{-1}$ and 2.094 Å) agrees very well with the bond length found by Kölmel and Ahlrichs (2.074–2.176 Å) and the formation energy found by Guido *et al* ($-382.3\ kJ\ mol^{-1}$). (The absolute value of the formation energy is a measure of the bond energy of the overall bond energy.) Moreover, for the dissociation of the gaseous $(CuCl)_2$ molecule (shown in figure 7.5(c)), Kölmel and Ahlrichs found a dissociation energy of 183 $kJ\ mol^{-1}$, whereas Guido *et al* reported on a value of 181 $kJ\ mol^{-1}$. The theoretically calculated dissociation energy for $(CuCl)_2$ is 223 $kJ\ mol^{-1}$, which is approximately 22% higher than the experimentally obtained one. However, the theoretically obtained geometry and intra-molecular distances ($r_{Cu-Cu} = 2.527$ Å versus $r_{Cu-Cl} = 2.351$ Å) agrees very well with the corresponding values found in [13] ($r_{Cu-Cu} = 2.432$–2.784 Å versus $r_{Cu-Cl} = 2.271$–2.380 Å).

7.3 Adsorption of Cu-containing growth species

In studying the ALD growth mechanism of Cu(s), from gaseous CuCl and H_2, one has to start with the adsorption of the growth species (i.e. CuCl) to the growing film, or to the substrate. There are four distinctly different adsorption sites on the Cu(111) surface (see figure 7.4); (1) a three-fold ccp site, (2) a three-fold hcp, (3) a two-fold bridge site, and (4) an on top site. Adsorption sites of types 1 and 2 are hollow sites, but of different types. As can be seen in figure 7.4, the adsorbed copper(I) chloride molecules are, for these sites, oriented normal and with the copper atom closest to the Cu surface. Furthermore, the CuCl monomer is being used here, even though it has been shown that CuCl most often exists as a dimer or a polymer in the gas phase [13–15]. This choice is motivated by the fact that the dimers or polymers are expected to dissociate when adsorbed onto the surface, or in the gas phase in the vicinity of the surface prior to adsorption. However, this assumption must be proved.

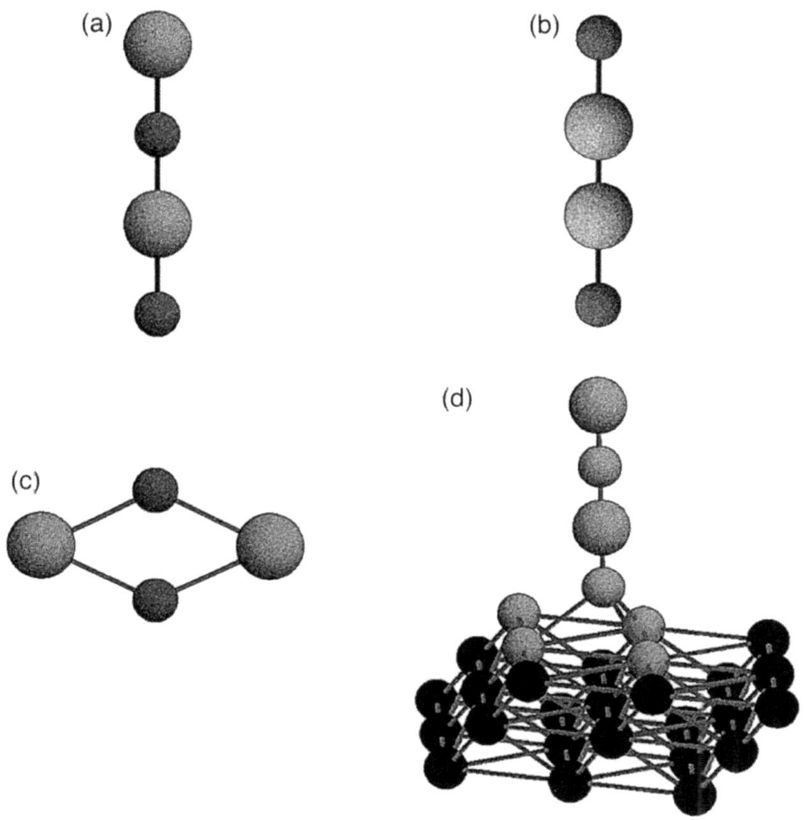

Figure 7.5. The three gaseous $(CuCl)_2$ molecules (a)–(c), and the adsorbed linear $(CuCl)_2$ used for the investigation of the disproportionation of CuCl. Atoms which were allowed to relax during the calculations are marked with grey, whereas atoms held fixed are black. The Cl atoms are represented by spheres that are larger than the Cu atoms. [11], reproduced with permission © Elsevier.

The adsorption energies, Cu–Cl intramolecular bond lengths, and $Cu_{surface}$–Cu_{CuCl} bond lengths are shown in table 7.3. As can be seen in this table, the adsorption energies for the ccp and hcp three-fold sites are identical (−152 kJ mol^{-1}), whereas it is 10 kJ mol^{-1} weaker for the adsorption onto the two-fold site (−142 kJ mol^{-1}). On the contrary, the adsorption on the on-top site is an endothermic process (+88 kJ mol^{-1}). Hence, CuCl does not prefer to adsorb at this type of surface site. For the other adsorption sites (three-fold ccp, three-fold hcp, and two-fold bridge sites) the various adsorption strengths are reflected in the $Cu_{surface}$–Cu_{CuCl} bond lengths. These bond lengths do decrease with an increase in adsorption strength. This can be expected as a higher adsorption energy (in absolute values) corresponds to a stronger and shorter chemical bond. Interestingly, the $Cu_{surface}$–Cu_{CuCl} bond lengths correspond (within 0.5%) to the inter-atomic distances found in metallic copper.

Table 7.3. Adsorption energies and bond distances for CuCl adsorbed on the available adsorption sites on a Cu(111) surface. The surface–Cu distance is the vertical distance from the Cu atom in CuCl to the top layer of the Cu surface.

Adsorption site	Adsorption energy ($kJ\ mol^{-1}$)	Cu–Cl bond length (Å)	Surface–Cu bond length (Å)
Three-fold (ccp)	−152	2.098	2.105
Three-fold (hcp)	−152	2.170	2.071
Two-fold bridge	−142	2.125	2.210
Ontop	+88	2.145	2.313

Table 7.4. The relative energies for CuCl at different distances from a copper (111) surface, showing the absence of an energy barrier for the adsorption of CuCl onto a two-fold bridge site. The relative total energies are related to the one obtained for the shortest $Cu_{surface}$–Cu_{CuCl} distance.

Cu_{CuCl}–$Cu_{surface}$ distance (Å)	Relative energy ($kJ\ mol^{-1}$)	Cu–Cl bond length (Å)
4.0	+1158	2.126
3.6	+978	2.131
3.2	+954	2.126
2.8	+665	2.171
2.4	+6	2.170
2.126	0	2.126

The findings presented above are also reflected in the electron density plots. From these, the electron density for the copper atom in adsorbed CuCl differs widely from that of gaseous CuCl. When CuCl is adsorbed on sites (a)–(c), the electron density of the copper atom closely resembles that of metallic copper. The copper atom in CuCl is also well-incorporated into the copper surface and the resulting template corresponds closely to a copper surface with an adsorbed chlorine atom. Finally, no changes in the electron density of the chlorine atom in CuCl could be observed upon adsorption on any of the sites studied.

7.3.1 Barrier energies for adsorption of Cu-containing species

The calculated relative total energies, for various distances between a surface Cu atom and the Cu atom in the CuCl adsorbate (4.0, 3.6, 3.2, 2.8, 2.4, and 2.13), are presented in table 7.4. The shortest distance (2.13 Å) is the one obtained from the geometry optimization calculations of CuCl being adsorbed onto a two-fold bridge site. For the other distances, the $Cu_{surface}$–Cu_{CuCl} distance has been fixed while the other atoms are being geometrically relaxed (except for the lower Cu layer in the

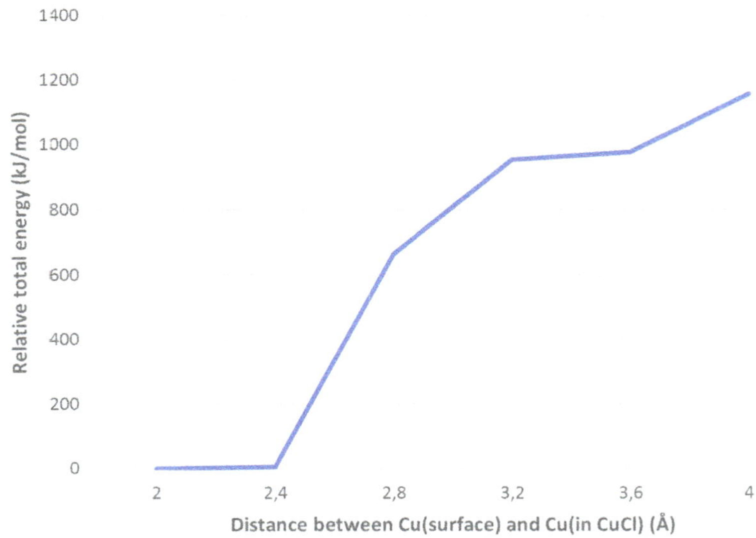

Figure 7.6. Total energy evolution for the adsorption process of CuCl onto a two-fold bridge site on Cu(111). The relative total energies are related to the one obtained for the shortest $Cu_{surface}$–Cu_{CuCl} distance.

surface slab). Moreover, the relative total energies are related to the one obtained for the shortest $Cu_{surface}$–Cu_{CuCl} distance. As can be seen in table 7.4, the relative energy for a $Cu_{surface}$–Cu_{CuCl} distance of 2.4 Å is very close to the value of the relaxed model. This implies that the global minimum is very wide and shallow. In addition, the results in table 7.4 do not give any indication of the presence of an energy barrier for this type of adsorption processes. This is, however, more clearly shown in figure 7.6.

It has not been found to be valuable to also continue with these barrier energy investigations for the two three-fold sites as the experimental results also suggest that there are no adsorption barriers for these types of adsorption steps.

7.4 Disproportionation of the copper(I)chloride molecule

Previous XPS studies have shown that $CuCl_2$ is present in the copper film after completion of the ALD deposition [1]. The growth rate per cycle was also found to be abnormally increased when using prolonged precursor pulses. There is, therefore, an urgent need to study the probability for a disproportionation process in the present example. This disproportionation reaction is expected to proceed as if the CuCl molecules are adsorbed on adjacent sites, or as if CuCl is formed in the gas phase and adsorbs intact. Since the volatility of $CuCl_2$ is lower than the volatility of CuCl, traces of $CuCl_2$ are expected to be found in the film if the deposition temperature is low enough. However, $CuCl_2$ will desorb from the surface at an elevated temperature, leaving a non-terminated adsorption site and an incorporated copper atom (within the growing film). Obviously, this process destroys the self-limitation in the ALD process and must, therefore, be avoided.

Table 7.5. The total energies for different combinations of surfaces and gaseous $(CuCl)_n$ molecules ($n = 1$ or 2).

Model system	Total energy (eV)	ΔE_{ads} (kJ mol^{-1})
Cu(surface) + 2CuCl(g)	−30 722.113 59	
Cu(surface) adsorbed with CuCl+CuCl(g)	−30 723.689 43	−152
Cu(surface) + $(CuCl)_2$(g)	−30 723.257 48	
Cu(surface) adsorbed with $(CuCl)_2$ (see figure 7.5(d))	−30 721.711 52	−149
2 CuCl(g)	−2883.592 82	
$(CuCl)_2$(g) (see figure 7.5(a))	−2883.736 71	
$(CuCl)_2$(g) (see figure 7.5(b))	−2883.514 31	
$(CuCl)_2$(g) (see figure 7.5(c))	−2885.918 04	

$$2CuCl(ads) \rightarrow CuCl_2(ads) + Cu(ads) \quad (7.2)$$

$$(CuCl)_2(ads) \rightarrow CuCl_2(ads) + Cu(ads) \quad (7.3)$$

In the present example, the disproportionation process is being studied under the assumption that it will take place on the growing Cu film. At first, the most stable gaseous $(CuCl)_2$ species must be identified. Three different arrangements are, thereby, considered (see figures 7.5(a)–(c)). As can be seen in table 7.5, the formation energies of $(CuCl)_2$ in the gas phase (i.e. from two gasoues CuCl molecules) are −110 kJ mol^{-1} and −224 kJ mol^{-1} for the $(CuCl)_2$ formations shown in figures 7.5(a) and (c), respectively. However, the formation process for the arrangement shown in figure 7.5(b) is endothermic. Since we are interested in a disproportionation on the copper surface, there are two demands on the $(CuCl)_2$ species. Its formation process must be exothermic (which outrules the arrangement shown in figure 7.5(b)) and it shall just be weakly exothermic (since it is expected to disproportionate on the surface). Based on these demands, the arrangement shown in figure 7.5(a) has been chosen for the adsorption of $(CuCl_2)$ onto the three-fold hcp site on the Cu surface (see figure 7.5(d)). As can be seen in table 7.3, this adsorption process is, however, energetically unfavourable, with an endothermic adsorption energy of +149 kJ mol^{-1}. As a conclusion, there are no indications of $(CuCl)_2$ disproportionations to take place on the Cu (111) surface.

7.5 Removal of Cl from the CuCl adsorbate

7.5.1 Introduction

The next step in the CVD growth mechanism of copper is the removal of Cl from the CuCl adsorbate, by the addition of hydrogen. This type of abstraction reaction is a reduction reaction since the Cu atom will go from the oxidation number +1 to 0. (The reason is that the resulting Cu adsorbate is incorporated in bulk Cu, and the oxidation number of a bulk metal is 0.) As is always the case, we have to look for the energetically most favourable reaction pathway for this reduction step, and there are two different surface reactions to choose from: (1) reactions between the CuCl adsorbate and gaseous atomic (or molecular) hydrogen. (2) Surface reactions between adsorbed CuCl and adsorbed hydrogen. (3) The reduction reaction takes place in the gas phase, before any adsorption has taken place (i.e. between CuCl(g) and H(g), or H_2(g)).

7.5.2 Gas phase reactions

In calculating the reaction energies for the reactions between CuCl(g) and H(g) (or H_2(g)), all molecules should be completely relaxed (i.e. geometry optimized). The following reactions are then assumed to take place:

$$2\mathrm{CuCl(g)} + \mathrm{H_2(g)} \rightarrow 2\mathrm{CuClH(g)},\ \text{or} \tag{7.4a}$$

$$\mathrm{CuCl(g)} + \mathrm{H(g)} \rightarrow \mathrm{CuClH(g)} \tag{7.4b}$$

$$\mathrm{CuClH(g)} \rightarrow \mathrm{Cu(g)} + \mathrm{HCl(g)} \tag{7.5}$$

For a gas phase reaction between H(g) and CuCl(g), the reaction is assumed to proceed as described in the following steps (and as shown in figure 7.7):

1. H(g) approaches the Cl atom in CuCl(g).
2. A Cu–Cl–H(g) intermediate is formed.
3. The Cu–Cl bond (in Cu–Cl–H(g)) breaks with a release of HCl(g).

For a gas phase reaction between H_2(g) and CuCl(g), the following steps are assumed to take place (and are shown in figure 7.8):

1. Two CuCl(g) molecules approach towards the H_2 (g) molecule with their Cl atoms pointing towards the H atoms.
2. A reaction intermediate is formed with 2CuCl(g) and H_2(g) in a linear arrangement.
3. The H–H bond (in the linear arrangement) breaks to form two H–Cl–Cu intermediate complexes.
4. The Cu–Cl bonds (in H–Cl–Cu) breaks with a release of two HCl(g) molecules.

As can be seen in table 7.6, the reaction between H_2(g) and CuCl(g) (see equation (7.4a)) is endothermic, with a positive reaction energy of +204 kJ mol^{-1}. However, it

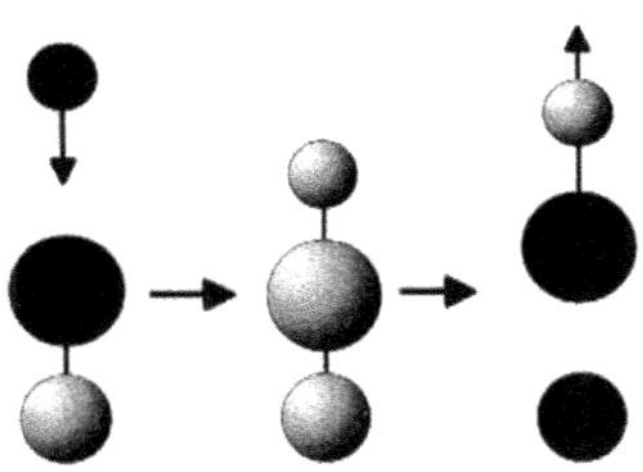

Figure 7.7. A schematic illustration of the reaction for gas phase elimination of chlorine by the addition of atomic hydrogen. [10], reproduced with permission © Elsevier.

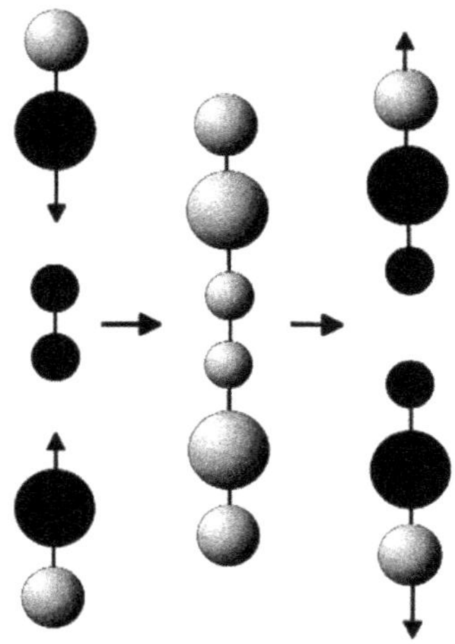

Figure 7.8. A schematic illustration of the reaction for gas phase elimination of chlorine through a hydrogen bridge formation. [10], reproduced with permission © Elsevier.

Table 7.6. Reaction energies for the reactions between CuCl and H_2 (or H) in the gas phase.

	Reaction energy (kJ mol^{-1})
$2CuCl(g) + H_2(g) \rightarrow 2CuClH(g)$	+204
$2CuCl(g) + H(g) \rightarrow 2CuClH(g)$	−204
$CuClH(g) \rightarrow Cu(g) + HCl(g)$	+5

is exothermic (−204 kJ mol^{-1}) when H(g) is, instead, used as the reducing agent (see equation (7.4b)). Moreover, the subsequent removal of HCl from the Cu–Cl–H intermediate (see equation (7.5)) requires only +5 kJ mol^{-1}. The low reaction energy in this elimination step is also reflected in the geometry of Cu–Cl–H where the Cu–Cl and the Cl–H bond lengths are 2.81 Å and 1.31 Å, respectively. These bond lengths are to be compared with the ones for single CuCl and HCl molecules (2.09 and 1.31 Å, respectively). Hence, it is clear that the Cu–Cl bond is considerably weakened in the intermediate state, while the final H–Cl bond has already been formed. In summary, out of the plausible CuCl-containing gas phase reactions (i.e. equations (7.4a) and (7.4b)), the reaction between CuCl(g) and atomic H(g) has been found to be the energetically preferred one.

7.5.3 Reactions between adsorbed CuCl and gaseous hydrogen

7.5.3.1 Reduction by molecular H_2

One possible mechanism for de-chlorination of CuCl adsorbates is by using gaseous H_2 in a direct elimination of the chlorine atoms via H_2 bridge formation between two adjacent CuCl molecules. This mechanism has previously been suggested by Seifert *et al* for the elimination of Cl (as HCl) in the CVD growth of GaAs [16]. As a first trial, the total reaction has here been divided into three steps (see equations (7.6)–(7.8)). They are, furthermore, illustrated in figure 7.1.

$$2CuCl(ads) + H_2(g) \rightarrow (CuClH)_2(ads) \quad (7.6)$$

$$(CuClH)_2(ads) \rightarrow 2CuClH(ads) \quad (7.7)$$

$$CuClH(ads) \rightarrow Cu(s) + HCl(g) \quad (7.8)$$

The initial arrangement, with the gaseous H_2 molecule at a long distance from the surface and centred between the two CuCl adsorbates, is shown in figure 7.1. As the first step in the total reaction, a reaction intermediate will be formed, with the hydrogen molecule forming a bridge between the two CuCl adsorbates (see figure 7.1(b)). As the second step, figure 7.1(c) shows the situation when the H–H bond has broken, leaving two neighbouring CuClH adsorbates on the Cu surface. Finally, figure 7.1(d) shows the third and last step, where the surface achieves two extra (and incorporated) Cu atoms after the breakage of the Cu–Cl bonds, and a following release of 2 HCl(g).

Unfortunately, the $(CuClH)_2$ adsorbate was in [9] found to be energetically unstable. Hence, we have here reduced these three steps to only two, where equation (7.9) is the sum of equations (7.6) and (7.7):

$$2CuCl(ads) + H_2(g) \rightarrow 2\,CuClH(ads) \quad (7.9)$$

$$CuClH(ads) \rightarrow Cu(s) + HCl(g) \quad (7.10)$$

As can be seen in table 7.7, the reaction process between the two neighbouring CuCl adsorbates and one gaseous hydrogen molecule (equation (7.9)) is slightly endothermic, with a reaction energy of just +15 kJ mol^{-1}. On the other hand, when gaseous atomic hydrogen is being used (see equation (7.4b)), the reaction is highly exothermic with a reaction energy of −299 kJ mol^{-1} (see table 7.7). This is to be compared with the energies for the corresponding gas phase reactions (+204 and −204 kJ mol^{-1} for molecular and atomic hydrogen, respectively). As presented above, the breaking of the Cu–Cl bond, after that the stable intermediate complex (see figure 7.1(c)) has been formed, requires very little energy, only 17 kJ mol^{-1}.

Table 7.7. Reaction energies for the reactions between two adsorbed CuCl species and a gaseous H_2, or H, species.

	Reaction energy (kJ mol^{-1})
$2CuCl(ads) + H_2(g) \rightarrow 2CuClH(ads)$	+15
$CuCl(ads) + H(g) \rightarrow CuClH(ads)$	−299
$CuClH(ads) \rightarrow Cu(s) + HCl(g)$	+17

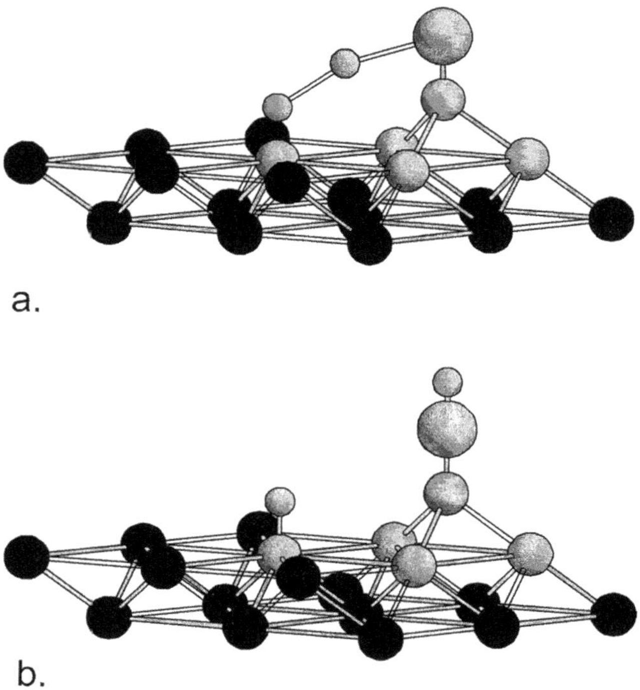

Figure 7.9. A schematic illustration of the reaction between gaseous H_2 and CuCl adsorbed on Cu(111). (a) The activated complex prior to the breakage of the H–H bond. (b) The copper surface with a stable reaction intermediate and an extra hydrogen atom on an adjacent adsorption site (after the breakage of the H–H bond). [9], reproduced with permission © Elsevier.

Another plausible reaction pathway for the reaction between the CuCl adsorbate and H_2(g) is through the formation of a bridge between the chlorine atom in CuCl and a surface copper atom. The following steps are assumed to take place, which are further shown in figure 7.9.

1. The approach of H_2(g) towards the CuCl adsorbate, ending up with the formation of a bridge between the chlorine atom and one of the surface copper atoms (figure 7.9(a)).
2. The breakage of the H–H bond, forming neighbouring CuClH and H adsorbates (figure 7.9(b)).
3. The breakage of the Cu–Cl bond in the CuClH adsorbate. HCl(g) will, thereby, be released from the surface which now has an extra copper atom and an H adsorbate (figure 7.9(c)).

The mechanism of the process shown in figure 7.9 can be expressed by equations (7.11)–(7.13), and the total reaction is presented in equation (7.14).

$$\mathrm{CuCl(ads)} + \mathrm{H_2(g)} \rightarrow \mathrm{CuCl - H - H(ads)} \tag{7.11}$$

Table 7.8. Reaction energies for the reactions between a CuCl adsorbate and an adsorbed H_2 species. The H_2 species is assumed to bind with one of its H atoms to a neighbouring surface site (hollow, bridge, or ontop) and the other to Cl in the CuCl adsorbate (see figure 7.9(a)).

	Reaction energy (kJ mol^{-1})
$2CuCl(ads) + H_2(g) \rightarrow 2CuClH(ads)$ (hollow site)	+64
$2CuCl(ads) + H_2(g) \rightarrow 2CuClH(ads)$ (bridge site)	+56
$2CuCl(ads) + H_2(g) \rightarrow 2CuClH(ads)$ (ontop site)	+29

$$CuCl - H - H(ads) \rightarrow CuClH(ads) + H(ads) \tag{7.12}$$

$$CuClH(ads) \rightarrow Cu(s) + HCl(g) \tag{7.13}$$

$$CuCl(ads) + +H_2(g) \rightarrow Cu(s) + HCl(g) + H(ads) \tag{7.14}$$

The reaction energy for the total reaction presented in equation (7.14) depends on which adsorption site the hydrogen atom prefers to adsorb to. As can be seen in table 7.8, the reaction processes for H binding to a threefold site, bridge site, and ontop site are all endothermic (+64, +56, and +29 kJ mol^{-1}, respectively). (These specific sites are the closest ones to the CuCl adsorbate site.)

7.5.3.2 Reduction by atomic H

For comparison, the possibility for a reaction between an adsorbed CuCl species and gaseous H should also be considered. The various reaction steps, in addition to the total reaction, are presented in equations (7.15), (7.16) and (7.17), respectively. The mechanism is also illustrated in figure 7.3.

$$CuCl(ads) + H(g) \rightarrow CuClH(ads) \tag{7.15}$$

$$CuClH(ads) \rightarrow Cu(s) + HCl(g) \tag{7.16}$$

$$CuCl(ads) + H(g) \rightarrow Cu(s) + HCl(g) \tag{7.17}$$

The situation is completely different when using gaseous H, instead of H_2, in the reduction process. As can be seen in table 7.5, the total reduction reaction is highly exothermic, with a reaction energy of −299 kJ mol^{-1}. This is to be compared with the energies for the corresponding gas phase reactions when molecular and atomic hydrogen were used (+204 and −204 kJ mol^{-1}, respectively) (see table 7.6). As noticed for also the gas phase reactions, the following breakage of the Cu–Cl bond (see equations (7.10) and (7.13)) requires very little energy (+17 kJ mol^{-1}).

7.5.4 Reaction between adsorbed CuCl and adsorbed H

Finally, the possibility for a reaction between CuCl and H adsorbates on the surface has also been investigated. Gaseous H_2 is then assumed, at first, to dissociate on the

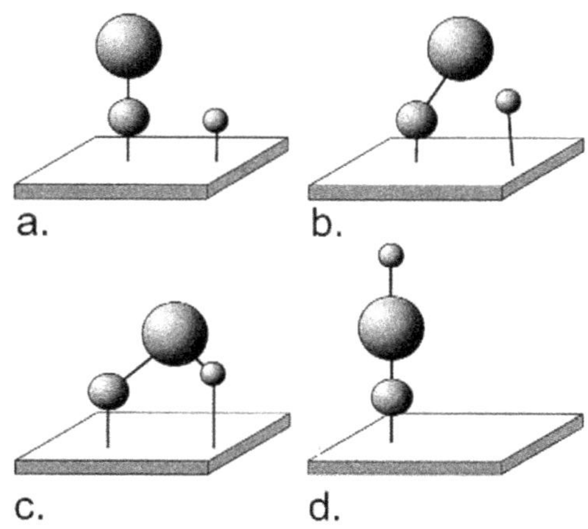

Figure 7.10. A sketch of a Cu surface with CuCl and H chemisorbed on two adjacent adsorption sites: (a) the initial arrangement. (b) Stretching of the Cu–Cl and $Cu_{surface}$–H bonds. (c) Formation of the bridge-like reaction intermediate. (d) Prior to release of HCl(g). [10], reproduced with permission © Elsevier.

copper surface. This type of surface-induced dissociation has earlier been found to end up with H adsorbed to different types of Cu surface sites [7, 8, 17]. As a next step, the reactions between CuCl (adsorbed on a ccp threefold surface site) with hydrogen (adsorbed on the respective hcp threefold, bridge, or on-top site) are being investigated (see figure 7.10). For all three of these adsorption sites, the reactions are assumed to proceed according to steps 1–6 below.

1. Adsorption of CuCl to a ccp threefold site.
2. Dissociative adsorption of H_2 on a position next to the already adsorbed CuCl species (see figure 7.10(a)).
3. Stretching of the Cu–Cl and $Cu_{surface}$–H bonds (figure 7.10(b)).
4. Formation of a bridge-like reaction intermediate including H, Cl, and Cu (figure 7.10(c)).
5. The breakage of the $Cu_{surface}$–H bond (figure 7.10(d)).
6. The breakage of the Cu–Cl bond, which results in the release of HCl(g) (figure 7.10(e)).

The final complete reaction for the reaction between the CuCl and H adsorbates is given in equation (7.18).

$$\mathrm{CuCl(ads)} + \mathrm{H(ads)} \rightarrow \mathrm{Cu(s)} + \mathrm{HCl(g)} \tag{7.18}$$

As can be seen in table 7.9, the surface reactions between adsorbed CuCl and H show reaction energies that are just somewhat dependent on type of H adsorption site. For H adsorbed onto a hcp threefold (i.e., hollow), bridge and ontop sites, the reaction energy for the overall reaction (see equation (7.18)) is +88, +80 and +53 kJ mol^{-1}, respectively. Hence, the H adsorption energies are very similar for the different H adsorption sites studied. This circumstance, together with the fact that hydrogen

Table 7.9. Reaction energies for the reactions between CuCl(ads) and H(ads). The H species is assumed to bind to a neighbouring surface site (hollow, bridge, or ontop) (see figure 7.3).

	Reaction energy (kJ mol^{-1})
CuCl(ads) + H(ads) → CuClH(ads) (hollow)	+88
CuCl(ads) + H(ads) → CuClH(ads) (bridge site)	+80
CuCl(ads) + H(ads) → CuClH(ads) (ontop site)	+53

readily dissociates with a rather low barrier energy on Cu(111), implies that there will be a high surface coverage (i.e. surface concentration) of hydrogen adsorbates on the surface. Another observation is that the calculated reaction energies, for each of the three different reaction paths (i.e. for the different H adsorption sites), are close to the experimental value for the reduction step of CuCl onto the Cu(111) surface (+85 kJ mol^{-1}) [7].

As presented and discussed in section 7.3, CuCl prefers to adsorb on a ccp threefold site on Cu(111), which, in fact, is its final incorporation site. It has also been found experimentally that the adsorption of CuCl is not limited by any adsorption barrier, and is, thereby, very fast [1]. This experimental study did also reveal that the subsequent reaction between hydrogen and adsorbed CuCl is kinetically controlled and requires an activation energy of approximately +85 kJ mol^{-1}. Interestingly, the calculated reaction energies for the reactions given in equations (7.11) and (7.17) are of the same order as the activation energy found in [1] (see tables 7.8 and 7.9). Consequently, they are all possible to occur. The only reaction that seems to have a very low probability to occur is the reaction between the adsorbed CuCl and gaseous H_2 through a bridge formation (see equation (7.9)).

7.6 Reaction barriers

7.6.1 Introduction

In order to unambiguously determine the rate determining reaction step in the CuCl/H_2 copper ALD process, reaction barriers for some selected reactions must also be calculated. These reactions are:

1. Gas phase reactions between CuCl and atomic H, or H_2 (see equations (7.4a) and (7.4b)).
2. Reaction between two adjacently adsorbed CuCl species and gaseous H_2 through a bridge formation (see equation (7.6)).
3. Surface reactions between adsorbed CuCl (ccp threefold site) and adsorbed H (adsorbed on the neighboring ontop site) (see equation (7.18)). The on top site has been chosen here, since it gives the smallest endothermic reaction energy.

The results presented in this chapter are discussed more in detail in [10].

7.6.2 Gas phase reactions

7.6.2.1 Gas phase reactions between CuCl(g) and H(g)

For the gas phase reaction between CuCl(g) and H(g), the energy barriers for both the formation of the reaction intermediate (figure 7.7, first step), and the subsequent elimination of HCl(g) (figure 7.7, second step), has also been calculated. For the identification of the reaction intermediate, the H–Cl distance was decreased in fixed steps, whereas Cu was allowed to completely relax. For the subsequent HCl elimination, the Cu–Cl distance was increased in fixed steps, whereas both Cu and H were allowed to completely relax.

As can be seen in figure 7.11, no reaction barrier is found for the gas phase reaction between H(g) and CuCl(g). This observation is also reflected in the change in Cu–Cl bond length, which continuously increases from 2.172 Å (for free CuCl) to 2.817 Å (in the linear CuClH intermediate). Hence, the formation of the CuClH intermediate is accompanied by a strong weakening of the Cu–Cl bond. For the second, and last, step in the reduction mechanism (i.e. the release of HCl from the CuClH intermediate), there is a minor reaction barrier of approximately 5 kJ mol^{-1}. This infinitesimal energy barrier is accompanied by a marginally increased H–Cl bond distance; from 1.313 Å in CuClH to 1.297 Å for molecular HCl.

7.6.2.2 Gas phase reaction between CuCl(g) and H_2(g)

Similarly, the reaction mechanism for the gas phase reaction between CuCl(g) and H_2(g) is demonstrated in figure 7.8. The reaction between CuCl and molecular

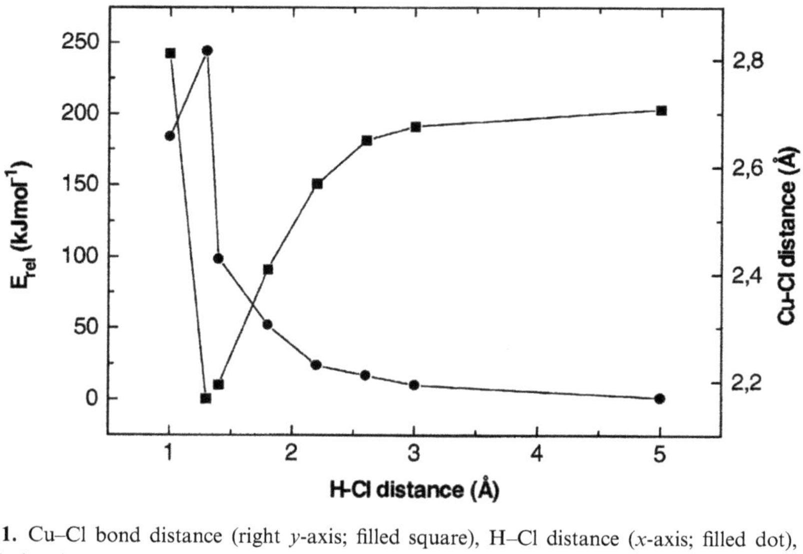

Figure 7.11. Cu–Cl bond distance (right *y*-axis; filled square), H–Cl distance (*x*-axis; filled dot), and total energy (relative the energy for the reaction intermediate) (left *y*-axis) for chlorine elimination by using atomic hydrogen in the gas phase. [10], reproduced with permission © Elsevier.

hydrogen is assumed here to take place by a hydrogen bridge formation between 2CuCl and H_2. The energy barriers for the formation of the reaction intermediate (see equation (7.6)), and the following breakage of the H–H bond (see equation (7.7)), should hence also be calculated. For the calculations of the reaction intermediate, the H–H bond distance has been fixed at the distance of molecular H_2, whereas the H–Cl distances have been shortened in fixed steps. On the other hand, for the breakage of the H–H bond, it has been increased in fixed steps, whereas the H–Cl distance was fixed at that of the reaction intermediate. In addition, the Cu atom was allowed to completely relax. Unfortunately, it was not possible to perform a total geometry optimization of the model due to severe SCF convergence problems. These types of convergence problems are usually due to very unlike structural geometries, e.g. highly strained structures that are far from a minimum in the potential energy surface. Consequently, one has to sequentially (and manually) optimize parts of the geometrical structure.

In contradiction to the situation with atomic hydrogen as a reducing agent, the formation of the reaction intermediate is limited here by a remarkably high energy barrier (of +270 kJ mol^{-1}). However, when increasing the H–H bond length simultaneously with a shortening of the H–Cl bond, the energy barrier will be eliminated (see figure 7.12). Thus, the only barrier energy required for the reaction to take place is the one determined by the total reaction energy (i.e. +209 kJ mol^{-1}). The reaction pathway for the former type of mechanism is illustrated by a solid line in figure 7.12, whereas the latter mechanism is marked with a dashed line. Upon formation of the reaction intermediate, the Cu–Cl bond distance increases from 2.171 Å for long H–Cl distances to 2.45 Å for shorter ones. Furthermore, the subsequent splitting of the H–H bond (to form two separate H–Cl–Cu gaseous intermediates) is limited by an energy barrier of approximately 36 kJ mol^{-1}. The release of HCl from

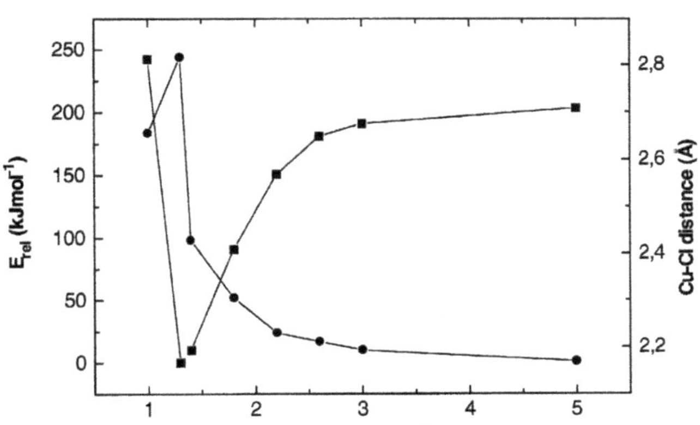

Figure 7.12. Total energy (relative the sum of the energies of the reactants) as a function of reaction coordinate for chlorine elimination of CuCl(g) by the addition of H_2(g) (with (solid line) and without (dashed line) a stretching of the H–H bond). [10], reproduced with permission © Elsevier.

these species has been presented and discussed above (see section 7.6.2). It is a somewhat endothermic process with a reaction energy of +5 kJ mol^{-1}.

7.6.3 Reactions between adsorbed CuCl and gaseous hydrogen

The calculations of an energy barrier for the reaction between two adjacent CuCl adsorbates and gaseous H_2 (through a hydrogen bridge formation) have been performed on models where, with one exception, all atoms were fixed at positions obtained from the corresponding gas phase reactions. The reason for this choice of model was also due to SCF convergence problems. However, some atoms were allowed to completely relax in the calculations of geometrical structure of the reaction intermediate shown in figure 7.2(b). Consequently, the calculated energy barriers might become too high, which is due to the restriction in the number of geometrical optimized atoms. As a result, the energy barrier for the reaction between the adsorbed CuCl species and molecular hydrogen has been found to follow the same pattern as in the corresponding gas phase reaction (see figure 7.12). The H–H distance in the intermediate (see figure 7.2(b)) was also kept fixed here at a distance identical to that of free H_2, while the H–Cl bond was shortened in fixed steps. As can be seen in figure 7.13, a very sharp increase in total energy takes place when shortening the H–Cl distance, with a magnitude close to that of the corresponding gas phase reaction. The largest energy barrier is +220 kJ mol^{-1}, which is to be compared with the one for the corresponding gas phase reaction (+270 kJ mol^{-1}). However, it was not possible to identify any low barrier pathways, i.e. a pathway corresponding to that illustrated by the dashed line in figure 7.12. The main reason is that the geometrical structure of the intermediate shown in figure 7.2(b) is extremely strained compared to the corresponding gaseous situation in figure 7.8. Consequently, this strained intermediate structure does also have a smaller probability to occur.

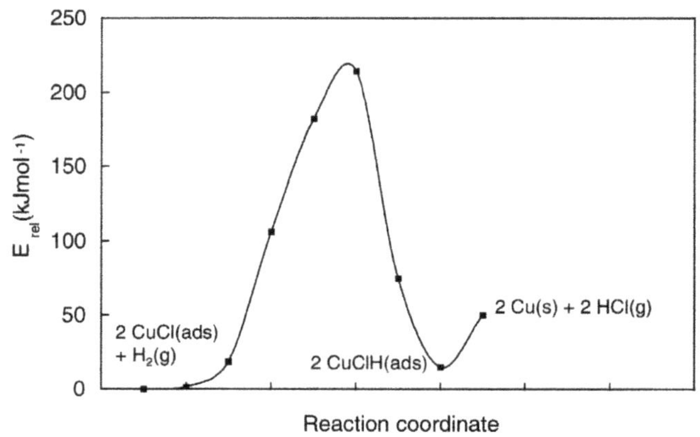

Figure 7.13. Total energy (relative the sum of the energies of the reactants) as a function of reaction coordinate for chlorine elimination from 2CuCl(ads) by the addition of H_2(g). [10], reproduced with permission © Elsevier.

7.6.4 Reaction between adsorbed CuCl and adsorbed H

The reaction between atomic H and CuCl, both adsorbed on adjacent sites on the Cu (111) surface, can proceed through a number of reaction pathways (presented in section 7.6.4). The differences between the different reaction pathways (i.e. for hollow, bridge, and on top positions of the neighbouring H adsorbate) are the degree of stretching of the $Cu_{surface}$–H bond prior to reaction, and the H–Cl bond formation. The important parameter here is not only the $Cu_{surface}$–H and H–Cl, distances, but a combination of those (see figure 7.14). If the $Cu_{surface}$–H distance is too short when the H–Cl bond is formed, a high energy barrier will be required for the process to take place (of the order of +100 to +120 kJ mol^{-1}). The same is also true if the $Cu_{surface}$–H length is increased to above 1.7 Å prior to the H–Cl bond formation. The high energy barriers are probably caused by two different mechanisms. In the former case, it is most probably caused by strong repulsions within the activated complex, whereas in the latter case, it is most probably a reflection of the high adsorption energy of H. However, for the situation when the stretching of the H–$Cu_{surface}$ bond and shortening of the H–Cl bond distance is occurring simultaneously, the barrier is dramatically $Cu_{surface}$ bond is increased to about 2 Å, while the H–Cl bond is simultaneously decreased to approximately 1.6 Å (i.e. prior to the completion of the total reaction). The energy barrier is, thus, found to be completely reduced for this specific reaction pathway. Hence, the required energy barrier corresponds to the reaction energy for this type of reaction (i.e. of about +53 kJ mol^{-1}).

The highest energy barrier for the combination of processes, when going from a starting point with both CuCl and H adsorbed in their preferred adsorption sites (i.e. both threefold hollow sites) to a neighboring ontop position of H, with the following surface reaction, is to be regarded as the energy barrier for the complete

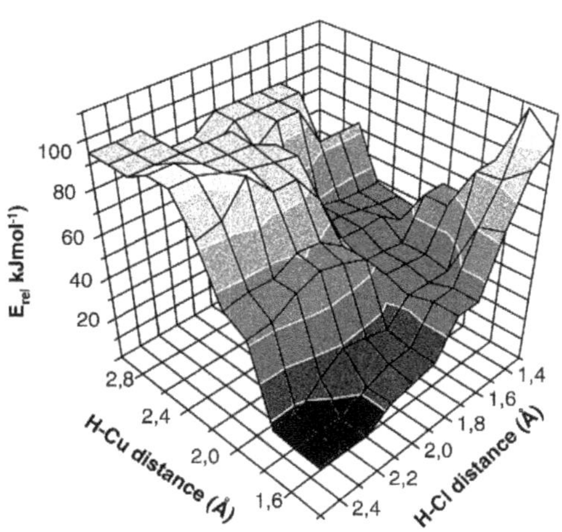

Figure 7.14. Total energy (relative the sum of the reactant energies) as a function of reaction coordinate for the chlorine elimination of 2CuCl(ads) by the addition of H_2(g). [10], reproduced with permission © Elsevier.

surface process. The numerical value of this activation energy is +83 kJ mol^{-1}. This value is in an extremely good agreement with the experimentally obtained one (+85 kJ mol^{-1}), and this reaction mechanism thus provides a good explanation to the question on how the hydrogen reduction of CuCl in the CuCl/H_2 copper ALD process proceeds.

References

[1] Martensson P and Carlsson J-O 1997 Atomic layer deposition of copper on tantalum *Adv. Mater.* **9** 45–50

[2] Kodas T and Hampden-Smith M J (ed) 1994 *The Chemistry of Metal CVD* (Weinheim: Wiley-VCH) and references therein

[3] Campbell J M and Campbell C T 1991 The dissociative adsorption of H_2 and D_2 on Cu (110): activation barriers and dynamics *Surf. Sci.* **259** 1–17

[4] Rasmussen P B, Holmblad P M, Christoffersen H, Taylor P A and Chorkendorff I 1993 Dissociative adsorption of hydrogen on Cu(100) at low temperatures *Surf. Sci.* **287–8** 79–83

[5] Dai J and Zhang J Z H 1995 Quantum adsorption dynamics of a diatomic molecule on surface: four-dimensional fixed-site model on H_2 on Cu(111) *J. Chem. Phys.* **102** 6280–9

[6] Hammer B, Scheffler M, Jacobsen K W and Nørskov J K 1994 Multidimensional potential energy surface for H_2 dissociation over Cu(111) *Phys. Rev. Lett.* **73** 1400–3

[7] Forni A, Wiesenekker G, Baerends E J and Tantardini G F 1994 The chemistry of hydrogen on Cu(111): a dynamical study *Int. J. Quant. Chem.* **52** 1067–80

[8] Gross A, Hammer B, Scheffler M and Brenig W 1994 *Phys. Rev. Lett.* **73** 3121–4

[9] Mårtensson P, Larsson K and Carlsson J-O 1999 Atomic layer epitaxy of copper: an *ab initio* investigation of the CuC/H_2 process. II. Reaction energies *Appl. Surf. Sci.* **148** 9–16

[10] Mårtensson P, Larsson K and Carlsson J-O 2000 Atomic layer epitaxy of copper: an *ab initio* investigation of the CuC/H_2 process. III. Reaction barriers *Appl. Surf. Sci.* **157** 92–100

[11] Mårtensson P, Larsson K and Carlsson J-O 1997 Atomic layer epitaxy of copper: an *ab initio* investigation of the CuC/H_2 process. I. Adsorption of CuCl on Cu(111) *Appl. Surf. Sci.* **136** 137–46

[12] Cotton F A and Wilkinson G 1980 *Advanced Inorganic Chemistry: A Comprehensive text* 4th edn (New York: Wiley)

[13] Kölmel C and Ahlrichs R 1990 An *ab initio* investigation of copper complexes with supershort copper–copper distances *J. Phys. Chem.* **94** 5536–42

[14] Guido M, Gigli G and Balducci G 1972 Dissociation energy of CuCl and Cu_2Cl_2 gaseous molecules *J. Chem. Phys.* **57** 3731–5

[15] Madar R, Bernard C, Palleau J and Torres J 1992 Termodynamic analysis of copper CVD using CuCl as precursor *Microelectron. Eng.* **19** 571–4

[16] Seifert W, Schwetlick S, Reinold J and Butter E 1984 A quantum chemical study of chlorine desorption by hydrogen in the VPE of GaAs *J. Cryst. Growth* **66** 333–7

[17] White J A and Bird D M 1993 Dissociation adsorption of H_2 on Cu(100) *Chem. Phys. Lett.* **213** 422–6

Chapter 8

Prerequisites for vapour phase growth of phase pure cubic BN

Boron nitride (BN) is a compound that occurs in several structures. In a similar way as carbon, BN can occur in hexagonal sp^2-bonded (h-BN), cubic sp^3-bonded (c-BN), and in nanotube [1] and fullerene [2, 3] structures. The geometrical structure of c-BN is very similar to diamond in that every other C atom has been replaced with B atoms, and the other ones with N atoms. This also makes the properties very similar. In fact, cubic boron nitride is a synthetic wide band gap crystalline compound with many excellent properties similar to diamond (e.g. extreme hardness, second only to diamond, high thermal conductivity, large electric resistivity, wide band gap, chemical inertness, low density, and transparency from near ultraviolet to infrared). It can also easily be n- and p-type doped. However, there are also large dissimilarities between c-BN and diamond. Cubic boron nitride possesses a higher thermal stability, both in oxidizing environments and in contact with Fe, Co, and Ni [4]. It is, hence, promising as a tool coating for machining of steel, cast iron and ferrous alloys. Furthermore, c-BN can be made as both n- and p-type semiconductors, suitable for p–n junction diodes. All these properties strongly suggest that c-BN is an extremely promising multifunctional material, which could be tailored for a very large range of advanced applications (e.g. micro- and opto-electronics, electron emission devices, radiation detection, biosensing, high temperature and/or radiation resistant devices). Most of these applications require the deposition of c-BN films in a large area. However, there are at present severe problems in the vapour phase synthesis of c-BN at low pressure. Compared to diamond (which can be relatively easily grown as thick polycrystalline films), the c-BN obtained at low pressure has always become a mixture with other polymorphic forms of BN. It is, therefore, of largest importance to outline the possibilities for a chemical route to c-BN growth by using more gentle chemical vapour deposition (CVD) techniques where the bombarding ions will not penetrate the surface and hence not detrimentally destroy the BN structure.

doi:10.1088/978-0-7503-3107-4ch8

The experimental problems with, and prerequisites for, c-BN growth are described in the present chapter, with the purpose of giving a good background to the theoretical studies of the phase-pure c-BN growth. It will then be shown that theoretical modeling is a necessary tool for a successful growth of phase-pure c-BN.

8.1 Energetical vapour phase deposition

A common feature is noticed in the literature about physical vapour deposited (PVD) c-BN films. That is, the c-BN phase is being created by using 'brute force', i.e. under the influence of surface bombardment with highly energetic ions of nitrogen and argon. 'Brute force' has also been found to be necessary for successful CVD routes [5–12]. Growth of c-BN does then occur by a mechanism in which the structural changes are accomplished by a high compressive stress of several GPa. A layered structure is most often the result, which consists of an amorphous (a-BN) interface layer closest to the substrate, followed by textured sp^2-bonded BN and a final layer of c-BN. The amorphous interface layer has a thickness that is of the order of the ion range, and the *c*-axis of the sp^2-bonded BN (h-BN or t-BN) is parallel with the substrate. Finally, c-BN starts to nucleate on top of the hexagonal BN layer. It has turned out that the parameter window for c-BN nucleation is totally different from the parameter of the continued growth, with a decisive parameter threshold for c-BN growth being lower than for nucleation [13, 14]. It has also been experimentally shown that the impact of energetic ions during the growth of BN films mainly has three effects [15, 16]. At first, it generates a large amount of stresses and defects, which favours the growth of the cubic phase relative the hexagonal one. Secondly, the induced high film stresses limit the thicknesses of the films. Finally, the generation of defects gives a poor crystalline quality, small grain size, and high density of defects and grain boundaries. Accordingly, a delamination from the substrates will be attributed to these high compressive stresses.

8.2 Gentle chemical vapour phase deposition

As presented above, plasma activation CVD techniques usually operate at the low energy regime [17–20]. The ions will normally not penetrate the surface upon impact when the ion energies are well below 50 eV. This raises the question why the bombarding ion velocity cannot be set to zero. It is at present an urgent need for a process development of a more gentle CVD method, where no bombardment of ions can detrimentally affect crystallinity and/or film stress content. Within traditional CVD techniques, chemical reactions are assumed to be the controlling mechanism on the growing material surface. For the situation with CVD growth of c-BN, B- and N-containing precursors are generally used in the gaseous mixture, which diffuse to and self-assemble on the growing c-BN surface without the impact of high-energy heavy ions. Surface dominated growth mechanisms are also assumed for those CVD techniques where the ion energies are well below 50 eV. The growth of a c-BN film from the vapour phase is a very complicated physical and chemical process, and the detailed growth mechanism is still not clear. The CVD techniques involve chemical reactions in both the gaseous phase and on the growing surface. To optimize the deposition processes of c-BN, it is very important to obtain a deeper knowledge (at the molecular level) of all surface processes in the overall growth process (e.g. surface vacancy generation, adsorption of growth species, reactions between the growth species adsorbates and gaseous species, etc). Other aspects that are of the largest importance to study is the effect of the substrate on the phase-pure growth of c-BN, and the combination of substrate and surface terminations on the geometrical structures of the first atomic layers of BN growth. In addition, comparison should be made with experimental observations. It is important to emphasize, though, that a comparison with the CVD growth of diamond shall always be in mind.

8.3 Termination of the c-BN surface

8.3.1 Introduction

Single bonds that form between the terminating H, F or Cl species and the c-BN surface result from s-orbital overlap between the surface dangling bond (i.e. unpaired electrons) and a corresponding orbital of the adsorbate. Fluorine has the highest electronegativity value (3.98) amongst the halogens, with chlorine (3.16) being less reactive towards other elements [44]. The situation is, however, reversed when comparing the covalent radii of F and Cl (0.57 versus 1.02 Å) [21]. Fluorine will, thus, form the strongest bonds with other atoms, and chlorine the weakest. Hydrogen has both the smallest radius and the lowest electronegativity value (2.2), giving it an intermediate position [21]. The differences in size and electronegativity of the different terminating species will not only yield diverse bond energies for the different terminating species, but most probably structural differences as well. One of the desirable roles of these adsorbates is to maintain the sp^3 hybridization of the surface atoms during growth of c-BN, thereby hindering the surface to collapse from a cubic to a hexagonal phase.

The F species is expected to be efficient as a cubic BN surface-stabilizing agent due to its large ability to form strong bonds. One of the main questions is, however, if surface termination is necessary to sustain the sp^3 hybridization of the surface atoms during c-BN growth. Is it possible that the surface nitrogen atoms, with their lone pair electrons, show sp^3 hybridization similar to NH (thereby inducing a cubic structure)? As a test, an N-covered c-BN surface has not been terminated but allowed to be geometry optimized at the same level of theory as the other calculations presented in this chapter. A totally collapsed geometrical structure was then obtained. The sp^3 hybridization of the original structure was completely converted to sp^2 hybridization. Hence, it is obvious that surface termination by a suitable species is of the highest importance for a continued growth of c-BN.

As discussed in section 6.3, one important role of H during CVD of diamond is to maintain the sp^3 configuration of surface carbons. As a matter of fact, it has earlier been theoretically observed that H (in addition to F) is very effective in stabilising the diamond (111) surface. As can be seen in table 8.1, the adsorption energies are almost identical for adsorption of H and F to the diamond (111) surface (−446 versus −447 kJ mol^{-1}). The $\nu_{C\text{–}C(surface)\text{–}C}$ angle does also show an effective

Table 8.1. Adsorption and abstraction energies (in kJ mol^{-1}) for the chemisorption of H (or F) onto a diamond (111) surface, B- or N-covered c-BN (111) and (110) surfaces, and B- or N-surface atoms in c-BN (100). The species in parenthesis is the gas phase species that is assumed to abstract the surface adsorbate.

Surface	$\Delta E_{ads,H}$ (kJ mol^{-1})	$\Delta E_{abs,H}$ (kJ mol^{-1})	$\Delta E_{ads,F}$ (kJ mol^{-1})	$\Delta E_{abs,F}$ (kJ mol^{-1})	$\Delta E_{ads,Cl}$ (kJ mol^{-1})
Diamond (111)	−446	+12 (H)	−447		
c-BN(111)					
B-covered	−442	+18 (H)+104 (F)	−653	+108 (H) +512 (F)	+262
N-covered	−465	+45 (H)	−251	+28 (H)	
c-BN(110)					
B-site	−418				
N-site	−416				
c-BN(100)					
B-covered	−198	−258 (H)−403 (F)	−471	−130 (H)+252 (F)	
N-covered	−431	−25 (H)−170 (F)	−283	−318 (H)+63 (F)	

sp^3-hybridization of the diamond surface C atoms (111.3° versus 111.2°). An H-terminated diamond (111), with an adsorbed growth species, is shown in figure 6.33. The question is now if H has a similar effect on the c-BN surface. It is also worth studying the effect by Cl. Is it similar to H and F? As stated above, a surface structural collapse (to the hexagonal phase) should thereby be hindered, which is a prerequisite for a CVD growth of phase-pure c-BN films.

The (111) surface of c-BN consists of alternate layers of B and N (figure 5.4(b)). This is also the situation for the (100) surface (figure 5.4(d)), which is one of the dominant c-BN growth surfaces under CVD conditions [31]. On the other hand, the (110) surface includes both B and N atoms (figure 5.4(c)). Before studying the effect by surface terminating species on the surface stabilities and geometries, one must find information about the reconstructions of the various c-BN surface planes. The non-terminated B- and N-covered c-BN(100) surfaces are known to reconstruct into a stable (2 × 1) surface structure and metastable (2 × 4), (2 × 4(3)), and c(2 × 2) structures (see figure 8.1). It is therefore of a great interest to investigate if an as-deposited c-BN surface, with a 100% surface coverage of H atoms, will undertake similar surface reconstructions. Any information about plausible surface reconstructions of the corresponding c-BN(111) and c-BN(110) surfaces has not been found. Hence, we have here to start from the ideal, bulk-like structures in the geometry optimization calculations of the respective models.

It is now of a large interest to study the effect of surface termination on the surface geometries (with their eventual reconstructions). But this is not enough. As stated in section 6.4, the surface stabilizing species must also be able to undergo abstraction reactions with gaseous species and, hence, leave room for an incoming B- or N-containing growth species. The abstraction process is, therefore, a central reaction step in the CVD growth of c-BN. The combination of binding energy for H (or F)

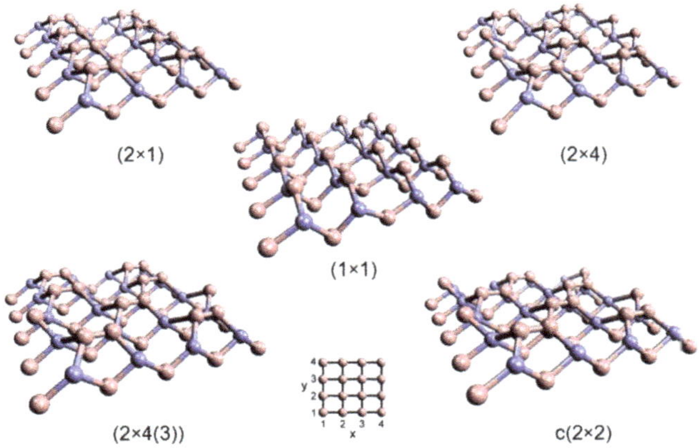

Figure 8.1. Models showing the non-reconstructed ((1 × 1)) and the reconstructed ((2 × 1), (2 × 4), (2 × 4(3)), and c(2 × 2)) B-covered surfaces of c-BN(100). (By interchanging the B and N atoms one obtains the corresponding N-terminated surfaces.) The coordinate system shows the non-reconstructed surface from above. In the text the surface atoms are referred to this coordinate system. [22], reproduced with permission © American Chemical Society.

adsorbed to the surface and binding energy of the molecular gaseous XY product (X = H or F, Y = H or F), determines the numerical value of these types of abstraction reaction energies (see equations (8.1) and (8.2)).

$$\text{Surface-X} + \text{Y} \rightarrow \text{Surface(rad)} + \text{X} - \text{Y} \quad (8.1)$$

$$E(\text{X} - \text{Y}) - E(\text{Surface-X}) = E(\text{Y}) - E(\text{Surface(rad)}) \quad (8.2)$$

where Surface* is the surface with a radical site.

Furthermore, the resulting radical surface sites should not collapse to sp^2 hybridization. Hence, the following aspects of terminated c-BN surfaces are of the largest importance to study more thoroughly:

(1) Adsorption processes of various types of terminating species,
(2) Influence of surface termination on surface geometries and reconstructions,
(3) Abstraction processes of various types of terminating species, and
(4) Degree of hybridization of the surface radical sites (i.e. for a non-perfect termination)

8.3.2 Adsorption processes of various types of terminating species

The surface stabilizing effect by different types of adsorbates is of great interest to study in the phase-pure growth of c-BN. In addition to H, halogens like F and Cl have been frequently used in the vapour phase deposition of BN materials.

If we start with H-termination, in comparison between diamond (111) and B-covered c-BN (111), we can see that H has an almost identical surface stabilizing effect on both type of surfaces (see table 8.1). The adsorption energies are -446 kJ mol^{-1} and -442 kJ mol^{-1}, respectively. However, F has a tendency to bind much more strongly to this type of c-BN surface (-442 (H) versus -653 (F) kJ mol^{-1}). In fact, the difference in adsorption energies between H and F is numerically the largest one for this specific c-BN surface. Another halogen that would be interesting to test is the Cl species. Both F and Cl, being the two most electronegative elements, experience a strengthening in the $B_{surface}$–F/Cl bonds by an ionic–covalent resonance. This means that there is either a double bond or a single bond between $B_{surface}$ and the terminating F/Cl species [23]. Termination by F or Cl would also satisfactorily mimic the heteropolarity of BN. However, it has not been found possible to terminate a B- or N-covered (111) surface to 100% with Cl adsorbates. The Cl adsorption energy for the B-covered surface was strongly endothermic (+262 kJ mol^{-1}), while it was not possible at all to bind the Cl species to the N-covered surface. The reason for these results is most probably the strong repulsions on the B-covered surface. Hydrogen, H, is the only terminating species that gives an almost nonpolar bond $B_{surface}$–H bond (which is due to similar electronegativity values of B and H). In comparison with the more polar B–N bonds in bulk material, this is expected to weaken the B–H bond to a certain extent.

The DFT calculations presented here have been performed at zero Kelvin, while the real CVD process is dynamic and takes place at a finite temperature. It is, hence, of greatest interest to also investigate the stability of the variously terminated B-covered c-BN(111) surface at an elevated temperature. *Ab initio*-based molecular dynamic (MD) calculations at 1200 K are, therefore, used to study the effect of temperature on the bond strength of the different terminating species. The force constants are, in these calculations, obtained from quantum-mechanical calculations [24]. A step length of 0.2 fs has been used for these simulations. The bond length between the adsorbate-binding surface atom and the terminating adsorbate is then regarded to be a measure of the corresponding strength. As a result of the MD simulations (at 1200 K), the $B_{surface}$–H and $B_{surface}$–F bonds have become slightly increased (by about 10% and 2%, respectively). Hence, the bond strengths of the terminating H and F species on the B-covered c-BN(111) surface are not appreciably affected by an increase in temperature from 0 to 1200 K.

As compared with the B-covered surface, it is very interesting to note that the N-covered c-BN(111) surface gives a comparable adsorption strength for H (-465 kJ mol^{-1}), but a much lower one for F (-251 kJ mol^{-1}) (see table 8.1). Thus, there is a much larger probability for H species to become chemisorbed to the N-covered surface (as compared with the F species). This rather weak F adsorption energy is strongly correlated with the reversed polarity of the $N_{surface}$–F bond (with respect to the N–B bonds in the bulk). The reason is the larger electronegative value of F as compared with N. On the contrary, the $N_{surface}$–H bond show the same type of polarity as the B–N bonds. It should also be stressed here that both N and F have very large electronegative values (3.04 versus 3.98) and are therefore not expected to easily interact with each other. In fact, they are both electron-pair donating bases. This is also the situation for an eventual binding between N and Cl. In addition to large sterical respulsions on a Cl-terminated N-covered surface, both N and Cl have very similar and large electronegativity values (3.04 versus 3.16). In fact (as stated above), it was not possible at all to form a 100% Cl-termination layer on the N-covered c-BN(111) surface.

Quantum-mechanical based MD calculations have also been performed here for surface termination with H or F species. As a result, the $N_{surface}$–H bond is only slightly increased (by approximately 15%). However, the $N_{surface}$–F bond is considerably prolonged (by about 42%). Hence, it has been found that an increase in temperature from 0 K to 1200 K will negatively affect the N–F bond strength (i.e. it becomes weakened). It is, thus, not advisable to use F species in the termination of N-covered c-BN(111) surfaces.

As can be seen in table 8.1, the adsorption energies for an atomic H, F, and Cl on a (100) monoradical B surface site were calculated to be −198, −471 and −17 kJ mol^{-1}, respectively. The situation with the c-BN(110) surface is, however, completely different. The H atom binds with an almost identical strength to both B and N surface sites (−418 versus −416 kJ mol^{-1}), which is somewhat weaker than the adsorption of H to c-BN(111). The result for the (110) surface is easily understood from the fact that each c-BN layer consists of both B and N atoms, which in turn leads to similar relaxation behaviour of the $B_{surface}$ and $N_{surface}$ radical sites (see section 6.4.2).

Both H and F have, hence, proven to be effective in stabilizing the three different types of c-BN surfaces. As presented in figure 8.1, the c-BN(100) surface can have several reconstructed geometrical structures, which will be discussed more in detail in the forthcoming chapters. All details in the computational treatments can also be made for c-BN (111) and c-BN(110), but this is up to the reader (as an exercise).

8.3.3 Influence of surface termination on surface geometries and reconstruction

8.3.3.1 Introduction

Widany *et al* performed MD simulations on clean (i.e. non-terminated) c-BN (111) surfaces at room temperature [25]. A transformation to sp^2 hybridization of the surface atoms was then observed. This was observed for both B- and N-covered surfaces. The influence on c-BN growth by the addition of hydrogen to the gas phase in an inductively coupled plasma CVD reactor was investigated by Kuhr *et al* [26]. It was found that H_2, in addition to the hydrogen introduced with the source gas trimethylborazine, prevents the formation of c-BN but not that of the sp^2 nucleation layer. The surface structures of single-crystal c-BN (100) and (111) have been exposed to argon-ion sputtering and then to a prolonged exposure of hydrogen plasma in a work by Loh *et al* [27, 28]. The low-energy argon-ion sputtering of c-BN (100) resulted in a local disrupture on the surface (from sp^3 to sp^2-type bonding). The following H-etching resulted in an H-terminated and restructured c-BN (100)–(2 × 1) surface with sp^3-type bonding. The situation for the (111) surface was that a prolonged exposure in H plasma leads to an H-terminated (111)–(1 × 1) surface with sp^3-type bonding. The influence of F_2 on c-BN in thermodynamic equilibrium was found to be similar to that of atomic hydrogen in a work by Bohr *et al* [29].

The challenge in the CVD process is now to find a way to stabilize the c-BN surface and to suppress the formation of h-BN, being equivalent to the use of atomic hydrogen (or fluorine) in diamond CVD. Stabilization of the c-BN surface during growth is one of the conditions that, according to Bartl *et al* [30], has to be fulfilled in order to achieve successful c-BN formation.

8.3.3.2 B-Covered c-BN(111) surfaces

The angles formed by the surface B atom and two of its binding N neighbors (ν(N–$B_{surface}$–N)), for both a 100% terminated surface and a monoradical surface, are presented in table 8.2 for the different termination situations. As can be seen in table 8.2, both investigated terminating species (H, F) sufficiently sustain the sp^3-hybridization of the $B_{surface}$ atoms.

The angles ν(N–$B_{surface}$–N), for a 100% terminated surface, are all ranging between 110.0° and 111.7° (H: 110.0°, F: 111.7°). The angle obtained for H is, thus, close to the tetrahedral angle. On the other hand, there is a minor influence of sp^2 hybridization for F. Moreover, both terminating species lead to interatomic B–X distances between 1.60 and 1.62 Å. This is to be compared with the bulk B–N value of 1.57 Å. Both the large adsorption energies for H and F, and the optimized structural geometries for the closest environments to these adsorbates, will, hence, support the conclusion that F, as well as H, will be effective in stabilizing the B-covered c-BN B(111) surface (i.e. protecting the sp^3-hybridized surface atoms from being sp^2-hybridized to any larger extent). However, the B-covered c-BN (111) surface has a more pronounced tendency to become terminated by F species due to the larger adsorption energy of F compared to H. As expected from the large difference in electronegativity of F on one hand, and H on the other, the fluorine atom will induce a much larger atomic charge rearrangement near the surface with a large depletion of charge from the surface borons and, hence, an appreciable ionic character of the B–F chemical bond. This charge rearrangement in the structural region of adsorption is reflected in the much larger adsorption energy of F to an F-terminated surface compared to the adsorption of H to an H-terminated surface.

Table 8.2. Angles (N–$B_{surface}$–N for 100% termination versus monoradical) for H- and F-termination of the B- and N-covered c-BN(111) surfaces.

c-BN (111) surface	Termination type	N–$B_{surface}$–N/B–$N_{surface}$–B angle (°)	N–$B_{surface}$–N/B–$N_{surface}$–B angle (radical B site) (°)
B-covered	H	109.2	118.6
	F	108.8	118.6
N-covered	H	113.0	112.9
	F	110.8	113.0

8.3.3.3 N-Covered c-BN(111) surfaces

The angles formed by the surface N atom and two of its binding B neighbors (ν(N–$B_{surface}$–N)), for both a 100% terminated surface and a monoradical surface, are presented in table 8.2 for the different termination situations. As can be seen in this table, the sp^2-hybridization of the N-covered c-BN(111) surface is, with one exception, found to be sufficiently sustained by these terminating species (H and F). The ν(B–$N_{surface}$–B) angle increases slightly to 110.8° when terminating the surface with F species. A partial transition from sp^3 to sp^2 hybridization was also observed for the H-termination. The ν(B–$N_{surface}$–B) angle increased with as much as 3.6° (up to 113.0°) for this termination type. This enlarged angle is consistent with the increasing double-bond character in the substrate with a decreasing N–H bond strength i.e. with an increasing 'surface-radical character'. Hence, hydrogen will most likely not sustain the sp^3 hybridization of the N-covered c-BN(111) surface. The interatomic distances between a surface N to its nearest B atom are also found to be 1.59, and 1.61 Å for H- and F-termination, respectively. These distances are all longer than the value of 1.57 Å for the corresponding bulk material [30].

8.3.3.4 c-BN(110) surfaces

The c-BN(110) surface consists of stoichiometric proportions of B and N atoms in each atomic layer. H- or F-termination studies for this specific surface show that the possibility to uphold the cubic structure is very similar for both the B and N atoms (see figure 8.2). This is expected since each surface atom is strongly influenced by its neighbouring ones. The similarity in surface geometrical structtures is strongly supporting the similarities in adsorption energy for both H and F to this specific surface type (see table 8.1).

8.3.3.5 B-Covered c-BN(100) surfaces

8.3.3.5.1 H-termination

The supercell model used for the terminated B-covered c-BN(100) surface is shown in figure 6.15. The calculations show that a 100% H-termination, on the initially ideal c-BN(100)–(1 × 1) surface, induces a (2 × 1) reconstruction of the B-covered surface (see figure 8.3), whereas the initially reconstructed (2 × 1), (2 × 4), (2 × 4(3)), and c(2 × 2) surfaces only exhibit a significant relaxation of the B–B dimer bonds. The relaxations of the dimer bonds on the H–terminated reconstructed surfaces are expected since chemisorption of H to each of the dimer B atoms will break the partial π bond character of the B–B bonds (which is similar to the behavior of the 100% H-covered diamond (100)–(2 × 1) surface) [31, 32]. Furthermore, the H-induced reconstruction of the initially ideal (1 × 1) surface, to any of the reconstructed surfaces ((2 × 1), (2 × 4), (2 × 4(3)), and c(2 × 2)), is also expected since a 100% surface coverage (at ontop positions) leaves one dangling bond per surface B atom. These dangling bonds will rapidly undergo pairwise reactions, forming B–B σ bonds. It might also be argued that each surface B atom will be terminated with 2 H atoms. However, this situation is not likely due to the sterical hindrances that would occur amongst the H adsorbates. Most

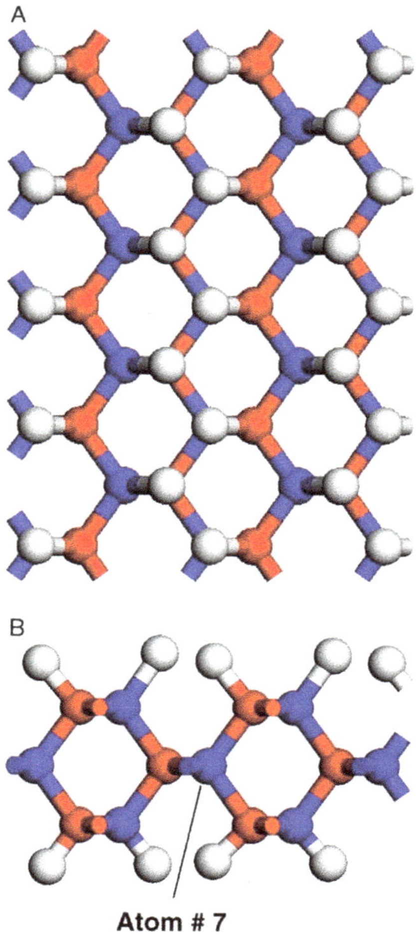

Figure 8.2. (a) Top view and (b) side view of the optimized c-BN(110) structure. The B, N, and H atoms are shown in red, blue, and white, respectively. The corresponding F termination showed a similar structural geometry. [33], reproduced with permission © Elsevier.

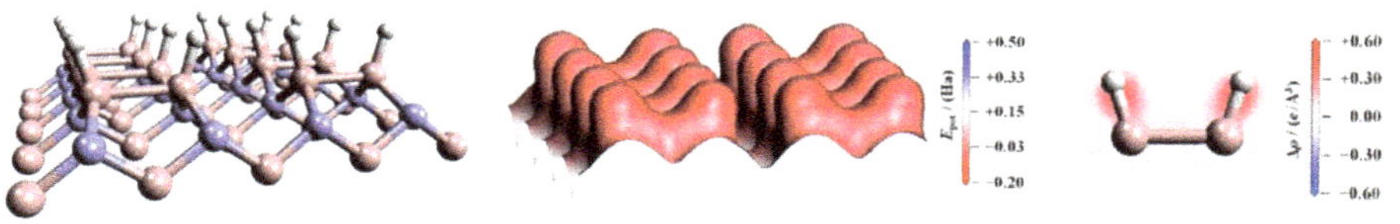

Figure 8.3. (a) 100% H-terminated B-covered c-BN(100)–(2 × 1) surface (optimized). (b) Electrostatic potential for the surface in (a). (c) Deformation density of a B–B dimer in a plane perpendicular to the surface in (a). [22], reproduced with permission © American Chemical Society.

interestingly, the H-terminated B-covered c–BN(100)–(2 × 1) structure has also been experimentally observed in a study by Loh *et al* [27]. It is therefore decided that this type of surface should be used here for further investigations about the phase–pure growth of c-BN.

For the H-terminated B-covered c-BN(100)–(2 × 1) surface, the surface B atoms are to a large degree sp^3 hybridized (i.e. have a tetrahedral environment), with an average surface N–$B_{surface}$–N bond angle and $B_{surface}$–N bond length of 110° and 1.56 Å, respectively. These values are very similar to the bulk c-BN values of 109.5° and 1.57 Å, which is also a result that is in accordance with the study by Loh *et al* [27]. It can, therefore, be concluded that a 100% surface coverage of H on the B-covered c-BN(100)–(2 × 1) surface is able to uphold a bulk-like bond angle and bond length for the surface B atoms. However, the H-termination is not able to up-hold the bulk-like structure for the atomic layers further below. We will, therefore, compare the optimized structure with the initial one (i.e. prior to the geometry optimization calculations). The interplanar B-N spacing between the first (B) and second (N) layer is 0.86 Å for the 2 × 1-reconstructed surface, which is within 4% of the value forinitial (i.e. non-reconstructed) surface (0.90 Å). The B–B dimer bond lengths are also 1.97 Å and 2.56 Å, respectively. The B–B dimer on the 2 × 1-reconstructed surface is significantly longer than the corresponding dimer on the non-terminated 2 × 1-reconstructed surface (1.97 Å versus 1.63 Å). This is not surprising since the dimer bond on the non-reconstructed surface has a partial π bond character, while it is only a σ bond on the H-terminated surface.

Continuing with the 2 × 1-reconstructed H-terminated surface, the B–B dimer rows are separated by troughs, approximately 3.14 Å apart (figure 8.3(a)). Moreover, the B–B bonds are tilted 17° from the surface normal, in the opposite direction with respect to their neighbouring B–H bond (in the same dimer). Also, the B–H bonds have a length of 1.20 Å and are positioned in a plane perpendicular to the surface (table 8.3). As a comparison, the experimentally determined B–H bond length in BH_3 is 1.19 Å [35]. As presented in section 8.4.1, the calculated adsorption energy for the H species gives an indication of a medium-sized covalent bond strength (−198 kJ mol^{-1}).

Table 8.3. Angles (N–$B_{surface}$–N), bond lengths ($B_{surface}$–N, B–B dimer, and B–H), and distances ($B_{1st\ layer}$–$N_{2nd\ layer}$) for various types of H-terminated B-covered c-BN(100) surfaces.

H-terminated B-covered c-BN (100) surface	N–$B_{surface}$–N angle (°)	$B_{surface}$–N length (Å)	$B_{1st\ layer}$–$N_{2nd\ layer}$ distance (Å)	B–B dimer length (Å)	B–H length (Å)
Optimized 2 × 1-reconstructed	110.0	1.56	0.86	1.97	1.20 1.19 (exp.)
Initial non-reconstructed	109.5	1.57	0.90	2.56	
Non-terminated optimized 2 × 1-reconstructed				1.63	
Bulk	109.5	1.57			

As can be seen in figure 8.3(b), calculated electrostatic potentials also show the dimer-formation in rows. The electrostatic potentials of the H adsorbates are negative, which is due to the difference in electronegativity values between B and H. The H atom has a higher electronegativity value than B (2.20 versus 2.04), which means that there is a partial transfer of electron from the B surface atoms to their terminating H atoms. This transfer of electronic charges can also be observed by calculating the deformation density (see figure 8.3(c)), which is the difference in electron density between the total density for the whole model and, in this situation, the sum of electron densities for all involving atoms (in their free states). There is also a third way to observe the partial electron transfer, and that is to calculate the H atomic charges. The calculations of Mulliken charges have demonstrated here that the average charge of the H adsorbates is −0.12 e. It should again be stressed that it is a very good idea to use all possible detection methods. The results will then become more trustworthy.

8.3.3.5.2 F-termination

As was the situation with H-termination, the calculations show that a 100% surface coverage with F species, on the initially non-terminated B-covered c-BN(100)–(1 × 1) surface, induces a (2 × 1)-reconstruction of the surface (see figure 6.21(a)). As is the situation with H-termination, this restructuring saturates the dangling bonds of the surface B atoms and, hence, yields a lower surface energy. As can be seen in table 8.4, the average B–B dimer bond length is 1.90 Å, while the corresponding B–B distance is 2.56 Å in the initial (1 × 1) structure (i.e. prior to geometry optimization calculations). The B–B dimer rows are also separated here by troughs, approximately 3.21 Å apart. Moreover, the surface B atoms exhibit tetrahedral coordination (i.e. a large degree of sp^3 hybridization), with an average surface N–B–N bond angle and B–N bond length of 109° and 1.57 Å, respectively. These values are in very good agreement with the bulk values of 109.5° and 1.57 Å. As was the situation with H-termination, a 100% F-termination is able to uphold bulk-like bond angles and

Table 8.4. Angles (N–$B_{surface}$–N), bond lengths ($B_{surface}$–N, B–B dimer, and B–F), and distances ($B_{1st\ layer}$–$N_{2nd\ layer}$) for various types of F-terminated B-covered c-BN(100) surfaces.

F-terminated B-covered c-BN (100) surface	N–$B_{surface}$–N angle (°)	$B_{surface}$–N length (Å)	$B_{1st\ layer}$–$N_{2nd\ layer}$ distance (Å)	B–B dimer length (Å)	B–F length (Å)
Optimized 2 × 1-reconstructed	109.0	1.57	0.87	1.90	1.35(exp. BF_3: 1.31)
Initial non-reconstructed	109.5	1.57	0.90	2.56	
Non-terminated optimized 2 × 1-reconstructed				1.63	
Bulk	109.5	1.57			

bond lengths for the surface B atoms on the B-terminated c-BN(100)–(2 × 1) surface. As is also the situation with an H-terminated B-covered c-BN(100) surface, the F-termination is not able to up-hold the bulk-like structure for the atomic layers further below in the lattice. We will also, therefore, compare the optimized structure here with the initial one (i.e. prior to the geometry optimization calculations). As can be seen in figure 6.21(b), the interplanar B–N spacing between the first (B) and second (N) layer is 0.87 Å for the 2 × 1-reconstructed surface, which is within 3% of the value for initial (i.e. non-reconstructed) surface (0.90 Å). Moreover, the B–F bonds are tilted 17° from the surface normal, in the opposite direction with respect to their neighbouring B–F bond (in the same dimer). Also, the B–F bond lengths are 1.35 Å, and these bonds are positioned in a plane perpendicular to the surface. (The experimentally determined B–F bond length in BF_3 is 1.31 Å [24].) As presented in section 8.4.1, the calculated adsorption energy of F presents an indication of a very strong covalent bond (−471 kJ mol^{-1}).

It can be concluded from these results that atomic F chemisorbs significantly stronger than atomic H to the B-covered c-BN(100)–(2 × 1) surface. This difference in adsorption energy is most probably due to the more ionic character of the B–F bond compared to that of the B–H bond. This conclusion is based on the atomic charge calculations of B and F in the B–F bond (q(B) = +0.40 e, q(F) = −0.25 e) and those of B and H in the B–H bond (qB = +0.19 e, qH = 0.12 e). This means that there is a larger attractive electrostatic contribution to the B–F bond energy compared to that for the B–H bond energy. The more ionic character of the B–F bond is also in accordance with calculated electronegativity differences ($\Delta\chi$), which characterize the B–F bond ($\Delta\chi$(B–F) = 1.94) as 'predominantly ionic', whereas the B–H bond ($\Delta\chi$(B–H) = 0.16) is 'predominantly covalent' [36].

As was the situation with H-termination, one should try to verify any conclusion by also performing other types of calculations. The results from electrostatic potential calculations and deformation density calculations are, together with the final geometrical structure, shown in figure 6.21. As can be se figure 6.21(b), calculated electrostatic potentials confirm the dimer-formation in rows, which is clearly visible in the geometrical structure given by figure 6.21(a). Also, the electrostatic potentials of the F adsorbates are negative, which is due to the difference in electronegativity values between B and F. As is the situation with H-termination, the F atom has a higher electronegativity value than B (2.20 versus 2.04), which means that there is a partial transfer of electrons from the B surface atoms to their terminating F atoms. This transfer of electronic charge can also be observed by calculating the deformation density (see figure 6.21(c)). In addition, the calculations of Mulliken charges show that the surface-covering F atoms have a negative value (q(F) = −0.25 e). They are, hence, nucleophilic and will thereby be highly susceptible towards an attack by electrophilic radical species.

In summary, both H and F species are highly efficient in upholding the cubic structure of a 100% terminated c-BN surface. However, F binds much more strongly than H to the growing c-BN surface, which may be negative for the necessary abstraction reactions in the c-BN CVD growth mechanism.

8.3.3.5.3 Cl-termination

In contradiction to the situation with a B-covered c-BN(111) surface, it has for the (100) surface shown to be possible to terminate the surface to 100% with Cl adsorbates (even though the adsorption energy for the last Cl is very small). As was the situation with H- and F-termination, the calculations show that a 100% surface coverage of Cl on the initially non-terminated B-covered c-BN(100)–(1 × 1) surface induces a 2 × 1-reconstruction of the surface structure (see figure 8.7(a)). As can be seen in table 8.5, the average B–B dimer bond length is 1.71 Å, which is to be compared with the B–B distance of 2.56 Å for the initial bulk-like structure of the B-covered c-BN(100)–(1 × 1) surface (i.e. before geometry optimization). The B–B dimer rows are separated by troughs, approximately 3.40 Å apart. Moreover, the surface B atoms are found to exhibit a tetrahedral coordination, i.e. a large degree of sp^3 hybridization, with an average surface N–$B_{surface}$–N bond angle and B–N bond length of 109° and 1.57 Å, respectively. These values are in very good agreement with the values of 109.5° and 1.57 Å for the initial c-BN(100)–(1 × 1) surface. Hence,

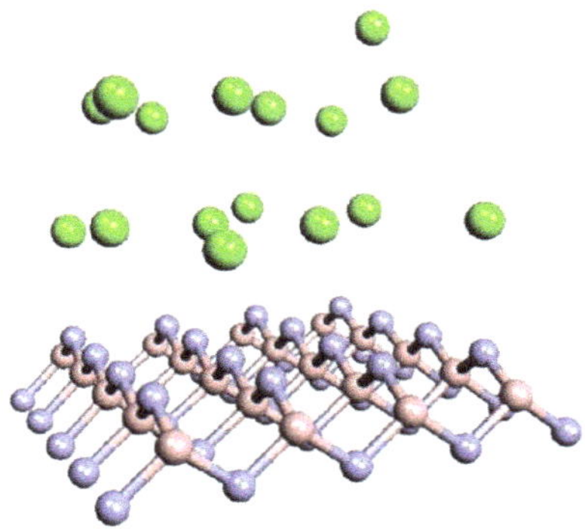

Figure 8.7. Desorption of Cl atoms and Cl_2 molecules from an initial 100% Cl-terminated N-covered c-BN (100)–(1 × 1) surface. [34], reproduced with permission © American Chemical Society.

Table 8.5. Angles (N–$B_{surface}$–N), bond lengths ($B_{surface}$–N, B–B dimer, and B–Cl), and distances ($B_{1st\ layer}$–$N_{2nd\ layer}$) for various types of Cl-terminated B-covered c-BN(100) surfaces.

Cl-terminated B-covered c-BN (100) surface	N–$B_{surface}$–N angle(°)	$B_{surface}$–N length (Å)	$B_{1st\ layer}$–$N_{2nd\ layer}$ distance (Å)	B–B dimer length (Å)	B–Cl length (Å)
Optimized 2 × 1-reconstructed	109.0	1.57	0.82	1.71	1.87(exp. BCl_3: 1.74)
Initial non-reconstructed	109.5	1.57	0.90	2.56	
Non-terminated optimized 2 × 1-reconstructed				1.63	
Bulk	109.5	1.57			

as is the situation with H- and F-termination, the Cl-termination is able to uphold a bulk-like structure of the upper **B** atomic layer, with its terminating species, on the c-BN(100)–(2 × 1) surface.

However, the Cl atoms are not able to uphold the ideal bulk-like (1 × 1) structure for the geometrical structure further below in the lattice. As can be seen in table 8.5, the interplanar B–N distance between the first (B) and second (N) atomic layer is 0.82 Å, which is within 9% of the bulk value of 0.90 Å. Moreover, the B–Cl bonds are tilted 17° from the surface normal, in the opposite direction with respect to their neighbouring B–F bond (in the same dimer). Also, the B–Cl bond lengths are 1.87 Å, and these bonds are positioned in a plane perpendicular to the surface (the experimentally determined B–Cl bond length in BCl_3 is 1.74 Å). As presented in section 8.4.1, the calculated adsorption energy of Cl presents an indication of a medium strong covalent bond (-471 kJ mol^{-1}). This is an adsorption energy for the chemisorption of Cl to a bare (i.e. non-terminated) surface. However, the adsorption energy for the chemisorption of Cl to an otherwise Cl-terminated surface is just -17 kJ mol^{-1}, which is much lower than the adsorption energy for F. Compared to F-termination, this lower adsorption energy is most probably due to the repulsive sterical repulsions between adjacent Cl adsorbate (the shortest distance between the Cl atoms on the surface is 2.54 Å, which is significantly smaller than the van der Waals distance of 3.50 Å). Hence, it is very difficult to terminate the surface to 100% with Cl species. As was the situation with H- and F-terminations, one should try to verify these conclusions by also performing other types of calculations (like electrostatic potential calculations and deformation density calculations). As can be seen in figure 8.4(b), calculated electrostatic potentials confirm the dimer-formation in rows, which is clearly visible in the geometrical structure given by figure 8.4(a). Also, the electrostatic potentials of the Cl adsorbates are negative, which is due to the difference in electronegativity values between B and Cl. As was the situation with H-termination, the Cl atom has a higher electronegativity value than B (2.20 versus 2.04), which means that there is an electronic charge transfer of electrons from the **B** surface atoms to their terminating Cl atoms. The effect of type of adsorbates on the electrostatic potential is clearly seen by comparing figures 6.23(b), 8.3(b), and 8.4(b) for F, H, and Cl, respectively. This transfer of electronic charge can also be observed by calculating the deformation density (see figure 8.3(c)). The effect of type of adsorbates on the electron transfer is, however, not clearly seen in figures. 6.23(c), 8.3(c), and 8.4(c) for F, H, and Cl, respectively.

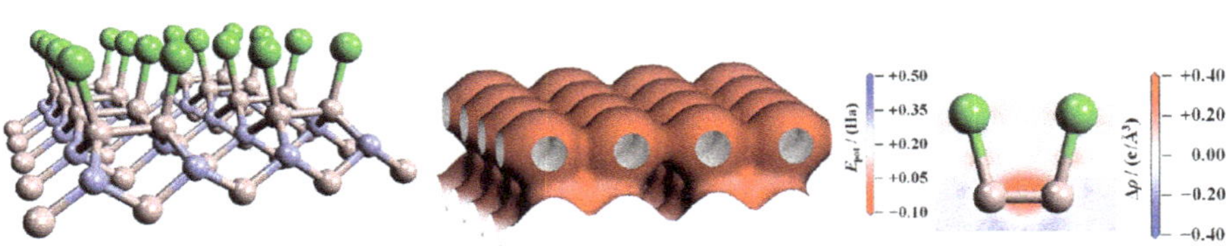

Figure 8.4. (a) 100% Cl-terminated B-covered c-BN(100)–(2 × 1) surface (optimized). (b) Electrostatic potential for the surface in (a). (c) Deformation density of a B-B dimer in a plane perpendicular to the surface in (a). [34], reproduced with permission © American Chemical Society.

8.3.3.6 N-covered c-BN(100) surfaces

8.3.3.6.1 H-termination

The calculations show that a 100% surface termination with H will de-reconstruct the N-covered c-BN(100)–(2 × 1) surface to the bulk-like (1 × 1) structure (see figure 8.5(a)). The calculated electron deformation density does not show any increase in electron density between the surface N atoms, which indicates that a complete surface de-reconstruction has taken place. The high stability of the 100% H-covered N-terminated c-BN(100)–(1 × 1) surface is most probably due to the σ bonding overlap between the 1s orbital on H and the partial occupied antibonding orbital of the N–N dimers. This overlap will increase the electron density in the antibonding orbital and thereby destabilize the already weak N–N dimer bond. Moreover, the surface N atoms on the H-terminated surface are found to exhibit a trigonal planar coordination, i.e. a large degree of sp^2 hybridization, with an average surface B–$N_{surface}$–B bond angle and $N_{surface}$–B bond length of 116° and 1.51 Å, respectively. Furthermore, the average surface N–N distance is 2.55 Å, which is to be compared with the bulk-like distance of 2.56 Å of the initial non-optimized structure (see table 8.6). The interplanar N–B spacing between the first (N) and second (B) layer is 0.81 Å, which is within 10% of the initial non-reconstructed value of 0.90 Å. Moreover, the N–H bond is parallel to the surface normal, with a length of 1.01 Å (the experimentally determined N–H bond length in NH_3 is 1.01 Å). As can be seen in table 8.1, the calculated H adsorption energy is –431 kJ mol^{-1}. Hence, it has been shown that H chemisorbs strongly to both the B-terminated c-BN(100)–

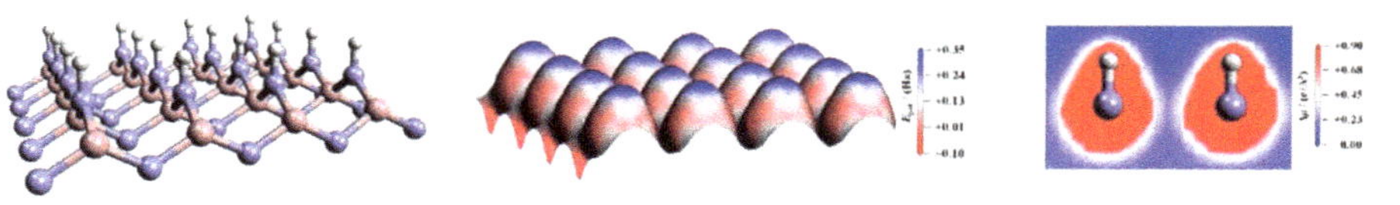

Figure 8.5. (a) 100% H-terminated N-covered c-BN(100)–(1 × 1) surface (optimized). (b) Electrostatic potential for the surface in (a). (c) Deformation density of B–H bonds in a plane perpendicular to the surface in (a). [22], reproduced with permission © American Chemical Society.

Table 8.6. Angles (B–$N_{surface}$–B), bond lengths ($N_{surface}$–B, N–N dimer, and N–H), and distances ($N_{1st\ layer}$–$B_{2nd\ layer}$) for various types of H-terminated N-covered c-BN(100) surfaces.

H-terminated N-covered c-BN (100) surface	B–$N_{surface}$–B angle (°)	$N_{surface}$–B length (Å)	$N_{1st\ layer}$–$B_{2nd\ layer}$ distance (Å)	N–N dimer length (Å)	N–H length (Å)
Optimized 2 × 1-reconstructed	116.0	1.51	0.81	2.55	1.01 (exp. NH_3: 1.01)
Initial non-reconstructed	109.5	1.57	0.90	2.56	
Bulk	109.5	1.57			

(2 × 1) surface ($\Delta E_{ads} = -198$ kJ mol^{-1}) and the N-terminated c-BN(100)–(1 × 1) surface ($\Delta E_{ads} = -431$ kJ mol^{-1}), with a preference for the N-terminated surface. This preference is also visible in the average B–H and N–H bond lengths, with a significantly shorter bond length for the N-terminated surface (1.01 Å) compared to the B-terminated surface (1.20 Å). Thus, the surface N sites show a larger reactivity than the surface B sites.

The N site preference can partially be explained by the fact that a surface radical N atom is nucleophilic (with an atomic charge, q(N), of −0.57 e), while a surface radical B atom is electrophilic (with an atomic charge, q(B), of +0.29 e). Since H is an electrophilic species, it prefers to bind to the N-terminated surface. Another factor that influences the adsorption energy is the more prominent polar character of the N–H bond (q(N) = −0.48 e, q(H) = +0.12 e) compared to the B–H bond (q(B) = +0.19 e, q(H) = −0.12 e), which increases the attractive electrostatic contribution to the N–H bond energy. This polar nature is in accordance with calculated electronegativity differences, which characterize the N–H bond as a 'polar covalent', while the B–H bond is 'mostly covalent' [36]. (The electronegativity differences are 0.84 and 0.16, respectively.)

As is clearly visible in the geometrical structure presented in figure 8.5(b), the calculated electrostatic potentials confirm the 1 × 1 structure of the H-terminated N-covered c-BN(100) surface. Also, as a direct consequence of the difference in electronegativity values, the electrostatic potentials of the H adsorbates are positive. Furthermore, the strong N–H bonds, and the breakage of the N–N dimer bonds, are strongly visible by the calculations of the electronic deformation density (see figure 8.5(c)).

8.3.3.6.2 F-termination

As was the situation with H-termination, the results from geometry optimization calculations show that a 100% surface termination with F will de-reconstruct the N-covered c-BN(100)–(2 × 1) surface to the bulk-like (1 × 1) structure (see figure 8.6(a)). The surface N atoms are also here exhibiting a trigonal planar coordination, with an average surface B–N$_{surface}$–B bond angle and N–B bond length of 117° and 1.50 Å, respectively (see table 8.7). These values are almost identical to the corresponding values for an 100% H-terminated surface (116° vs. 1.51 Å). However, these surface values are different from the corresponding bulk values of 109.5° versus 1.57 Å. Moreover, the interplanar N–B spacing between the

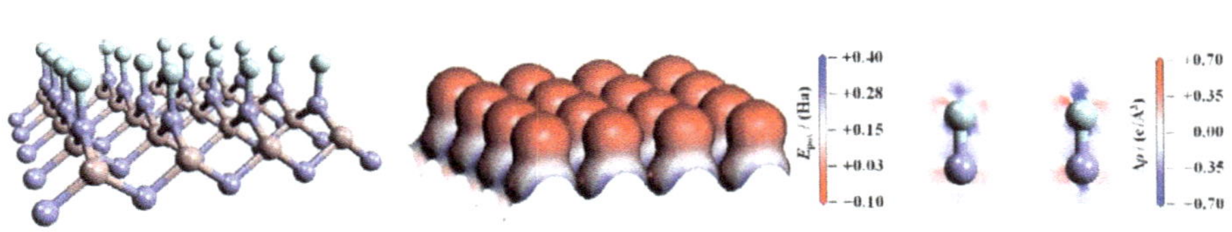

Figure 8.6. (a) 100% F-terminated N-covered c-BN(100)–(1 × 1) surface. (b) Electrostatic potential for the surface in (a). (c) Deformation density of N–F bonds in a plane perpendicular to the surface in (a). [34], reproduced with permission © American Chemical Society.

Table 8.7. Angles (B–$N_{surface}$–B), bond lengths ($N_{surface}$–B, N–N dimer, and N–F), and distances ($N_{1st\ layer}$–$B_{2nd\ layer}$) for various types of F-terminated N-covered c-BN(100) surfaces.

F-terminated N-covered c-BN (100) surface	B–$N_{surface}$–B angle (°)	$N_{surface}$–B length (Å)	$N_{1st\ layer}$–$B_{2nd\ layer}$ distance (Å)	N–N dimer length (Å)	N–F length (Å)
Optimized 2 × 1-reconstructed	117.0	1.50	0.79	2.55	1.38 (exp. NF_3: 1.35)
Initial non-reconstructed	109.5	1.57	0.90	2.56	
Bulk	109.5	1.57			

first (N) and second (B) atomic layer is 0.79 Å, which is within 12% of the bulk value (0.90 Å). These differences in angles, bond lengths, and interplanar distances are most probable caused by the surface relaxation of the N-covered c-BN surface. The degree of surface reconstruction can also be evaluated by studying the distances between the upper N surface atoms (in relation to the corresponding bulk values); 2.55 Å versus 2.56 Å. In addition, the N–F bond is parallel with the surface normal and has a length of 1.38 Å (the experimentally determined N–F bond length in NF_3 is 1.35 Å). Hence, the F- (or H-) terminated N-covered c-BN(100) surface is an ideal ground for a continued cubic growth of c-BN.

The adsorption energy for F chemisorbed to a monoradical surface site, on an otherwise completely F-terminated N-covered c-BN(100) surface, has been calculated to be −283 kJ mol^{-1} (see table 8.1). This adsorption process is, hence, rather strongly exothermic. From these results it can be concluded that atomic F chemisorbs strongly to both the B-covered c-BN(100)–(2 × 1) surface (ΔE_{ads} = −471 kJ mol^{-1}) and the N-covered c-BN(100)–(1 × 1) surface (ΔE_{ads} = −283 kJ mol^{-1}). However, the F species binds significantly stronger to the B-covered surface compared to the N-covered one. The weaker bond strength of the N–F bond, compared to that of the B–F bond, is most probably caused by the repulsions between non-bonding electrons in N and F. In addition, by calculating the atomic charges (here the Mulliken charges) the partial charges on the N and F atoms (in N–F) are both negative; −0.17 e vs. −0.20 e. Also, the Mulliken charges of B and F (in B–F) are +0.40e and −0.25 e, respectively. This results in a repulsive (destabilizing) electrostatic contribution to the N–F bond energy, whereas the B–F bond will be stabilized by an attractive electrostatic contribution. It has also been shown that atomic H (ΔE_{ads}=−431 kJ mol^{-1}) chemisorbs significantly stronger than atomic F (ΔE_{ads} = −283 kJ mol^{-1}) to the N-covered c-BN(100)–(1 × 1) surface (see table 8.1). This is reasonable since (i) H has no non-bonding electron and (ii) the N–H bond is electrostatically stabilized (q(N) = −0.48 e, q(H) = +0.12 e). The preferential adsorption of H onto the N-covered surface, and of F onto the B-covered surface, is also in agreement with experimental studies by Zhang *et al* [37].

8.3.3.6.3 Cl-termination

As was the situation with the c-BN(111) surface, when starting with a 100% coverage of Cl adsorbates on an initially clean N-covered c-BN(100)–(1 × 1) surface, it was found that the Cl atoms desorbs from the surface during the course of the geometry optimization calculations (see figure 8.7). The (1 × 1)-surface was chosen for these calculations since it is the most stable structure for both H- and F-terminations. The desorption of Cl atoms is most probably caused by the repulsive sterical repulsions between the Cl adsorbates. The distance between the Cl adsorbates on the initial non-reconstructed surface is 2.56 Å, which is significantly smaller than the corresponding van der Waals distance of 3.50 Å. Such interactions were also found to be present on the 100% Cl-terminated B-covered surface. However, in contrast to the polar covalent B–Cl bond (with a difference in B and Cl electronegativity of 1.12), the N–Cl bond (with a difference in N and Cl electronegativity of 0.12) is almost non-polar and therefore less electrostatically stable. The low stability of the N–Cl bond is also in accordance with the limited stability of NCl_3. Furthermore, interactions between adjacent Cl atoms has also been observed in the early stages of the Cl desorption process from the N-covered c-BN(100) surface. This is a strong indication of a recombinative molecular desorption of Cl_2 from the surface.

From these results (i.e. for the Cl adsorption on the B- and N-covered c-BN(100)–(1 × 1) surfaces), it is reasonable to assume that adsorption of Br and I will be highly unfavorable. This assumption is based on the fact that:

- (i) Br and I have larger van der Waals radii (1.85 Å vs. 1.98 Å) compared to Cl (1.75 Å). Thus, there will be more pronounced sterical hindrances amongst the adsorbates on the Br- and I-terminated surfaces.
- (ii) The $B_{surface}$–Cl ($N_{surface}$–Cl), $B_{surface}$–Br ($N_{surface}$–Br), and $B_{surface}$–I ($N_{surface}$–I) bonds are formed by the overlap of a hybridization orbital of B (N) with a 3p, 4p, and 5p orbital of Cl, Br, and I, respectively. Since the average electron density is lower in a 4p and 5p orbital, compared to a 3p orbital, the $B_{surface}$–Br ($N_{surface}$–Br), and $B_{surface}$–I ($N_{surface}$–I) bonds are expected to be weaker than the $B_{surface}$–Cl ($N_{surface}$–Cl) bonds [38].
- (iii) The polar covalent B–Br bond (with a difference in B and Br electronegativity of 0.92) and polar covalent B–I bond (with a difference in B and I electronegativity of 0.62) are less electrostatically stabilized than the B–Cl bond (with a difference in B and Cl electronegativity of 1.12).
- (iv) There is no significant electrostatic contribution to the N–Br (with a difference in N and Br electronegativity of 0.08) and N–I (with a difference in N and I electronegativity of 0.38) bond energies. These bonds are 'mostly covalent' [38].
- (v) The NBr_3 and NI_3 molecules are less stable than NCl_3.

8.3.4 Abstraction of surface terminating species

8.3.4.1 B-covered c-BN(111) surfaces

The abstraction of an adsorbed H, F, or Cl species results in a radical B surface site that is almost totally collapsed into sp^2 configuration (see table 8.8). (The

Table 8.8. Angles (N–$B_{surface}$–N) for H-, F-, and Cl-terminated B-covered c-BN(100) surfaces; monoradical and 100% terminated.

B-covered c-BN (111) surface	N–$B_{surface}$–N angle H (°)	N–$B_{surface}$–N angle F (°)	N–$B_{surface}$–N angle Cl (°)
Monoradical	118.6	118.6	118.2
100% terminated	109.2	108.8	110.1

Table 8.9. Angles (B–$N_{surface}$–B) for H-, F-, and Cl-terminated N-covered c-BN(100) surfaces; monoradical and 100% terminated.

N-covered c-BN (111) surface	B–$N_{surface}$–B angleH(°)	B–$N_{surface}$–B angleF(°)	B–$N_{surface}$–B angleCl(°)
Monoradical	112.9	113.0	113.1
100% terminated	113.0	110.8	109.6

abstraction of a Cl adsorbate from a 100% termination with Cl species is included here even though the probability of such a layer formation is very low.) The neighboring adorbates are, thus, not able to sustain the sp^3 hybridization of the radical B surface atom. The angle ν(N–$B_{surface}$–N) increases from approximately 109°–110° to about 118°–119° for all these adsorbates. This difference gives evidence of the transition from sp^3 hybridization (i.e. 109.4°) to predominantly sp^2 hybridization (i.e. 120°). The radical surface site has then been stabilized by a hyperconjugation interaction [39]. This is a resonance interaction in which the unpaired electron in a p-type orbital on the surface radical B atom interacts with an adjacent B–N bond (being aligned with the radical B atom). This gives a partial double bond character to the $B_{surface}$–N bond and weakens the aligned B–N bonds. Several features in the present c-BN structure support such a resonance stabilization. At first, the interatomic $B_{1st\ layer}$–$N_{2nd\ layer}$ distance shortens, with an average shortening of 0.10 Å for the H, F, and Cl adsorbates. Secondly, the aligned B–N bonds become lengthened with an average increase of 0.06 Å, while the other B–N bond lengths do not change at all.

8.3.4.2 N-covered c-BN(111) surfaces

A terminating species X (X = H, F or Cl) has also been removed for the N-covered c-BN(111) surface from the surface, thereby forming a monoradical site. However, this was found to yield just a local partial transition to sp^2 hybridization (see table 8.9). The angle ν(B–$N_{surface}$–B) increased to only about 113° for all terminating species (i.e. H, F, and Cl). If we compare this result with the corresponding result for the B-covered c-BN(111) surface (with ν(N–$B_{surface}$–N) angles within the range

118.2–118.6° and, hence, a collapse of the cubic (111) structure), we can conclude that radical sites on N-covered surface are only partially converted to sp^2. It is, thus, likely that the cubic phase of c-BN will be sustained even for situations with a not perfectly terminated nitrogen surface.

8.3.4.3 c-BN(110) surfaces

As was the situation with the 100% terminated c-BN(110) surfaces, the abstraction of a terminating species, X (X = H, F, or Cl), results in large similarities in structural geometries for the radical B and N surface sites. The underlying reason is the same, namely that each surface atom is strongly influenced by its neighbouring ones.

8.3.4.4 B-covered c-BN(100) surfaces

8.3.4.4.1 H-termination

As can be seen in figure 8.3(b), the H adsorbates on the c-BN(100)–(2 × 1) surface are nucleophilic (i.e. they will attract electron-deficient species). This is further verified by the calculations of atomic charges (here Mulliken charges), which show that the H adsorbates have a negative charge of −0.12e. This result is in accordance with the higher electronegativity value of H (2.20), as compared with B (2.04). These nucleophilic surface-terminating H atoms are, hence, highly susceptible toward the attack by electrophilic radical species (i.e. the process of H abstraction with gaseous H atoms is expected to be energetically favourable). As can be seen in table 8.1, this is in agreement with the strongly exothermic H abstraction process (with an abstraction energy of −258 kJ mol^{-1}). Similar results have been obtained for the corresponding process of H abstraction by using gaseous F radicals. This process is significantly more exothermic (ΔE_{abs} = −403 kJ mol^{-1}) than the abstraction process using gaseous H (ΔE_{abs} = −258 kJ mol^{-1}).

The B–B dimer bonds on the 2 × 1-reconstructedc-BN(100) surface are also susceptible toward electrophilic attacks. Hence, it might at first seem plausible to de-reconstruct the (2 × 1) surface, to a (1 × 1) dihydride one by adding H to the surface. However, such a high surface coverage of H atoms gives an energetically unstable surface, which is due to repulsive sterical interactions, as well as dipole-dipole interactions, between adjacent H adsorbates.

The removal of an adsorbate from the 100% H-terminated B-covered c-BN(100)–(2 × 1) surface results in a pronounced, but very local, collapse of the corresponding surface B radical site (see figure 8.8(a)). As can be seen in figure 8.1, these atoms are

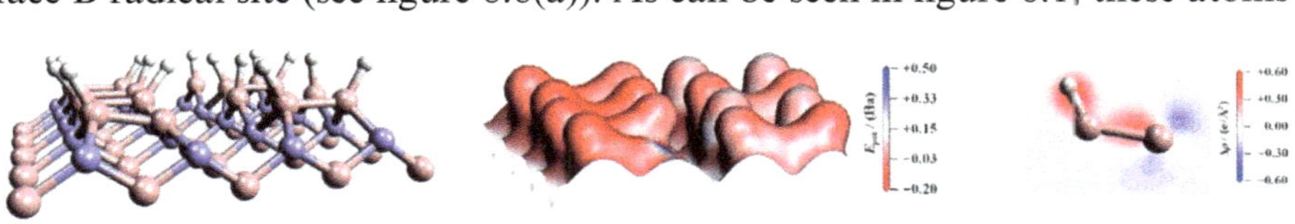

Figure 8.8. (a) A surface radical site on an otherwise H-terminated B-covered c-BN(100)–(2 × 1) surface. (b) Electrostatic potential for the surface in (a). (c) Deformation density of the radical B–B dimer in a plane perpendicular to the surface shown in (a). [22], reproduced with permission © American Chemical Society.

denoted H(2,1) and B(2,1), respectively. The B(2,1) atom relaxes down by −0.26 Å to a more trigonal planar coordination (ν(N–B(2,1)–N) = 115°) Furthermore, this collapse is accompanied by a buckling of the B(2,1)–B(1,1) dimer. The buckled dimer bond length and buckling height in the z-direction were 1.71 and 0.28 Å, respectively. The change in bonding coordination for the radical surface B atom (i.e. B(2,1) indicates a change in hybridization, from sp^3 to sp^2). This change is most probably due to an electron withdrawal from the less electronegative B(2,1) atom (χ (B) = 2.04) toward the more electronegative second layer N atoms (χ(N) = 3.04). Hence, B(2,1) is electron-deficient. By adopting an sp^2 hybridization, the electron-deficiency of the surface B atoms will be reduced by the orbital-overlap between the unhybridized 2p orbital on B(2,1) and adjacent σ bonds. Such an hyperconjugation interaction has earlier been observed to stabilize a B radical site on the H-terminated B-covered c-BN(111)–(1 × 1) surface [40]. It is also in accordance with the work by Brown *et al* who reported on a hyperconjugation stabilization of a C surface site on the H-terminated diamond (100)–(2 × 1) surface [41].

As can be seen in figure 8.8(a), it should also be noted that some of the B–B dimers have become broken as a result of the geometry optimization of the monoradical H-terminated B-covered c-BN(100) surface. This is most probably due to the high stability of the trigonal planar B. That is, the B element has three valence electrons and, thereby, prefers to bind with three atomic neighbours in a trigonal planar form.

8.3.4.4.2 F-termination

On the other hand, the process of F abstraction from the surface, with gaseous atomic F, has been found to become strongly endothermic with an abstraction energy of +252 kJ mol^{-1} (see table 8.1). The abstraction process has, however, been found to be exothermic if H radicals are used as the abstracting species (with an abstraction energy of −130 kJ mol^{-1}). The explanation to this exothermic process is most probably the nucleophilic properties of the F adsorbates, and thereby their high susceptibility toward attacks by electrophilic radical species (like H). The removal of an F adsorbate from the 100% F-terminated B-covered c-BN (100)–(2 × 1) surface have also resulted in a pronounced, but very local, collapse of the resulting B radical site. This is demonstrated for the F(2,1) adsorbate in figure 6.22(a). The corresponding surface B radical (B(2,1)) has sank down along the z-direction by 0.35 Å, forming a more trigonal planar coordination (ν(N–B (2,1)-N) = 116°). As was the situation with the corresponding H-termination, this type of collapse is accompanied by a buckling of the B(2,1)–B(1,1) dimer. The buckled dimer bond length and buckling height in the z-direction are 1.71 Å and 0.32 Å, respectively. The change in bonding coordination for the radical B surface site also indicates here a change in hybridization, from sp^3 to sp^2, for the B(2,1) atom. As was the situation for H-termination, this is most probably due to (i) an electron withdrawal from the less electronegative B(2,1) atom (with an electronegativity value of 2.04) towards the more electronegative second layer N atoms (3.04), and (ii) a hyperconjugation interaction between the unhybridized 2p orbital on B(2,1) and adjacent σ bonds. This behavior of a monoradical F-terminated B-covered c-BN(100)–(2 × 1) surface is, thus, very similar to the behavior of a monoradical H-terminated B-covered c-BN(100)–(2 × 1) surface.

8.3.4.4.3 Cl-termination

Even though there are repulsive interactions between the adsorbed Cl atoms on the 100% Cl-terminated B-covered c-BN(100)–2 × 1 surface, the B-Cl bonds are strong enough to hinder any atomic desorption of Cl (i.e. the adsorption energy of Cl to this type of surface is positive, +252 kJ mol^{-1} (see table 8.1)). The shortest distance between the Cl atoms on the surface is 2.54 Å, which is too large for a recombinative molecular desorption of Cl_2 to take place. (The calculated and experimentally determined Cl-Cl bond lengths in Cl_2 are 2.02 and 1.99 Å, respectively.) The process of Cl abstraction from the surface, with gaseous atomic Cl, has been calculated to be strongly exothermic (with an abstraction energy of −503 kJ mol^{-1}). The abstraction process is even more exothermic if H radicals are used as abstracting species (ΔE_{abs} = −696 kJ mol^{-1}). This is most probably due to the higher bond strength of the more polar H-Cl bond (+434 kJ mol^{-1}), as compared to the Cl-Cl bond (+241 kJ mol^{-1}). (The experimentally determined bond energies for H-Cl and Cl-Cl, in HCl and Cl_2, are +431 and +242 kJ mol^{-1}, respectively.)

It is usually possible to follow the changes in geometries during the course of the geometry optimization process, and this is the geometrical trajectory obtained from the optimization of the monoradical Cl-terminated B-covered c-BN(100)–2 × 1 surface. As could be seen from this trajectory, the removal of the Cl(2,1) atom from the surface does at first result in a prominent, but local, collapse of the B(2,1) radical surface site. The B(2,1) atom relaxes 0.32 Å down into the c-BN lattice, towards an almost perfect trigonal planar coordination (∠(N–B(2,1)-N) =121°). That is, the B (2,1) atom becomes approximately sp^2 hybridized. This behaviour of the mono-radical Cl-terminated B-covered c-BN(100)–(2 × 1) surface is similar to the behaviours of the corresponding H-, and F-terminated surfaces. However, this specific surface relaxation/reconstruction does not stop here. The electron-deficiency of the surface B(2,1) radical is being eliminated by a donation of the lone pair electrons in Cl(3,1) to the unhybridized 2p orbital in B(2,1), which gives rise to a chloride bridge bond between B(2,1) and B(3,1) (see figure 8.9(a)). Such a high order of coordination is common for halogen atoms. The B–Cl–B bridge formation is tilted 51° from the surface normal and has B–Cl bond lengths of 1.96 Å. Moreover,

Figure 8.9. (a) Monoradical Cl-terminated B-covered c-BN(100)–(2 × 1) surface. (b) Electrostatic potentials for the surface in (a). (c) Deformation density of the N–Cl–N formation in a plane perpendicular to the surface in (a). [34], reproduced with permission © American Chemical Society.

the hybridization of the B(2,1) atom has now changed from sp^2 to sp^3, with a tetrahedral coordination and an elevation of the B(2,1) atom in the z direction (with +0.21 Å). Compared to 100% surface termination with Cl adsorbates, both of the B(2,1) and B(3,1) surface atoms exhibit a small collapse in the *z*-direction (Δz (B(2,1)) = −0.11 Å; Δz(B(3,1)) = −0.09 Å). They thereby form a somewhat distorted tetrahedral coordination (ν(N–B(2,1)-N) = 114°; ν(N–B(3,1)-N) = 114°). Furthermore, the collapse of the B(2,1) and B(3,1) surface atoms is accompanied by an +0.10 Å elevation of B(1,1) and B(4,1) in the z direction, which results in a buckling of the B(1,1)–B(2,1) and B(3,1)–B(4,1) dimer bonds. The average buckled dimer bond length and buckling height in the *z*-direction are 1.76 Å and 0.20 Å, respectively.

It can, thus, be observed that the radical B site is stabilized by an adjacent Cl atom, whereas this doesn't take place for an H- or F-terminated surface. This is most probably due to the longer B–Cl bond length (1.87 Å), as compared with the B–H (1.20 Å) and B–F (1.35 Å) lengths.

8.3.4.5 N-covered c-BN(100) surfaces

8.3.4.5.1 H-termination

Since the surface-terminating H atoms are electrophilic with an atomic charge of +0.12 e (which is in accordance with the lower electronegativity of H (2.20) as compared with N (3.04)), they exhibit a lower susceptibility for an electrophilic H radical attack. This is completely different to the H-terminated B-covered surface, for which the H adsorbates are nucleophilic with an atomic charge of −0.12 e. This agrees well with the calculated H abstraction energy (ΔE_{abs} = −25 kJ mol^{-1}), which indicates that the process of H abstraction is significantly less exothermic than the corresponding process for the B-covered surface (ΔE_{abs} = −258 kJ mol^{-1}). This difference in abstraction energy is also in agreement with the larger strength of the N–H bond (ΔE_{ads} = −431 kJ mol^{-1}), as compared with the B–H bond (ΔE_{ads} = −198 kJ mol^{-1}) (see table 8.1). The less favorable abstraction process for the H-terminated N-covered c-BN(100)–(1 × 1) surface is unfortunate since non-terminated sites are a necessary prerequisite for growth to occur. However, it is generally assumed that weakly endothermic processes at 0 K are plausible ones at elevated temperatures and at the supersaturation condition of H in the CVD growth of c-BN thin films. In other words, the process of H abstraction from the H-terminated N-covered c-BN(100)–(1 × 1) surface is most probable to occur during CVD growth of c-BN.

As can be seen in figure 8.10(a), the removal of the H(2,1) atom from the 100% H-terminated N-covered c-BN(100)–(1 × 1) surface results in a medium-sized collapse of the N(2,1) radical site (with a relax by −0.14 Å in the z-direction). However, the relative collapse is very small since the other surface N atoms also relax downward (with approximately 0.10 Å). This minor collapse of the radical site is most probably due to the fact that the very electronegative surface radical N(2,1) atom is able to withdraw electrons from the second-layer B atoms (with electronegativity values of 3.04 and 2.04, respectively). Thus, the N radical does not need to

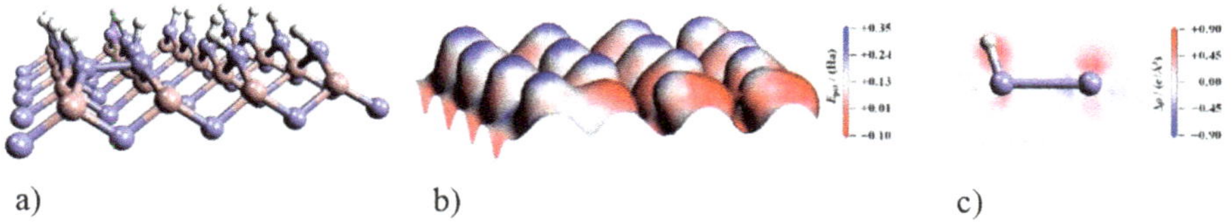

Figure 8.10. (a) A surface radical site on an otherwise H-terminated N-covered c-BN(100)–(2 × 1) surface. (b) Electrostatic potential for the surface in (a). (c) Deformation density of an N–N dimer in a plane perpendicular to the surface in (a). [22], reproduced with permission © American Chemical Society.

relax downward to increase its surrounding electron density. Furthermore, the relaxation of the radical site is accompanied by a dimerization of N(2,1) and N(1,1), with a resulting bond length of 1.58 Å. For comparison, the average N-N dimer bond length on the non-terminated N-covered c-BN(100)–(2 × 1) surface is 1.55 Å [25]. However, the calculated electron deformation density shows that an antibonding orbital on the dimer is occupied, which weakens the dimer bond. Nevertheless, the dimerization has induced a dramatic change in bonding coordination for N(2,1) and N(1,1); going from a trigonal planar coordination for the 100% H-terminated surface (ν(B–N(2,1)–B) = 116°; ν(B–N(1,1)–B) = 116°) to a more tetrahedral coordination for the monoradical H-terminated surface (ν(B–N(2,1)–B) = 108°; ν(B–N(1,1)–B = 109°)). This change in bonding coordination indicates a change in hybridization for the N(2,1) and N(1,1) atoms; from sp^2 to sp^3. The major change in the B–N–B angle for the N(2,1) atom is most probably also due to the electron-rich dangling bond. In order to minimize the electrostatic repulsions between this dangling bond and the bonding electron pairs, the angle between these two types of electron densities will be maximized, which results in a decreased B–N(2,1)–B angle.

There are also indications of electrostatic interactions between N(2,1) and the neighboring H(3,1) adsorbate (see figure 8.10(b)). The calculated atomic charges of N(2,1) and H(3,1) are −0.57e and +0.13e. As can be seen in figure 8.10(a), this results in a significant change in the H-to-surface normal angle for H(3,1) and H (4,1), which are both electrostatically attracted to N(3,1). The calculated atomic charges of N(3,1) and H(4,1) are −0.62e and +0.13e. The H-to-surface normal angle goes from 0° for the 100% H-terminated surface to 48° for the corresponding monoradical surface. This significant change in angle is accompanied by a decrease in surface B–N–B bond angle for N(3,1) and N(4,1), from 116° to 114°. Both of these changes indicate a change in bonding coordination for N(3,1) and N(4,1), from trigonal planar (i.e. a large degree of sp^2 hybridization) for the 100% H-terminated surface to a more tetrahedral coordination (i.e. a large degree of sp^3 hybridization) for the monoradical H-terminated surface. This change in H-to-surface normal angle also makes N(3,1) and N(4,1) highly susceptible toward electrophilic attacks. In summary, it can be concluded that a strong correlation exists between the H adsorption energy and degree of radical surface site collapse (i.e. the endothermic adsorption energy corresponds to a more severe radical collapse).

8.3.4.5.2 F-termination

Even though the F adsorbates on the 100% F-terminated N-covered c-BN(100)–(1 × 1) surface are nucleophilic (with an atomic charge of +0.17 e), the process of F abstraction, with gaseous atomic F, is found to be endothermic (ΔE_{abs} = +63 kJ mol^{-1}). However, the abstraction process from the N-covered surface is less endothermic than the abstraction process for the of B-covered surface (ΔE_{abs} = +252 kJ mol^{-1}). This is in agreement with the smaller bond adsorption energy of F onto the N-covered surface (ΔE_{ads} = −283 kJ mol^{-1}) as compared with the adsorption energy of F for the binding to the B-covered surface (ΔE_{ads} = −471 kJ mol^{-1}). For the B-terminated surface, the abstraction process was found to be significantly more favorable if H radicals were used as abstracting species (ΔE_{abs} = −318 kJ mol^{-1}) (see table 8.1). Similar results were obtained for the process of H abstraction from the 100% H-terminated N-covered c-BN(100)–(1 × 1) surface. The abstraction with gasous F (ΔE_{abs} = −170 kJ mol^{-1}) was found to be significantly more exothermic than abstraction with gaseous H (ΔE_{abs} = −25 kJ mol^{-1}). This is most probably due to the significantly more ionic character of the H-F bond (as compared with the H-H and F-F bonds). The difference in electronegativity values for the H-F bond is 1.78, while the electronegativity differences for H-H or F-F is zero. The unequal sharing of electrons in the H-F bond makes the F end of the bond partially negative (q(F) = −0.35 e) and the H end partially positive (qH = +0.35 e) (see figure 6.24(b)). Thus, the attraction between these partial charges increases the energy required to break the bond. On the other hand, the F–F bond is weakened by strong repulsions between the non-bonding electrons on the F atoms. The experimentally determined bond energies for H–F, H–H, and F–F in molecular HF, H_2, and F_2 are +570, +436, and +159 kJ mol^{-1}, respectively. The exchange of the H-H, or F-F, bond with the H-F bond will, thereby, provide a large energy gain in the gas phase abstraction of F, or H. These observations indicate that the use of only F as the surface stabilizing species is highly unfavorable for the growth of c-BN (100). A much better choice is, though, to only use H. However, the best alternative is to use a mixture of H and F. Such a mixture of gases will make it possible to grow c-BN exclusively through surface reactions (i.e. there is no need for surface damaging ion bombardment). This allows the production of electronic grade, single-crystal c-BN wafers.

However, it has been experimentally observed that energetic particle bombardment is a necessity for the occurrence of F abstraction from the B-terminated c-BN (100) surface [37]. As can be seen in figure 6.24(a), the removal of the F(2,1) atom from the 100% F-terminated N-covered c-BN(100)–(1 × 1) surface results in a medium-sized collapse of the N(2,1) radical site (i.e. the radical site relaxes 0.16 Å downwards into the lattice). However, the relative collapse is very small since the other surface N atoms relax downward with approximately 0.11 Å. Furthermore, the relaxation of the radical site is accompanied by a dimerization of the surface N (2,1) and N(1,1) atoms, with a bond length of 1.45 Å. The dimerization induces a dramatic change in bonding coordination for N(2,1) and N(1,1), going from a trigonal planar coordination for the 100% F-terminated surface (ν(B–N(2,1)-B) = 117°; ν(B–N(1,1)-B) = 117°) to a more tetrahedral coordination for the monoradical F-terminated surface (ν(B–N(2,1)-B) = 109°; ν(B–N(1,1)-B) = 107°). There are also indications of interactions between N(2,1) and the neighboring F(3,1) adsorbate, which results in a significant change in F-to-surface normal angle for F (3,1) (i.e. going from 0° for the 100% F-terminated surface to 30° for the monoradical F-terminated surface). Hence, the behaviour of the monoradical F-terminated N-covered c-BN(100)–(1 × 1) surface is very similar to the behaviour of the monoradical H-terminated N-covered c-BN(100)–(1 × 1) surface. This is most probably due to (i) the electron withdrawal from the second layer B atoms towards the surface radical N(2,1) atom, and (ii) the electron-rich dangling bond on N(2,1). (The electronegativity values of B and N are 2.04 and 3.04, respectively, and the resulting charge differences upon electron transfer are +0.69 e and −0.43 e, respectively.)

8.3.4.6 Summary

Different abstraction reactions have been presented in the present chapter. As can be seen in table 8.1, the energetically most favorable abstraction reactions involve surface-terminating H species on the B-covered c-BN(100)–2 × 1 surface. An abstraction of these H adsorbates by gaseous F has, thus, been found to be the most exothermic process (−403 kJ mol^{-1}). This is to be compared with the endothermic abstraction of surface-terminating F species with gaseous F (+252 kJ mol^{-1}). Hence, the energetically least favorable abstraction reaction includes no hydrogens. Furthermore, the abstraction energy for an H-terminated N-covered c-BN(100) surface, using gaseous H, was found to be slightly exothermic (−25 kJ mol^{-1}), while it was not possible at all to abstract these H adsorbates by using gaseous F. Moreover, the abstraction reactions involving both H and F (either as an adsorbate or as a gaseous species) were all found to be endothermic for the c-BN (111) surface (see table 8.1).

However, there are some very weakly endothermic processes for both the B- and N-covered c-BN(111) surface: (i) abstraction of an H adsorbate with gaseous H (B-covered), and (ii) abstraction of an F adsorbate with gaseous H (N-covered). The abstraction energies are for these processes +18 and +28 kJ mol^{-1}, respectively. It must be stressed, though, that these very weak abstraction energies are calculated at 0 K, and might, thus, occur at higher temperatures and/or at a supersaturation condition of gaseous species in the CVD reaction chamber. Based on these energetic investigations, it is possible to draw the conclusion that H-terminated c-BN(100) surfaces, both B-covered and N-covered, are the energetically most favorable ones for gas phase abstraction reactions including gaseous F (or H) and H species, respectively. It is then recommended to use the diamond (100) surface as a substrate for the heteroepitaxial growth of c-BN(100), and to use a gas phase that contains only H (i.e. no F species). These calculations have, thus, turned out to be highly valuable for further experimental activities in the strives to grow phase-pure c-BN by using CVD.

However, to be certain about the conclusions drawn from the thermodynamic investigation above, one has to complement them with corresponding kinetic studies. Since Cl-termination, with its corresponding abstraction reactions, was found to be not that effective, we will in the next chapter continue studying the kinetics of the abstraction reactions for only H- and F-terminated surfaces. Both H- and F-terminated surfaces are then assumed to undergo abstraction reactions including gaseous H, or F, species. The surface plane that will be used for this type of study is the c-BN (100) surface. However, similar calculations can easily also be done for the c-BN(111) and (110) surfaces.

8.3.5 Kinetics of gas phase abstraction reactions

What is most important for a complete understanding of the abstraction processes is to perform a kinetic study, where both the geometrical structure and the total energy of the system are carefully followed during the approach of H (or F) to the surface-binding H (or F). The results presented here do strongly emphasize the importance with kinetic calculations when studying the abstraction processes. The results presented above (see section 8.4.4) were obtained from solely thermodynamic calculations where only the reaction energies were calculated. On the contrary, kinetic calculations involve the location of transition states and the calculations of barrier energies. The methods for the calculations of barrier energies are presented in section 4.5.3.

The kinetics of the gas phase abstraction of H- versus F-adsorbates from the c-BN (100) surface have already been presented in great detail in the present book (see section 6.4.2). As a result from these studies, the following similarities with the thermodynamic results can be observed:

(i) It is not possible to chemically abstract a surface-binding F, by using gaseous F radicals, from neither the N- or B-covered c-BN(100) surface. However, the abstraction of F from the B-covered surface, with gaseous H radicals, is found to be highly probable from both a thermodynamic and kinetic point of view. This is also, in a sense, the situation for the F-terminated N-covered surface (with just a minor energy barrier of +8 kJ mol^{-1}).

(ii) It is shown to be possible to chemically abstract H from the surfaces (B- or N-covered) by using gaseous H. However, a minor energy barrier is observed for the N-covered surface (+13 kJ mol^{-1}), which is not expected to be of any major issue at realistic CVD deposition temperatures.

(iii) The major difference between the thermodynamic and kinetical studies concerns the abstraction of H from the B- or N-covered c-BN(100) surface, using gaseous F. The thermodynamic study (chapter 10) shows a clear exothermic reaction for both types of surface coverage (N; −403 kJ mol^{-1} versus B; −170 kJ mol^{-1}). On the other hand, the kinetic study shows that neither of these will ever take place. The underlying reasons for these observations are twofold. At first, the incoming F species does not react with the surface-binding H species on the B-covered surface. It is, instead, binding to the surface B atom and, thereby, introducing a local surface restructuring. Secondly, the incoming F species reacts with the surface-binding H species on the N-covered surface, forming a more complex adsorbate with an enhanced difficulty to desorb from the surface.

8.4 Adsorption of growth species on the c-BN surface

8.4.1 Introduction

As presented in section 8.1, there are at present severe problems in the synthesis of phase pure c-BN. Compared to diamond, which can be easily grown by a HPHT catalytic conversion of graphite to homogeneous single diamond crystals of up to cm^3 in size, the largest c-BN single crystals reported so far are of a few mm^3 in size [43]. The difference is even more striking when considering the vapour growth at low pressure. Compared to diamond (which can be relatively easily grown as thick polycrystalline films containing mm sized crystallites), the c-BN obtained at low pressure has always become a mixture with other polymorphic forms of BN.

Physical vapour deposition (PVD) has often been used in the strives to grow phase-pure c-BN. Hence, the aim has been to deposit the c-BN phase by using 'brute force' under the influence of surface bombardment with highly energetic ions of nitrogen and argon. 'Brute force' has also been found to be necessary for successful CVD routes [5–12]. By using these methods, c-BN growth takes place by a mechanism in which the structural changes are accomplished by several features such as high compressive stress of several GPa, and growth of a layered structure consisting of an amorphous (a-BN) interface layer (with a thickness in the order of the ion range), followed by textured sp^2-bonded BN (with the *c*-axis parallel to the substrate) and a final layer of c-BN, which starts to nucleate on top of the hexagonal nucleation layer.

It has turned out that the parameter windows for c-BN nucleation and growth (i.e. the continuation of growth on a c-BN nucleus) are totally different [13, 14], with a decisive parameter threshold for c-BN growth being lower than for nucleation. It has experimentally [15, 16] been shown that the impact of energetic ions, during the growth of BN films, is predominantly three-fold. It generates a large amount of stresses and defects, and it favours the growth of the cubic phase relative to the hexagonal one. Moreover, the induced high film stresses limit the thicknesses of the films. The generation of defects will in addition give poor crystalline quality, small grain size, and high density of defects and grain boundaries. Accordingly, a delamination from the substrates will occur which can be attributed to these high compressive stresses.

The c-BN properties described in section 8.1 make c-BN an extremely promising multifunctional material, which can be tailored for a very large range of advanced mechanical, tribological, thermal, electronic, and optical applications. However, to exploit these applications, a successful route for large area CVD of c-BN films is required. Thus, the possibility for a chemical route c-BN growth by using more gentle CVD techniques, where the bombarding ions will not penetrate the surface and thus not detrimentally destroy the BN structure, is then needed.

Common to all ordinary CVD methods are dilute mixtures of simple growth species in an excess of a reactive surface terminating species. As discussed in section 6.3, H and F have generally been considered as very important reactants for diamond growth since one of its dominant roles during growth is to maintain the sp^3 configuration of the surface carbon atoms. As stated above, this is also the situation for the c-BN surfaces. Thus, the role of H and/or F species is to maintain the sp^3 hybridization of the c-BN surface atoms by completely hydrogenate, or fluorinate, the growing surface. However, surface radical sites will appear during the growth process as a result of atomic H, or F, abstraction reactions. Since the gas-phase is saturated with H and/or F species, such abstractions are usually followed by the recapture of an H, or F, atom [44]. However, a surface radical site is also expected to attract a B- or N-containing growth species. It must be stressed here that the cubic BN growth is more complicated than the growth of diamond, which is due to the binary nature of c-BN material. This circumstance has been shown to lead to severe restrictions in the thin film growth of phase-pure c-BN. The main demand for the growth of c-BN is that a B-containing species is chemisorbed to the N surface site, while the N-containing species are chemisorbed to the B surface sites [37]. This is the situation for all c-BN planes. This situation is not that severe for the c-BN (111) and (100) surface planes, where a layer-by-layer growth of B- versus N-containing growth can take place. However, the situation is more severe for the c-BN (110) surface, which consists of a mixture of B and C atoms. (The dominant c-BN surfaces under CVD conditions are the (100), (110), and (111) surfaces.)

Mixtures of a boron-containing precursor and N_2 (or NH_3) are generally used for c-BN deposition [45]. Diborane, B_2H_6, has been widely used as the boron-containing precursor. Other examples include BCl_3 [30, 46] and BF_3 [47]. The $BF_3/N_2/H_2$ gas mixture was successfully used to deposit B_xN_x films under CVD conditions in a work by Hentschel *et al* [48]. Single-source precursors have also been used [45]. One major difference between a single-source precursor and, e.g. NX_3 (X=H, F) is that the former species contains both B and N species, being a growth species for growth of two layers simultaneously. When using NX_3 as a growth species, it is only possible to deposit an N(111) surface. A boron-containing species must also be introduced into the CVD reactor in order to deposit a second layer of B(111).

The adsorption of H- and F-containing growth species on H- or F-terminated (111) surfaces of diamond have been studied earlier in a series of investigations [42, 49–53]. It has also been assumed that CH_3 and the C_2H fragment of acetylene are the major contributing precursors in the growth of diamond [54]. It is, thus, of a large interest to also study the effect of the corresponding B- and N-containing species on the growth of c-BN(111). In addition to H and F, Br will then be used as ligands in some of the growth precursors. Also, we will assume here that we are using CVD techniques, and this means that we mix the B- and N-containing growth precursors in the reactor gas. These precursors include BH_3(g), BF_3(g), or BBr_3(g), together with NH_3, or NF_3(g). The HBNH(g) precursor will also be tested. Other growth precursors might also be of great interest to study, and this can easily be done by following the methodology used for the examples presented here.

An equilibrium condition is not expected in the CVD reaction chamber, which is mainly due to the prominent excitations of growth precursors in the gas mixture (see chapter 2). Most of these molecules are, thereby, expected to become decomposed to various extents. Consequently, some of these fragments are expected to approach the growing thin film surface. Thus, the following species has been theoretically tested here as growth species for the deposition of phase-pure c-BN materials: B, BH, BH_2, BH_3, BF, BF_2, BF_3, BBr, BBr_2, BBr_3, N, NH, NH_2, NH_3, NF, NF_2, and NF_3. In addition, the non-fragmented precursors and the terminating species H, F, and Br have been included in the list of plausible gaseous adsorbates (the former as growth precursors).

8.4.2 Adsorption of growth species onto the B-covered c-BN(111) surface

As presented in section 6.6, it has been shown that CH_2 is efficient as a surface species in the growth of diamond. It binds very strongly to the diamond surface with an adsorption energy of -364 kJ mol^{-1}. It is then of a large interest to also study the corresponding influence of BH_2 and NH_2 in the growth of c-BN. Moreover, it was also shown in section 6.4.2 that a mixture of F and H species in the gas phase will result in energetically favorable abstraction reactions on the c-BN surface. Thus, it is also of great interest to study the effect of F (or H) on the adsorption reactions during thin film growth of c-BN. Let us first study the possibility for c-BN growth when BH_2, BF_2, NH_2 and NF_2. Consider the absolute values of the adsorption energies (as has been done earlier in the book): NH_2 (H) NF_2 (H) >NH_2 (F) >NF_2 (F). The parentheses indicate the type of surface termination. The NH_2 species adsorbed onto an H-terminated surface is, in fact, the most strongly adsorbed one (-459 kJ mol^{-1}). It is even somewhat more strongly bonded to the surface than the H species (-442 kJ mol^{-1}). On the contrary, the NF_2 species adsorbed onto an F-terminated surface is found to be the weakest adsorbed one (-17 kJ mol^{-1}). One reason for this very small adsorption energy is most probably the sterical hindrances induced by the larger F atoms (both in NF_2 and on the surface) (see figure 8.11).

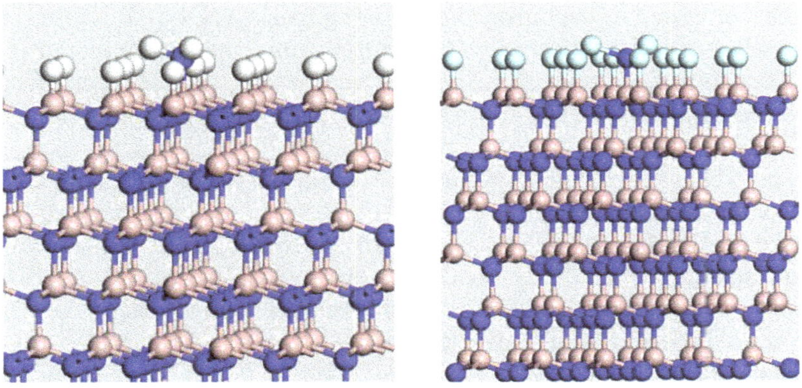

a) b)

Figure 8.11. Geometrical structure of an (a) NH_2 and (b) NF_2 adsorbate on an otherwise H- and F-terminated B-covered c-BN(111) surface, respectively.

Table 8.10. Adsorption energies for the chemisorption of NX_2, or BX_2, species to a monoradical site on otherwise completely terminated H-, or F-terminated B-covered c-BN(100) surfaces. X=H or F. The unit is kJ mol^{-1}.

B-covered c-BN (111) surface	H-termination	F-termination
NH_2	−459	−201
NF_2	−201	−17
BH_2	−394	−353
BF_2	−412	−327

As can be seen in table 8.10, the bonding situation, in which gaseous species BX_2 is bonded to the B-covered c-BN(111) surface, results in a much smaller range of adsorption energies as compared with the corresponding adsorptions of NX_2 species (−412 to −327 kJ mol). The following order of adsorption energies is obtained for the BX_2 species: BF_2 (H) < BH_2 (H) < BH_2 (F) < BF_2 (F) with the values (−412 < −394 > −353 < −327 kJ mol^{-1}); and for the absolute value of the adsorption energies: BF_2 (H) > BH_2 (H) > BH_2 (F) > BF_2 (F). As is clear from table 8.9, the adsorption of BX_2 to an F-terminated surface results in much larger bond strengths, as compared to the adsorption of NX_2 to an F-terminated surface. On the contrary, the adsorption of BX_2 to an H-terminated surface results in somewhat smaller adsorption energies, when being compared to the adsorption of NX_2 to an H-terminated surface. As a consequence, a mixture of boron- (BX_2) and nitrogen- (NX_2) containing species in the gas phase (during chemical vapor deposition of c-BN thin films) will most probably result in a mixture of these species on the B-covered c-BN(111) surface as well. This is an unfavorable result since only nitrogen-containing species are expected growth precursors on this type of surface. The situation is even worse for an F-terminated B-covered surface, for which the boron-containing species (BX_2) is found to be energetically much more strongly bonded to the surface (compared to the nitrogen-containing species (NX_2)). It is also interesting to observe that the adsorption of NH_2 and BH_2, onto the H-terminated B-covered c-BN(111) surface, corresponds to adsorption energies that are numerically very similar to the adsorption of CH_2 onto an H-terminated diamond (111) surface: −459 (NH_2), −394 (BH_2), and −416 (diamond) kJ mol^{-1}.

Let us now compare the adsorption of NH_2 and BH_2 with the adsorption of NF_2 and BF_2 to a B-covered c-BN(111) surface (i.e. to study the importance of ligands in the molecular adsorbates). As can be seen in figure 8.12, the NH_2 species adsorbs somewhat stronger than BH_2 (with an energy difference of 63 kJ mol^{-1}). The opposite is, however, the case for the fluorinated growth species (i.e. BF_2 binds stronger than NF_2 to the B-covered surface). It might be tempting to use the strongest binding growth species for, e.g. the H-terminated B-covered surface, which are the BF_2 and NH_2 species. However, the difference in adsorption energy for these species is only 47 kJ mol^{-1}. Hence, we will then most probably end up with a B-covered surface that is terminated with a mixture of BF_2 and NH_2 adsorbates.

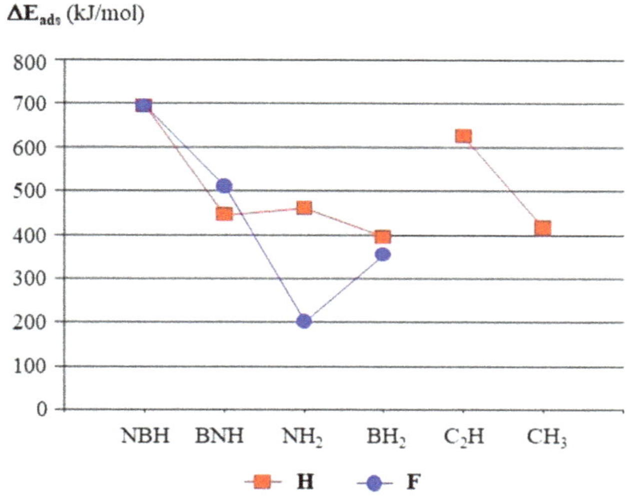

Figure 8.12. Calculated adsorption energies for plausible H- and F-terminated growth species to an H- and F-terminated B-covered c-BN (111) surface, respectively. The adsorption energies are given in absolute values. Red: H is present as both ligands and terminating species. Blue: F is present as both ligands and terminating species (and is, thus, replacing H in the adsorbates given in the figure). [53], reproduced with permission © Elsevier.

This is also the situation if one uses a mixture of BH_2 and NH_2 species in the gas phase. The difference in adsorption energy is only 63 kJ mol^{-1} for these species, which is not big enough to preclude any co-existence of BH_2 and NH_2 on the B-covered c-BN(111) surface.

The possibilities for co-adsorption of B- and N-containing species onto the B-covered c-BN(111) surface can, however, be eliminated by using an atomic layer deposition (ALD) technique instead of ordinary CVD (i.e. since the latter method uses a mixture of growth precursors). One instead has to use an ALD process with alternating B- and N-containing gas pulses. More specifically, one has to look for a preferential adsorption of N-containing adsorbates to a B-covered c-BN(111) surface, and the adsorption of B-containing adsorbates onto the corresponding N-covered surface.

Another option would be to use a growth precursor that includes both B and N atoms; like HBNH. Plausible fragments are then the linear radicals HBN and BNH. As shown in table 8.10 and figure 8.12, the adsorption energy of HBN (with N binding to $B_{surface}$), when attached to the H-terminated B-covered c-BN(111) surface, is 250 kJ mol^{-1} larger than the corresponding one for NH_2 (in absolute values). What is most interesting is the observation that there is a large energetical preference for adsorption of NBH to the B-covered c-BN surface (with a difference of about 270 kJ mol^{-1} with respect to BNH- with B binding to $B_{surface}$). Hence, there is just a minor probability that one also finds HNB (with B binding to $B_{surface}$) adsorbates (in addition to HBN) on the B-covered c-BN(111) surface. One major drawback with the HNBH precursor is, however, that it is highly explosive so one must find another alternative.

The possibility for CVD growth of phase-pure c-BN films can also be investigated by using other types of growth precursors, with their fragments. If possible, one should use as much information as possible from correlating experimental studies. This is, in fact, what has been done in the next example. Attempts were made to experimentally to grow c-BN by using laser-assisted ALD and alternate gas pulses with BBr_3 and NH_3, with Ar purge pulses in between [55], and optical emission spectroscopy (OES) analysis of the gas directly above the growing film showed the presence of NH, B, and BBr species. Since a laser-induced fragmentation of the initial NH_3 and BBr_3 growth precursors had taken place in the different gas pulses, it was also possible that the N, NH_2, BBr_2, H, and/or H_2 species could be found above the growing film. This is also the situation with the non-fragmented BBr_3 and NH_3 species. Unfortunately, no c-BN was deposited in this study. Instead, a smooth nano-crystalline turbostratic film was formed, which was stoichiometric with respect to B and N. Theoretical DFT calculations were, therefore, made to explain these findings. A c-BN(111) surface (B- versus N-covered) was used for these calculations, in addition to the following growth species/termination adsorbates; BBr_3, BBr_2, BBr, B, NH_3, NH_2, NH, N, H_2, H, and Br. The gas pulse for the growth of the N layer should include NH_3, NH_2, NH, N, H_2, and/or H species, and the growth of the B layer should include BBr_3, BBr_2, BBr, B, and/or Br. Thus, the N-covered surface is expected to be H-terminated, and the B-covered one is assumed to be Br-terminated. To highlight the necessity to use an ALD technique, instead of CVD (i.e. with mixed B- and N-containing precursors), all B- and N-containing species have been allowed to adsorb on both the B- and N-covered surfaces of c-BN(111).

The adsorption energies of these gaseous species will at first be presented for the B-covered c-BN(111) surface. As can be seen in table 8.11 and figure 8.13, the more

Table 8.11. Adsorption energies for the B- versus N-covered c-BN(111) surface ($kJ\ mol^{-1}$).

Adsorbate	Adsorption energy B-covered ($kJ\ mol^{-1}$)	Adsorption energy N-covered ($kJ\ mol^{-1}$)
N	−515	−622
NH	−426	−475
NH_2	−451	−295
NH_3	−236	+30
H	−437	−465
B	−368	−531
BBr	−331	−307
BBr_2	−338	−479
BBr_3	+96	−241
Br	−436	−392

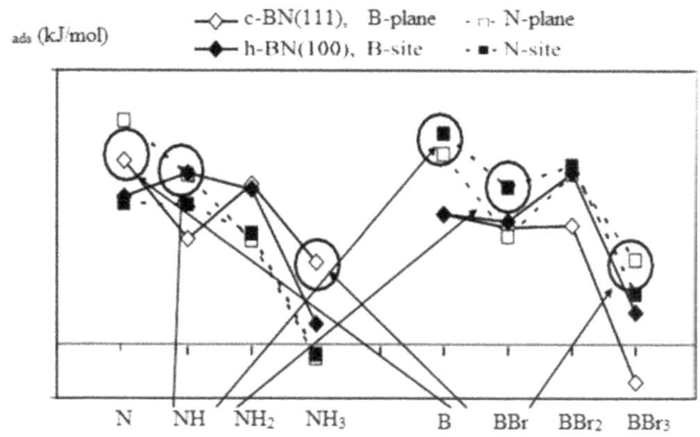

Figure 8.13. Calculated adsorption energies for adsorption of various B- and N-containing species to the B and N edge atoms of h-BN versus the B and N atoms on the (111) surfaces of c-BN. Absolute values of the adsorption energies are presented in kJ mol^{-1}. The arrows (with corresponding circles) show the points at the situations where the respective growth species gives either a cubic or hexagonal growth of BN. [53], reproduced with permission © Elsevier.

Table 8.12. Angles and distances for the adsorbates on the c-BN(111) surfaces.

Adsorbate	B-covered c-BN(111) surface		N-covered c-BN(111) surface	
	$r_{surface-adsorbate}$(Å)	$\nu_{N-B(surface)-N}$(°)	$r_{surface-adsorbate}$(Å)	$\nu_{B-N(surface)-B}$(°)
N	1.32	104.8	1.32	103.8
NH	1.41	106.0	1.44	106.6
NH_2	1.45	106.0	1.48	107.5
NH_3	1.62	109.6	3.01	111.3
H	1.20	107.7	1.03	108.3
B	1.76	108.7	1.35	104.4
BBr	1.73	108.0	1.40	104.3
BBr_2	1.86	104.7	1.41	100.9
BBr_3	1.96	103.6	1.61	101.9
Br	2.01	107.7	2.31	112.1

dissociated precursor, the stronger it adsorbs. The numerical order of the calculated adsorption energies is: N (−515) < NH_2 (−451) < H (−437) ~ Br (−436) < NH (−426) < B (−368) < BBr_2 (−338) ~ BBr (−331) < NH_3 (−236) < BBr (+96). The numbers within parentheses show the calculated adsorption possibility to uphold the sp^3-hybridization of the surface B atoms. The ν(N–$B_{surface}$–N) angles are all between 103.68° and 109.68°. These angles, together with the corresponding surface-to-adsorbate distances, are shown in table 8.12. As presented and discussed above (see section 8.4.4), a non-terminated surface B site will more or less completely collapse to a ν(N–$B_{surface}$–N) angle of 117.98° (i.e. to an almost completely sp^2-hybridized surface radical atom). Hence, the surface B atom, with its three valence electrons, forms a triangular coordination polyhedra, as predicted by the valence shell electron-pair repulsion theory.

8.4.3 Adsorption of growth species onto the N-covered c-BN(111) surface

As was the situation in section 8.5.2, the same gaseous species have been allowed here to adsorb to the B-covered c-B(111) surface. The results obtained for these species indicate that the more dissociated the precursor adsorbs the strongest to the N(111) surface of c-BN (see table 8.11 and figure 8.13). The numerical order of the calculated adsorption energies is:

N (−622) < B (−531) < BBr_2 (−484) < NH (−475) < H (−465) < Br (−392) < BBr (−307) < NH_2 (−295) < BBr (−241) < NH (+30).

With one exception, the adsorption energies were all found to be strongly exothermic. The exception is the gaseous NH species that was found to not have any possibility to bind to the N-covered c-BN(111) surface (i.e. the adsorption reaction was endothermic). Moreover, the corresponding ν(B–$N_{surface}$–B) angles were, with one exception, found to be in the region 101.98°–107.58°. These adsorption species are, thus, very promising for the stabilization of the N-covered c-BN(111) surface. The exception is the adsorption of Br species, which results in a slightly sp^2-hybridized surface N atom (i.e. ν(B–$N_{surface}$–B) = 112.1°).

On the contrary to the situation with the B-covered c-BN(111) surface, the abstraction of a surface-terminating H species from the N-covered surface results in an almost maintained sp^3-hybridization of the surface N atom (see table 8.9). The ν (B–$N_{surface}$–B) angle for the non-terminated N atom is 112.9°. It is, thus, very likely that the cubic phase of the N-covered c-BN(111) surface will be sustained even for the situation with a not perfectly terminated surface. The nitrogen atom has five valence electrons, and the valence shell of the surface N atoms will be filled by forming bonds to three B atoms in atomic layer 2. It is, hence, favorable for the N atom to maintain the sp^3-hybridization and, thereby, a tetragonal coordination according to the valence shell electron-pair repulsion theory.

8.4.4 Adsorption of growth species onto the c-BN(110) surface

The adsorption energies for H and NH_x ($x = 1$–3), chemisorbed to B and N surface sites on c-BN(110), have also been theoretically investigated earlier [33]. As can be seen in figure 8.14, it is obvious that both NH and NH_2 strongly favour the B surface site. Another observation is that both NH and NH_2 bind somewhat more strongly than H to the B surface sites. This feature is very important since the incoming growth species must be able to replace a terminating H atom for growth to occur. As can further be seen in figure 8.14, the gaseous NH_3 species does not have any tendency to bind to either of the surface sites on c-BN(110) (i.e. B or N). Hence, a decomposition of NH_3 in the CVD reactor is very crucial for adsorption and, thereby, growth to occur. As discussed in chapter 2, this can experimentally be

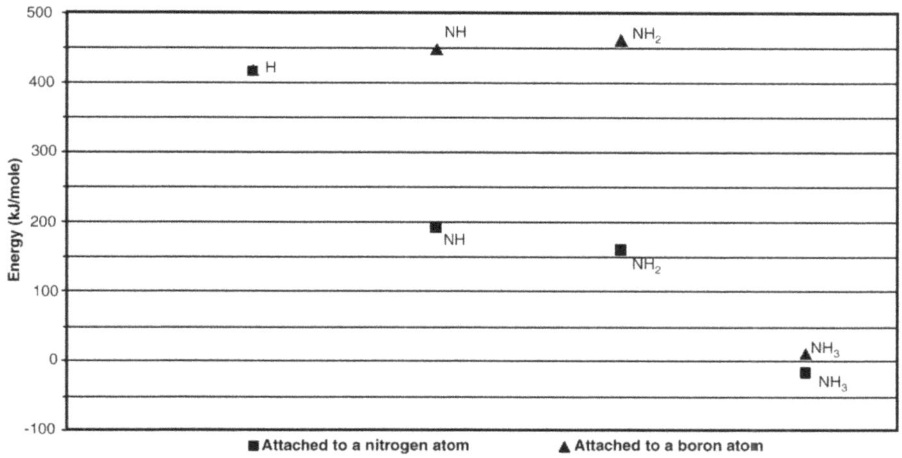

Figure 8.14. Calculated adsorption energies for plausible N-containing growth species, adsorbed to B surface sites on an H-terminated c-BN (110) surface. Presented as absolute values in kJ mol^{-1}. [34], reproduced with permission © Elsevier.

achieved by using various means of gas phase activation (e.g. thermally, hot filament, laser or plasma activation).

When comparing the adsorption energies for NH_3 chemisorbed to (110) and (111) surfaces, it is clear that the non-decomposed NH_3 species adsorbs easier to a B site on the (111) surface (−236 kJ mol^{-1} compared to −12 kJ mol^{-1} for a corresponding site on the (110) surface). (The adsorption energies are presented as absolute values in figures 8.13 and 8.14.) This circumstance implies that it is more important with decomposition for growth of (110), as decomposed to NH, NH_2 and BBr, these fragments will experience larger binding energies to the (110) surface (compared to (111)).

The corresponding adsorption energies for H and BH_x (x = 1–3), to N surface sites on c-BN(110), are shown in figure 8.15. As shown in this figure, some obvious trends can be drawn when considering the relative adsorption energies. The B-containing adsorbate with the largest number of H ligands (i.e. BH_3) is found to have no tendency to bind to either of the surface B- or N-sites on c-BN(110). As shown in figure 8.14, this was also the situation for NH_3. Another observation is that there is only one adsorbate, BH_2, that energetically favours a binding to a surface N site, and, hence, creates a B–N bond. The difference in adsorption energies for surface B and N sites is 110 kJ mol^{-1} (−310 (B) versus −420 (N) kJ mol^{-1}). The conclusion that can be drawn from figure 8.15 is that a gas mixture including BH will most probably result in BH species adsorbed to N surface sites, and with a preference for BH on the corresponding B sites. The only type of B-containing adsorbate that has the possibility to compete with H (considering adsorption reactions on surface N sites) is BH_2. The corresponding adsorption energies are −416 eV (H) and −420 eV (BH_2). As presented above, gaseous BH_2 does experience a rather strong discrimination between the various surface sites (in favour of the N

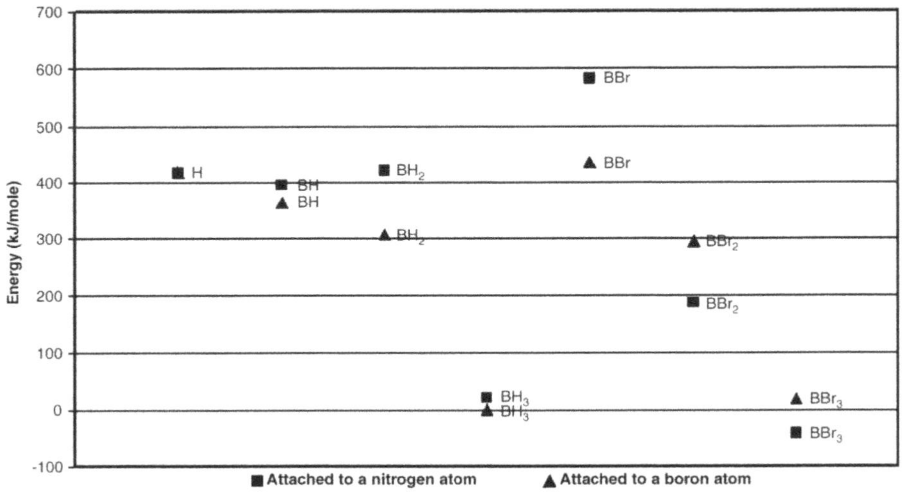

Figure 8.15. Calculated adsorption energies for plausible B-containing growth species to N surface sites on an H-terminated c-BN (110) surface. Presented as absolute values in kJ mol^{-1}. [34], reproduced with permission © Elsevier.

site). In addition, since it has an adsorption energy of a similar strength compared to an H adsorbate, it cannot be ruled out as a plausible contributor to the c-BN growth of the (110) plane.

Several striking observations are made when studying figures 8.14 and 8.15. First of all, non-decomposed BH_3 and NH_3 species do not have any tendency to become chemically bonded to the cBN(110) surface (which strongly supports the need for a gaseous phase activation and/or a use of controlled chemical gas phase reactions). The plausible H-containing growth species proposed here (BH_2, NH, NH_2) are all within a specific range of adsorption energies (−420 to −460 kJ mol^{-1}), being somewhat larger than for the corresponding H species (−416 kJ mol^{-1}). However, other species than the plausible growth species also bind very strongly to the surface B sites. This is especially the situation with the adsorption of the B-containing adsorbates BH and BH_2, for which the adsorption energies are −364 and −306 kJ mol^{-1}. These species will, hence, compete with our preferred N-containing growth species (if present in the gas mixture during the film deposition of c-BN). This problem can, however, be circumvented by using an ALD method instead of ordinary CVD.

8.4.5 Adsorption of growth species onto the c-BN(100) surface

8.4.5.1 Introduction

As shown in sections 8.5.2 and 8.5.3, the BH_x and NH_x species ($x = 0, 1, 2, 3$) have been found to be promising as ALD growth species for the stoichiometric (110) surface and for the B- or N-covered c-BN(111) surface. We should, therefore, also study the possibility for BH_x, BF_x, and NH_x to act as growth species in the vapour phase growth of B- and N-covered c-BN(100) surfaces. Let us then assume that the surfaces are either H or F-terminated. Interestingly, the (100) surface of c-BN is then found to behave somewhat differently compared to (111) and (110). Namely, the (100) surface shows characteristically different reconstructions when exposed to H or F species; the 100% H- or F-terminated B-covered c-BN(100) surface will be (2×1)-reconstructed, while the 100% H- or F-terminated N-covered c-BN(100) surface will be (1 × 1)-reconstructed (see section 8.4.3).

8.4.5.2 Adsorption of BH_x or BF_x species onto an H-terminated c-BN(100) surface

The resulting BH_x and BF_x ($x = 0, 1, 2, 3$) adsorption structures, for chemisorption onto a monoradical surface site on the H-terminated N-covered c-BN(100)–(1 × 1) surface, are shown in figures 8.16 and 8.17. The corresponding adsorption energies

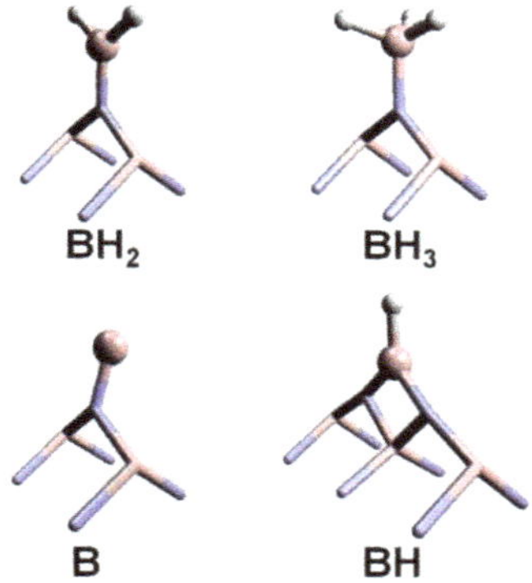

Figure 8.16. The BH_x ($x = 0, 1, 2, 3$) adsorption structures for chemisorption onto a monoradical surface site on the H-terminated N-covered c-BN(100) surface (balls = adsorbate species; sticks = surface atoms). [56], reproduced with permission © American Chemical Society.

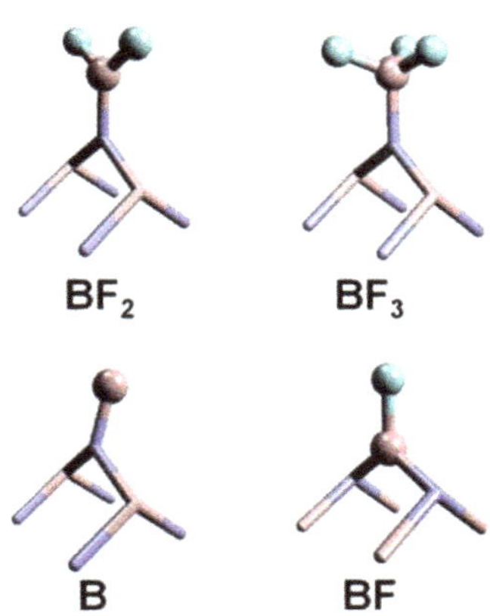

Figure 8.17. The BF_x ($x = 0, 1, 2, 3$) adsorption structures for chemisorption onto a monoradical surface site on the H-terminated N-covered c-BN(100) surface (balls = adsorbate species; sticks = surface atoms). [56], reproduced with permission © American Chemical Society.

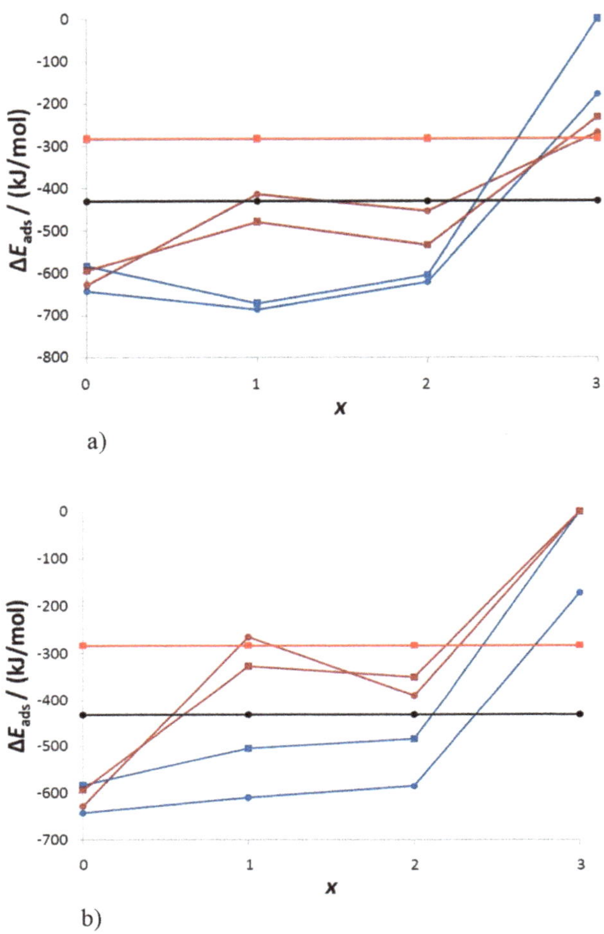

Figure 8.18. Adsorption energies for the (a) BH_x and (b) BF_x species ($x = 0, 1, 2, 3$) on a monoradical surface site on the H- (blue ▲) or F-terminated (blue ■) N-covered c-BN(100)–(1 × 1) surface, and the (b) H- (brown ▲) or F-terminated (brown ■) B-covered c-BN(100)–(2 × 1) surface. The corresponding adsorption energies for H (black ▲) and F (red ■) are also shown. (Points at 0 kJ mol^{-1} do not represent adsorption energies, but indicate that the species did not chemisorb, or became dissociatively chemisorbed, onto the surface.) [56], reproduced with permission © American Chemical Society.

for both H-terminated N-covered c-BN(100)–(1 × 1) surface, and the H-terminated B-covered c-BN(100)–(2 × 1) surface, are shown in figure 8.18 and table 8.13. The following orders of adsorption energies (in absolute values) are: BH (−686) < B (−642) < BH_2 (−621) < BF (−609) < BF_2 (−585) < BH_3 (−179) ~ BF_3 (−172) kJ mol^{-1} for the H-terminated N-covered surface. Moreover, the order of adsorption energies for the corresponding B-covered surface is: B (−627) < BH_2 (−454) < BH (−414) < BF_2 (−390) < BF (−264) kJ mol^{-1}. The adsorption process of BF_3 is endothermic, which means that this species does not chemisorb to the H-terminated B-covered surface. As can be understood from these results, in principle all BH_x and

Table 8.13. Adsorption energies for the chemisorption of BH_x, BF_x, and NH_x species (x = 0, 1, 2, 3) onto a monoradical surface site on the H- or F-terminated N-covered.

ΔE_{ads}/(kJ mol^{-1})

	N-termination		B-termination	
	Gas-phase		Gas-phase	
Growth species	H-covered	F-covered	H-covered	F-covered
B	−642	−584	−627	−594
BH	−686	−671	−414	−479
BH_2	−621	−604	−454	−533
BH_3	−179	Dissociative chemisorption	−270	−233
BF	−609	−505	−264	−327
BF_2	−585	−484	−390	−350
BF_3	−172	No chemisorption	No chemisorption	No chemisorption
N	−514	−416	−502	−704
NH	−491	−436	−700	−813
NH_2	−338	−212	−594	−550
NH_3	−106	−115	−192	−233
H	−431	—	−198	—
F	—	−283	—	−471

BF_x species adsorb rather strongly to both types of surfaces (i.e. H-terminated N- and B-covered surface). However, it is obvious that the BH, BH_2, BF, and BF_2 species have a more pronounced preference for the N-covered surface. This preference is largest for BF, followed by BH, BF_2, and BH_2 (with a difference of 345, 272, 195, and 167 kJ mol^{-1}, respectively, with respect to the B-covered surface). It must also be stressed that these species also bind significantly more strongly than H ($\Delta E_{ads,H}$ = −431 kJ mol^{-1}). This feature is very important in thin film growth since the incoming growth species must be able to replace a surface-terminating H atom for growth to occur. The B species is also found to adsorb significantly stronger than H onto the H-covered N-terminated surface.

However, it does not exhibit any significant preference for the H-covered N-terminated surface (the difference is only 15 kJ mol^{-1} with respect to the H-terminated B-covered surface). The situation with the gaseous BF_3 species is, however, completely different. It will only adsorb onto the H-terminated N-covered surface, and then significantly weaker compared to H. The only B-containing species that becomes adsorbed onto the H-terminated B-covered surface is the BH_3 species (with a difference of 91 kJ mol^{-1} with respect to the corresponding N-covered surface). The adsorption of BH_3 will, thereby, result in B-B bonds in the growing film. However, such bonds are not present in an ideal c-BN structure. It can from these calculations be concluded that the BH, BH_2, BF, and BF_2 species are energetically more favorable than the B, BH_3, and BF_3 species for a successful and continuous CVD growth of c-BN onto the H-terminated N-covered (100) surface.

8.4.5.3 Adsorption of BH_x or BF_x species onto an F-terminated c-BN(100) surface

The resulting BH_x and BF_x ($x = 0, 1, 2, 3$) adsorption structures, for chemisorption onto a monoradical surface site on the F-terminated N-covered c-BN(100)–(1 × 1) surface, are shown in figures 8.19 and 8.20. The corresponding adsorption energies for both the F-terminated N-covered c-BN(100)–(1 × 1) surface, and the

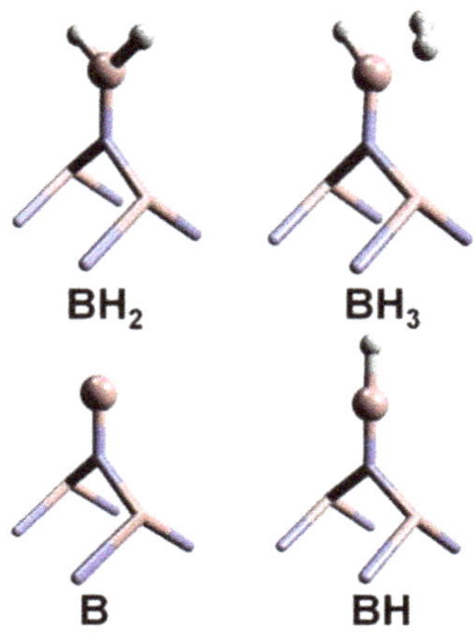

Figure 8.19. The BH_x ($x = 0, 1, 2, 3$) adsorption structures for chemisorption onto a monoradical surface site on the F-terminated N-covered c-BN(100)–(1 × 1) surface (balls = adsorbate species; sticks = surface atoms). [56], reproduced with permission © American Chemical Society.

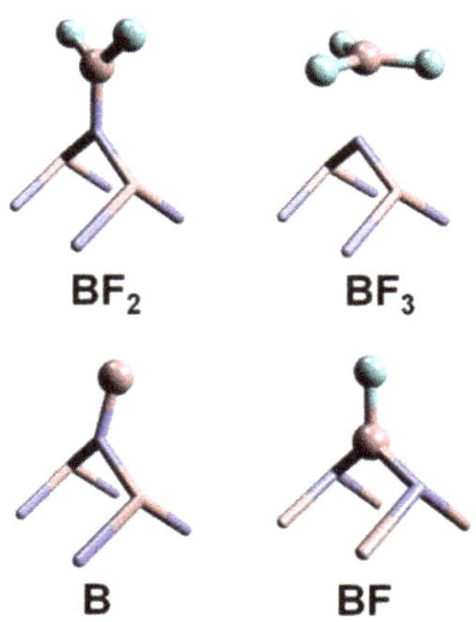

Figure 8.20. The BF_x ($x = 0, 1, 2, 3$) adsorption structures for chemisorption onto a monoradical surface site on the F-terminated N-covered c-BN(100)–(1 × 1) surface (balls = adsorbate species; sticks = surface atoms). [56], reproduced with permission © American Chemical Society.

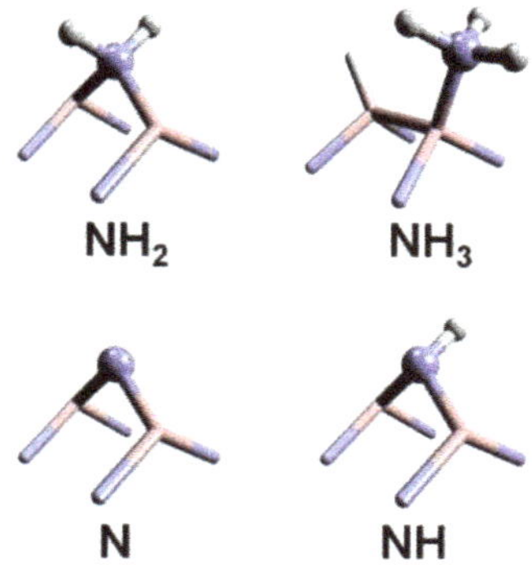

Figure 8.21. The NH$_x$ ($x = 0, 1, 2, 3$) adsorption structures for chemisorption onto a monoradical surface site on the H-terminated B-covered c-BN(100)–(2 × 1) surface (balls = adsorbate species; sticks = surface atoms). [56], reproduced with permission © American Chemical Society.

F-terminated B-covered c-BN(100)–(2×1) surface, are shown in figure 8.21 and table 8.13. The following order of (absolute) adsorption energies for the H-terminated surface is: BH (−671) < BH_2 (−604) < B (−584) < BF (−505) < BF_2 (−484). Moreover, the BH_3 species have become dissociatively chemisorbed and the BF_3 species did not bind at all to the surface (i.e. an endothermic adsorption process). Also, the order of (absolute) adsorption energies for the F-terminated N-covered surface is: B (−594) < BH_2 (−533) < BH (−479) < BF_2 (−350) < BF (−327) < BH_3 (−233). As was the case for the c-BN(100)–(1×1) surface, and onto a H- or F-terminated B-covered c-BN(100)–(1 × 1) surface, adsorption of BH_3 to the H-terminated surface, the BF_3 species did not chemisorb to the F-terminated N-covered c-BN surface. Furthermore, it was found here that the BH, BH_2, BF, and BF_2 species do all energetically favour the F-terminated N-covered surface. This preference is largest for BH, followed by BF, BF_2, and BH_2 (with a difference of 192, 178, 134, and 71 kJ mol^{-1}, respectively, with respect to the results obtained for the corresponding B-terminated surface). It must also be stressed that these species bind significantly more strongly than F ($\Delta E_{ads,F} = -283$ kJ mol^{-1}). It must also be noted that B adsorbs significantly more strongly than F. However, the B species is calculated to bind almost as strongly to the F-terminated N-covered surface as to the F-terminated B-covered surface (the difference is as small as 10 kJ mol^{-1}). Furthermore, the BH_3 species is found to dissociatively chemisorb onto this specific N-terminated surface, forming a BH adsorbate and gaseous H_2. Finally, the BF_3 species has no tendency to chemisorb to the surface. It can from these findings be concluded that the adsorption processes of BH, BH_2, BF, and BF_2 are energetically more favorable than the corresponding ones for B, BH_3, and BF_3 when adsorbed onto the F-terminated N-covered surface. The more favorable adsorption energies for the smaller fragments of the inlet growth precursors (being chemisorbed onto either the H- or F-covered N-terminated surface), are most probably due to the combination of a smaller degree of repulsive interactions among the adsorbates, and the ability of these species to form π bonds with the surface N atoms. This π-bond

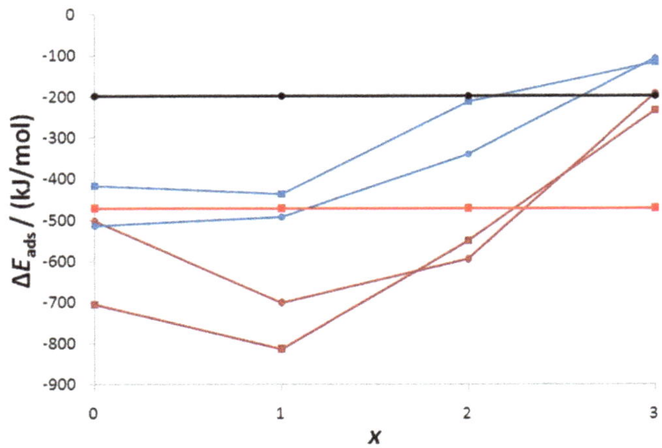

Figure 8.22. Adsorption energies for the NH_x species ($x = 0, 1, 2, 3$) on a monoradical surface site on the H- (blue ▲) or F-terminated (blue ■) B-covered c-BN(100)–(2 × 1) surface, and the H- (brown ▲) or F-terminated (brown ■) N-covered c-BN(100)–(1 × 1) surface. The corresponding adsorption energies for H (black ▲) and F (red ■) are also shown. [56], reproduced with permission © American Chemical Society.

formation is caused by the donation of electrons from the unhybridized 2p-orbital on the N atom to a vacant 2p-orbital on the B atom in the chemisorbed B-containing growth species. Furthermore, the adsorption reactions for the B, BH_3, BF, BF_2, and BF_3 species are more favourable for the H-terminated N-covered surface than for the corresponding adsorption onto the F-terminated one. However, the BH and BH_2 species form bonds of similar strength to both the H- and F-terminated N-covered surfaces. Moreover, the situation for the B-covered surface is the following. The BH, BH_2, and BF species adsorb less strongly to the H-terminated surface (as compared with the F-terminated one). On the contrary, the B, BH_3, and BF_2 species bind stronger to the H-terminated surface. In summary, the variations in bond strengths indicate a larger preference for the adsorption of almost all BH_x and BF_x species onto the H-terminated N-covered surface (with respect to the F-terminated one). The only exceptions to this behaviour include the B and BH_3 species.

8.4.5.4 Adsorption of NH_x species onto an H-terminated c-BN(100) surface

The resulting NH_x ($x = 0, 1, 2, 3$) adsorption structures, for chemisorption onto a monoradical surface site on the H-terminated B-covered c-BN(100)–(2 × 1) surface, are shown in figure 8.21. The corresponding adsorption energies for both the H-terminated B-covered c-BN(100)–(2 × 1), and the H-terminated N-covered c-BN (100)–(1 × 1) surface, are shown in figure 8.22 and table 8.11. The following orders of absolute adsorption energies have been obtained: NH (−700) < NH_2 (−594) < N (−502) < NH_3 (−192) for the H-terminated B-covered surface, and N(−514) < NH (−491) < NH_2 (−338) < NH_3 (−106) kJ mol^{-1} for the H-terminated N-covered surface. All adsorption processes are strongly exothermic, i.e. all of the investigated

NH_x species have the capacity to chemisorb onto both the B- and N-covered c-BN (100) surfaces. It is of an outmost importance that, for a successful c-BN growth to occur, the NH_x species energetically prefer the B-covered surface relative the N-covered one. This is, indeed, the case for the NH and NH_2 species. The preference is somewhat stronger for NH_2 relative NH (with a difference of 256 and 209 kJ mol^{-1}, respectively, with respect to the corresponding N-covered surface). These species also have the capability to replace a surface-terminating H atom ($\Delta E_{ads,H} = -198$ kJ mol^{-1}) on the H-terminated B-covered surface. The N species was also found to adsorb significantly stronger than H onto the H-terminated B-covered surface. However, the adsorption energy is almost identical for both the H-terminated B-and N-covered surfaces (the difference is only 12 kJ mol^{-1}). Also, the NH_3 species favours the H-terminated B-covered surface (with a difference of 86 kJ mol^{-1} with respect to the corresponding N-covered one). However, it has an adsorption energy of similar strength as the adsorption of H onto the H-terminated B-covered surface. It can be concluded from these calculations that the NH and NH_2 species are energetically more favourable than N and NH_3 for a successful and continuous CVD growth of c-BN onto the H-terminated B-covered surface.

8.4.5.5 Adsorption of NH_x species onto an F-terminated c-BN(100) surface

The resulting NH_x ($x = 0, 1, 2, 3$) adsorption structures, for chemisorption onto a monoradical surface site on the F-terminated B-covered c-BN(100)–(2 × 1) surface, are shown in figure 8.23. The corresponding adsorption energies are shown in figure 8.22 and table 8.11. The following orders of adsorption energies have been obtained: NH (−813) < N (−704) < NH_2 (−550) < NH_3 (−233) kJ mol^{-1} for the F-terminated B-covered surface, and NH (−436) < N (−416) < NH_2 (−212) < NH_3 (−115) kJ mol^{-1} for the F-terminated N-covered surface. All adsorption processes are exothermic and, thereby, energetically favourable for both the F-terminated B- and N-covered surfaces. It can also be concluded that all NH_x species exhibit a significant preference for the F-terminated B-covered surface. This preference is the strongest for NH, followed by NH_2, N, and NH_3 (with a difference of 377, 338, 288,

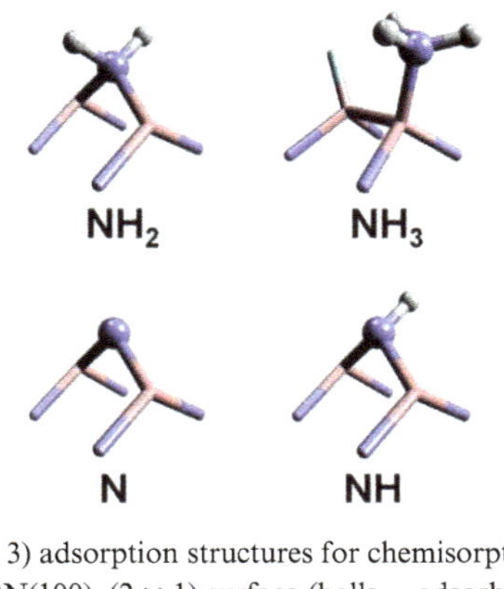

Figure 8.23. The NH_x ($x = 0, 1, 2, 3$) adsorption structures for chemisorption onto a monoradical surface site on the F-terminated B-covered c-BN(100)–(2 × 1) surface (balls = adsorbate species; sticks = surface atoms). [56], reproduced with permission © American Chemical Society.

and 118 kJ mol^{-1}, respectively, with respect to the F-terminated N-covered surface). However, only N, NH, and NH_2 are associated with an adsorption process that is more exothermic than adsorption of F ($\Delta E_{ads,F}$ = −471 kJ mol^{-1}) onto the F-terminated B-covered surface. Hence, the adsorption process for the N, NH, and NH_2 species are, for the F-terminated B-covered surface, more favourable when compared with the NH_3 species. The less favourable adsorption energy for the NH_3 species, onto both H- and F-terminated B-covered surfaces, is most probably due to the fact that NH_3 remains in the on-top position (as a result of geometry optimization). The other NH_x species become inserted into a B-B dimer bond, thereby forming two N–B bonds. It should be stressed that such insertion reactions are a necessary condition for a continued growth of c-BN. In addition to the different adsorbate-to-surface bond formations, the less hydrogenated NH_x species exhibit smaller sterical hindrances on the surface. When comparing the preferential adsorptions onto the F- and H-terminated B-covered surfaces, the N, NH, and NH_3 species are found to bind more strongly to the F-terminated B-covered surface. An exception to this behaviour is the NH_2 species, which binds somewhat more weakly to the F-terminated B-covered surface. Moreover, the N, NH, and NH_2 species adsorb less strongly to the F-terminated N-covered surface compared to the corresponding H-terminated one. Furthermore, the NH_3 species form surface-to-adsorbate bonds of similar strength to both F- and H-terminated N-covered surfaces. These variations in bond strength point at a larger B preference for all NH_x species when adsorbed onto an F-terminated c-BN(100) surface (with respect to the H-terminated one).

8.4.5.6 Adsorption of BH_x, BF_x, and NH_x from an H/F-saturated gas phase
Based on the results presented above, it is possible to conclude that a continued growth on the B-covered c-BN(100) surface is more favorable in an F-saturated gas phase, i.e. which forms F-termination on the surface. On the other hand, a continued growth on the N-covered surface is more favourable in an H-saturated environment. In addition, the process of H abstraction from the H-terminated N-covered surface, with gaseous H species, was in section 8.4 shown to be only slightly exothermic ($\Delta E_{abs,H\ with\ H}$ = −25 kJ mol^{-1}). Moreover, the process of F abstraction from the F-covered B-terminated surface, with gaseous F species, was shown to be strongly endothermic ($\Delta E_{abs,F\ with\ F}$ = +252 kJ mol^{-1}). The abstraction processes were, however, found to be strongly exothermic if F radicals were used for abstraction of H from the H-terminated N-covered surface ($\Delta E_{abs,H\ with\ F}$ = −170 kJ mol^{-1}), and H radicals were used for abstraction of F from the F-terminated B-covered surface ($\Delta E_{abs,F\ with\ H}$ = −130 kJ mol^{-1}). It is therefore highly probable that the optimal gas phase for CVD growth of c-BN consists of a mixture of H and F species (if it is not possible to instead use the ALD technique with its sequential pulses). On the basis of the calculated values of adsorption energies, a B-covered c-BN surface is expected to become preferentially F-terminated ($\Delta E_{ads,H}$ = −258 kJ mol^{-1}, $\Delta E_{ads,F}$ = −403 kJ mol^{-1}), while the N-covered surface is expected to be mainly H-terminated ($\Delta E_{ads,H}$ = −431 kJ mol^{-1}, $\Delta E_{ads,F}$ = −283 kJ mol^{-1}). Moreover, all BH_x and BF_x species, with the exception of BH_3, do energetically favour the H-terminated N-covered

surface (see table 8.13 and figure 8.22). This preference is largest for BF_3 ($\Delta E_{ads,H\text{-}term.} = -172$ kJ mol^{-1} and $\Delta E_{ads,F\text{-}term.} = >0$ kJ mol^{-1}), followed by BF, BF_2, BH, BH_2, and B (with a difference of 282, 235, 207, 88 versus 48 kJ mol^{-1} compared to the F-terminated B-covered surface). However, only the B, BH, BH_2, BF, and BF_2 species show an adsorption process that is more exothermic than the H adsorption ($\Delta E_{ads,H} = -431$ kJ mol^{-1}) to the otherwise H-terminated N-covered surface. Furthermore, all NH_x species exhibit a significant preference for the F-terminated B-covered surface (see table 8.13 and figure 8.22). This preference is largest for NH, followed by NH_2, N, and NH_3 (with a difference of 322, 212, 190 vs 127 kJ mol^{-1} compared to the H-terminated N-covered surface). However, only the N, NH, and NH_2 species show an adsorption process that is more exothermic than the adsorption of F ($\Delta E_{ads,F} = -471$ kJ mol^{-1}) to the otherwise F-terminated B-covered surface. Based on all these calculated results, it can be concluded that the adsorption of B, BH, BH_2, BF, BF_2, N, NH, and NH_2 are energetically more favourable than the adsorption of BH_3, BF_3, and NH_3 (in a H/F-saturated gas-phase) for a successful and continuous CVD growth of c-BN to occur.

As presented in chapter 2, vapour phase deposition techniques like CVD generally uses some type of high energy activation of the gas phase in order to decompose the incoming gaseous growth precursors (like BH_3, BF_3, and NH_3). It is, thereby, extremely difficult to control the formation of specific gaseous fragments and a mixture of all type of fragments is expected to be formed. It is, hence, very difficult to completely avoid an unwanted result where the growth species adsorb onto non-ideal surface sites (like BH_x adsorption onto a surface B atom). This problem can, however, be circumvented by using a vapour phase deposition method where the incoming BH_3, BF_3, and NH_3 growth species are separately introduced into the reactor, e.g. by using an ALD type of method (see chapter 2). It can now also be concluded that the H/F-saturated gas-phase is the most optimal one since it exhibits: (i) favorable abstraction reactions, (ii) more exothermic adsorption reactions, and (iii) the largest range of favorable growth species. It is now possible to suggest the following (simplified) ALD growth sequence of c-BN, starting from an 100% F-terminated B-covered c-BN(100) surface in an H- versus F-saturated gas-phase and with BF_x versus NH_x growth precursors (see figure 8.24):

1. Abstraction of surface-terminating F species by gaseous H radicals (i.e. thereby forming surface B radical sites).
2. Formation of an 100% H-terminerad N-covered c-BN(100) surface by the adsorption of NH_x growth species to the B radical sites.
3. Abstraction of surface-terminating H atoms by gaseous F radicals (i.e. thereby forming surface N radical sites).
4. Formation of an 100% F-terminated B-covered c-BN(100) surface by the adsorption of BF_x growth species to the N radical sites.

The steps 1–4 are thereafter repeated. These results are in very good agreement with the experimental results presented by Zhang *et al* [37]. However, energetic

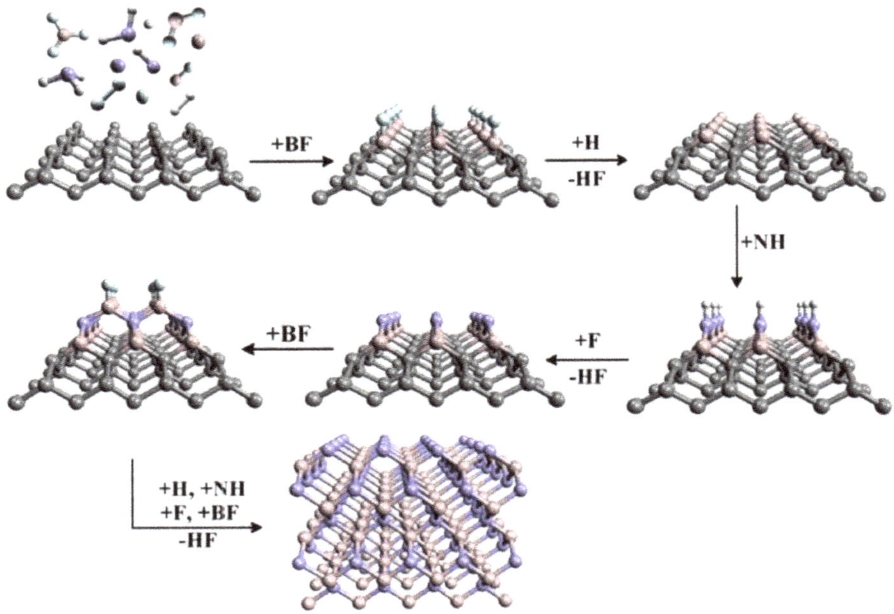

Figure 8.24. Proposed ALD growth mechanism for a phase-pure growth of c-BN. C: grey; B: pink; N: blue; H: white; F: light blue. [56], reproduced with permission © American Chemical Society.

particle bombardment, in addition to an increased temperature, was found to be necessary in that study. Hence, the c-BN growth mechanism seems to be driven by a surface kinetic control.

8.4.6 Growth of c-BN versus h-BN

Boron nitride thin films have been grown earlier by using a laser-assisted ALD (LALD) technique at temperatures in the interval 250 °C–750 °C [55]. The alternating growth species pulses consisted of NH_3 and BBr_3, respectively, with a purging gas in between. These precursors were dissociated by using an ArF excimer laser, and the resulting film consisted of hydrogen-terminated t-BN/h-BN nanocrystallites. No cubic phase of BN was detected. Theoretical modelling was also used to identify those precursors that favour the cubic phase of c-BN (instead of the hexagonal one). The adsorption of NH_3 and BBr_3, and their decomposed fragments, onto c-BN(111) surfaces and h-BN(10–10) edges were then of a special interest to study. As can be seen in figure 8.13, almost all adsorption reactions were found to be exothermic. The endothermic reactions include NH_3 adsorption to the N sites of either c-BN or h-BN, as well as BBr_3 adsorption to the B-covered c-BN(111) surface. The most weakly bonded species to the cubic phase of BN are the non-fragmented NH_3 and BBr_3 species. Furthermore, the N, NH, and BBr_2 species do not prefer adsorption sites that are energetically favourable for a phase-pure growth of c-BN to occur. The N species prefers an N surface site on c-BN, whilst the NH species prefers either a cubic N site or a hexagonal B site. Furthermore, BBr_2 favours almost all surface site types (with an exception of the cubic B site). Hence, there is an apparent risk for N –N and B –B bond formations with these species. Moreover, adsorptions of NH_2, B, or BBr were predicted to preferably form bonds to the hexagonal N sites, with a large probability for NH_2 to adsorb to a cubic N site. The present results suggest that there is a large probability to obtain a mixture of B–N, N–N, and B–B bonds, when using activation of a NH_3/BBr_3 gas mixture within a CVD reactor. Separate introductions of precursors into the reactor (e.g. by using ALD) can prevent such a situation. The most optimal precursor fragments to use for c-BN growth is N in addition to the parent molecules NH_3 and BBr_3. It is very interesting to note that the experimental ALD results do strongly support the experimental results in [55].

The calculated adsorption energies for the N fragment, in addition to BBr_3 and NH_3 are especially marked in figure 8.13. Based solely on the adsorption energies, it is then obvious that these species favour a hexagonal growth instead of a cubic one. It can also be seen in figure 8.13 that the species that favour a c-BN growth were not experimentally detected in the ALD chamber. These species (the N fragment and the non-fragmented BBr_3 and NH_3 molecules) would, hence, be ideal in discrimination between c-BN and h-BN. Unfortunately, a high degree of radical surface terminating species is needed to stabilize the growing c-BN surface, and any type of fragmentation is, therefore, unavoidable.

The example above has, thus, taught us that the theoretical simulations should generally be done prior to further experimental studies. If that had been the case in [55], one had most probably tuned the laser to achieve the species that favours a c-BN growth and disfavours an h-BN growth.

8.5 Surface migration during growth of c-BN

8.5.1 General

As presented in section 6.7.5, a stepped morphology on the surface has been shown to be characteristic of chemical vapor deposited diamond [53]. Earlier experiments have also indicated that surface diffusion can take place during CVD growth of diamond [57]. However, before this discovery, the growth of diamond was usually discussed without including the possibility for surface diffusion. Past mechanistic studies have usually presumed that the growth of diamond takes place by a direct Eley–Rideal type of mechanism [58]. As presented in section 3.3.3, a gaseous precursor then undergoes a chemical reaction immediately upon its collision with a reactive surface site or, if the reaction cannot take place, scatters away from the surface. The main reason for this assumption is that the carbon-containing growth species are forming very strong covalent bonds with the surface, and the probability for surface diffusion is, hence, expected to become very low. However, with the discovery of surface steps and surface diffusion, growth of diamond should occur through propagation of steps over the diamond surface. In addition, the complex dynamic processes occurring during chemical vapor deposition of diamond is, also assumed to include migration of surface-terminating species (e.g. H, F). Due to the large chemical and physical similarities between diamond and c-BN, these surface migrations assumptions are also valid for the CVD growth of c-BN. As discussed by Bartl *et al* the challenge in the CVD of c-BN is to grow the stable phase by either suppressing the nucleation of h-BN or to favour the nucleation of c-BN [30]. A high mobility of the growth precursors (e.g. NH_2) is then needed for a continuous growth of the stable c-BN phase.

A surface migration process assumes that there is a neighboring surface radical site next to the migrating species. As discussed in section 6.7.4, the migration process is then assumed to occur as a single jump from the first occupied site to the neighbouring radical site. We will in the present chapter demonstrate the importance of surface migration on the B- and N-covered c-BN(111) surfaces. It is, for the reader, easy to use this knowledge for corresponding migration studies on the c-BN (100) and c-BN(110) surfaces.

8.5.2 Surface migration of growth species on the c-BN(111) surface

As was the situation for a diamond (111) surface, the surface migration on the c-BN (111) surface has an anisotropic character since every surface atom has six symmetrically equivalent neighbouring sites on the B- versus N-covered surface. The energy barriers for migration of NH_2 and BH_2 species onto the H-terminated B- and N-covered c-BN(111) surfaces are shown in table 8.14. The smallest energy barrier is attributed to the migration of BH_2 species onto the B-covered surface. Its value is +120 kJ mol^{-1}, which should be compared with the corresponding energy barrier for NH_2 (+276 kJ mol^{-1}). These barriers of energies are relatively large and will most probably prevent an adsorbed NH_2 (or BH_2) species to move more or less freely over the surface at moderate temperatures. As can be seen in table 8.14, the energy barriers of both BH_2 and NH_2 are, however, considerably more energetically favorable as compared with the desorption processes. It is well known that the desorption energy for a chemisorbed species is always larger than, or equal to, its adsorption energy (see figure 8.25). Thus, since the adsorption energies for BH_2 and NH_2 are −394 and −459 kJ mol^{-1}, respectively, an addition of energy will preferentially cause surface migrations instead of desorption of the actual adsorbates. The geometries of the transition states for the surface migration processes of

Table 8.14. Adsorption energies in addition to energy barriers for the migration of NH_2 and BH_2 species onto the H-terminated B- and N-covered c-BN(111) surface, respectively. The adsorption and migration of CH_2 on the diamond (111) surface is presented for comparison.

H-terminated (111) surfaces	B-covered c-BN (kJ mol^{-1})		N-covered c-BN (kJ mol^{-1})		Diamond (kJ mol^{-1})	
(kJ mol^{-1})	ΔE_{barr}	ΔE_{ads}	ΔE_{barr}	ΔE_{ads}	ΔE_{barr}	ΔE_{ads}
BH_2	+120	−394	+101	−621		
NH_2	+276	−459	+232	−338		
CH_2					+52	−261

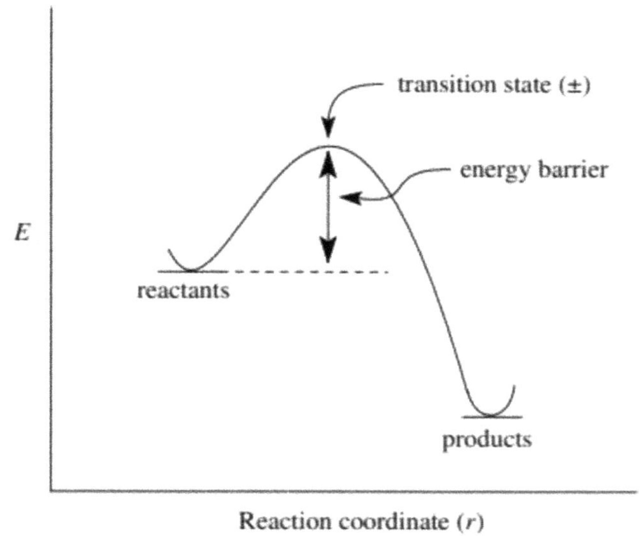

Figure 8.25. Sketch of the evolution of energy during the course of a chemical reaction. E_a = activation energy when going from the reactants to the product; ΔE = reaction energy.

BH_2 and NH_2 (i.e. for the jumping between two surface sites) are shown in figure 8.26. In these transition states, the distances between the N atom (in NH_2) and the involving radical B sites, and between the B atom (in BH_2) and the involving radical B sites, are 2.04 versus 2.26 Å. These distances are longer than ordinary N-B and B-B covalent bonds (1.565(c-BN) versus 1.4 (free B_2) Å), thus indicating the absence of bonding interactions at the respective transition state. On the contrary, the corresponding energy barrier obtained for the CVD growth of diamond differs considerably from those obtained for the CVD growth of c-BN [59]. As shown in table 8.14, the barrier energy obtained for the migration of CH_2 on the H-terminated diamond (111) surface is much smaller than for the migration of BH_2 or NH_2 on the H-terminated B-covered c-BN(111) surface (+52, +120, and +276 kJ mol^{-1}, respectively). The calculations show that the CH_2 species binds strongly to the diamond surface at the transition state (figure 6.30(c)). Two single bonds are then formed between the triplet state of CH_2 and each of the two radical surface carbons by overlapping dangling bonds (i.e. non-paired electrons). This is not the case for the corresponding transition states for the migrating BH_2 and NH_2 species on the c-BN surface. The reason is that both BH_2 and NH_2 have only one non-paired electron, whilst CH_2 has two. The stabilized CH_2 transition state is, furthermore, identified by the short bond lengths between C (in CH_2) and each of the two radical surface sites on the diamond (111) surface (1.35 Å), which is to be compared with the average C–C bond length in covalent compounds; 1.54 Å. This stabilization explains the very low barrier energy and, thereby, fast rate of the migrating CH_2 adsorbate on the diamond (111) surface.

The energy barrier for migration of NH_2 has also been calculated for the N-covered c-BN (111) surface and is +232 kJ mol^{-1}. This value is to be compared with

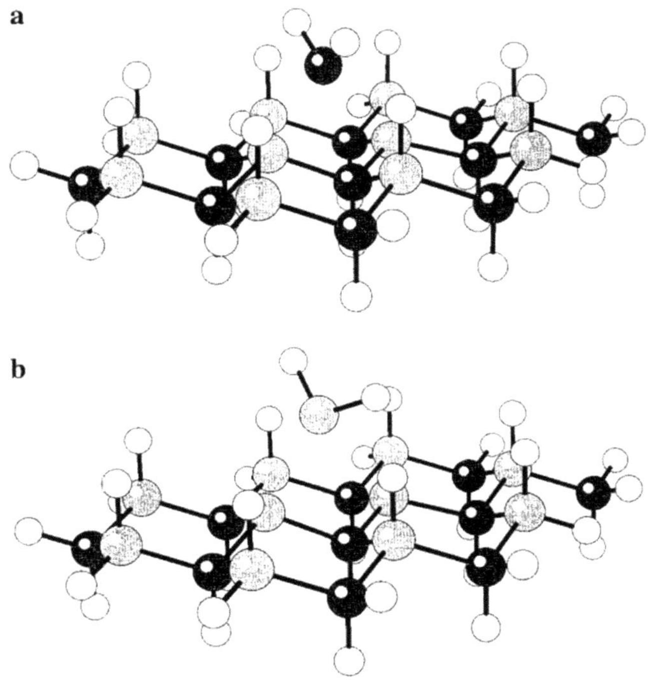

Figure 8.26. Optimized transition state geometries for a migrating (a) NH_2 and (b) BH_2 species on a B-covered c-BN(111) surface. B: grey; N: black. [49], reproduced with permission © American Chemical Society.

the energy barrier for migration on the corresponding B-covered surface: +276 kJ mol^{-1}. Furthermore, the energy barriers for migration of BH_2 on the N- versus B-covered surfaces are +101 and +120 kJ mol^{-1}, respectively. Thus, the migration of both the NH_2 and BH_2 species seems to be faster on the N-covered surface (as compared to the B-covered on). Since the energy barriers for the migrations on the B-covered surface are much smaller than the corresponding adsorption energies, and thereby the respective desorption energy, the most probable scenarios are surface migrations on this specific surface.

For the growth precursors, low migration barriers, in combination with large adsorption energies, is generally expected to result in a large probability for a smooth and low-defect growth of thin films. More specifically, the present calculated result does also speak in favor for a preferred adsorption (and migration) of BH_2 growth species onto the N-covered c-BN(111) surface, and a preferred adsorption (and migration) of NH_2 growth species onto the B-covered one.

8.5.3 Surface migration of terminating species on the c-BN(111) surfaces

The probability for surface migration of the various terminating species is also of large interest to study. The reason is that the formation of surface radical sites is needed for a continuous CVD growth of thin films to occur. As discussed in section

Table 8.15. Adsorption energies, abstraction energies, and energy barriers for the migration of F and H species onto the H-terminated B- and N-covered c-BN(111) surface, respectively. The values in parenthesis are the gas phase species involved in the abstraction processes.

Adsorbate	B-covered c-BN(111)			N-covered c-BN(111)		
($kJ\ mol^{-1}$)	ΔE_{barr}	ΔE_{ads}	ΔE_{abs}	ΔE_{barr}	ΔE_{ads}	ΔE_{abs}
F	+330	−653	+512 (F)	+182	−251	+28 (F)
H	+252	−442	+18 (H)	+312	−465	+45 (H)

6.4.2, gas phase abstraction of the terminating species creates these radical sites. However, the probability for the creation of a specific surface radical site, S, increases if the surface mobility of the terminating species is high. In other words, a surface radical site further away can 'move fast' to site S. Hence, site S will either be formed by a direct abstraction or by surface migration of the terminating species that is originally positioned on S. For the B-covered c-BN(111) surface, it has been found that F has not only a very pronounced adsorption energy to the surface ($\Delta E_{ads,F} = -653$ kJ mol^{-1}), but also a very high abstraction energy with gaseous F ($\Delta E_{abs,F\ with\ F} = +512$ kJ mol^{-1}). In addition, the calculated energy barrier for the terminating F species has been calculated to also be very high $\Delta E_{mig,F} = +330$ kJ mol^{-1} (see table 8.15). Thus, the surface terminating F species seems to block the possible growth surface sites on surface sites on the B-covered surface. Contrary to the results obtained for F-terminated surfaces, the surface stabilizing H atoms ($\Delta E_{ads,H} = -123$ kJ mol^{-1}) are found to be much more mobile, either by a very low abstraction energy ($\Delta E_{abs,H\ with\ H} = +18$ kJ mol^{-1}) or by migration of the H adsorbates on the surface ($\Delta E_{mig,H} = +252$ kJ mol^{-1}). Both of these energies are much lower compared to the corresponding situation with an F-terminated surface. However, it seems that a desorption is preferable over a migration for this terminating species.

The probability of surface migration of H, or F, on the N-covered c-BN(111) surface has also been calculated. As a result, there is a very large tendency for F to undergo surface abstraction reactions with gaseous F species ($\Delta E_{abs,F\ with\ F} = +28$ kJ mol^{-1}), in addition to a relatively weak adsorption energy ($\Delta E_{ads,F} = -251$ kJ mol^{-1}), is strongly correlated with the relatively small energy barrier for the surface migration of F adsorbates ($\Delta E_{mig,F} = +182$ kJ mol^{-1}). The barrier energies for migration of the H adsorbates is, however, much higher ($\Delta E_{mig,H} = +312$ kJ mol^{-1}), which correlates to a pronounced adsorption energy ($\Delta E_{ads,H} = -465$ kJ mol^{-1}). However, the abstraction energy for the removal of adsorbed H species, by gaseous H, is unexpectedly low ($\Delta E_{abs,H\ with\ H} = +45$ kJ mol^{-1}). $A^{-1}s^{-1}$ a result, the H species is, compared to F, very efficient as a terminating species since it has a larger probability of (i) binding to and, hence, stabilizing the surface, and (ii) leaving room for an incoming growth species to the N-covered c-BN(111) surface.

References

[1] Chopra N G, Luyken R J, Cherry K, Crespy V H, Cohen M L, Louie S G and Zettle A 1995 Boron nitride nanotubes *Science* **26** 966–7

[2] Goldberg D, Bando Y, Stephan O and Kurashima K 1998 Octahedral boron nitride fullerenes formed by electron beam irradiation *Appl. Phys. Lett.* **73** 2441–3

[3] Alexandre S S, Mazzoni M S C and Chacham H 1999 Stability, geometry, and electronic structure of the boron nitride $B_{36}N_{36}$ fullerene *Appl. Phys. Lett.* **75** 61–3

[4] Yarbrough W A 1991 Current research problems and opportunities in the vapor phase synthesis of diamond and cubic boron nitride *J. Vac. Sci. Technol.* A **9** 1145–52

[5] Okamoto M, Utsumi Y and Osaka Y 1992 Formation and properties of cubic boron nitride films on tungsten carbide by plasma chemical vapor deposition *Jpn. J. Appl. Phys.* **31** 3455–60

[6] Kester D J and Messier R 1992 Phase control of cubic boron nitride thin films *J. Appl. Phys.* **72** 504–13

[7] Friedmann T A, Mirkarimi P B, Medlin D L, McCarty K F, Klaus E J, Boehme D R, Johnsen H A, Mills M J, Ottesen D K and Babour J C 1994 Ion-assisted pulsed laser deposition of cubic boron nitride films *J. Appl. Phys.* **76** 3088–101

[8] Yamashita N, Wada T, Ogawa M, Kobayashi T, Tsukamoto H and Rokkaku T 1992 Adhesion and structure of c-BN films produced by ion-beam-assisted deposition *Surf. Coat. Technol.* **54/55** 418–22

[9] Kester D J, Ailey K S, Lichtenwalner D J and Davis R F 1994 Growth and characterization of cubic boron nitride thin films *J. Vac. Sci. Technol.* A **12** 3074–81

[10] Mirkarimi P B, Medlin D L, McCarty K F and Barbour J C 1995 Growth of cubic BN films on β-SiC by ion-assisted pulsed laser deposition *Appl. Phys. Lett.* **66** 2813–15

[11] Hahn J, Richter F, Pintaske R, Röder M, Schneider E and Welzel T 1997 Formation of c-BN thin films under reduced ion impact *Surf. Coat. Technol.* **92** 129–34

[12] Gimeno S, Andujar J L and Lousa A 1997 Cubic boron nitride thin films by tuned r.f. magnetron sputtering *Dia. Relat. Mater.* **6** 604–7

[13] Yoshida T 1996 Vapour phase deposition of cubic boron nitride *Diam. Relat. Mater* **5** 501–7

[14] McKenzie D, McFall W, Smith H, Higgins B, Bosewell R, Durandet A, James B and Falconer I 1995 High pressure phases produced by low energy ion implantation with reference to cubic boron nitride *Nucl. Instrum. Methods* **106** 90–5

[15] Jime´nez I, Jankowski A, Terminello L J, Carlisle J A, Sutherland D G J, Doll G L, Mantese J V, Tong W M, Shuh D K and Himpsel F J 1996 Near-edge x-ray absorption fine structure study of bonding modifications in BN thin films by ion implantation *Appl. Phys. Lett.* **68** 2816–18

[16] Jime´nez I, Jankowski A, Terminello L J, Sutherland D G J, Carlisle J A, Doll G L, Tong W M, Shuh D K and Himpsel F J 1997 *Phys. Rev.* B **55** 12025–37

[17] Okamoto M, Yokoyama H and Osaka Y 1990 Formation of cubic boron nitride film on Si with boron buffer layers *Jpn. J. Appl. Phys.* **29** 930–3

[18] Weber A, Bringmann U, Nikulski R and Klages C P 06, 1993 Growth of cubic boron nitride and boron-carbon-nitrogen coatings using N-trimethylborazine in an electron cyclotron resonance plasma process *Dia. Relat. Mater.* **2** 201–2

[19] Kulish W and Ulrich S 2003 Parameter spaces for the nucleation and the subsequent growth of cubic boron nitride films *Thin Solid Films* **423** 183–195

[20] Chayahara A, Yokoyama H, Imura T and Osaka Y 1987 Function of substrate bias potential for formation of cubic boron nitride films in plasma CVD technique *Jpn. J. Appl. Phys.* **26** L1435–6
[21] Shriver D F, Atkins P W and Langford C H 1994 *Inorganic Chemistry* (Oxford: Oxford University Press) pp 377–543
[22] Karlsson J and Larsson K 2010 Hydrogen-induced de/reconstruction of the c-BN(100) surface *J. Phys. Chem.* C **114** 3516–21
[23] Cotton F A and Wilkinson G 1988 *Advanced Inorganic Chemistry* (New York: Wiley) pp 162–205
[24] CASTEP Program System of Molecular Simulation/Biosym Technologies, Inc., San Diego
[25] Widany J, Weich F, Köhler T, Porezag D and Frauenheim. T 1996 Dynamic properties and structure formation of boron and carbon nitrides *Diam. Relat. Mater* **5** 1031–41
[26] Kuhr M, Freudenstein R, Reinke S, Kuhlisch W, Dollinger G and Bergmaier A 1996 Hydrogen incorporation during nucleation and growth of c-BN films *Diam. Relat. Mater.* **5** 984–9
[27] Loh K P, Sakaguchi I, Nishitani-Gamo M, Taniguchi T and Ando T 1998 Hydrogen-induced surface structuring of a cubic boron nitride (100) face studied by low-energy electron diffraction and electron spectroscopic techniques *Phys. Rev.* B **57** 7266
[28] Loh K P, Sakaguchi I, Nishitani-Gamo M, Taniguchi T and Ando T 1997 Surface structure of single-crystal cubic boron nitride (111) studied by LEED, EELS, and AES *Phys. Rev.* B **56** R12791
[29] Bohr S, Haubner R and Lux B 1995 Comparative aspects of c-BN and diamond CVD *Dia. Relat. Mater.* **4** 714–9
[30] Bartl A, Bohr S, Haubner R and Lux B 1996 A comparison of low-pressure CVD synthesis of diamond and c-BN *Int. J. Refract. Mater.* **14** 145–57
[31] Steckel J A, Phung T, Jordan K D and Nachtigall P 2001 Concerted use of slab and cluster models in an *ab initio* study of hydrogen desorption from the Si(100) surface *J. Phys. Chem.* B **105** 4031–8
[32] Verwoerd W S 1981 The adsorption of hydrogen on the (100) surfaces of silicon and diamond *Surf. Sci.* **108** 153–68
[33] Arvidsson I and Larsson K 2007 Adsorption of H, NH_x, BH_x and BBr_x on a (110) surface of c-BN: A quantum-mechanical DFT study *Diam. Relat. Mater.* **16** 131–7
[34] Karlsson J and Larsson K 2011 Halogen-induced reconstruction of the c-BN (100) surface *J. Phys. Chem.* C **115** 22910–16
[35] Lide D R 2009 *CRC Handbook of Chemistry and Physics* 90th edn (Boca Raton, FL: Internet Version 2010; CRC Press/Taylor and Francis)
[36] Silberberg M S 2003 *Chemistry: The Molecular Nature of Matter and Change* 3rd edn (New York: McGraw-Hill)
[37] Zhang W J, Chan C Y, Meng X M, Fung M K, Bello I, Lifshitz Y, Lee S T and Jiang X 2005 The mechanism of chemical vapor deposition of cubic boron nitride films from fluorine-containing species *Angew. Chem.* **44** 4749
[38] Bruice P Y 2001 *Organic Chemistry* 3rd edn (New York: Prentice-Hall)
[39] Morrison R I and Boyd R N 1973 *Organic Chemistry* (Boston, MA: Allyn and Bacon)
[40] Mårlid B, Larsson K and Carlsson J-O 1999 Theoretical investigation of hydrogen- and halogen-terminated *c*-BN (111) clusters *Phys. Rev.* B **60** 16065–72

[41] Brown R C, Cramer C J and Roberts J T 1997 An *ab initio* study of hydrogen abstraction from cluster models for the diamond surface *J. Phys. Chem.* B **101** 9574–80
[42] Kern G, Hafner J and Kresse G 1996 Atomic and electronic structure of diamond (111) surfaces I. Reconstruction and hydrogen-induced de-reconstruction of the one dangling-bond surface *Surf. Sci.* **366** 445–63
[43] Kagamida M, Kanda H, Akaishi M, Nukui A, Osawa T and Yamaoka S 1989 Crystal growth of cubic boron nitride using Li_3BN_2 solvent under high temperature and pressure *J. Cryst. Growth* **94** 261–9
[44] Pederson M R, Jackson K A and Pickett E 1991 Local-density-approximation-based simulations of hydrocarbon interactions with applications to diamond chemical vapor deposition *Phys. Rev.* B **44** 3891–9
[45] Konyashin I, Bill J and Aldinger F 1997 Plasma-assisted CVD of cubic boron nitride *Chem. Vap. Dep.* **3** 239–55
[46] Komatsu S, Mariiiyoshhi Y, Kassamatsu M and Yamada K 1994 U.S. Patent 286553
[47] Mendez J M, Muhl S, Farias M, Soto G and Cota-Araiza L 1991 Preparation of cubic boron nitride films by plasma-enhanced chemical vapour deposition of BF_3, N_2 and H_2 gas mixtures *Surf, Coat, Technol.* **41** 422–6
[48] Allendorf M D and Melius C F 1997 Thermochemistry of molecules in the B–N–Cl–H system: *ab initio* predictions using the BAC-MP4 method *J. Phys. Chem.* A **101** 2670–80
[49] Larsson K and Carlsson J-O 1999 Surface processes in cubic boron nitride growth: a theoretical study *J. Phys. Chem.* B **103** 6533–8
[50] Larsson K, Lunell S and Carlsson J-O 1993 *Ab initio* study of hydrocarbons on diamond (111) *Dia. Relat. Mater* **2** 949–51
[51] Larsson K, Lunell S and Carlsson J-O 1993 Adsorption of hydrocarbons on a diamond (111) surface: an *ab initio* quantum mechanical study *Phys. Rev.* B **48** 2666–74
[52] Larsson K, Carlsson J-O and Lunell S 1995 Nearest-neighbour influence on hydrocarbon adsorption on diamond (111): studied by *ab initio* calculations *Phys. Rev.* B **51** 10003–12
[53] Larsson K 2006 CVD growth of cubic boron nitride: a theoretical/experimental approach *Thin Solid Films* **515** 401–6
[54] Kalish R 2001 The search for donors in diamond *Diam. Relat. Mater.* **10** 1746–55
[55] Olander J and Larsson K 2006 CVD growth of cubic boron nitride: a theoretical/ experimental approach *Thin Solid Films* **515** 401–6
[56] Karlsson J and Larsson K 2011 Adsorption of growth species on the c-BN(100) surface *J. Phys. Chem.* C **115** 16977–83
[57] Frenklach M and Skokov S J 1997 Surface migration in diamond growth *Phys. Chem.* **B 101** 3025–36
[58] Enckewort W J P 1994 *Synthetic Diamond: Emerging CVD Science and Technology* ed K E Spear and J P Dismukes (New York: Wiley) pp 307–53
[59] Hentschel F, Wochnowski C and Benndorf C 1997 Presented at the 8th European Conf. on Diamond, Diamond-Like, and Related Materials, Edinburgh, Scotland

Chapter 9

Effect of substrates on the vapour phase growth of thin film materials

There is at present an urgent need to develop more gentle vapor phase methods whereby it is possible to grow defect-free c-BN films without noncubic interlayers. Generally, the initial nucleation stage during vapor phase deposition of thin films is expected to have a major effect on the quality of the film. This is especially the situation for c-BN growth, for which the initial nucleation stage of the cubic phase is much more difficult to achieve than the continuous film growth (i.e. the parameter window for the initial nucleation stage is much more narrow than for the later stage of growth) [1–3]. The role of the substrate material has been found earlier to be very crucial. It must not only help improve the binding between the substrate material and the film, but should also be able to control the phase ratio (within the BN film) in favour of the cubic phase [4]. The lattice mismatch is in this sense a factor of ultimate importance for heteroepitaxial growth of c-BN. The most common substrate materials of today fall within the following categories: hard (diamond, β-SiC, Si), soft (metals), and ionic (KBr, KI) [2, 5–10].

The search for an optimal substrate material is, thus, regarded to be one of the most important research directions for a successful vapour phase growth of phase-pure c-BN. Theoretical modeling is a necessary tool to be used in these strives.

9.1 Substrate effect on the vapour phase growth of c-BN

9.1.1 Diamond, Si versus SiC substrates

The three hard substrates mentioned above (diamond, β-SiC, and Si) have been theoretically tested for their usage as optimal substrates in the vapour phase growth of phase-pure c-BN thin films. The geometrical structures of these substrates are very similar, with a cubic zinc blende structure. In addition, the elements C and Si are also very similar since they have identical electronic structures in their respective

doi:10.1088/978-0-7503-3107-4ch9

valence shell (s^2p^2). The main difference, though, is their different unit cell sizes, which, in turn, cause various degrees of lattice mismatch between c-BN and these substrates. The cubic cell sizes ($a=b=c$) of diamond, β-SiC, and Si are 3.556 Å, 4.348 Å, and 5.430 Å, respectively, which correspond to the following substrate/c-BN lattice mismatches: 1.7%, 16.9%, and 33.4%, respectively. The diamond surface was not only used as a substrate due to its large similarity with the diamond lattice (i.e. a minor lattice mismatch) but also because of recent successful growth results that have shown that it is possible to grow the cubic phase of BN directly onto the diamond surface [1, 11, 12].

Test-calculations have been performed for all three types of cubic structures, and only the diamond substrate was found to be useful in heteroepitaxially depositing a thin c-BN adlayer. The usage of both SiC and Si substrates resulted in the formation of an amorphous BN adlyer. This is a result that can be directly attributed to the large lattice mismatch between c-BN and β-SiC (or Si).

If the lattice mismatch between the substrate and the adhered film (in this case c-BN) is too large, it is advisable to use a model where the c-BN film is modelled by using a large cluster. The advantage is then that there will not be any effect of different crystalline cell dimensions in the x- and y-dimensions (i.e. parallel to the interface), as is the case for the 2D modeling of a substrate/c-BN slab. As is the case for the latter situation, the usage of two 2D slabs in the construction of the interface puts a restriction on the supercell borders. To be more precise, the two borders in the x-, y- and z-directions are identical. This fact will directly influence the modelled unit cell size of the two involving 2D slabs. It is only for the situation with a perfect lattice mismatch that this problem will not occur. Also, the larger the supercell size, the smaller the resulting lattice mismatch. Remember here that it is also best to base the supercell size on the crystalline cell size of the substrate since this substrate is much thicker than the grown adlayer film (i.e. it is generally more difficult to stretch a thick substrate material than the much thinner adlayer).

We do not have this problem if we model the adlayer by using a large cluster model. However, the cluster, which is a 0D material, must be large enough to be able to represent a 2D material. Various test-calculations can be performed to make sure that this is the case. One of these involves the calculations of partial density of states (pDOS) (see section 4.6.3). The pDOS of a free-standing c-BN slab is then compared with the pDOS of a c-BN cluster with the same thickness in the z-direction.

It is very difficult in many cases to make the decision if the lattice mismatch is small enough or not. It is then recommended to compare the results from calculations on a large enough cluster and the corresponding film for a carefully chosen problem (i.e. the effect of H-termination on the adhesion strength between c-BN and diamond).

9.1.2 Diamond substrate-effect of surface plane

9.1.2.1 Introduction

As presented above, in comparison to β-SiC and Si, the diamond substrate was found to be the best choice for a phase-pure growth of c-BN. We will, thus, continue to use this substrate in the following calculations. This is because, when using vapor phase deposition methods, the resulting quality of the films will be completely dependent on the initial nucleation and growth process, which in turn completely depends on the degree of lattice mismatch when growing heteroepitaxial films of, e.g. c-BN on diamond substrates. In addition to the almost perfect lattice mismatch in the diamond/c-BN interface, a careful layer-by-layer growth study can bring some light in the search for a successful recipe for c-BN growth. The reason is that a lot of questions still remain in this specific area [13–15]. Therefore, a careful geometrical and energetic-study of this layer-by-layer growth of c-BN (100) onto diamond (100) will be presented here. As is the case with all other examples in the book, this choice of materials is just an example of what can be done, and how it can be done for similar types of material systems.

In the present example, neither H nor F have been used to terminate the growing c-BN adlayer on the diamond (100) surface. The main reason is that here we want to study the pure effect of the substrate on the phase-pure growth of c-BN. Also, many PVD processes (like ion-beam-assisted evaporation) use growth precursors that are free from these types of terminating species in the UHV environment [16, 17]. The reason is that the total gas phase pressure is so low that termination of the BN surface with species like H, or F, is not expected to take place to any larger extent (at least much less than 100%). Boron carbide and elemental boron are examples of precursors that are used in these experimental setups.

9.1.2.2 A monolayer of N (or B) attached to diamond (100)

The energetically most stable configuration of a monolayer of N atoms adsorbed onto the 2×1-reconstructed diamond (100) surface, is shown in figure 9.1. Various adsorption sites for these N atoms were at first modelled and geometry optimized. The final structure corresponds to the very large interfacial adhesion energy of −780 kJ mol^{-1} per C–N bond, which, in fact, is a measure of the largest interfacial bond energy amongst all diamond/c-BN interfaces investigated (see figure 9.2).

For the corresponding situation with a monolayer of B atoms onto diamond (100), the most stable B layer formation was found to be the situation with B atoms in bridge formations along the C–C dimer chain (see figures 9.1(b) and (d)). As can be seen in figure 9.2, the adhesion energy for this most stable configuration (−540 kJ mol^{-1} per C–B bond) is more favourable, by 80 kJ mol^{-1}, than the adhesion energy for a heterostructural continuation of the underlying diamond (100)–2 × 1 structure (−620 kJ mol^{-1} per C–B bond). (The negative adhesion energies in figure 9.2 are presented as positive binding energies in the unit eV.) The calculated interfacial adhesion energy for the preferred B position is, thereby, smaller than for an N monolayer (−540 versus −780 kJ mol^{-1} per monolayer atom). This weaker interfacial C–B bond strength is further visualized in the structural geometry of the interface. The much weaker C–B bond has not the capacity to counteract the 2 × 1 reconstruction of the diamond (100) surface, which, on the other hand, N has (see figure 9.1). However, the most prominent difference between the N and B monolayers is that N can form a heteroepitaxial continuation of the c-BN(100) lattice, whilst B cannot (at least not in its most stable form).

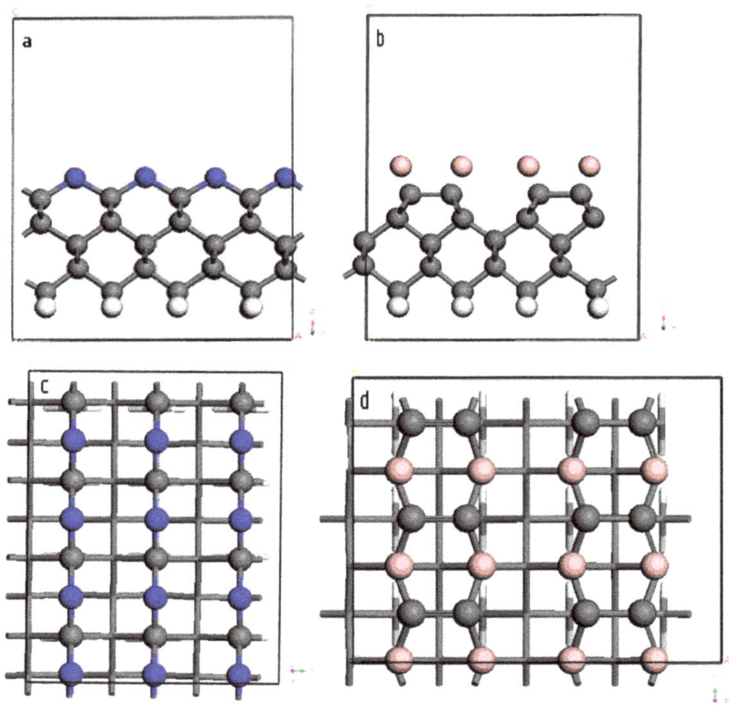

Figure 9.1. Periodic slabs containing the upper five layers of C in a 2×1-reconstructed diamond (100) surface. The bottom C layer is hydrogen terminated, and all structures are geometry optimized. Side views of (a) a heteroepitaxial adlayer of N, and (b) an adlayer of B (side views). The corresponding ontop views are shown in (c) and (d). [18], reproduced with permission © American Chemical Society.

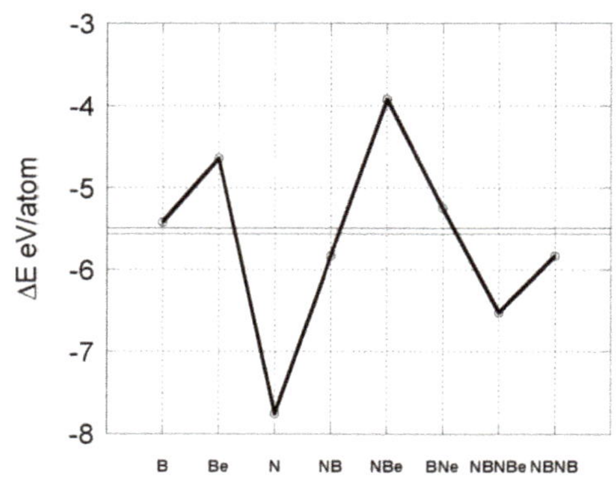

Figure 9.2. Interfacial binding energies (ΔE (eV) per binding adlayer atom) for geometry-optimized structures. The suffix *e* (in e.g. Be) means that the adlayer is heteroepitaxially attached to the diamond (100) surface. The values for the thicker BN adlayers are inserted as reference horizontal lines. The lower and upper horizontal lines show the binding for an adlayer with N (lower) versus B (upper) binding to the diamond substrate. [18], reproduced with permission © American Chemical Society.

9.1.2.3 A BN adlayer attached to diamond (100)

The most stable surface configurations of one monolayer of N (or B) on the diamond (100) substrate here is the starting structure onto which one additional layer of B (or N) is added. Two different interfacial structures were initially constructed for the situation with B attached to the first monolayer of N: one with a cubic phase of BN heterostructurally positioned on diamond (100) (i.e. with aligned *x*-axes) (see figure 9.3(a)) and another where the *x*-axis of c-BN is aligned with the *y*-axis of the diamond lattice (see figure 9.3(b)). As a result of the geometrical relaxation, the heteroepitaxially modelled c-BN layer stays perfectly cubic while the other c-BN structure, which was rotated with respect to diamond, became more amorphous-like (see figures 9.4(a)–(d)). Both BN adlayer structures show relatively strong interfacial adhesion energies to the underlying diamond substrate (−390 and −580 kJ mol^{-1} per binding N atom). However, the amorphous-like BN structure was found to be much more energetically favourable with a difference of 190 kJ mol^{-1} per binding N atom (see figure 9.3). When adding a second layer of N to the most stable conformation of the first monolayer of B, it results in a detachment of B from the surface (i.e. no BN adlayer can be formed on the diamond surface). It must here be stressed that the starting geometry for the B atoms were not heterostructural with diamond since that type of structure does not correspond to the most stable monatomic layer. However, it is of an interest to also study the possibility for a BN formation for these less stable heterostructural positions of B atoms. As a result of geometry optimization, the initial heteroepitaxial BN adlayer is found to stay heteroepitaxially cubic with a surface reconstructed N layer (see figures 9.4(e) and (f)) and with an interfacial adhesion energy of −530 kJ mol^{-1} per binding B atom. As can be seen figure 9.2, this interfacial adhesion energy is similar to the corresponding situation with a thicker c-BN adlayer (−550 kJ mol^{-1} per binding B atom).

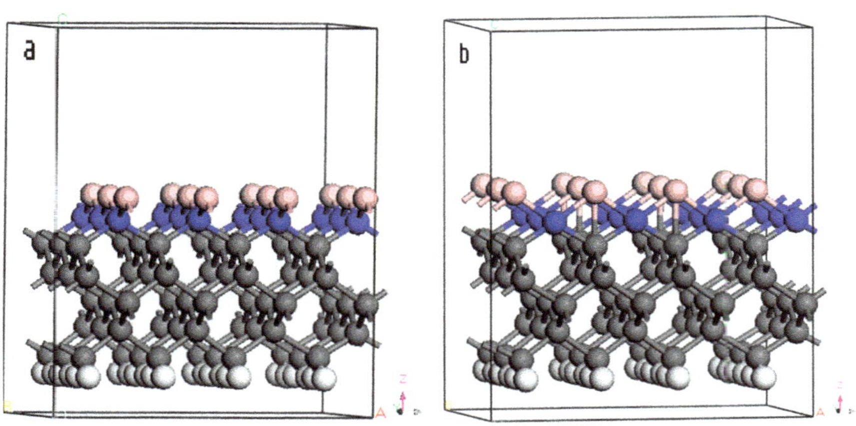

Figure 9.3. Two-atomic c-BN adlayers on diamond (100)–2 × 1 with N attached to the diamond substrate. The *x*-axis of the c-BN layer is either (a) in parallel with the *x*-axis of the diamond substrate, or (b) in parallel with the *y*-axis of the substrate. These model structures are initial ones, prior to geometry optimization. [18], reproduced with permission © American Chemical Society.

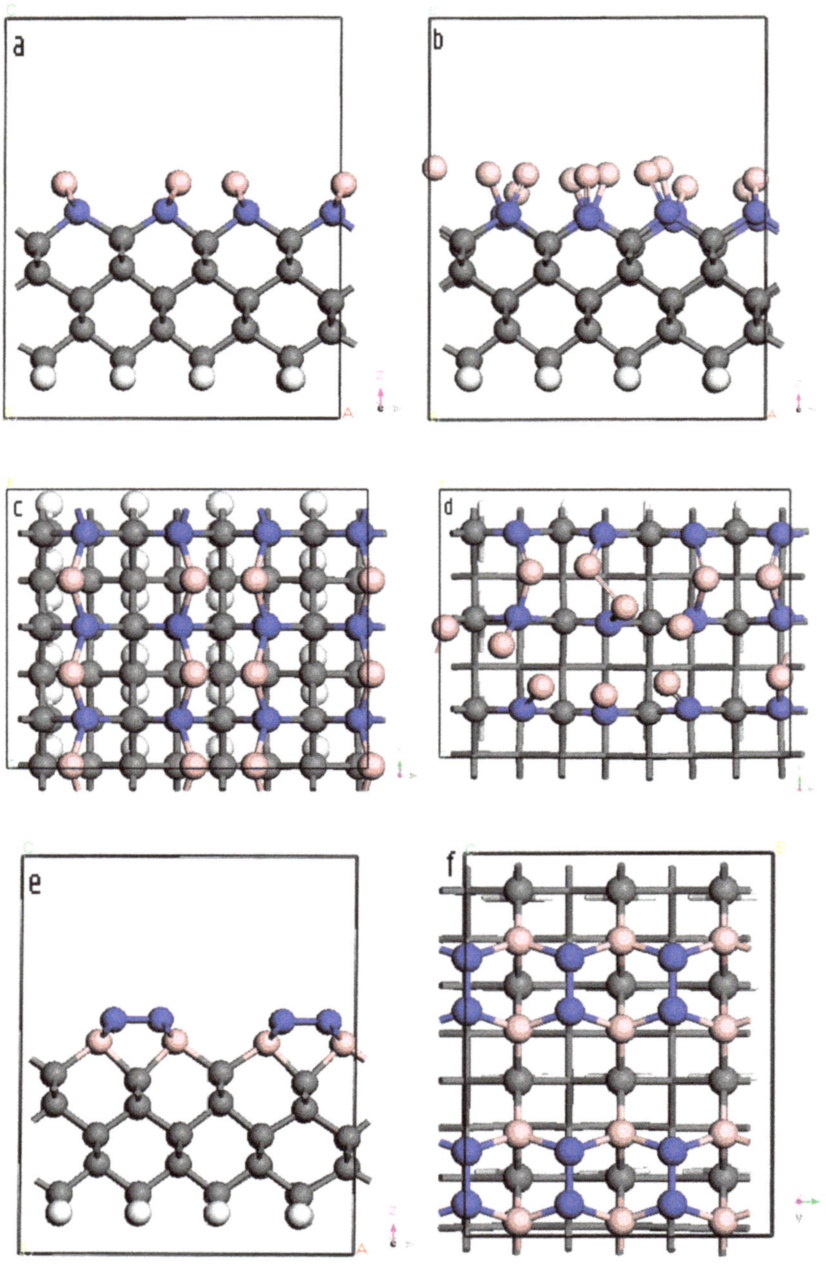

Figure 9.4. Geometry optimized diamond/BN interfaces: (a), (b) side views of the two-atomic thick BN layer (with N attached to the diamond substrate) for an initially heteroepitaxially built adlayer structure (a), and an initially non-heteroepitaxially built adlayer structure (b); (c), (d) corresponding top views; (e) side view and (f) top view of the initially built heteroepitaxially constructed adlayer with B attached to the diamond substrate. [18], reproduced with permission © American Chemical Society.

As presented above, the difference in total energy for the two different B monolayers is only 80 kJ mol^{-1} per binding B atom (in favour of the non-heterostructural adlayer) (see figure 9.2). During a layer-by-layer growth of BN, and starting from the most stable non-heterostructural configuration of B atoms, an energy barrier of ~80 kJ mol^{-1} per binding adlayer atom, hence, has to be overcome in order to grow a heteroepitaxial adlayer of c-BN. This is possible to occur, especially at higher deposition temperatures. However, even though it is somewhat energetically feasible for a two-layer c-BN layer to form with B attached to a diamond (100) substrate, it is less stable than the situation with an amorphous-like BN adlayer with N attached to diamond (with an adhesion energy of −530 versus −580 kJ mol^{-1} per binding adlayer atom). Furthermore, it is interesting to note that the heteroepitaxial c-BN structure with B attached to the diamond substrate is much more stable than for the heteroepitaxial structure with N attached to diamond (with an adhesion energy of −530 versus −390 kJ mol^{-1} per binding atom). These results point at the necessity to use an ALD-technique in the deposition of phase-pure c-BN films, whereby it is possible to choose the right atomtype to bind to the diamond substrate.

9.1.2.4 Two BN adlayers attached to diamond (100)

A BN adlayer is finally built on top of the two-atomic-layer thick BN adlayer, thereby, forming a four-atomic-layer thick BN adlayer on diamond (100). The purpose of this model is now to see if the geometrical structure and interfacial bond strength will dramatically change in the continuation of the layer-by-layer growth of BN onto diamond (100). For the initially heteroepitaxial twoatomic-layer BN structure (from here on denoted as BN1), with nitrogen attached to the diamond surface, the upper B atomic layer becomes reconstructed when adding the second BN layer to it (see figures 9.5(a) and (b)). Despite this tendency for reconstruction, an overall high degree of heteroepitaxy (with the underlying diamond (100) structure) is sustained. For this four-atomic layer thick BN adlayer, the resulting interfacial adhesion energy (to diamond (100)) becomes improved when compared to the situation with a two-atomic thick BN adlayer (−650 versus −390 kJ mol^{-1} per binding N atom). For a structure where the adlayer is initially built as a cubic BN structure, but rotated 90° with respect to the diamond lattice (from here on denoted as BN2), the addition of an extra BN layer results in an obvious amorphous structure (see figures 9.5(c) and (d)). For this amorphous BN adlayer, the interfacial adhesion energy becomes somewhat less energetically favourable when compared to the heteroepitaxial four-atomic thick adlayer described above (−580 versus − 650 kJ mol^{-1} per binding N atom). It is also interesting to note that the interfacial adhesion energies for the amorphous two- and four-atomic BN layers are identical, namely −580 kJ mol^{-1} per binding N atom.

The interfacial binding energies presented here have been compared for two main types of models for thick heteroepitaxial c-BN adlayers on diamond (100): one adlayer with N closest to the diamond substrate and one adlayer with B closest to the diamond substrate. As shown in figure 9.2, oscillations in the interfacial binding energy are observed when starting from N attached to the substrate in the initial

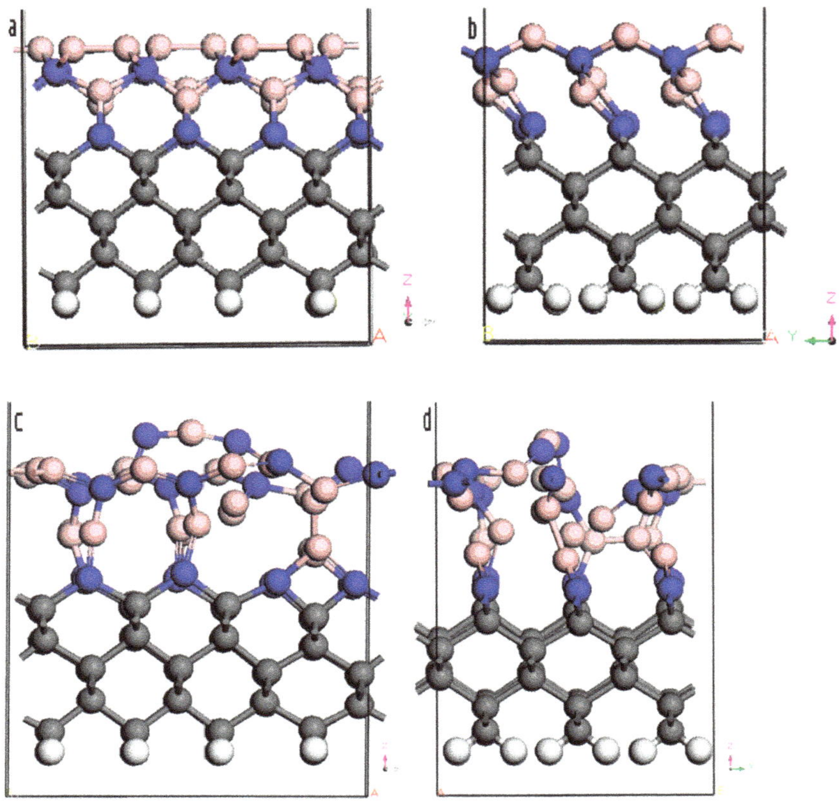

Figure 9.5. Geometry optimized diamond/BN interfaces: Side views of the four-atomic thick BN layer (with N attached to the diamond substrate) for initially (a), (b) heteroepitaxially and (c), (d) non-heteroepitaxially built adlayer structures. Two different side views are used, either along the *y*-axis (a), (c) or along the *x*-axis (b), (d). [18], reproduced with permission © American Chemical Society.

layer-by-layer growth of c-BN(100) on diamond (100). The heterostructural N monolayer binds extremely strongly to the substrate (with an interfacial adhesion energy of −780 versus −560 (thicker c-BN) kJ mol^{-1} per binding N). Depending on the following growth pathways, the next atomic layer (consisting of B atoms) will lead to either a heterogeneous c-BN structure or an amorphous BN structure. For the heterostructural two-atomic-layer structure, the addition of the B layer onto the N layer results in a marked weakening in the interfacial binding to the diamond substrate (with an adhesion energy from −780 to −390 kJ mol^{-1} per binding N). However, an adhesion strength of −390 kJ mol^{-1} is still strong, it is even somewhat stronger than an ordinary C–C bond (−346 kJ mol^{-1}). A^{-1} four-atomic layer of heteroepitaxial c-BN, on the other hand, is found to bind much stronger to the diamond (100) substrate: with an adhesion energy of −650 kJ mol^{-1} per binding N. In fact, this value is even larger than for the thicker c-BN adlayer with N closest the diamond substrate (with an adhesion energy of −560 kJ mol^{-1} per binding N).

On the other hand, the corresponding amorphous two- and four-layer structures are found to bind strongly with almost identical interfacial adhesion energies (−580 kJ mol^{-1} per binding N), which are also very similar to the interfacial adhesion energy for the thicker c-BN with N closest the diamond substrate. Hence, it is very important to go into the right (i.e. heterostructural) pathway when depositing c-BN.

9.1.2.5 Electron bond population versus adhesion strength during a layer-by-layer growth of c-BN(100) on diamond (100)

As presented in section 4.7.4, the electron bond population (i.e. electron density between two atoms) is generally regarded to be a measure of the covalent bond strength. The results of the calculated electron bond populations for the diamond/c-BN interfaces are shown in figure 9.6. Except for the C–N (C–B) bonds in the interfaces, the N–N (B–B) interatomic interactions within the first atomic layer in the c-BN adlayer are also included in this figure. As can be seen in figure 9.6, all diamond/c-BN interfaces show a numerically large electron bond population for the bonds between the surface C atoms and their binding N (or B) atoms. When building a BN adlayer sequentially, by starting from one monolayer of N and continuing to a thicker BN adlayer, no strict correlation between adlayer binding energy and C–N bond population was observed. The main reason for this observation is most probably the structural environment to the substrate-binding N atoms. It is possible to identify two different types of structural environments: (i) sp^3 hybridization of N

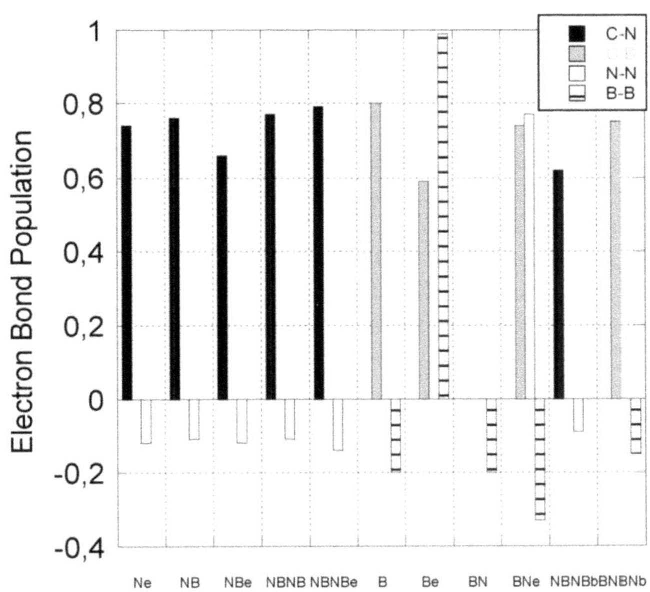

Figure 9.6. Presentation of electron bond populations for the bonds across the diamond/BN interface. The suffix *e* (in, e.g. Ne) is a notation for the adlayer that is heteroepitaxially built on the diamond (100) surface, and the suffix *b* (in, e.g. NBNBb) is a notation for the thicker heteroepitaxial BN adlayer. [18], reproduced with permission © American Chemical Society.

where no electron is filling antibonding states, and (ii) sp^3 hybridization where electrons occupy antibonding states (and thereby weaken the C–N bonds). Members of the first type of N structure are monatomic N and four-atomic BN adlayers. The interfacial adhesion energies for these adlayers (−780 vs −650 kJ mol^{-1} per binding N), together with the C–N bonding populations (+0.75 versus +0.78), are to be compared with the corresponding values for the second type of N structural environment: two-atomic and thicker BN adlayers. Based on the interfacial adhesion energies for these latter adlayers (−390 versus −550 kJ mol^{-1} per binding N), together with the C–N bonding populations (+0.65 and +0.62), it is clear that these adlayer formations form a separate group. The underlying reason for these observations is most probably bond weakening effects due to partial filling of antibonding N–C orbitals.

The calculated bonding populations for the interfaces with B binding to diamond show a much clearer trend for the heteroepitaxial interfaces. The following numerical order of interfacial adhesion energies has been obtained: −460 kJ mol^{-1} (heteroepitaxial B monolayer), −530 kJ mol^{-1} (heteroepitaxial BN layer), and −520 kJ mol^{-1} (thicker c-BN adlayer) per binding B. The corresponding C–B bond populations are +0.59, +0.74, and +0.75. It is obvious that when the bonding population is high, the interfacial binding energy is strong.

9.1.2.6 Degree of electron transfer versus adhesion strength during a layer-by-layer growth of c-BN(100) on diamond (100)

The degree of electron transfer between atoms in the diamond/BN interface can be used to explain the underlying causes for geometrical structures and interfacial binding energies. It can be obtained from calculated atomic charges and is generally regarded to be a strong indication of electrostatic interactions and, hence, ionic bonding. It is, thereby, a complement to calculations of electron bond populations, which indicate the degree of covalent bonding in the bonds.

The two material surfaces in an interface usually interact by forming overlapping orbitals. The factors that are governing the interaction strengths and bond types are (i) degree of spherical orbital overlap, and (ii) orbital energy difference (see figure 9.7). Depending on these factors, the interaction spans all the way from perfectly covalent bonds over polar covalent bonds to ionic bonds. This is also the situation with degree of electron transfer. Perfect covalent bonds do not show any

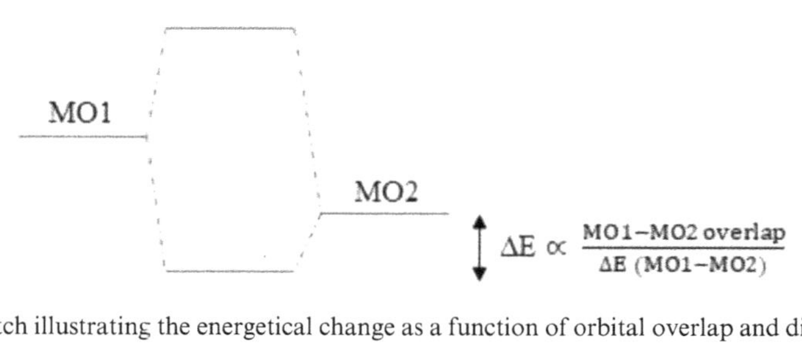

Figure 9.7. A sketch illustrating the energetical change as a function of orbital overlap and difference in orbital energy.

degree of electron transfer (and the calculated electron bond populations show the true covalency in the bonds). However, the more polar the bond is, the larger the degree of electron transfer. A complete electron transfer is reflected in a true ionic bond, with zero electron bond populations. The electron transfer over the diamond/c-BN interface is also affected by the circumstance that N is a Lewis base and B is a Lewis acid. This means that the N atoms are donating their filled sp^3 orbital to the empty sp^3 orbital of B. The N atoms will, thereby, contribute with one electron pair to the bond formed with the electron-deficient B atoms. In addition to this donation of electrons, the difference in electron negativity will induce a withdrawal of these binding electrons toward the N atoms.

9.1.2.6.1 c-BN adlayers with N closest to diamond

For the situation with just a monolayer of N in heteroepitaxy with the underlying diamond (100)–2×1 surface (being the energetically most preferred structure), all three indicators for interfacial bond strength show the same results; adhesion energy, electron bond populations, and degree of electron transfer. As can be seen in table 9.1 and figures 9.2 and 9.6, the adhesion energy of the interfacial bonds, the electron population within these bonds, and degree of electron transfer along the bonds indicate a very strong interfacial C–N bonding situation. As can be understood from table 9.1, the degree of electron transfer is the largest (the surface C atom in diamond donates approx. 0.25e to its binding N atom) of all interfaces in the present example, indicating a strong interaction between the diamond substrate and the c-BN adlayer. The driving force for this electron transfer is the difference in electronegativity between the elements C and N. The resulting lone pair orbitals on the attached N atoms, which are each more than half-filled with electrons, now have the possibility to overlap with each other and thereby form very weak antibonding N–N interactions.

As can be seen in figure 9.6, the indication of overlapping lone-pair orbitals among these N atoms is observed in the form of negative N–N electron bond populations. When adding a monatomic B layer on top of the first attached and geometry optimized N adlayer, a strong correlation between interfacial bond strength, electron bond population, and degree of electron transfer is also observed

Table 9.1. Variation in atomic charges as a result of electron transfer across the diamond (100)/BN interface, with N binding to the diamond substrate.

Atom layer	Ne	NB	NBe	NBNB	NBNBe	NBNB (bulk)
C charge (e)	+ 0.25	+ 0.22 to + 0.24	+ 0.21	+ 0.21 to + 0.25	+ 0.21	+ 0.19
N charge (e)	− 0.25 to − 0.30	− 0.41 to − 0.46	+ 0.22 to + 0.24	− 0.39	− 0.37 to − 0 46	− 0.40 to − 0.48
B charge (e)		+ 0.14 to + 0.24	+ 0.16	+ 0.15 to + 0.55	+ 0.54 to + 0.72	+ 0.66

here (both for the resulting heteroepitaxial BN adlayer and for the more amorphous BN adlayer). As can be seen in table 9.1 and figures 9.2 and 9.6, the somewhat weaker interlayer bonding observed for the heteroepitaxial BN adlayer (with an adhesion energy of -580 kJ mol^{-1}) is accompanied by a smaller degree of electron transfer (0.22e) from C to N. However, the substrate-binding N layer will receive electrons from both the underlying diamond and from the upper B layer (i.e. due to differences in electronegativity). Since the N element has five valence electrons but only four bonds, anti-bonding states must be filled. This is observed here as weaker C–N binding interactions (with electron bond populations of +0.66) and as a more pronounced tendency for anti-bonding N–N interactions (with electron bond populations of -0.12). A similar trend is also observed for the amorphous BN counterpart. Furthermore, larger variations in degree of electron transfer are observed for these types of BN adlayers on diamond (111). This is due to the larger variations in C–N and B–N bond lengths. It is, hence, obvious that B prefers three valence electrons and N five valence electrons in the BN adlayer. Consequently, the surface B layer reconstructs and form pairwise B–B binding interactions (see figure 9.4(a)).

The situation with a four atomic BN adlayer on diamond (100) is very similar to the two-atomic BN scenario. However, for the situation with a heteroepitaxial interface, the preference for three valence electrons, and a more planar structure, is more apparent for the B atoms in the second atomic layer (i.e. positioned in between the attached N layer and the second N layer). As a result, the substrate-attached N atoms will bind to only three other atoms, with no tendency for filling the C–N antibonding orbitals (see figures 9.5(a) and (b)). The four-atomic BN adlayer interfacial binding energy will thereby become larger when compared to the corresponding scenario for the two-atomic BN adlayer (with adhesion energies of -650 versus -390 kJ mol^{-1} per binding N atom) (see figure 9.2). Moreover, the more pronounced amorphous structure of the four-atomic BN adlayer can be explained by the large variation in degree of electron transfer from B to its surrounding atoms (0.17e–0.39e).

9.1.2.6.2 c-BN adlayers with B closest to diamond

For the situation with a B monolayer attached to the diamond (100) surface, the direction of the electron transfer is the opposite to the one for the N monolayer. Hence, the B atoms will partially donate electrons to the C atoms. For the most stable, non-heteroepitaxial B adlayer, the electron transfer for each B is 0.32e. This value is high and correlates well with both the interfacial adhesion energy (see figure 9.2) and the electron bonding populations (figure 9.6). It is also apparent from figure 9.6 that there is an appreciable amount of anti-bonding interactions between the B atoms. The predominant degree of electron transfer from the B atoms, in combination with the B–B antibonding interactions, speaks in favor of the formation of a three-electron bond situation over the C–B–C formation. This phenomenon is more common for electron-deficient atoms that are very similar with respect to electronegativity (identical to or less than 0.5) [20]. The difference in electronegativity for B and C is just 0.5. Hence, one of the B electrons will be

Table 9.2. Variation in atomic charges as a result of electron transfer across the diamond (100)/BN interface, with B binding to the diamond substrate.

Atom layer	B	Be	BN	BNe	BNBN (bulk)
C charge (e)	− 0.32	− 0.15 to − 0.35	− 0.01 to − 0.03	− 0.14 to − 0.39	− 0.21
B charge (e)	+ 0.30 to + 0.32	+ 0.02 to + 0.33	+ 0.93 to + 0.96	+ 0.54	+ 0.52
N charge (e)			−0.93 to − 0.96	−0.35	−0.63

involved in a C–B–C bond with the substrate, and the rest of the electrons will form a lone pair, which overlaps with neighboring B lone pairs.

As can be seen in table 9.2, there is a larger spread in degree of electron transfer (from the B atoms to the C atoms) for the heteroepitaxially built diamond/B interface: 0.15e–0.35e. The major explanation is the variation in orbital overlap, caused by the different B–C bond lengths: 1.52–2.49 Å. On average, the degree of electron transfer within the interface is much smaller compared to the energetically more stable non-heteroepitaxial structure (0.24e versus 0.32e). This smaller value of electron transfer correlates strongly with the somewhat weaker interfacial adhesion energy (−464 versus −542 kJ mol^{-1} per binding B) (see figure 9.2) and smaller electron bond populations +0.59 versus +0.80 (figure 9.6). As can also be seen in figure 9.6, there are large positive electron bond populations between the B atoms in the adlayer, which strongly support the pairwise B–B binding observed in the structure (see figures 9.1(b) and (d)). Hence, there is also a strong preference for three valence electrons within the B atoms for this interfacial structure.

The heteroepitaxial BN adlayer obtained by adding a monatomic N layer on top of the first attached heteroepitaxial B layer shows an electron transfer of 0.54e from the B atoms to surrounding C and N atoms (see table 9.2). Due to differences in electronegativity, the substrate-binding B atoms will partially donate electrons to both the underlying C atoms and their binding N atoms. As can be seen in figures 9.4(e) and (f), the 'capture' of B atoms between the diamond substrate and the surface N atomic layer will force the B atoms to stay sp^3-hybridized. Every B atom binds to four other atoms (two N and two C), with C–B and B–N bonds that are very strong (with bonding population of +0.74 and +0.73, respectively). Boron is an element that has only three valence electrons. However, the obtained geometrical structure, in addition to the high electron bond population in each of the four bonds, prove that N has donated an electron pair to form a fourth B–N bond (although strongly polarized). This can be explained by the fact that N is a Lewis base and B is a Lewis acid. The remaining valence electrons on N will, after the formation of two N–B bonds, have the possibility to form strong N–N bonds. As can be seen in figure 9.6, calculated N–N electron bond populations (+0.77) give evidence of strong N–N bond formations, which in turn explains the observed surface reconstruction (see figures 9.4(e) and (f)).

9.2 Combined effect of substrate and terminating species on the vapour phase growth of c-BN

9.2.1 Introduction

It is also of the largest interest to add the influence of terminating species in the study of substrate-induced c-BN growth. As was shown in section 8.4, terminating species influence the surface geometrical structure of c-BN and are, thereby, highly important for the trapping of the cubic structure between the substrate and the terminating adlayer (especially closest to the substrate). The problem with vapour phase growth of phase-pure c-BN will, hence, boil down to the problem to find the most optimal substrate/terminating species combination. Practically seen, the substrate/c-BN adhesion energy and structural stability of c-BN thin films must therefore be carefully investigated for various candidates of substrates and terminating species. DFT-calculations are for these tasks a highly valuable technique to use. The other reason to include the effect of surface terminating species on the initial c-BN growth (i.e. first atomic layer growth of c-BN) is the more recent usage of low substrate biased plasma CVD setups (in addition to 'brute force' PVD methods). Since these methods use a much higher gas phase pressure it is expected that the growth c-BN film becomes terminated with gaseous species. As presented in chapter 8, a rather successful c-BN growth has earlier been obtained by using rough and faceted diamond (100) substrates at low substrate bias (−20 V) and by using an electron-cyclotron-resonance microwave plasma CVD setup [21]. The gas mixture contained N_2, BF_3, and H_2 as reactant gases, with a pressure kept at about 2×10^{-3} Torr. This resulted in 2 μm thick c-BN films with a low film stress of about 2 GPa. Even though rather good results have been obtained with CVD methods based on low-energy ion bombardment, it has never been shown to be possible to use more gentle CVD (or ALD) methods in the growth of phase pure c-BN thin films. Especially the ALD method should be highly versatile for the development of a precursor design for c-BN growth. As presented in chapter 2, the ALD technique is a thin film deposition method that sequentially introduces different gaseous precursors into the reaction chamber, thereby depositing a monatomic layer-by-layer thin film of desired form and with a high and even thickness control. Experimentally, it would be highly feasible to tailor-make the necessary surface termination design by using this specific deposition technique in the growth of phase-pure c-BN.

For the same reason as presented in section 9.1, diamond (100) will also be used here as the substrate for phase-pure c-BN growth. Moreover, the terminating species are H and F, which were found to be very efficient in the c-BN growth calculations described in chapter 8. Because of previous experimental results, only H is used here as terminating specie on the N atoms, while both H and F species are used for the B atoms [22].

9.2.2 Structural geometries and surface stabilization

9.2.2.1 A monolayer of N (or B) on diamond (100)

For one monolayer of N, perfectly heteroepitaxially positioned on top of a diamond (100) surface, surface termination by H does not render any major structural changes when compared with the non-termination scenario. As can be seen in table 9.3, the

Table 9.3. Interfacial adhesion energies ΔE (kJ mol^{-1} per binding adlayer atom) for geometry optimized structures. The suffix e (in, e.g. Ne) means that the BN adlayer is heteroepitaxial with respect to the diamond (100) substrate.

Adlayer (kJ mol^{-1})	No termination	H-termination	F-termination
N_e	−780	−740	
B_e	−460	−580	−610
B	−540	−380	−430
NB_e	−390	−610	−600
NB	−580	−490	−420
BN_e	−520	−530	
BN		−500	
NBNB	−580	−580	−500
$NBNB_e$	−650	−640	−630
$NBNB_e$	−680	−590	
BNBN	−410	−450	
$NBNBNB_e$	−330	−940	
NBNBNB	−440	−460	
$NBNBNBNB_e$	−470	−520	
NBNBNBNB	−450	−460	

interfacial adhesion energy for a non-terminated N surface is −780 kJ mol^{-1} per binding N atom, which is to be compared with a value of −740 kJ mol^{-1} per binding N atom for the situation with H-termination. The corresponding results for the heterostructural B monolayer are, however, quite different. Stronger interfacial adhesion energies were obtained when terminating by H (or F) species (see table 9.3). For the situation where no terminating species is used, the non-heteroepitaxial interfacial structure is calculated to be energetically more favored than the heterostructural counterpart. Hence, surface termination induces stabilization of the interfacial adhesion strength for the heteroepitaxial structure. However, the various types of terminating species stabilize differently (with adhesion energies of −610 kJ mol^{-1} (H-termination) versus −580 e (F-termination) per B atom). Except for the role of the terminating species to induce a strong heteroepitaxial adhesion to the surface, it is also noteworthy that sp^3-hybridization of the upper surface atoms will take place (i.e. an induced cubic structure) (see figures 9.8(a)–(d)). It is also interesting to note that F-termination (compared to H-termination) will, for both non- and heteroepitaxial structures of the B monolayer formation, result in slightly improved interfacial bond strengths (see table 9.3).

Various changes in surface geometrical structure are, however, obtained, depending on the matching of the BN adlayer to the underlying diamond substrate. For the heteroepitaxial interface, the F-terminated B monolayer becomes slightly surface reconstructed (but still crystalline). The F adsorbates are slightly moved in the negative y-direction for every other row along the x-axes, resulting in a (2×1)-reconstruction (see figures 9.8(e) and (f)). The H-terminated N monolayer does also

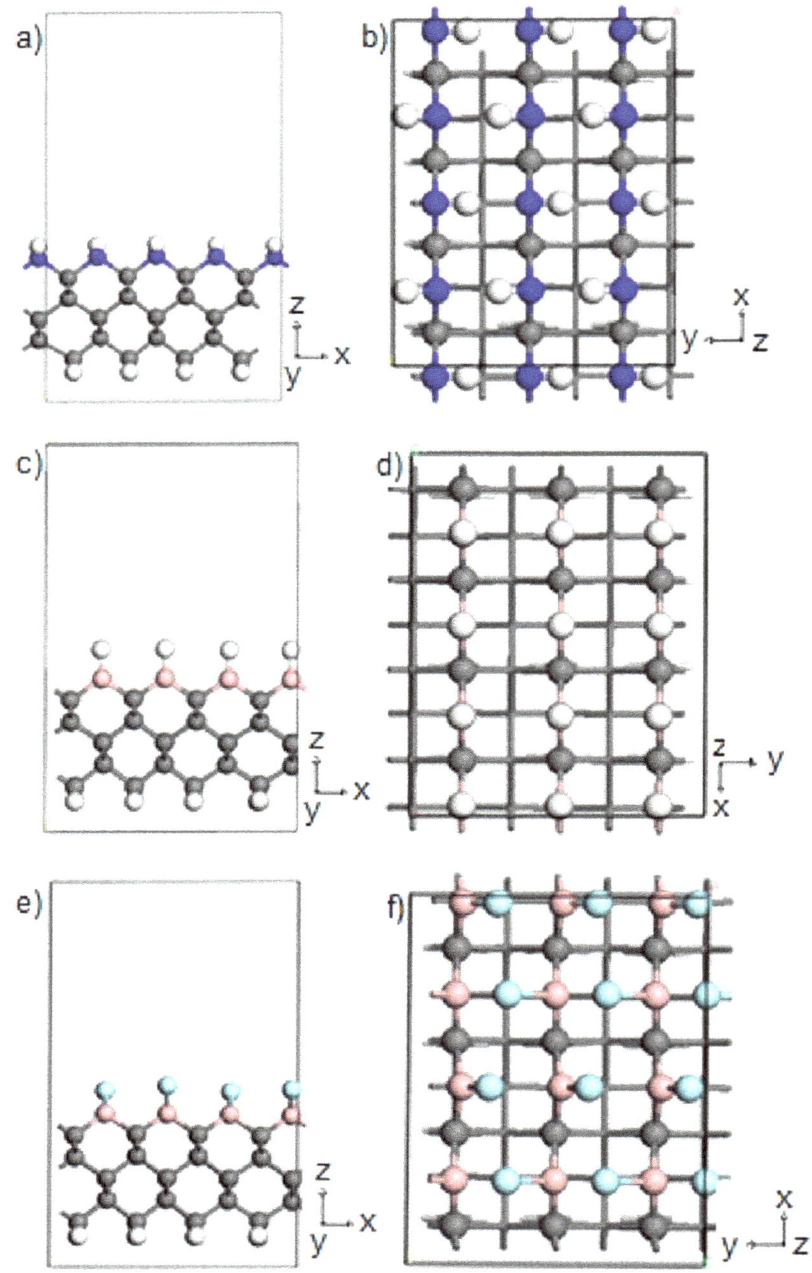

Figure 9.8. Periodic slabs containing five atomic layers of C in a diamond (100)−2×1 substrate. The bottom C layer is H-terminated, and all structures are geometry optimized. (a), (b) An H-terminated monolayer of heteroepitaxial N. (c), (d) An H-terminated monolayer of heteroepitaxial **B**, and (e), (f) an F-terminated monolayer of heteroepitaxial **B**. Two different side views are presented: either along the *y*-axis (a), (c), (e) or from above (b), (d), (f). [19], reproduced with permission © American Chemical Society.

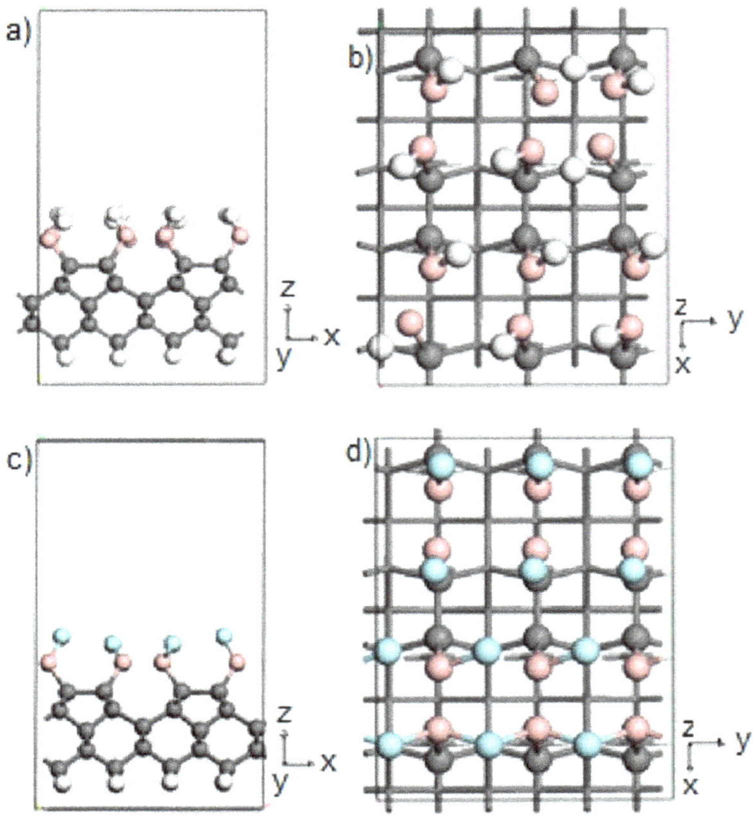

Figure 9.9. Periodic slabs containing five atomic layers of C in a diamond (100)–2×1 substrate. The bottom C layer is H-terminated, and all structures are geometry optimized; (a), (b) an H-terminated monolayer of non-heteroepitaxial B. Two different side views are presented: either along the *y*-axis (a), (c) or from above (b), (d). [19], reproduced with permission © American Chemical Society.

undergo this type of (2×1)-reconstruction (see figures 9.8(a) and (b)). However, the F-terminated B-monolayer is much more ordered than the H-terminated one for the non-heteroepitaxial interface, forming a (4×1) reconstruction (see figures 9.9(a)–(d)).

9.2.2.2 A two-atomic BN adlayer on diamond (100)

An additional monatomic layer of N (or B) is now added to the initial surface configuration of one monolayer of N (or B) on the diamond (100) substrate. For the situation where a B layer is added heteroepitaxially onto the first optimized N layer, the geometrical optimization results show a clear difference between the non-terminated and terminated adlayer (see figure 9.9). As can be seen in table 9.3, the terminated heteroepitaxial BN adlayers were found to attach much more strongly to the diamond substrate (compared to the non-terminated situations). As a result of the geometry optimizations, both the H- and F-terminated B surfaces became (2×1)-reconstructed. As can be seen in figure 9.10, the H-terminated

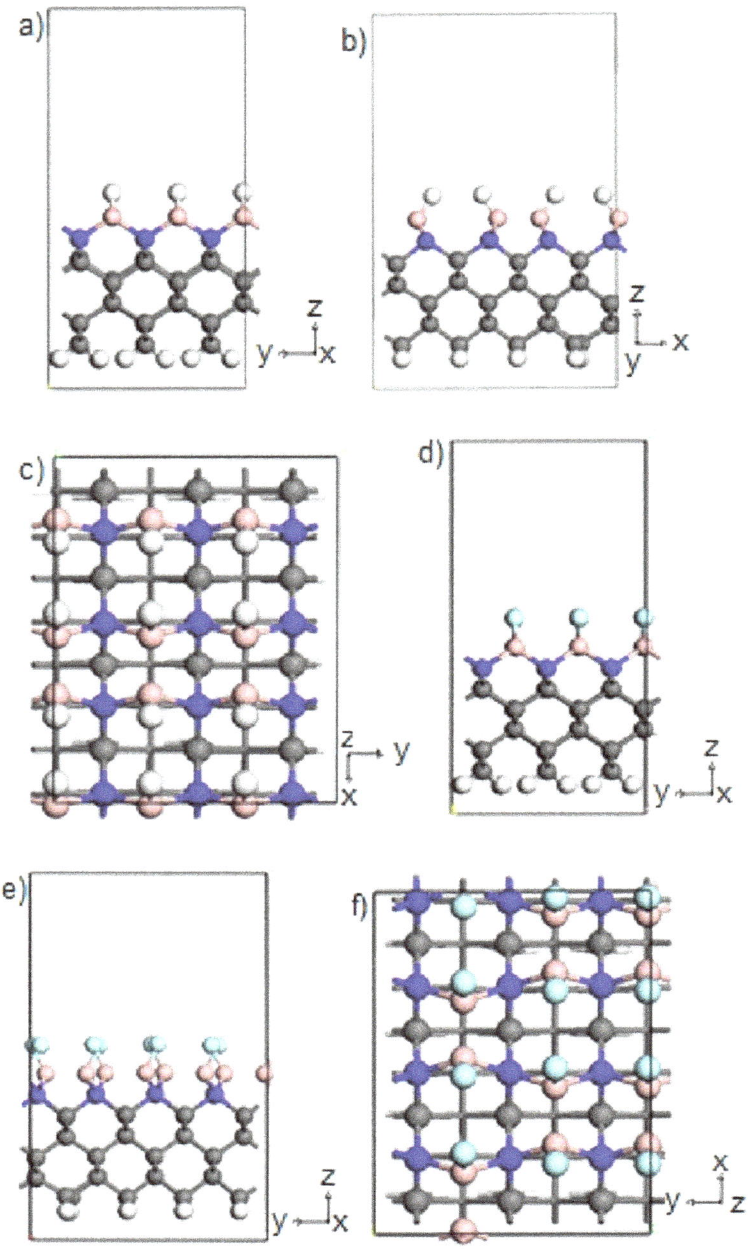

Figure 9.10. A geometry-optimized two-atomic layer of initially heteroepitaxial BN attached to diamond (100) (with N closest to the substrate). (a)–(c) Three different side-views of the H-terminated interface (either along the *y*-axis (a), along the *x*-axis (b), or from the top (c)). (d)–(f) Corresponding geometrical structures for F-terminated structures. [19], reproduced with permission © American Chemical Society.

surface shows B–B dimer rows, while the F-terminated surface shows B–B dimer formations in zigzag patterns. The most probable explanation for this difference in dimer formation is the more pronounced electron repulsion between the chemisorbed F species (as compared with H adsorbates). On the contrary, a completely different result is obtained for the initially non-heteroepitaxial interfacial structure. As compared to the H- and F-terminated c-BN surfaces, the non-terminated BN adlayer is found here to adhere stronger to the underlying substrate (see table 9.3). Furthermore, the F-terminated BN surface becomes amorphous-like, while the H-terminated one still shows a crystalline structure (although somewhat disordered; see figure 9.11).

As a conclusion, the use of surface-termination will steer the c-BN growth towards an ideal heterostructural adlayer onto diamond (100). Since this conclusion is solely based on thermodynamic considerations, it is highly likely that the preference for a cubic, but non-heteroepitaxial, growth of a H-terminated c-BN thin film will be enhanced at higher temperatures (i.e. since energy barriers must be

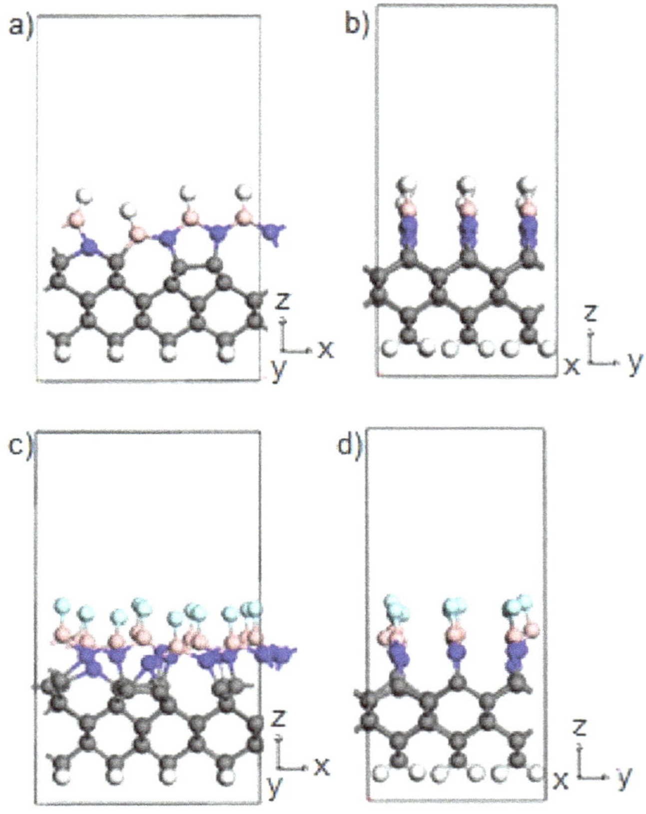

Figure 9.11. Geometry-optimized BN/diamond interfaces, including a non-heteroepitaxial BN adlayer with N closest to the diamond substrate. (a), (b) An H-terminated interface, using two different side-views (either along the y-axis (a) or along the x-axis (b)). (c), (d) Corresponding geometrical structure for the F-terminated BN adlayer. [19], reproduced with permission © American Chemical Society.

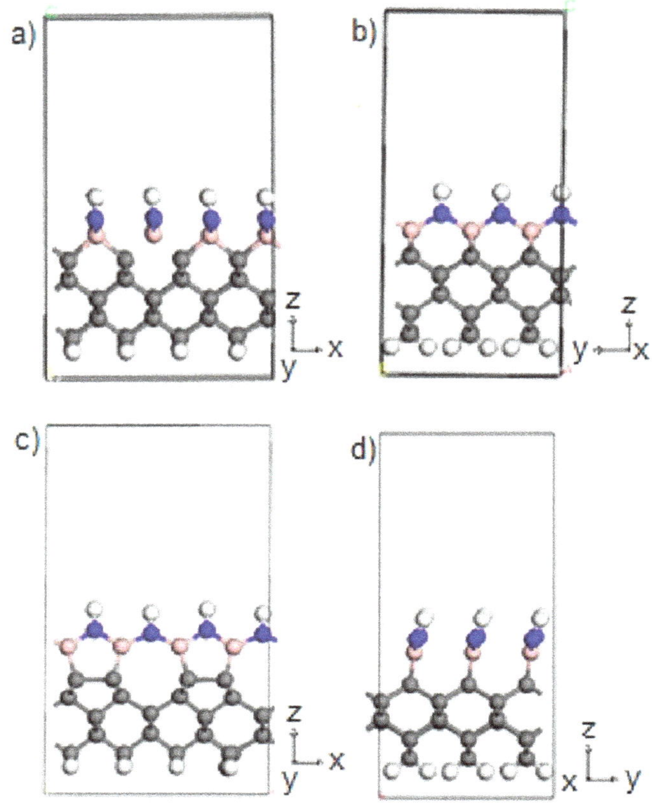

Figure 9.12. Geometry-optimized H-terminated BN/diamond interfaces, including heteroepitaxial and non-heteroepitaxial BN adlayers, with B closest to the diamond substrate. (a), (b) The heteroepitaxial structure, using two different side views (either along the *y*-axis (a) or along the *x*-axis (b)). (c), (d) The non-heteroepitaxial structure, using two different side views (either along the *y*-axis (c) or along the *x*-axis (d)). [19], reproduced with permission © American Chemical Society.

overcome). When adding a second layer of N onto the first attached non-heterostructural B monolayer (being the energetically most stable form), the calculated interfacial adhesion energy is much higher for H-termination (compared to the non-termination scenario; see figures 9.23(c) and (d)). In fact, a non-terminated adlayer is not found to bind to the diamond substrate at all. As can be seen in table 9.3, H-termination will result in a two-atomic BN adlayer that is even more strongly attached to the diamond substrate than the first attached B layer. When adding a second monolayer of N onto the first attached heterostructural B monolayer (see figures 9.11(a) and (b)), there will only be a minor difference in interfacial binding for the situations of H-termination versus non-termination (see table 9.3). Hence, it is possible to draw the conclusion that surface termination with H will not, for this very early stage of growth, control the growth toward the preferred cubic phase. Instead, a mixture of phases is expected to be formed.

9.2.2.3 A four-atomic BN adlayer on diamond (100)

Heteroepitaxial and non-heteroepitaxial BN adlayers are, thereafter, built on top of the two-atomic thick hetero- versus non-heteroepitaxial BN adlayer (i.e. thereby forming four-atomic thick BN adlayers). H- and F-terminated BN/diamond interfaces are also compared here with the corresponding non-terminated situations. For the situation with an initial heteroepitaxial four-atomic BN adlayer, with N closest to the diamond substrate, there is no significant F-termination (−600 to −630 kJ mol^{-1}) per binding N. Hence, an increased adlayer thickness from a two-atomic to a four-atomic BN adlayer will only induce small changes in the interfacial adhesion energy when using surface termination. However, a completely different situation occurs when using non-termination. As can be seen in table 9.3, a four-atomic thick BN adlayer will then bind with approximately a factor of two when compared with a two-atomic thick BN adlayer. Moreover, as mentioned above, there is no significant difference in interfacial adhesion energy when comparing the non-terminated surface within the H-, or F-, terminated one.

The differences in geometry optimized structures are, however, much more significant when comparing a two-atomic thick c-BN adlayer with a four-atomic thick one (see figure 9.13). The terminated BN adlayer becomes much more ordered, though reconstructed, in the form of a cubic BN structure for the heterostructural four-atomic thick interface. The terminating H (or F) atoms do apparently succeed in upholding the cubic structure of the c-BN surface atoms. On the other hand, the corresponding non-terminated BN adlayer shows a less ordered structure. The B atoms on the bare surface reconstruct since they strive toward a three-valence environment and, hence, a more planar structure. The strive by the B atoms to flatten out and bind to only three N neighbors is even more visible for the second B layer. In addition, the N atoms prefer a bonding situation with five valence electrons. Both of these tendencies for B and N will jointly cause the disrupton of some B–N bonds in the BN structure.

The non-heteroepitaxial structures show slightly lower interfacial adhesion energies when compared to the corresponding heteroepitaxial ones. In addition, the interfacial adhesion energies for the non-terminated and terminated non-heteroepitaxial structures are very similar.

Moreover, the resulting geometrical structures of the H- and F-terminated c-BN surfaces are both amorphous-like, and very similar (see figure 9.14). Hence, the formation of a heteroepitaxial BN adlayer, with N closest to the diamond substrate, seems to be the energetically most preferred result when using non-terminated BN adlayers (under the assumption of a layer-by-layer growth process). Moreover, the formation of heteroepitaxial cubic BN adlayers does also seem to be energetically favored when terminating with either H or F species.

A four-atomic thick c-BN adlayer, with B closest to the diamond substrate, was also initially built as either a heteroepitaxial or non-heteroepitaxial (but cubic) continuation of the diamond lattice. As a result, the H-termination does not induce any difference in interfacial adhesion energy when going from a two-atomic to a four-atomic thick c-BN adlayer (see table 9.3). These results are to be compared with

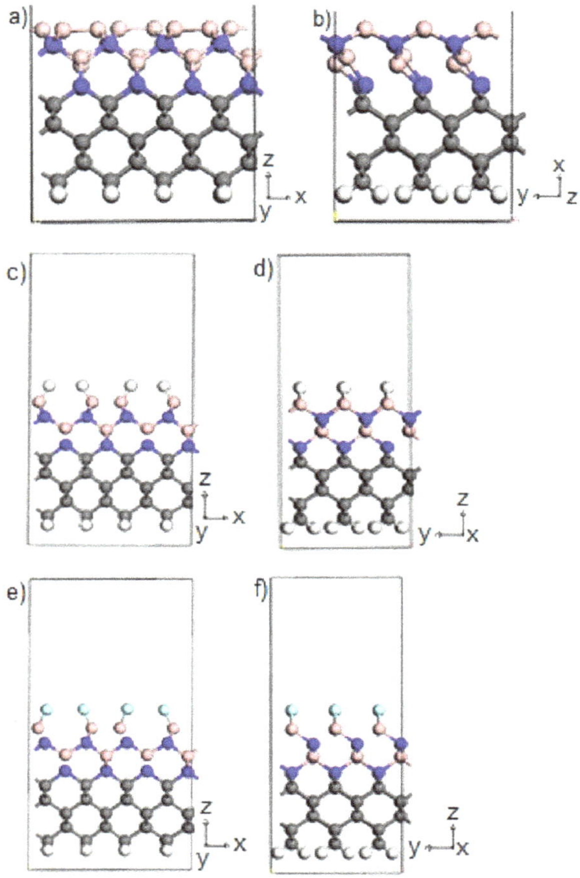

Figure 9.13. Geometry-optimized heteroepitaxial interfaces of a four-atomic thick adlayer of BN, with N attached to the diamond (100) substrate. (a), (b) The non-terminated interfaces, using two different side views (along the *y*-axis (a) and along the *x*-axis (b)). (c), (d) The H-terminated interfaces, using two different side views (along the *y*-axis (c) and along the *x*-axis (d)). (e), (f) The F-terminated interfaces, using two different side views (along the *y*-axis (e) and along the *x*-axis (f)). [19], reproduced with permission © American Chemical Society.

a corresponding non-terminated heteroepitaxial structure, for which a slightly stronger interfacial binding energy is observed.

This is also the situation for the non-heteroepitaxial structures: -500 kJ mol^{-1} per binding B (two-atomic) versus -450 kJ mol^{-1} per binding B (four-atomic) (see table 9.3). However, the non-heteroepitaxial cubic adlayer structure shows a completely different result for a non-terminated surface, in which case the two-atomic thick BN adlayer does not bind at all to the diamond (100) surface. The four-atomic thick c-BN adlayer is observed to bind with an interfacial adhesion energy of -410 kJ mol^{-1} per binding B. As can be seen in figure 9.15, both the heterostructural and non-heterostructural structures will become crystalline as a result of the

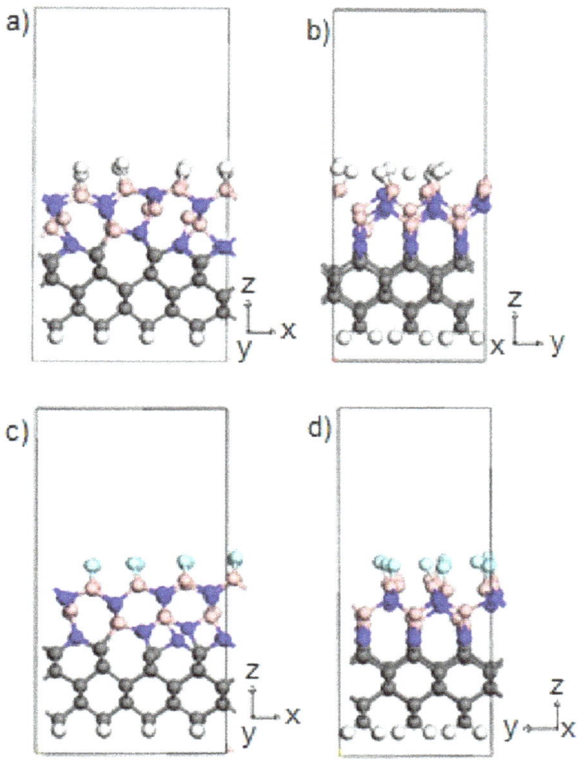

Figure 9.14. Geometry optimized non-heteroepitaxial interfaces of a four-atomic adlayer layer of BN, with N attached to the diamond substrate. (a), (b) The H-terminated interfaces, using two different side views (along the y-axis (a) and along the x-axis (b)). (c), (d) The F-terminated interfaces, using two different side views (along the y-axis (c) and along the x-axis (c)). [19], reproduced with permission © American Chemical Society.

geometry optimization. From all of the energy values presented in table 9.3, one can also draw the conclusion that only small differences in adlayer adhesion energies are observed for the H-terminated and non-terminated non-heteroepitaxial surfaces. For the heteroepitaxial c-BN structure, the non-terminated surface shows a slightly stronger adhesion energy.

9.2.2.4 Six- and eight-atomic BN adlayers on diamond (100)

When continuing up to six and eight atomic BN layers attached onto the diamond substrate, the non-terminated scenario is also of interest here to study in comparison with the terminated ones. However, only H-terminated surfaces are being studied since this termination type is found to give very similar BN adlayer adhesion energies, as compared to F-termination, for the situations with one, two, and four atomic layers, respectively. Only interfaces with N closest to the diamond substrate have then been calculated for adlayers built from three BN

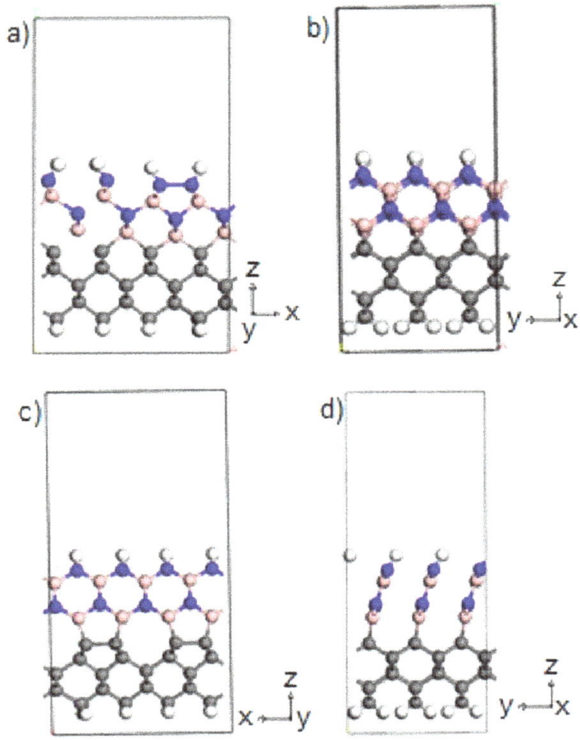

Figure 9.15. Geometry optimized interfaces of BN/diamond with B attached to the diamond substrate. All interfaces are H-terminated. (a), (b) A structure obtained from an initial heteroepitaxially constructed four-atomic BN adlayer (along the *y*-axis (a) and along the *x*-axis (b)). (c), (d) A structure obtained from an initial non-heteroepitaxially constructed four-atomic BN adlayer (along the *y*-axis (c) and along the *x*-axis (d)). [19], reproduced with permission © American Chemical Society.

layers (i.e. six atomic layers). As a result of the geometry optimizations, the initially heteroepitaxial H-terminated interface keeps its cubic structure (see figure 9.16). Furthermore, the interfacial adhesion energy is strong. It is even stronger than for the corresponding situations with one, two, and four BN layers (with N closest to the diamond substrate) (see table 9.3). Furthermore, all of these interfacial adhesion energies are much stronger than the one for a non-terminating interface built by three BN layers.

For the non-heteroepitaxial three BN layer thick H-terminated adlayer, an appreciably weaker interfacial adhesion energy is obtained (as compared with the heteroepitaxial counterpart (see figure 9.17)). However, the adhesion energy is almost identical to the corresponding non-terminated BN adlayer. Both the initial non-heteroepitaxial non-terminated and H-terminated surfaces result in crystalline cubic BN adlayers with a minor tendency for B to bind to only three N (due to the preferred number of valence electrons: three) and N to also bind to three B (and thereby fulfill a stable five-valence electron shell) (see figure 9.17).

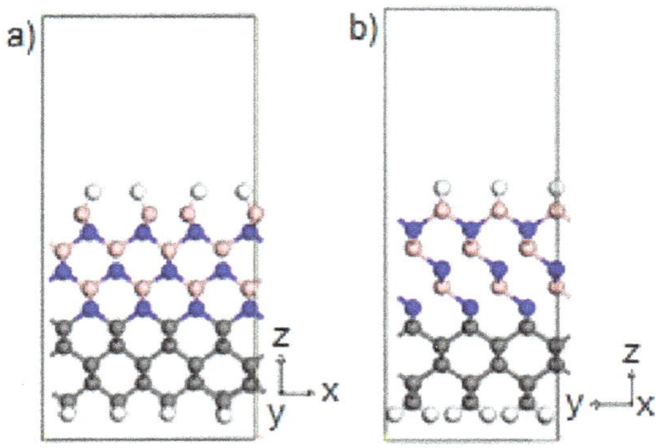

Figure 9.16. Geometry optimized interfaces of BN/diamond obtained from heteroepitaxially constructed six-atomic BN layers (with N attached to the diamond substrate). (a), (b) The H-terminated interface, using two different side views (along the *y*-axis (a) and along the *x*-axis (b)). [19], reproduced with permission © American Chemical Society.

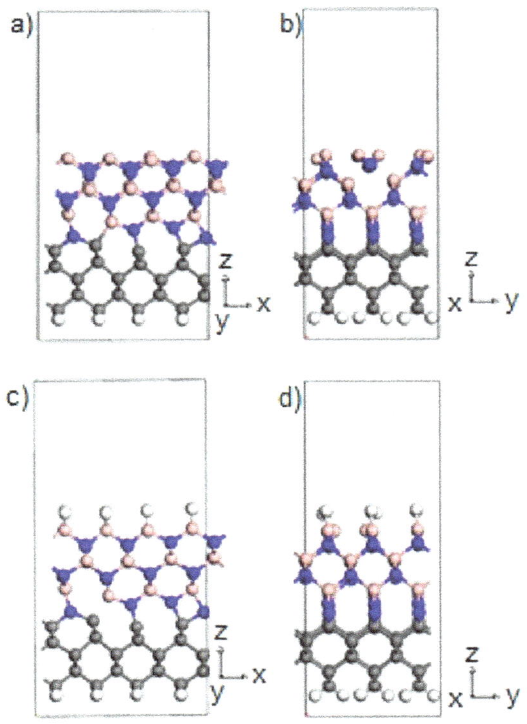

Figure 9.17. Geometry-optimized interfaces of BN/diamond obtained from non-heteroepitaxially constructed six-atomic BN adlayers (with N attached to the diamond substrate). (a), (b) The non-terminated interface, using two different side views (along the *y*-axis (a) and along the *x*-axis (b)). (c), (d) The H-terminated interface, using two different side views (along the *y*-axis (c) and along the *x*-axis (d)). [19], reproduced with permission © American Chemical Society.

It is possible from these calculations to draw the conclusion that the most energetically stable diamond/c-BN interfacial configuration is for the H-terminated BN adlayer, being heteroepitaxially attached to the diamond substrate. In addition, the H-termination will induce an even stronger interfacial adhesion energy, as compared to the non-terminated scenario. When approaching a BN film thickness of four heteroepitaxial BN layers (resulting in an eight-atomic-thick adlayer; see figure 9.18), the calculated interfacial adhesion energy becomes −520 kJ mol^{-1} per binding N atom. As can be seen in table 9.3, this bond strength is much weaker than the corresponding six-atomic thick adlayer (i.e. with three BN layers). However, it is at the same time somewhat stronger than the situation with a corresponding non-terminated c-BN adlayer. Both the initial heteroepitaxial non- and H-terminated BN surfaces show here an even larger tendency to create a crystalline cubic BN structure, although somewhat reconstructed, with more or less perfect sp^3-hybridization of both surface B and N atoms (see figure 9.19). For the non-heteroepitaxial diamond/BN interface situation, the

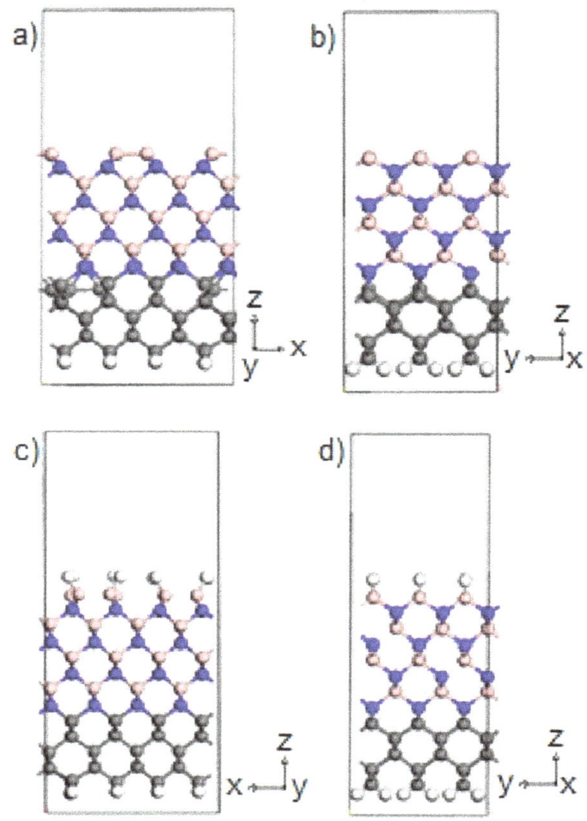

Figure 9.18. Geometry-optimized hetero-epitaxial BN/diamond interfaces with an eight-atomic thick BN adlayer (with N attached to the diamond (100) substrate). (a), (b) A non-terminated interface, visualized by using two different side views (along the *y*-axis (a) and along the *x*-axis (b)). (c), (d) An H-terminated interface, visualized by using two different side views (along the *y*-axis (c) and along the *x*-axis (d)). [19], reproduced with permission © American Chemical Society.

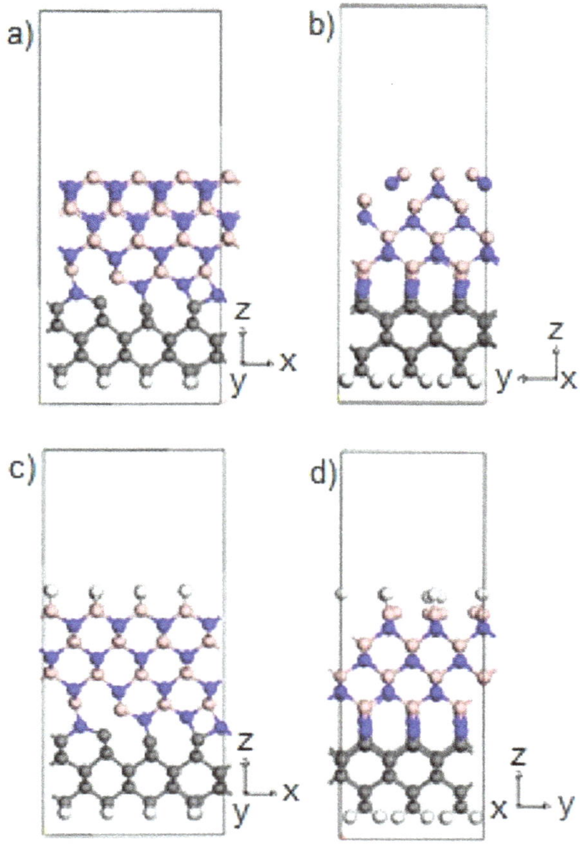

Figure 9.19. Geometry-optimized non-heteroepitaxial BN/diamond interfaces with an eight-atomic thick BN adlayer, and N attached to the diamond (100) substrate. (a), (b) A non-terminated interface in two different side views (along the *y*-axis (a) and along the *x*-axis (b)). (c), (d) An H-terminated interface in two different side views (along the *y*-axis (c) and along the *x*-axis (d)). [19], reproduced with permission © American Chemical Society.

H-terminated adlayer result in an interfacial adhesion energy of −460 kJ mol^{-1} per binding N atom. As can be seen in table 9.3, this result is identical to the result obtained for the six-atomic adlayer interface. It is also almost identical to the situation with no terminating species. Both the non-terminated and H-terminated surfaces did also result here in crystalline cubic BN adlayers with a minor degree of surface reconstruction, in addition to an otherwise more or less perfect sp^3-hybridization of surface B and N atoms. Even though the differences in binding energy are somewhat small, it can also be concluded here that the most energetically stable diamond/c-BN interfacial configuration is obtained for the H-terminated BN adlayer, being heteroepitaxially attached to the diamond substrate. In addition, the H-termination induces an even stronger interfacial binding energy, as compared to the non-terminated scenario.

9.3 Electron bond populations

9.3.1 Introduction

As presented and discussed in section 4.6.4, calculation of electron bond populations is generally expected to give an indication of covalent bond strengths. The electron bond population describes the electron density between two atoms and are for the present example presented as the calculated values for the bonds between the diamond surface C atoms and the first atomic layer of either N (or B) attached to the diamond substrate. These values are all numerically large (see tables 9.4 and 9.5, and figures 9.20 and 9.21).

9.3.1.1 A two-atomic BN adlayer with N attached to the diamond surface H- or F-terminated non-heteroepitaxial BN adlayers

Since the bond population is a clear indication of covalent bond strength, a strong correlation to covalent bond lengths is generally expected. For H-terminated BN

Table 9.4. Electron bond population values between the diamond surface C atoms and their binding N and C atoms in geometry-optimized diamond/BN interface of various BN thicknesses.

	Heteroepitaxial				Non-heteroepitaxial			
	H-terminated		F-terminated		H-terminated		F-terminated	
Atomic BN layers	C–N	C–C	C–N	C–C	C–N	C–C	C–N	C–C
1	0.67	0.80						
2	0.67	0.82	0.67	0.80	0.77		0.67	0.80
4	0.67	0.81	0.66	0.80	0.79	0.77	0.66	0.81
6	0.62	0.80			0.80	0.92		
8	0.58	0.80			0.81	0.92		

Table 9.5. Electron bond population values between the diamond surface C atoms and their binding B and C atoms in geometry-optimized diamond/BN interface of various BN thicknesses.

	Heteroepitaxial				Non-heteroepitaxial			
	H-terminated		F-terminated		H-terminated		F-terminated	
Atomic BN layers	C–B	C–C	C–B	C–C	C–B	C–C	C–B	C–C
1	0.89	0.73	0.92	0.77	0.89	0.79	0.90	0.77
2	0.69	0.84			0.92			
4	0.80	0.84			0.93			

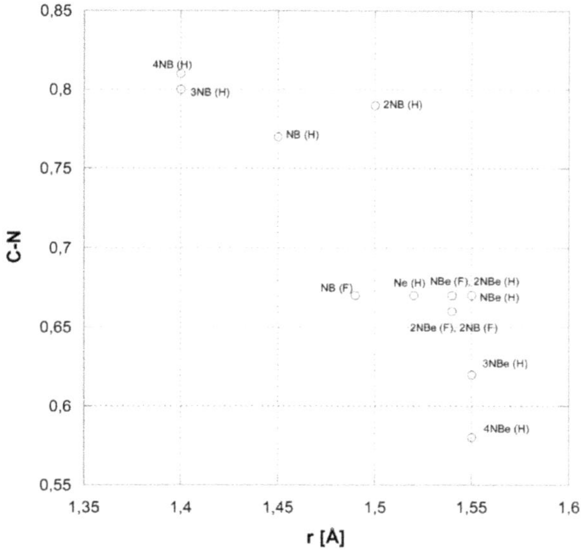

Figure 9.20. Electron bond population versus bond length between the diamond surface C atoms and the directly attached N atoms in the attached BN adlayer. The suffix e (in, e.g. N_e) means that the adlayer is heteroepitaxial with respect to the diamond (100) substrate. [19], reproduced with permission © American Chemical Society.

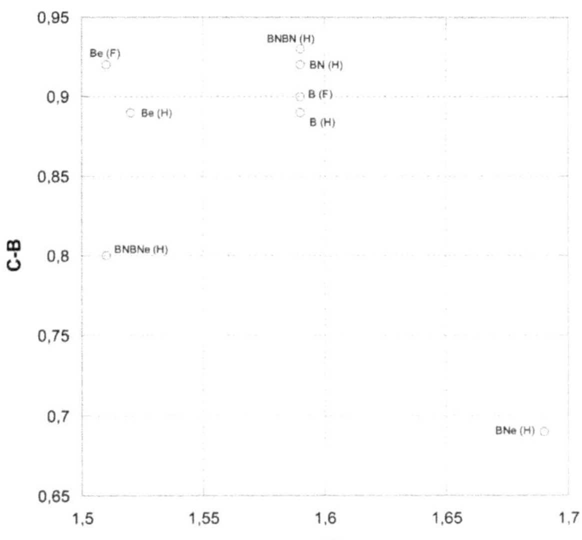

Figure 9.21. Electron bond population versus bond length between the diamond surface C atoms and the directly attached B atoms in the attached BN adlayer. The suffix e (in, e.g. B_e) means that the adlayer is heteroepitaxial with respect to the diamond (100) substrate. [19], reproduced with permission © American Chemical Society.

adlayers, with N being directly attached to the diamond substrate, the calculated results for the heteroepitaxially and non-heteroepitaxially built interlayers have been observed to be completely different. As can be seen in figure 9.20 and table 9.4, the non-heteroepitaxial interfaces show large C–N bond populations (around +0.8), with corresponding bond lengths of 1.4–1.5 Å. The bond populations do apparently increase with adlayer thickness, with an overall quite acceptable correlation with the bond lengths. An exception to the otherwise very good correlation between bond population and bond length is observed for the four-atomic BN adlayer. However, there are different plausible explanations for this exception. One is the fact that the calculated bond populations will only give information about the covalent nature of the bond, and other contributions are, hence, not included (e.g. ionic bond strengths). In addition, the geometrical structures of these interfaces are very complex in nature, showing very irregular patterns when increasing the BN adlayer thickness.

The F-terminated non-heteroepitaxial interfaces do not, however, show the same type of trend as the H-terminated ones, with an increasing electron bond population for an increase in BN adlayer thickness. In fact, the bond populations for F-terminated non-heteroepitaxial interfaces are in principle identical (+0.67 to +0.66), while the bond lengths varies (from 1.49 to 1.54 Å; see figure 9.20). The most plausible explanation for this observation is also here a combination of non-perfect covalent bonds and a too irregular bond formation. When comparing the results for F-termination with H-termination, it is clear though that the F-terminated interfaces show smaller electron bond population values and longer bond lengths (see figure 9.20 and table 9.4). This difference in results can be explained by the larger electronegativity of F compared to H (3.98 versus 2.20). This conclusion is further supported by the observation that the interfacial bond populations do increase with an increase in BN adlayer thickness. The explanation for this difference in bond population (for F and H) is that more electronegative F species will have the capacity to withdraw more electrons from the C–N bond, thereby weakening it. This is an effect that will be further diminished for an increased BN film thickness, for which the terminating species will be given a smaller influence on the interfacial bond region. When trying to correlate interfacial binding adhesion energies, various factors come to play. For instance, the adhesion energies are calculated for the whole supercell and contain contributions from various bonding types. Hence, it is not straightforward to project it to specific bonds. We have, hence, chosen here to correlate electron bond population with bond length, since both can be strictly located to a specific bond and since both are strongly coupled to bond strengths.

9.3.1.1.1 H- or F-terminated heteroepitaxial BN adlayers

The heteroepitaxially built H-terminated diamond/c-BN interfaces do all show much smaller electron bond populations in comparison to the non-heteroepitaxial counterparts (+0.58 to +0.67 versus ~+0.8). As expected, the corresponding bond lengths are also longer (1.52–1.55 Å versus 1.4–1.5 Å; see figure 9.20). In addition, the trends for these heteroepitaxially built interfaces are opposite to those of the non-heteroepitaxially built ones; the electron bond populations are observed to decrease with an increase in adlayer thickness (see table 9.4). Despite the difficulties to project the total adhesion energy to individual bonds, here we can just conclude that, with one exception, these results correlate quite well with the interfacial binding energies; a decrease in interfacial binding energy is accompanied by a decrease in electron bond population. As can be seen in figure 9.12, the F-terminated heteroepitaxial interfaces are also found to follow the same trend (although rather weak), with a decrease in bond population with an increase in BN adlayer thickness. The heteroepitaxial two-atomic BN adlayer shows an identical electronic bond population (+0.66) as the H-terminated counterpart, but with a slightly shorter C–N bond length (1.54 Å).

As a conclusion, the non-heteroepitaxial adlayer structures show slightly larger electron bond populations within the interfacial C–N bonds (as compared with the corresponding heteroepitaxially built structures). In addition, these bond populations do increase with adlayer thickness, which is not the case for the heteroepitaxially built structures (for which these values will decrease with increasing adlayer thicknesses). Furthermore, the bond lengths correlate perfectly with the bond populations, with longer bond lengths for smaller values of electron bond populations.

9.3.1.2 A two-atomic BN adlayer with B attached to the diamond surface

9.3.1.2.1 H- or F-terminated non-heteroepitaxial adlayers

For the situation with H-terminated non-heteroepitaxial diamond/BN interfaces, where B atoms are directly attached to the diamond substrate, large bond populations are observed between the C and B atoms across the interface. Also, the electron bond population increases with BN adlayer thickness, which is identical to the corresponding results for N directly attached to the diamond substrate. The electron bond populations are shown in table 9.5. As is the situation with an F-terminated non-heteroepitaxial adlayer where N is being attached to the diamond surface, there is no correlation observed between these bond population values and the corresponding bond lengths. Instead, there is a constant C–B bond length of 1.59 Å (see figure 9.21). It must again be stressed that the electron bond population results only give information about the covalent contribution to the bond and other additions, such as ionic, are not included. In addition, the geometries of these non-heteroepitaxial structures show very irregular patterns when the adlayer thickness increases and the bond length distributions are, therefore, likely to follow these irregular patterns.

The results for the F-terminated non-heteroepitaxial diamond/BN interfaces are very similar to the H-terminated ones (see figure 9.21). The corresponding C–B bond lengths are 1.59 Å for both types of terminations, which is identical to all non-heteroepitaxially built interfaces with B directly attached to the diamond surface (i.e. H- and F-terminated two- and four atomic BN adlayers).

9.3.1.2.2 H- or F-terminated heteroepitaxial adlayers

As was the situation for H-terminated non-heteroepitaxial diamond/BN interfaces, with B atoms directly attached to the diamond surface, corresponding heteroepitaxial interfaces do also have high electron bond populations (see table 9.5). As can be seen in figure 9.21, a good correlation between electron bond population and bond length is observed, with shorter bond lengths for larger bond population values. However, there is no correlation between the electron bond population and adlayer thickness. On the other hand, there is an almost perfect correlation between the interfacial binding energy and bond populations, and between the interfacial binding energy and bond lengths. When comparing the results for different types of atoms in the closest vicinity to the diamond substrate, the electron interfacial bond populations are found to be larger for the structures with B directly binding to the diamond substrate than with N binding to the diamond substrate. In addition, both interface types (i.e. C–B and C–N) show a clear correlation between electron bond populations and bond lengths.

Calculations on adlayers that start with B closest to the diamond substrate here have only taken place for up to four atomic BN layers. Depending on the purpose of the investigation, it is very easy to continue to even thicker BN slabs. Moreover, in the present study, only the B layer has been terminated with F atoms, and we start here by showing the results for F-terminated heteroepitaxial interfaces with B closest to the diamond substrate-the monatomic B adlayer. It shows an electron bond population of +0.92 for the C–B bonds, which is close to the situation for the corresponding H-terminated structure (+0.89; see figure 9.21). The corresponding C–B bond length for the F-terminated B monolayer is also very similar to the corresponding H-terminated adlayer. The minor difference in bond length is, however, expected since the bond population for the F terminated B monolayer is somewhat larger than for the H-terminated one. In addition, a comparison between the electron bond populations and interfacial binding energies shows a very clear trend for these types of bonds. This is also the case for the bond lengths versus interfacial binding energies.

When comparing with the situation with a BN adlayer with N closest to the diamond substrate, the calculated C–B bond populations for an F-terminated interface is found to be larger for the situation with B closest to the diamond substrate. The bond lengths are, on the other hand, very similar (see figure 9.21).

In summary, all of the heteroepitaxially built diamond/BN interfaces show good correlation between the electron bond populations and the interfacial binding energies. However, this is not the situation for the non-heteroepitaxially built structures. As discussed above, the most plausible explanation for this discrepancy is that other bond types (than covalency) are also present in these bonds. Or alternatively, there are more general electrostatic interactions over the diamond/BN interface. It must be stressed here that the non-heteroepitaxial structures show very irregular patterns and complex structures, which is not at all the situation for the heteroepitaxial structures.

9.4 Degree of electron transfer

9.4.1 Introduction

The degree of electron transfer over the interface, between the diamond substrate and the attached BN adlayer, can be estimated from calculated atomic charges.

This electron transfer is a strong indication for a chemical interaction over the interface and will, for instance, give an indication of degree of ionic bonding. It is, thereby, very useful as a complement to the electron bonding population calculations, from which it is possible to get information about covalent bond strengths. Both of these tools will, hence, be invaluable in achieving a deeper understanding about the underlying causes to the differences in interfacial binding energies, as well as for the resulting geometrical structures. The interaction between diamond surface C atoms and attached N (or B) atoms is dependent on the degree of spherical orbital overlaps and orbital energy differences. Perfect covalent bonds do not show any electron transfer at all, while the more polar bonds show a much higher degree of electron transfer. The interfacial binding interactions of interest in the present example have a certain degree of polarized contribution, to the otherwise covalent bonds, since the electronegativity values for C, N, and B are 2.55, 3.04, and 2.04, respectively [23].

9.4.2 A two-atomic BN adlayer with N attached to the diamond surface

9.4.2.1 H- or F-terminated non-heteroepitaxial adlayer

For the non-heteroepitaxial H-terminated BN adlayer, an almost constant degree of electron transfer is observed for a somewhat thicker BN adlayers (see figure 9.22). On the contrary, the very thin two-atomic BN adlayer shows an electron transfer of only 1.05 e. Hence, the different types of terminating species (H or F) are observed to have a more pronounced effect for thinner BN adlayers only. Thus, these results indicate that the degree of electron transfer over the interface will cease to a constant value for a BN thickness of at least four atomic layers. The F-terminated interfaces follow the same trend, although the electron transfer for the two atomic adlayer is higher than for the corresponding H-terminated one. As can be seen in figure 9.22, the four-atomic adlayer shows an electron transfer of 2.11e (as compared with 2.03e for the H-terminated interface). Since the electronegativity value of F is larger than H, the F-terminated adlayers will also show a larger degree of electron transfer here, which becomes more dominant for the thinner adlayers. The difference in electron transfer for F- and H-termination is 0.70 for a two-atomic adlayer, and 0.08 for the four-atomic one. It is, hence, obvious that the terminating species do have a much higher influence on the degree of electron transfer for non-heteroepitaxial adlayers, as compared to heteroepitaxial ones.

9.4.2.1.1 H- or F-terminated heteroepitaxial adlayers

For the situation with an H-terminated mono-atomic N adlayer, being hetero-epitaxially attached onto a diamond (100) substrate, there is a partial electron transfer from the diamond surface to the N adlayer of 2.61e per supercell. This is the expected direction of electron transfer since N has a higher electronegativity compared to C (3.04 versus 2.55). When increasing the number of BN adlayers, the electron transfer is, with some exceptions, found to decrease (see figure 9.22). The more general picture is, hence, that the binding energy will decrease with adlayer thickness, which correlates quite well with the values of partial electron

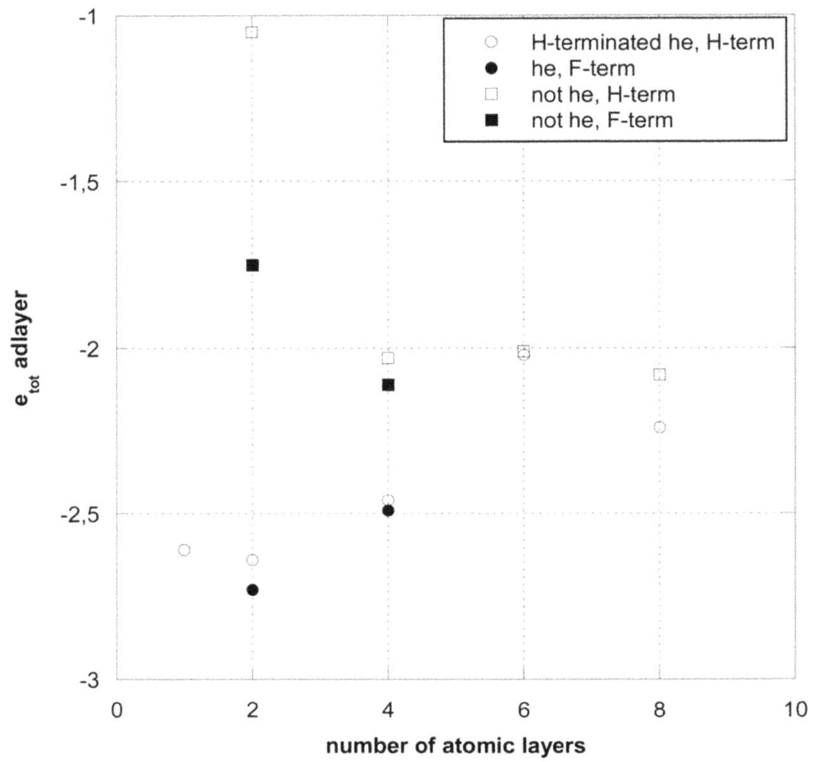

Figure 9.22. Electron transfer, with the direction from the BN adlayer to the diamond substrate versus number of atomic layers in the attached BN adlayer. Nitrogen atoms are directly attached to the diamond (100) substrate. The notation 'he' means that the adlayer is heteroepitaxial with respecti to the substrate. [19], reproduced with permission © American Chemical Society.

transfer. This is an expected result since a larger interaction between two materials in an interface usually implies a larger adhesion energy and a larger degree of electron transfer (due to overlapping surface orbitals between the two materials in the interface). For the situation with the F-terminated and heteroepitaxially built adlayers, with N directly attached to the diamond (100) surface, similar values for partial electron transfer, as for the H-terminated interfaces, has been observed (see figure 9.22). As can be seen in this figure, the F-termination electron transfer values are only slightly larger than the corresponding H-termination ones. The corresponding adlayer adhesion energies are −600 versus −610 kJ mol^{-1} per binding N atom for a two-atomic adlayer and −630 versus −640 kJ mol^{-1} per binding N atom for a four-atomic adlayer. The values of the electron transfer clearly show that the effect by the terminating species is the much more pronounced H for a thinner BN adlayer. The larger effect by the F-terminated interface, though, is expected since the F atoms are much more electronegative than atoms (3.98 versus 2.20). However, this influence by the terminating species becomes rapidly smaller when increasing the adlayer thickness to a four-atomic one. The difference in electron transfer for F versus H is 0.09 for the two-atomic adlayer, and 0.03 for the four-atomic one.

9.4.2.2 A two-atomic BN adlayer with B attached to the diamond surface

9.4.2.2.1 H- or F-terminated non-heteroepitaxial adlayers

The H-terminated non-heteroepitaxial diamond/BN interfaces, with B atoms binding to the diamond substrate, show partial electron transfers that correlate well with the interfacial binding energies (see table 9.3 and figure 9.23). As clearly demonstrated, the electron transfer results also converge here towards a fairly constant value for thicker adlayers. This convergence occurs very quickly, already for a two-atomic thick adlayer. This is also the situation for the interfacial binding energies. These results indicate that the terminating species are able to influence the electron transfer and binding situations for adlayer thickness as thin as one atomic layer only. As is the situation with a mono-atomic heteroepitaxial B adlayer, the electron transfer from the F-terminated adlayer to the diamond substrate is larger than for

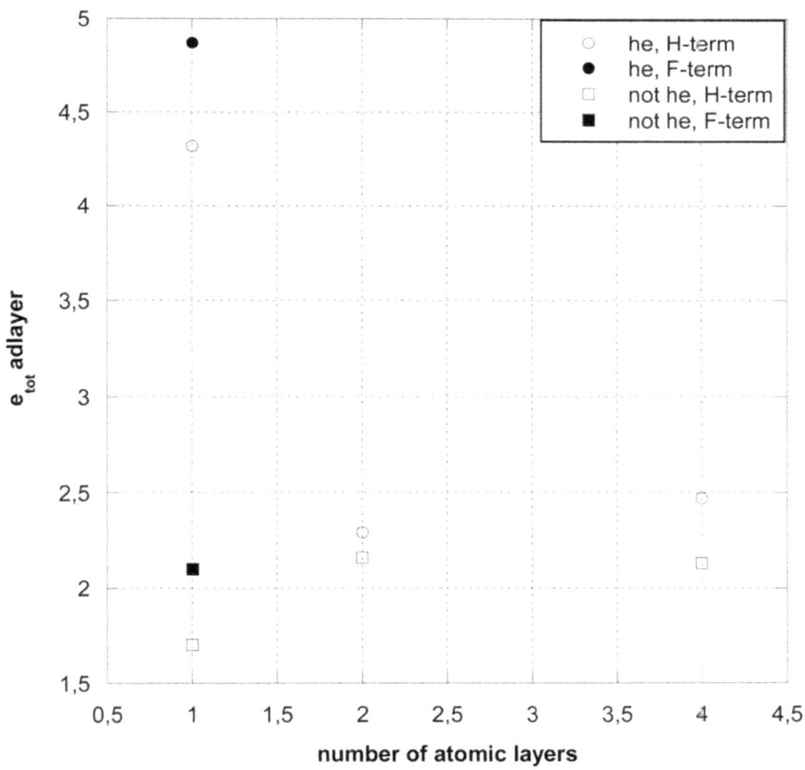

Figure 9.23. Electron transfer, with the direction from the BN adlayer to the diamond substrate versus number of atomic layers in the attached BN adlayer. Boron atoms are directly attached to the diamond (100) substrate. The notation 'he' means that the adlayer is heteroepitaxial with respective to the substrate. [19], reproduced with permission © American Chemical Society.

the corresponding H-terminated interface. There are two plausible explanations for this observation. One possible explanation is that the geometry optimized H- and F-terminated structures are structurally very different, which also results in different adhesion energies. And the second plausible explanation is, as was also discussed for heteroepitaxial interfaces, the repulsions between the electron density-rich F adsorbates. The degree of electron transfer for the H- and F-terminated interfaces are 1.70 e and 2.10 e, respectively. The corresponding interfacial adhesion energies are −430 versus −380 kJ mol^{-1} per binding B atom. This result is expected since a large degree of electron transfer results in a more pronounced adhesion energy (of ionic character).

9.4.2.2.2 H- or F-terminated heteroepitaxial adlayers

For the situation with B closest to the diamond surface in the interface, the direction of electron transfer is the opposite to the situation with N atoms being attached to the diamond surface. Due to the difference in the electronegativity between C (2.55) and B (2.04), the B atoms will partially donate electrons to the C atoms in the diamond substrate. The partial electron transfers (per supercell) for the H-terminated heteroepitaxial adlayers are shown in figure 9.23. It is hard to notice any clear trend with so few values, but there is a tendency for the electron transfer values to decrease when the adlayer thickness increases (from one atomic layer to four atomic layers). For the situation with only a one-atomic thick adlayer (i.e. a B layer), the influence by the terminating species is most probably the dominating factor for the degree of electron transfer between the diamond substrate and the B layer. This can, for a thicker BN adlayer, be explained by the difference in electronegativities between the terminating species H (2.20) and the N atoms (3.04). Hence, the smaller electronegativity value of the H atom allows a larger portion of electrons to be donated from the B atoms to the substrate C atoms, as compared with the N atoms (for a thicker BN adlayer). To be more precise, the withdrawal of electrons towards the N atoms will counteract the withdrawal of electron towards the substrate C atoms.

As can be seen in figure 9.23, the corresponding F-termination situation show an electron transfer result that is very similar to the H-termination one. According to the difference in electronegativity values between H (2.20) and F (3.98), the degree of electron transfer from the adlayer (with B closest to the diamond surface) to the diamond substrate is expected to be much smaller for the F-terminated adlayer than for the H-terminated one. As can be seen figure 9.18, this is not the case. One plausible explanation for the observation that F-termination also renders an electron transfer from B to C (over the interface) is a strong indication of sterical repulsions between the terminating F species. As can be seen in figure 9.18(f), sterical repulsion among the highly electron density rich F adsorbates causes them to undergo surface reconstructions. For this very thin adlayer, this repulsion is also expected to result in an induced electron transfer from the thinnest BN adlayers to the diamond surface C atoms.

9.5 Conclusions

As shown in sections 9.1–9.2, a lot of calculations of different kinds are needed in the strives to learn more about the effect of a substrate (here diamond (100)) on the hetereoepitaxial growth of phase-pure c-BN. However, these calculations are very important since the vapour phase growth of c-BN is still very difficult to accomplish experimentally. This is especially the case of the initial growth stage. The initial BN layers often result in a mixture of phases, such as hexagonal, amorphous, turbostratic, and cubic. These BN phases lead to films with higher compressive stress, poor crystalline quality, small grain size, and a high defect density. The film will most often peel off from the substrate material, even at lower thicknesses. Moreover, the initial growth process has been observed to be very important for the film adhesion to the substrate material. The main purpose with the calculations presented in section 9.2 was to outline the possibility for capturing the initial cubic BN phase directly onto the substrate, by combining the choice of substrate material with terminating species. The first choice of substrate material was naturally diamond since it has been shown to be the only substrate to date onto which it is possible to grow c-BN directly without any amorphous, or hexagonal, phase in between.

The results in the chapters above show that nitrogen is energetically the most favorable precursor to start the growth processes with. The reason is that a first atomic layer of N will directly bind to the diamond substrate in heteroepitaxial positions—both with and without terminating species like H and F. A single atomic layer of B, directly attached to the diamond substrate, was also found to energetically prefer a heteroepitaxial continuation of the diamond substrate. But this was the situation only for an H- or F-terminated B adlayer. The energetically preferred structure will otherwise (i.e. in the absence of terminating species) become amorphous-like. For a two-atomic-thick BN adlayer, with N closest to the diamond substrate, it was found that surface termination with either H or F will steer the growth towards a heterostructural cubic phase of BN. There was a very similar result obtained for the two-atomic thick BN adlayer, with B closest to the diamond substrate. An H-terminating heteroepitaxial diamond/c-BN interface was found to be the energetically most preferred structure. For the four-atomic-thick BN adlayer, with N closest to the diamond substrate, the crystalline heteroepitaxial structure is the most energetically preferred one. It is also possible for this adlayer thickness to draw the conclusion that the terminating species will uphold the cubic structure and steer the c-BN growth toward a crystalline, heterostructural adlayer. For the four-atomic-thick BN adlayer, with B closest to the diamond substrate, the crystalline heteroepitaxial structure is the most energetically preferred one, with and without terminating species. When continuing to six and eight atomic thick BN adlayers, with N closest to the diamond substrate, all structures became crystalline after the geometry optimization. However, there is an energetic preference for the H-terminated heterostructural diamond/c-BN interface.

All results have indicated that it is most important to use surface termination for the initial growth of the first four BN atomic layers in order to avoid amorphous and non-cubic phases of BN. After four atomic layers, a non-perfect surface termination will also render a phase-pure growth of c-BN directly onto the diamond substrate. In addition, at these thicker layers, the tendency for three-valence for B and five-valence for N is no longer apparent. This means that the strive by B and N to bind to three atoms each is not so dominating at these larger thicknesses. Hence, these calculations have verified the experimental observation that the growth of c-BN is more difficult in its initial stage, compared to later. Moreover, the effect of the terminating species H (or F) was also clearly present for the thinnest adlayers, showing a chemical influence on both the electron transfer over the interface and on the adhesion energy. This was an influence that was diminished for thicker atomic layers (as compared to two to four layers).

References

[1] Zhang W, Bello I, Lifshitz Y, Chan K, Wu Y, Chan C, Meng X and Lee S 2004 Thick and adherent cubic boron nitride films grown on diamond interlayers by fluorine-assisted chemical vapor deposition *Appl. Phys. Lett.* **85** 1344–6

[2] Mirkarimi P, McCarty K and Medlin D 1997 Review of advances in cubic boron nitride film synthesis *Mater. Sci. Eng.* R **21** 47–100

[3] Zhang W J, Bello Y, Lifshitz Y and Lee S T 2003 Recent advances in cubic boron nitride deposition *MRS Bull.* **28** 184–8

[4] Zhang F, Guo Y, Song Z and Chen G 1994 Deposition of high quality cubic boron nitride films on nickel substrates *Appl. Phys. Lett.* **65** 971–3

[5] Friedmann T, Mirkarimi P, Medlin D, McCarty K, Klaus E, Boehme D, Johnsen H, Mills M, Ottesen D and Barbour J 1994 Ion assisted pulsed laser deposition of cubic boron nitride films *J. Appl. Phys.* **76** 3088–101

[6] Yamamoto K, Keunecke M, Bewilogua K, Czigany Z and Hultman L 2001 Structural features of thick c-boron nitride coatings deposited via a graded B–C–N interlayer *Surf. Coat. Technol.* **142** 881–8

[7] Kester D, Ailey K, Lichtenwalner D and Davis R 1994 Growth and characterization of cubic boron nitride thin films *J. Vac. Sci. Technol.* **12** 3074–81

[8] Mirkarimi P, Medlin D, McCarty K and Barbour J 1995 Growth of cubic BN films on ß-SiC by ion-assisted pulsed laser deposition *Appl. Phys. Lett.* **66** 2813–5

[9] Hofsass H, Ronning C, Griesmeier U, Gross M, Reinke S, Kuhr M, Fischer R and Zweck J 1995 Characterization of cubic boron nitride films grown by mass separated ion beam deposition *Nucl. Instrum. Methods* B **106** 153–8

[10] Cameron D C and Karim M Z 1993 Properties of mixed-phase BN films deposited by r.f. PACVD *Thin Solid Films* **236** 96–102

[11] Zhang W, Bello I, Lifshitz Y, Chan K, Meng X, Wu Y, Chan C and Lee S 2004 Epitaxy on diamond by chemical vapor deposition: a route to high-quality cubic boron nitride for electronic applications *Adv. Mater.* **16** 1405–8

[12] Zhang W, Meng X, Chan C, Chan K, Wu Y, Bello I and Lee S 2005 Interfacial study of cubic boron nitride films deposited on diamond *J. Phys. Chem.* B **109** 16005–10

[13] Nose K, Yang H and Yoshida T 2005 Defect-induced electronic conduction of tBN thin films *Dia. Rel. Mater.* **14** 1297–301

[14] Zhang W, Chong Y, Bello I and Lee S J 2007 Nucleation, growth and characterization of cubic boron nitride (cBN) films *Phys. D: Appl. Phys.* **40** 6159–74

[15] Keunecke M, Wiemann E, Weigel K, Park S and Bewilogua K 2006 Thick c-BN coatings: preparation, properties and application tests *Thin Solid Films* **515** 967–72

[16] Djouadi M, Vasin A, Nouveau C, Angleraud B and Tessier P 2004 Deposition of boron nitride films by PVD methods: transition from h-BN to c-BN *Surf. Coat. Technol.* **180–81** 174–7

[17] Pauling L 1960 *The Nature of the Chemical Bond and the Structure of Molecules and Crystals: An Introduction to Modern Structural Chemistry* (Ithaca, NY: Cornell University Press)

[18] Pallas A and Larsson K 2010 Initial growth of c-BN on diamond substrates: a theoretical approach *J. Phys. Chem.* **114** 11448–55

[19] Pallas A and Larsson K 2014 Effect of terminating species on the initial growth of c-BN on diamond substrates *J. Phys. Chem.* **118** 3490–503

[20] Segall M D 1996 Population analysis in plane wave electronic structure calculations *Mol. Phys.* **89** 571–7

[21] Segall M D, Shah R, Pickard C J and Payne M C 1996 Population analysis of plane-wave electronic structure calculations of bulk materials *Phys. Rev.* B **54** 16317–20

[22] Aylward G and Findlay T 1994 *SI Chemical Data* 3rd edn (Milton: Jacaranda Wiley Ltd)

Chapter 10

Construction of growth reaction pathways

Experimental work has indicated that PEALD, using a BF_3–H_2–NH_3–F_2 pulse cycle and a diamond substrate, is a promising method for the deposition of c-BN films. The gaseous species will decompose in the plasma and form BF_x, H, NH_x, and F species ($x = 0, 1, 2, 3$). The H and F radicals will uphold the cubic structure by completely hydrogenating or fluorinating the growing surface. However, surface radical sites will appear during the growth process as a result of atomic H or F abstraction reactions. The addition of NH_x growth species ($x = 0, 1, 2$) to B radical sites, and BF_x growth species ($x = 0, 1, 2$) to N radical sites, will then result in a continuous growth of c-BN. However, energetic particle bombardment has been found necessary to promote the growth of c-BN.

The question is now, why isn't it enough to only use an increased temperature? Why do we need both an elevated temperature and a low-energetic particle bombardment? Theoretical modelling is a necessary tool in the search for an answer to these questions.

10.1 Simulation of an experimentally suggested c-BN growth mechanism

10.1.1 Introduction

How to calculate the various elementary growth steps in a layer-by-layer growth of c-BN has been presented in detail in chapter 8. These steps include adsorption to the surface, desorption from the surface, surface migrations, etc. However, the next step is to put these elementary steps together, forming a proposed layer-by-layer growth mechanism. As a guideline in the search for the most optimal growth mechanism, it is advisable to compare with experimental results. We will focus here on the growth mechanism derived by Zhang *et al* [1]. Their mechanism (see figure 10.1) is based on a combination of high-resolution transmission electron microscopy (HRTEM),

doi:10.1088/978-0-7503-3107-4ch10

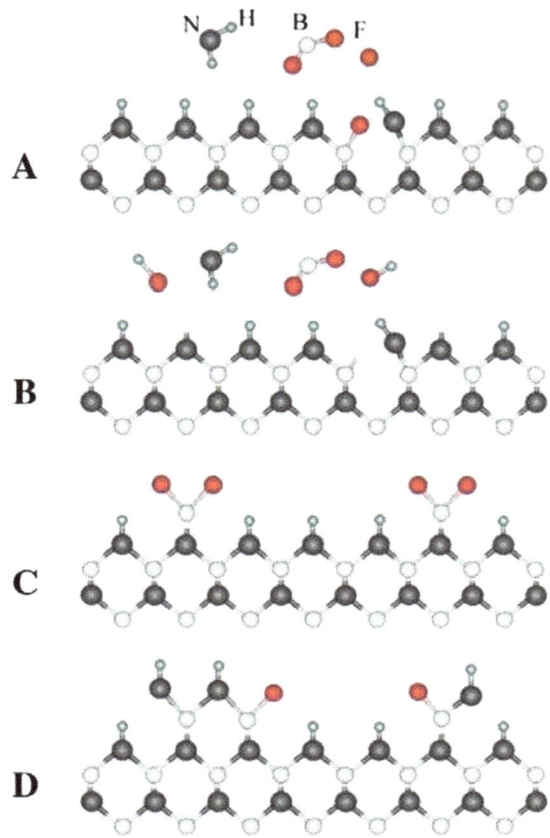

Figure 10.1. Suggested c-BN growth mechanism by using an F-based ECR-MWCVD reactor fed by a gas mixture of He(Ar)/N_2/BF_3/H_2 [2].

optical emission spectroscopy (OES), and x-ray photoelectron spectroscopy (XPS) on F-based electron cyclotron resonance-assisted microwave CVD (ECR-MWCVD) deposition of c-BN from a gas-mixture of He(Ar)/N_2/BF_3/H_2. In such a gas-mixture, the B-covered surfaces are expected to become F-terminated while the N-covered surfaces are H-terminated. This is an assumption that is based on the calculated results presented and discussed in chapter 8. Furthermore, the c-BN surfaces will be activated by surface abstraction by the impingement of plasma species, thus providing radical sites for further growth. However, it is a necessity that the B-containing growth species are chemisorbed to the N surface radical sites, while N-containing growth species are chemisorbed to the B surface radical sites. The results presented in chapter 8 show that the BFx species are good candidates for the chemisorption to the N-covered c-BN surface, while the NHx species can be used for the chemisorption to the B-covered surface. The adsorption energies have been obtained in a so-called thermodynamic study. In addition, what is most important for the final understanding of the adsorption processeses is to perform a kinetic study, where both the geometrical structure and the total energy of the system will be carefully followed for the situation with an BFx (NHx) species approaching an N (or B) surface radical site. The same is the situation for the gas phase abstraction reactions. The total growth rate of c-BN is completely governed by the slowest elementary step in the growth mechanism. It is, hence, very important to identify the energy barrier for this step, which can be done by performing a kinetic study. The advantage with this knowledge is that it might be possible to improve the rate-determining step, thereby speeding up the complete growth process.

10.1.2 Adsorption of BF_x in an H-saturated gaseous atmosphere

According to the results presented in chapter 8, the adsorption of non-fragmented BF_3, and its fragmented counterparts (BF_2, BF, and B), onto the H-terminated N-covered c-BN(100) surface, is simulated here. The BF_x (x = 0–3) growth species is initially fixed 3.0 Å directly above an N surface radical site, and is thereafter sequentially moved towards the radical site in steps (n) of 0.2 Å (i.e. the z-coordinate changes according to $z_n = 3.0\ n{\bullet}(-0.2)$), and the system is subsequently allowed to relax. The B atom in the adsorbate is, however, set fixed in the optimization calculations (which is also the situation for the lowest BN atomic layer, with its terminating H species). The interaction energy between the incoming BF_x species and the surface N radical site (process 1) is thereafter calculated according to equation (10.1).

$$\mathrm{BN}(94\%H) + \mathrm{BF}_x(z_n = 3.0 + n(-0.2)) \tag{10.1a}$$

$$(x = 0,\ 1,\ 2,\ or3;\ n = 0,\ 1,\ 2,\ \ldots..,\ n_{\max})$$

$$\Delta E_{\mathrm{rel}} = E_{\mathrm{BN(94\%H)\,+\,BF}_x(Z_n\,=\,3.0+n(-0.2))} - E_{\mathrm{BN(94\%H)\,+\,BF}_x(Z_n=3.0)} \tag{10.1b}$$

where $E_{\mathrm{BN(94\%H)\,+\,BF}_x(Z_n\,=\,3.0+n(-0.2))}$ and $E_{\mathrm{BN(94\%H)\,+\,BF}_x(Z_n\,=\,3.0)}$ are the total energies for the monoradical H-terminated surface with a BF_x species at 3.0 and $3.0 + n{\bullet}(-0.2)$ Å above the surface, respectively. The potential energy curve for equation (10.1) has, thereafter, been obtained by plotting ΔE_{rel} as a function of a reaction coordinate, ξ (see equation (10.2)).

$$\xi = n\,/\,n_{\max} \tag{10.2}$$

As a result, the adsorption of atomic B exhibits an energy that progressively decreases in the region $0 < \xi < 0.33$ (see figure 8.83(a)). In the region $0.33 < \xi < 0.42$, a bond is formed between the incoming B species and the N surface radical. This bond formation is accompanied by a bond breakage of the N(2,1)–N(1,1) dimer (see figure 10.2(a)). The energy does, thereafter, continue to decrease towards a stable minimum of -482 kJ mol^{-1} at $\xi = 0.67$ (corresponding to the distance of 1.35 Å between N and B). The total energy does slowly decrease in the region $0 < \xi < 0.47$ for the situation with a BF chemisorption (see figure 10.2(b)). This decrease in energy is accompanied by a gradual bond formation between the incoming BF species and the

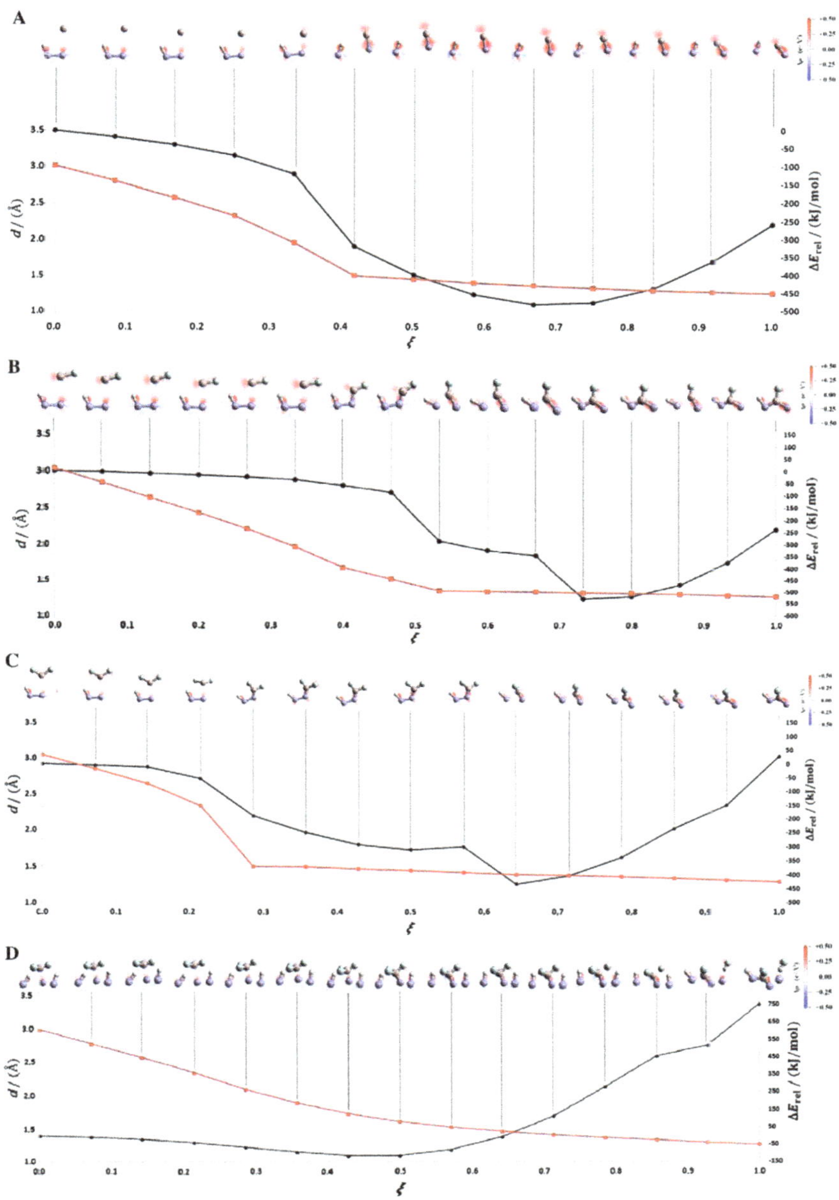

Figure 10.2 Potential energy curves, ΔE_{rel}, for the chemisorption of BF_x ($x = 0, 1, 2, 3$) onto a monoradical surface site on the H-covered N-terminated surface (marked in black). The distance (d) between the surface radical site is also given (marked in red). (a) B; (b) BF; (c) BF_2; (d) BF_3. [3], reproduced with permission © Elsevier.

N surface radical. The N(2,1)–N(1,1) dimer bond is thereafter broken in the region 0.47 < ξ < 0.53, and the BF is thereafter, inserted into this broken dimer (in the region 0.67 < ξ < 0.73). It must be stressed here that such an insertion reaction is a necessary condition for a continued growth of c-BN. The total energy continues, thereafter, to significantly decrease towards a stable minimum of -530 kJ mol^{-1} (at $\xi = 0.73$).

The potential curve for the BF_2 chemisorption can been in figure 10.2(c), where it slowly decreases in the region 0 < ξ < 0.21. Furthermore, a bond is formed between the incoming BF_2 species and the N surface radical in the region 0.21 < ξ < 0.29. The total energy is, thereafter, decreasing progressively until it reaches the region 0.57 < ξ < 0.64, where the N(2,1)–N(1,1) dimer bond will be broken. As the next step in the chemisorption process, the energy reaches a stable minimum of -433 kJ mol^{-1} at $\xi = 0.64$, which corresponds to a N$-BF_2$ distance of 1.39 Å.

Finally, the energy curve for chemisorption of BF_3 is slowly decreasing towards a stable minimum of -116 kJ mol^{-1} at $\xi = 0.43$ (see figure 10.2(d)). This minimum corresponds to a N-B bond length of 1.73 Å for the chemisorbed BF_3 species and a situation where the N(2,1)–N(1,1) dimer has been broken.

10.1.3 Adsorption of NH_x in an F-saturated gaseous atmosphere

The chemisorption of atomic N onto a radical surface B atom, on an otherwise F-terminated c-BN(100) surface, results in a total energy that slowly decreases in the region 0 < ξ < 0.21 (see figure 10.3(a)). This decrease in energy is accompanied by a gradual bond formation between the incoming N species and the B surface radical. The B(2,1)–B(1,1) dimer bond is then broken in the region 0.21 < ξ < 0.29. The energy does, thereafter, decrease towards a stable minimum of -503 kJ mol^{-1} at $\xi = 0.64$ (corresponding to a B–N distance of 1.34 Å).

10.1.4 Thermodynamically driven reaction pathway

According to the calculated deformation densities, all energy minima in figures 10.2 and 10.3 correspond to the formation of a covalent bond between the incoming BF_x (NH_x) species and the N (B) surface radical site. In addition, all adsorption processes have been found to be energetically barrierless. Thus, adsorption of NH_x growth species onto a monoradical F-terminated B-covered c-BN(100)-(2 × 1) surface is expected to be a highly facile process. This is also the situation for the adsorption of BF_x growth species onto a monoradical H-terminated N-covered c-BN(100)-(1 × 1) surface. This is most probably due to the fact that the B atom in the BF_x growth species is positively charged ($+0.49e \leqslant q(B) \leqslant +0.88e$) while the N surface radical is negatively charged ($-0.68e \leqslant q(N) \leqslant -0.10e$). Also, the N atom in the NH_x growth species is negatively charged ($-0.50e \leqslant q(N) \leqslant -0.06e$) charged while the B atom in the B surface radical is positively charge ($+0.20 \leqslant q(B) \leqslant +0.54$) charged. This to a stabilizing electrostatic contribution that lowers the total energy of the system. As can also be seen in figures 10.2 and 10.3, the less fluorinated (or hydrogenated) growth species exhibit a more exothermic adsorption reaction. This is most probably due to the reduced sterical hindrances (on the surface) for these species.

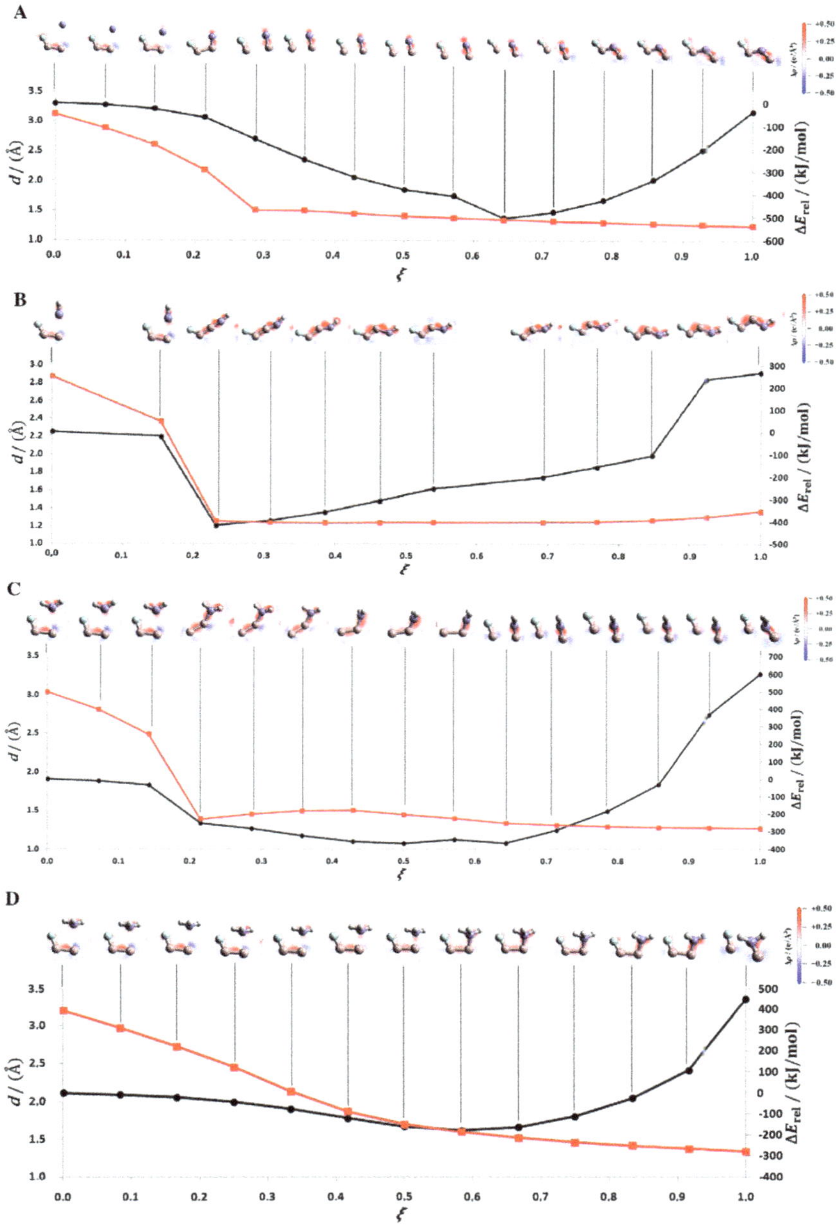

Figure 10.3. Potential energy curves, ΔE_{rel}, for the chemisorption of NH_x ($x = 0, 1, 2, 3$) onto a monoradical surface site on the B-covered F-terminated surface (marked in black). The distance (d) between the surface radical site is also given (marked in red). (a) N; (b) NF; (c) NF_2; (d) NF_3. [3], reproduced with permission © Elsevier.

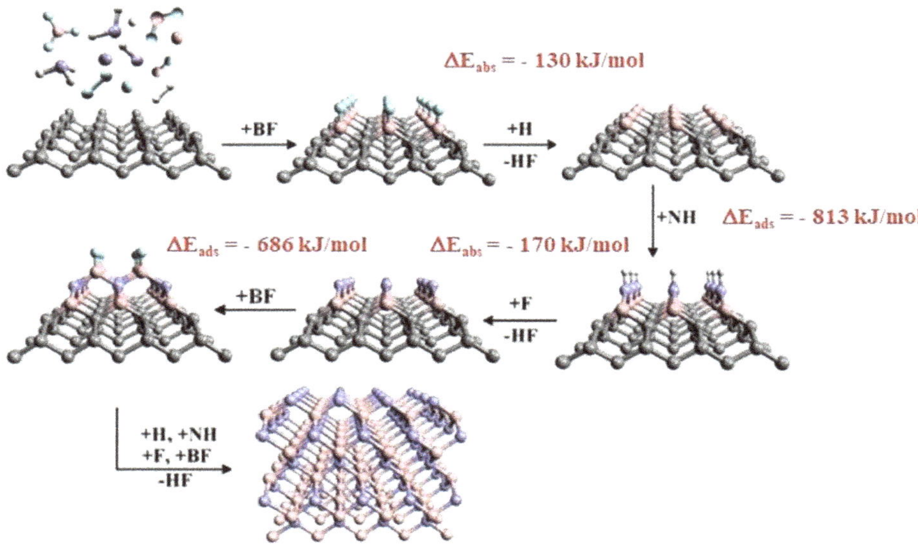

Figure 10.4. Thermodynamic reaction pathway for the heteroepitaxial layer-by-layer CVD (also called ALD) growth of phase-pure c-BN on a diamond(100) substrate (in kJ mol^{-1}). B: pink; N: blue; H: white; F: green; C: grey.

Decomposition of the incoming BF_3 and NH_3 growth species is, thus, very favorable for CVD growth of c-BN.

Gas phase abstraction of terminating species is very important for the c-BN growth process. The reason is that these species must leave room for an incoming growth species. Based on the energetic investigation presented in section 8.4.3, it is possible to draw the conclusion that the most energetically favorable abstraction reactions takes place with gaseous F removing an H adsorbate on the N-covered c-BN(100) surface, and with gaseous H removing an F adsorbate on the B-covered c-BN(100) surface. The corresponding abstraction energies are −170 and −130 kJ mol^{-1}, respectively (see table 8.1).

Based on the adsorption and abstraction reaction energies presented in tables 8.1 and 8.11, the thermodynamic reaction pathway for the layer-by-layer growth of phase-pure c-BN is shown in figure 10.4, The energetically most favourable adsorption and abstraction processes have been used in the construction of this growth pathway. However, by using instead N, NH_2, NH_3, B, BF_2, and BH_3 as plausible growth species, a similar overall exothermic growth process will take place. As can be seen in figure 10.4, it is clear that the whole growth process is thermodynamically driven since all adsorption and abstraction processes are exothermic. Based on these results, the layer-by-layer growth of c-BN should even be able to take place at 0 K! But, as we know, there are very large problems in growing c-BN by using vapour phase techniques. Thus, the growth process must be kinetically driven (i.e. controlled). The barrier energies for each step in the reaction pathway must, therefore, be taken into consideration.

10.1.5 Kinetic reaction pathway

As demonstrated in section 8.5.5, the adsorption processes of BF_x and NH_x ($x = 0$, 1, 2, 3) onto the N-covered and B-covered c-BN(100) surface, respectively, are all exothermic processes. However, the less fluorinated/hydrogenated growth species exhibit smaller sterical hindrances on the surface and, thereby, show more exothermic adsorption reactions. The gaseous B, BF, BF_2, N, NH, and NH_2 growth species have even the capability to replace a surface-terminating H or F atom. From this, it can be concluded that decomposition of the incoming BF_3, and NH_3 growth species is very crucial for CVD growth of c-BN. Moreover, it has also been shown that the H and F atoms are very promising surface-stabilizing species on the N- and B-covered c-BN(100) surfaces, respectively (see section 8.4.2). Also, abstraction of F (H) adsorbate with gaseous H (F) species is thermodynamically highly favourable. These results speak in favour for a CVD method where the incoming growth species are separately introduced into the reactor, e.g. by using an ALD type of method. As discussed in section 10.1.4, an ALD growth mechanism with the following sequential growth steps is then proposed:

$$B_{surface} + NH_x(g) \rightarrow B_{surface} - NH_x \text{ (will become } N_{surface} - H) \tag{10.3}$$

$$N_{surface} - H + F(g) \rightarrow N_{surface} + HF(g) \tag{10.4}$$

$$N_{surface} + BF_x(g) \rightarrow N_{surface} - BF_x \text{(will become } B_{surface} - H) \tag{10.5}$$

$$B_{surface} - \ + xH(g) \rightarrow B_{c-BN}(s) \ + \ xHF(g) \tag{10.6}$$

where $B_{c\text{-}BN}(s)$ and $N_{c\text{-}BN}(s)$ are B- and N-covered c-BN(100) surfaces. As discussed in section 10.1.4, this mechanism is shown to be thermodynamically feasible. In addition, the usage of an ALD method prevents the adsorption of growth species to non-ideal surface sites (like BF_x onto a surface B atom and NH_x onto a surface N atom). In addition, all of these adsorption processes are barrierless (i.e. concerning the adsorption of growth and surface terminating species). However, the kinetic studies presented in section 8.4.5 show that the processes of abstraction of the surface terminating species correspond to energy barriers of various extents (5 kJ mol^{-1} for the F abstraction from the B-covered surface, and 20 kJ mol^{-1} for the H abstraction from the N-covered surface). This is especially the situation for the latter type of abstraction.

The addition of energy is, thus, necessary to promote these abstraction reactions. The combined thermodynamic and kinetic reaction pathway for the phase-pure growth of c-BN can be seen in figure 10.5. It is very clear there that the ALD growth of c-BN is thermodynamically driven, but not kinetically. This means that although the complete growth process is exothermic (with a total reaction energy of $\Delta E = -1799$ kJ mol^{-1} per cycle), the overall growth rate will be very low (if not infinitesimally low). Thus, there is a strong need for the addition of extra energy to the system. This might be the reason why Zhang *et al* have succeeded to grow c-BN (although not phase-pure) [2, 4].

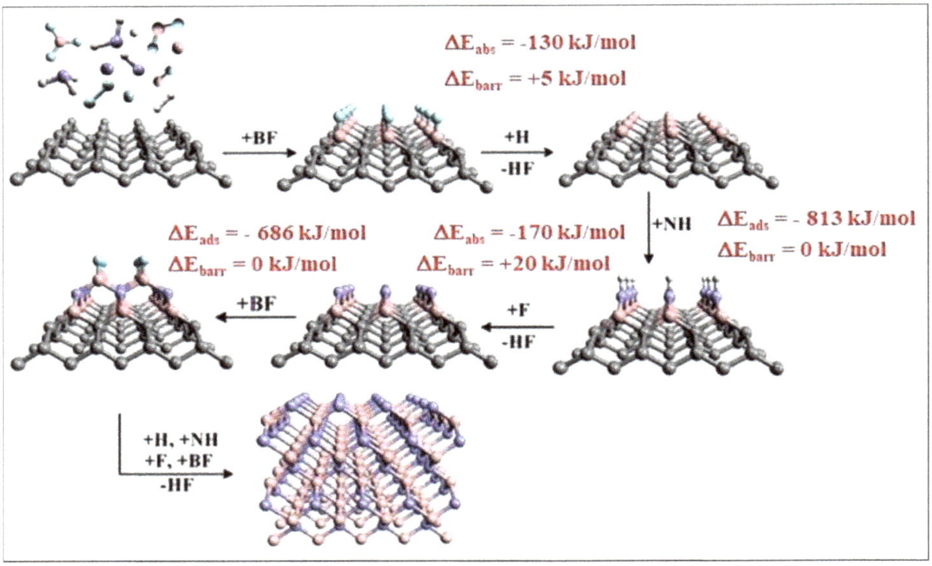

Figure 10.5. The combined thermodynamic and kinetic reaction pathway for the heteroepitaxial layer-by-layer CVD (also called ALD) growth of phase-pure c-BN on a diamond(100) substrate (in kJ mol^{-1}). B: pink; N: blue; H: white; F: green; C: grey.

Zhang *et al* used electron cyclotron resonance-assisted microwave CVD (ECR-MWCVD) in the deposition of c-BN in a gas-mixture of $Ar/N_2/BF_3/H_2$. This is a method where the growth species are approaching the growing c-BN surface with a specific velocity. Hence, the gaseous species approaches the surface with an increased energy. Even though Zhang *et al* used a CVD method with a gas mixture, instead of a layer-by-layer growth technique (such as ALD), they assumed that the following growth mechanism took place (see figure 10.1).

$$B_{surface} + NH_x(g) \rightarrow B_{surface} - NH_x \text{ (will become } N_{surface} - H) \quad (10.7)$$

$$N_{surface} - H + F(g) \rightarrow N_{surface} + HF(g) \quad (10.8)$$

$$N_{surface} + BF_x(g) \rightarrow N_{surface} - BF_x\text{(will become } B_{surface} - H) \quad (10.9)$$

$$B_{surface} - \; + xH(g) \rightarrow B_{c-BN}(s) + xHF(g) \quad (10.10)$$

This is a mechanism that is identical to the one presented above (see equations (10.3)–(10.6)). The theoretical results presented in figure 10.5 do, hence, strongly support the experimental findings by Zhang *et al* (and the other way around). It is quite understandable, then, that an addition of extra energy is needed for the achievement of phase-pure c-BN growth (more specifically, for the abstraction of surface-terminating H species from the N-covered surface (with gaseous F)), and to some extent also for the surface-terminating F species from the B-covered surface (with gaseous H). As a conclusion, one has to add an extra amount of energy by some means (larger velocities of the incoming species or by increasing the temperature), or by using other terminating and gaseous species (than H and F).

References

[1] Zhang W, Bello I, Lifshitz Y, Chan K, Meng X, Wu Y, Chan C and Lee S 2004 Epitaxy on diamond by chemical vapor deposition: a route to high-quality cubic boron nitride for electronic applications *Adv. Mater.* **16** 1405–8

[2] Zhang W, Chong Y, Bello I and Lee S 2007 Nucleation, growth and characterization of cubic boron nitride (cBN) films *J. Phys. D: Appl. Phys.* **40** 6159–74

[3] Karlsson J and Larsson K 2014 Kinetic considerations of gas-phase adsorption of growth species on the c-BN (100) surface *Thin Solid Films* **564** 246–52

[4] Zhang W, Meng X, Chan C, Chan K, Wu Y, Bello I and Lee S 2005 Interfacial study of cubic boron nitride films deposited on diamond *J. Phys. Chem.* B **109** 16005–10

Chapter 11

Other types of material growth in a CVD reactor

A monolayer of graphite (i.e. graphene) has recently become a very popular material with very promising electronic properties [1–3]. It is constituted of a two-dimensional hexagonal sheet, in which all C carbons are sp^2-hybridized. Different types of synthesis techniques for graphene exist; out of which chemical vapor deposition (CVD) of graphene onto Cu (or Pt) surfaces have recently been developed [4, 5]. Other examples of techniques include mechanical exfoliation and exfoliation from silicon carbide substrate [6, 7].

Diamond is a cubic phase of carbon, with quite different properties when compared to graphene (e.g. high break down field, high thermal conductivity, band gap of 5.47 eV). It has theoretically been discussed that the diamond (111) surface will be graphitized at a temperature around 2500 K. There will be an almost perfect lattice matching between an eventually formed graphene structure and the underlying diamond (111) structure. Thus, the diamond (111) plane is a most suitable surface texture for thermal phase transformation of the upper surface layer to graphene. The possibility for a diamond-to-graphene formation would be interesting for many applications. For example, a graphene layer adsorbed on a wide band gap diamond substrate can be applied in field effect transistors, which can be designed for various environment working conditions (i.e. bio-compatible sensing, high corrosion condition) but also for RF communication applications.

Theoretical modelling is a highly valuable tool in the study of the possibility for a diamond-to-graphene transformation.

11.1 Diamond-to-graphene transformation

11.1.1 Introduction

Another most common allotrope of carbon is diamond. It is a cubic phase of carbon, with quite different properties when compared to graphene (e.g. high break down field, high thermal conductivity, band gap of 5.47 eV). It has theoretically been discussed that a diamond (111) surface can be graphitized at a temperature around

doi:10.1088/978-0-7503-3107-4ch11

2500 K [8]. Since the surface atoms of a material have more freedom to move and also exhibit a higher energy relative the bulk material, the upper surface is expected to undergo more severe reconstructions and relaxations with an increased temperature. For the situation with the graphitization of a diamond (111) surface, there will be an almost perfect lattice matching between an eventually formed graphene structure and the underlying diamond (111) structure. *This is, in fact, a prerequisite for a more, or less, perfect transformation from the cubic phase to the hexagonal phase of carbon.* By modeling an interface based on diamond (111) and the basal plane of graphite, a mismatch of less than 2% can be found. Thus, the diamond (111) plane is a most suitable surface texture for a thermal phase transformation of the upper surface layer to graphene. A question may arise if, for well-controlled temperature and pressure parameters, this graphitization process can be controlled in the way to produce epitaxial graphene from the upper surface of diamond (111). The possibility for a diamond-to-graphene formation would be interesting for many applications. For example, a graphene layer adsorbed on a wide band gap diamond substrate can be applied in field effect transistors, which can be designed for various environment-working conditions (i.e. bio-compatible sensing, high corrosion conditions) but also for RF communication applications [9, 10].

A CVD reactor would be most suitable for the experimental diamond (100)-to-graphene transformation, with its possibilities to work at increased temperatures, low pressures, and with specific gas mixtures, pressures, and gas flow conditions.

In the present example, the condition for forming a monolayer of graphene on the diamond (111) surface has theoretically been investigated using DFT calculations under periodic boundary conditions. *Ab initio* molecular dynamics (MD) annealing processes (from 0 K to 2000 K) were then used to simulate the geometrical restructuring of the upper diamond (111) surface. The structure evolution process was studied under thermal equilibrium condition for each temperature, and set up with time steps of 0.15 fs for a total duration of 0.30 ps. The graphitization process was considered under the assumption of three different experimental assumptions. Firstly, a pure thermally induced transformation for diamond (111) to graphene was considered to take place under a strict vacuum condition (i.e. without the presence of gaseous species). Secondly, a sufficient amount of H radicals was introduced into the system during the annealing process. Thirdly, the system was saturated with H radicals prior to the annealing process. The purpose with the present study was also to elucidate the binding condition over the diamond–graphene interface, by calculating the adhesion energy between the developed graphene-like adlayer and the diamond surface. The electronic properties of the, hereby, constructed epitaxial graphene-like adlayers, have also been of great interest to investigate, and were in this example calculated and compared with ideal and free-standing graphene.

11.1.2 Geometrical restructuring of the diamond (111) surface during annealing at low temperatures

The possibility for graphitization of the upper part of the diamond (111) surface was, at first, simulated under strict vacuum conditions (i.e. no other gaseous species were

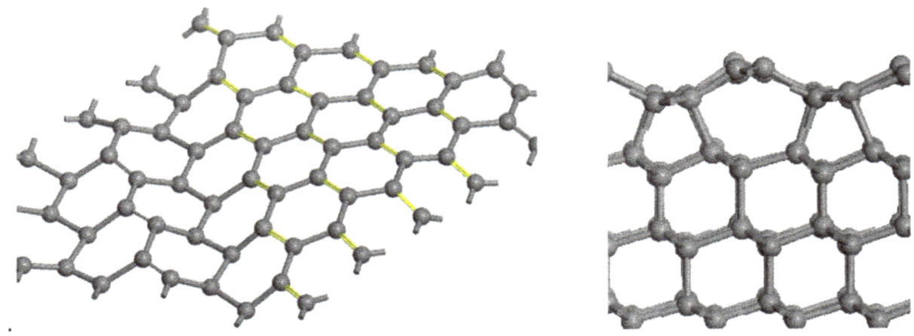

Figure 11.1. Surface reconstruction obtained at the temperature of 500 K. (a) Top and (b) side views of the transition to a 2 × 1 Pandey-chain structure on the non-terminated diamond (111) surface [12].

present in the models). Two MD calculations (at 300 K and 500 K) were initially performed with the purpose to relax and reconstruct the upper part of the surface. As a result of the geometry reconstruction, the upper two C layers within the diamond (111) surface underwent a 2 × 1 reconstruction towards a Pandey-chain π-binding structure [11]. The process had just started at 300 K and was completely finished at about 500 K (see figure 11.1). This type of initial reconstruction of the diamond (111) surface was expected and has earlier been observed in several experimental and theoretical [13] investigations. It can hereby be concluded that a surface reconstruction to a complete Pandey-chain structure is the preferred one at temperatures above (and close to) 500 K, and that this type of reconstruction is likely to completely fill the surface area of the diamond (111) surface.

The Pandey-chain reconstruction of the diamond (111) surface was, thereafter, used as an initial structure when studying the structural evolution of this system for temperatures above 500 K. For all MD simulations in the present investigation, the total simulation time were chosen to guarantee a stabilized constant of motion (and system total energy) for the part of the simulation where observations were made. To be more specific, the systems had achieved thermal equilibrium at a simulation time of approximately 0.15 ps (see figure 11.2).

11.1.3 Thermally-induced transformation of diamond (111) to graphene under strict vacuum conditions

A simulation of the diamond-to-graphene formation was performed here under the assumption that there are no gaseous species that can interact with the newly formed surface (i.e. under strict vacuum condition). The Pandey-chain reconstructed diamond (111) surface was found to be quite stable, and remained in the same geometry until the temperature was increased to about 800 K. Compared to the results from the *ab initio* MD simulations at 500 K, a geometrical transformation of the upper diamond C layer to graphene-like channels was observed at a temperature of 1000 K. Two nearby-positioned and parallel Pandey chain structures started to merge together, and thereby form this type of single layer graphene channel (with a

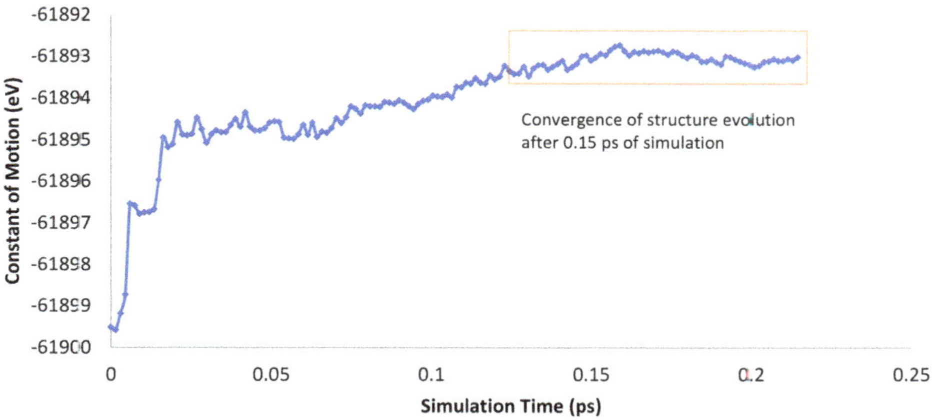

Figure 11.2. Evolution of the constant of motion during the annealing of a diamond (111) surface to 2000 K [12].

width of 7 Å and height of 2.5 Å). Hence, an obvious trend was observed in that an increase in temperature (from 500 K to 1000 K) resulted in a decrease in number of covalent $C_{diamond}$–$C_{grephene\text{-}like}$ bonds between the underlying diamond surface and the graphene-like adlayer. By increasing the temperature to 1500 K, and later on to 2000 K, the graphene-like channels became even more pronounced and well-shaped, and the number of $C_{diamond}$–$C_{grephene\text{-}like}$ bonds between the two materials decreased even further. In fact, the number of $C_{diamond}$–$C_{grephene\text{-}like}$ bonds (in the super cell) that were left between the two structures, decreased from originally 36 (0 K) to 18 (500 K), 14 (1000 K), and 13 (2000 K), respectively. The observed influence of temperature on the diamond (111) surface restructuring, is demonstrated in figure 11.3.

For comparison, the corresponding situations for the diamond (100)–2 × 1 and (110) surfaces were calculated as a reference to the (111) surface. The results were found to be completely different. When annealing the diamond (100)–2 × 1 surface to a temperature of 1000 K, the surface carbon atoms were found to move along their (2 × 1)-reconstruction direction, thereby changing the C–C dimer pattern (as demonstrated in figure 11.4). There were otherwise no major distortions observed, either for the carbon atom on the surface or further below in the lattice. These observations indicate that the thermal stability of the (2 × 1)-reconstructed diamond (100) surface is quite high. Similar observations were made for the diamond (110) surface. There was no indication of carbon atom distortion, or C–C bond breakage, up to a temperature of 1000 K. When reaching a temperature of 1500 K, few C–C bond breakages were found in the diamond structure, but not in the direction to form a graphene-like adlayer. In summary, the diamond (100)–2 × 1 and (110) surfaces do not show any tendency for a surface transformation to graphene. This is expected from the fact that there is very severe lattice mismatch between these diamond surface structures and graphene.

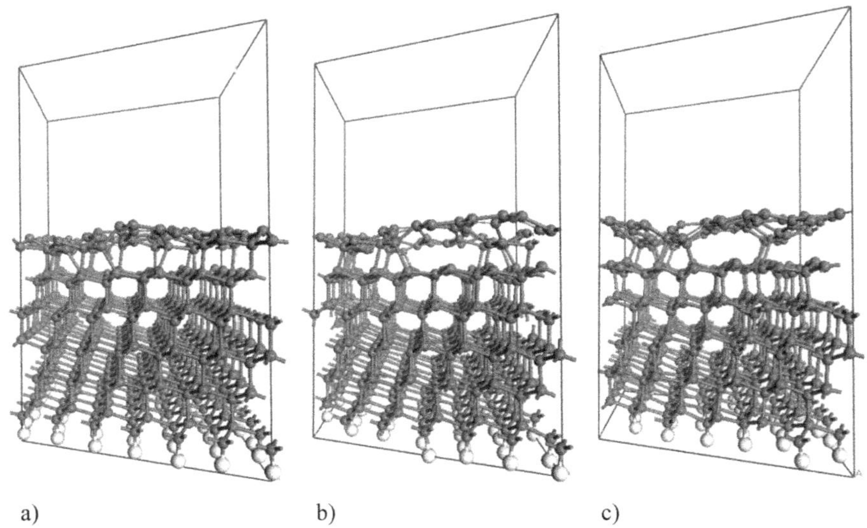

Figure 11.3. Influence of temperature on surface restructuring for a flat diamond (111) surface: (a) 500 K, (b) 1000 K, and (c) 2000 K [12].

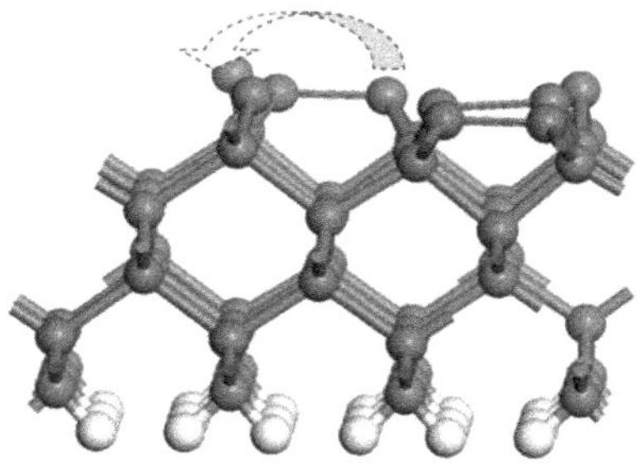

Figure 11.4. Model showing how C–C dimer bonds on diamond (100)–2 × 1 are shifted to neighboring positions during the annealing process at 1000 K [12].

11.1.4 Introduction of radical H species during the annealing processes

11.1.4.1 Introduction

In this part of the investigation, a simulation of the diamond-to-graphene formation was performed under the assumption that H is introduced into the system during the annealing process. When performing MD simulations of the diamond//graphene system at higher temperatures, oscillations of breakages and re-combinations of C–C bonds were observed between the diamond (111) surface and the graphene-like adlayer, and also within the respective system (i.e. the upper diamond surface and graphene-like adlayer, respectively). It is interesting to note that the frequency of C–C bond breakage and recombination within the different systems, were much

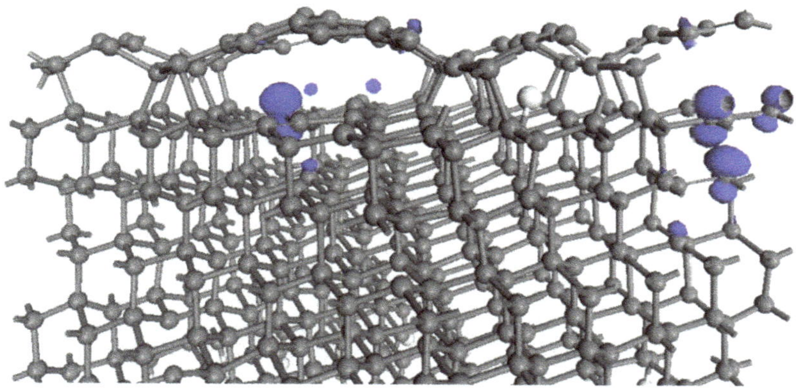

Figure 11.5. Radical reactivity of C atoms (shown in blue), as calculated by the Fukui functions $f^{(0)}$ [12].

faster than the corresponding processes in between these materials. In order to explain the reason to this phenomenon, Fukui function, $f^{(0)}$, calculations were performed for a diamond (111) surface at a temperature of 500 K, 1000 K and 1500 K, respectively. These functions will show the interfacial susceptibility towards a radical attack to the surface. The results of these calculations made it clear that the C–C bond breakages will result in very reactive non-binding diamond (111) carbon atoms.

As can be seen in figure 11.5, the most reactive carbon atoms were found in the upper C layers of the newly formed diamond (111) surface, which is the 2nd C atomic layer in the initial structure (i.e. before annealing). Since these highly reactive C atoms are also very susceptible to a C–C recombination across the diamond/graphene interface, the addition of gaseous radical species is expected to prevent this recombination by terminating the reactive C radicals and, hence, to make the whole system more thermodynamically stable. It should be stressed here that gaseous H radicals are small enough to diffuse into the diamond/graphene interface, and to chemisorb onto the highly reactive surface C atoms on the newly formed, diamond surface.

In order to investigate the structural effect by the H atoms that has been introduced into the diamond/graphene system during the annealing process, *ab initio* MD simulations have been performed with H radicals positioned interstitially within the interface. The aim here was to model the evolution of the geometrical structure when increasing the temperature, and at the same time introducing more and more H radicals to the interfacial system. Figure 11.6 shows the result when introducing only one H radical at 500 K. As a result of the *ab initio* MD simulations at this temperature, the number of remaining C–C bonds between the diamond surface and the graphene-like adlayer, have significantly decreased compared to the corresponding situation with no H atoms in the system.

For *ab initio* MD simulations at 1000, 1500, and 2000 K, and with 2, 3, and 4 interstitially positioned H radicals for each temperature, it was observed that all H radicals will aid in the breakage of $C_{diamond\ surface}$–$C_{graphene}$ bonds in the interface,

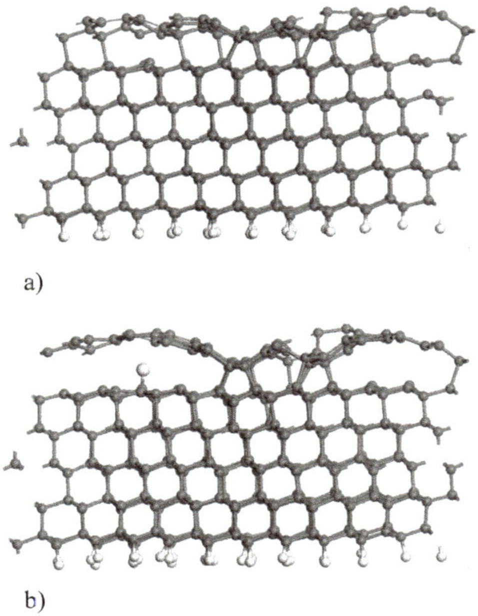

Figure 11.6. The structural changes of the upper diamond (111) surface at 500 K, (a) without any H radicals, and (b) with one H radical in the vicinity to a radical C site within the interface [12].

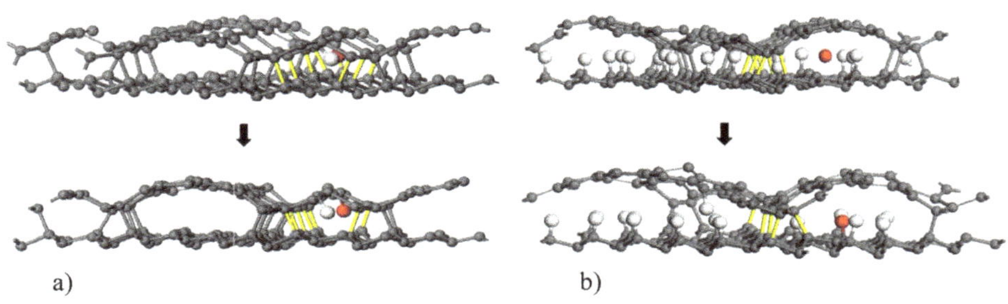

Figure 11.7. The effect of interstitially positioned H radicals (marked with red color) in an (a) unsaturated, and (b) H-saturated diamond–graphene interface. The upper figures show the starting structures, and the lower ones show the resulting structures at thermal equilibrium (T = 1000 K) [12].

and thereby bind to radically active carbon atoms on the newly formed diamond (111) surface. The bond breaking effect by the H radicals was found to be very efficient for the addition of the first H radical into the system (as compared with adding more H radicals to the system). For comparison, an *ab initio* MD simulation was performed at the same temperature, but with all radical C atoms saturated with H radicals. An extra H radical was also positioned within the interface. For this highly saturated situation, this extra H radical showed an efficiency to break more C–C bonds than just a single H does (see figure 11.7).

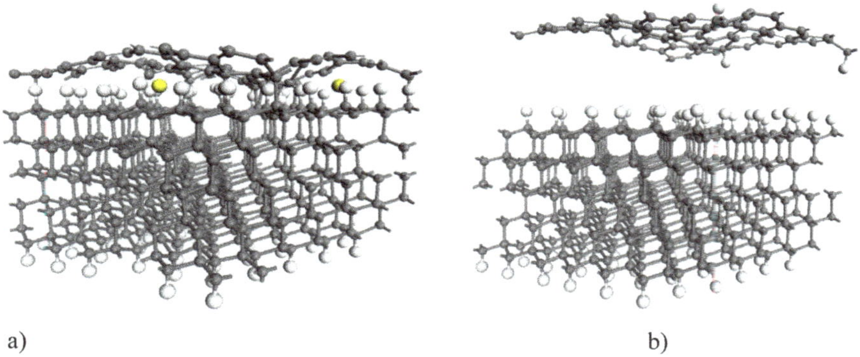

Figure 11.8. (a) Initial and (b) resulting structure after annealing at 1000 K. The initial diamond (111) surface is fully H-terminated with an extra H in the interface [12].

The efficiency of an even larger number of radical H atoms, during the course of the annealing process, has been tested at a temperature of 1000 K. This number of H atoms (33 H atoms) is assumed to be large enough to saturate the developed C radicals in the annealing process. As can be seen in figure 11.8, these H atoms are extremely efficient in causing a complete lift-off of a graphene sheet (although somewhat H-terminated) at 1000 K, with no sign of any binding between the graphene sheet and the resulting diamond (111) surface.

11.1.4.2 Introduction of a larger concentration of H radicals in the initial part of the annealing process

In this part of the investigation, a simulation of the diamond-to-graphene formation has been performed under the assumption that a larger concentration of H has been introduced to the system in the initial part of the annealing process. A sequentially increased number of H radicals (9, 12 and 17) were then introduced into the diamond (111) upper surface (i.e. between C layer 2 and 3). For a temperature at 1000 K, a quite dramatic result was observed when going from 9 to 17 H interstitials. The remaining $C_{diamond}$–$C_{graphene}$ bonds were only seven when adding nine H radicals to the lattice. However, this number decreased to 0 when introducing 17 H radicals (see figure 11.9).

11.1.4.3 Adhesion energy between a graphene-like sheet and a diamond surface

For all different temperatures that have been used in the present study, the adhesion energy between the diamond surface and the graphene-like sheet has also been calculated by following an *ab initio* MD simulation by geometry optimization at 0 K. The average adhesion energies for the C–C bonds at the various temperatures were in the range −370 kJ mol^{-1} (at 500 K) to −340 kJ mol^{-1} (at 2000 K). These values correspond to the average bond length ranging from 1.46 Å to 1.66 Å. As expected, the interface bond strengths decrease, with longer diamond-to-graphene bond distance, for larger temperatures. The corresponding values for a tabulated C–C sp^3 bond are −360 kJ mol^{-1} and 1.54 Å. This comparison gives an indication that, in

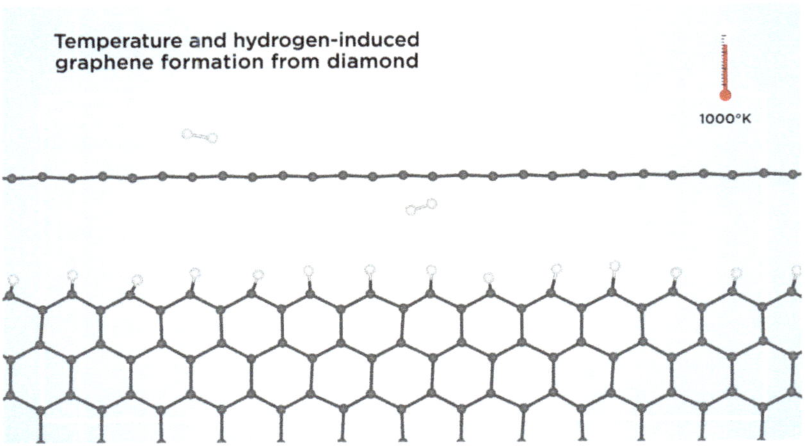

Figure 11.9. Graphene evolution from the diamond (111) surface, with 17 H radicals in the system at 1000 K [12]. Animation available at http://doi.org/10.1088/978-0-7503-3107-4.

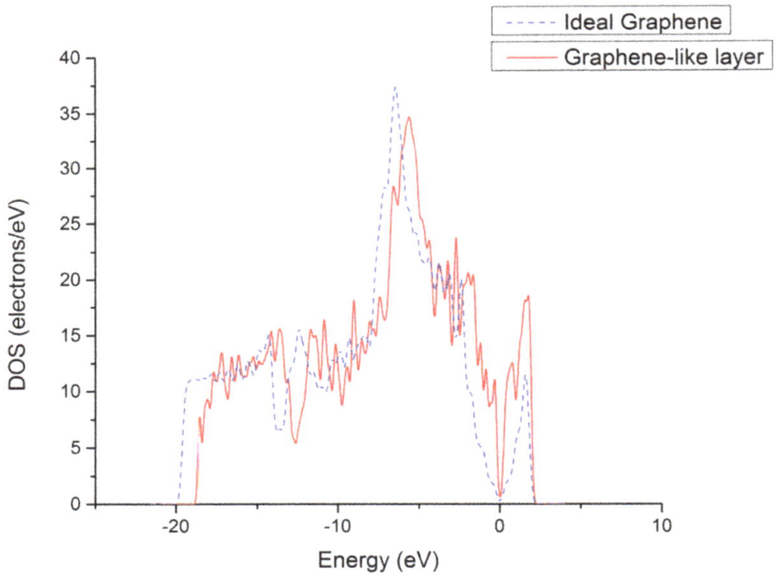

Figure 11.10. Comparison of DOS spectra for a graphene-like adlayer onto diamond (111) and ideal graphene sheet [12].

addition to covalent C–C bonds, there are also dispersion interactions between the diamond surface and the graphene-like adlayer (at least at lower temperatures).

11.1.4.4 Electronic states of the graphene-like adlayer

The electronic property of two types of graphene-like adlayers, produced without H at 500 K and with 18 $C_{diamond}$–$C_{graphene}$ bonds left in the interface, was in the present study compared with the one for an ideal graphene sheet. As can be seen in figure 11.10, the density of states (DOS) pattern for the graphene-like adlayer (solid line) has a very similar shape as the one for ideal graphene (non-solid line). A zero-level DOS at 0 eV (the Fermi level) can be found for both structures. This result shows that the graphene-like adlayer also has a cone-shaped band cross at the Fermi level and will thereby show similar electronic properties as a single layer graphene.

The present example shows that a single-layer graphene can be formed from a flat non-terminated diamond (111) surface by introducing an excess of H radicals in the CVD chamber. This type of non-terminated surface is easily obtained in the annealing process (at around 400 K, depending on the pressure in the CVD chamber) when starting from the H-terminated counterpart. The electronic properties of the graphene-like adlayer (formed at the rather low temperature of 500 K, and, thereby, prior to the formation of a freestanding graphene sheet) were very similar to the electronic properties of a free-standing graphene.

References

[1] Novoselov K S, Geim A K, Morozov S V, Jiang D, Zhang Y, Dubonos S C, Grigorieva I V and Firsov A A 2004 Electric field effect in atomically thin carbon films *Science* **306** 666–9

[2] Novoselov K S, Geim A K, Morozov S V, Jiang D, Katsnelson M I, Grigorieva I V, Dubonos S V and Firsov A A 2005 Two-dimensional gas of massless dirac fermions in graphene *Nature* **438** 197–200

[3] Geim A K and Novoselov K S 2007 The rise of graphene *Nat. Mater.* **6** 183–91

[4] Land T A, Michely T, Behm R J, Hemminger J C and Comsa G 1992 STM investigation of single layer graphite structures produced on Pt (111) by hydrocarbon decomposition *Surf. Sci.* **264** 261–70

[5] Nagashima A, Nuka K, Itoh H, Ichinokawa T, Oshima C and Otani S 1993 Electronic states of monolayer graphite formed on TiC (111) surface *Surf. Sci.* **291** 93–8

[6] Berger C *et al* 2004 Ultrathin epitaxial graphite: 2D electron gas properties and a route toward graphene-based nanoelectronics *J. Phys. Chem.* B **108** 19912–6

[7] Ohta T, Bostwick A, Seyller T, Horn K and Rotenberg E 2006 Controlling the electronic structure of bilayer graphene *Science* **313** 951–54

[8] DeVita A, Galli G, Canning A and Car R 1996 A microscopic model for surface-induced diamond-to-graphite transitions *Nature* **379** 523–26

[9] Ekinci K L and Roukes M L 2005 Nanoelectromechanical systems *Rev. Sci. Instrum.* **76** 061101

[10] Bakker E and Qin Y 2006 Electrochemical sensors *Anal. Chem.* **78** 3965–84

[11] Pandey K C 1981 New-bonded chain model for Si(111)–(2×1) surface *Phys. Rev. Lett.* **47** 1913–16

[12] Song Y and Larsson K 2018 Formation conditions for epitaxial graphene on diamond (111) surfaces *Adv. Nanosci. Nanotechnol* **2** 1–8

[13] Vanderbilt D and Louie S G 1984 Total energies of diamond (111) surface reconstructions by a linear combination of atomic orbitals method *Phys. Rev.* B **30** 6118–30

www.ingramcontent.com/pod-product-compliance
Ingram Content Group UK Ltd.
Pitfield, Milton Keynes, MK11 3LW, UK
UKHW051907100926
8556IPUK00005B/791